AF324676

Intelligent Early Warning of Risks in Complex Systems of Oil and Gas Production

Theory, Method and Application

Intelligent Early Warning of Risks in Complex Systems of Oil and Gas Production

Theory, Method and Application

Jinqiu Hu
China University of Petroleum-Beijing, China

Laibin Zhang
China University of Petroleum-Beijing, China

Shengnan Wu
China University of Petroleum-Beijing, China

World Scientific

NEW JERSEY · LONDON · SINGAPORE · BEIJING · SHANGHAI · HONG KONG · TAIPEI · CHENNAI · TOKYO

Published by

World Scientific Publishing Co. Pte. Ltd.

5 Toh Tuck Link, Singapore 596224

USA office: 27 Warren Street, Suite 401-402, Hackensack, NJ 07601

UK office: 57 Shelton Street, Covent Garden, London WC2H 9HE

Library of Congress Control Number: 2025009780

British Library Cataloguing-in-Publication Data
A catalogue record for this book is available from the British Library.

《油气生产复杂系统风险早期智慧预警理论、方法及应用》
Originally published in Chinese by Petroleum Industry Press
Copyright © Petroleum Industry Press, 2022

INTELLIGENT EARLY WARNING OF RISKS IN COMPLEX SYSTEMS OF OIL AND GAS PRODUCTION
Theory, Method and Application

ISBN 978-981-98-1012-3 (hardcover)
ISBN 978-981-98-1013-0 (ebook for institutions)
ISBN 978-981-98-1014-7 (ebook for individuals)

For any available supplementary material, please visit
https://www.worldscientific.com/worldscibooks/10.1142/14219#t=suppl

Desk Editors: Nambirajan Karuppiah/Julio Hong/Amanda Yun

Typeset by Stallion Press
Email: enquiries@stallionpress.com

Preface

The oil and gas industry is an industrial system that includes exploration, development, production, storage, transportation, processing, and utilization of resources, so it not only has a long industrial chain, spans a wide range of areas, and is strongly linked to other industries but is also susceptible to the influence of complex factors such as geology, environment, climate, and society. This industry is characterized by hazards such as high temperatures and high pressures, flammability and explosiveness, and toxicity and hazards. It is internationally recognized as a high-risk industry, characterized by a high incidence of frequently occurring major accidents, such as the "12-23" Chongqing Kaixian blowout accident, the "11-22" Qingdao Donghuang oil pipeline explosion, and the Penglai 19-3 oil spill accident in China the Gulf of Mexico "deep" oil spill accident and the "Deepwater Horizon" drilling platform blowout in the Gulf of Mexico in the United States. The consequences of such accidents are serious, the losses are huge, and the social impact is extremely bad.

At present, the characteristics of China's oil and gas production are deep-water dredging, unconventional mining, and oil and gas pipeline transportation. Offshore oil is gradually moving toward deepwater and offshore locations (maximum operating depth of 3,000 m, drilling depth of 10,000 m). The exploration and development of unconventional oil and gas resources (high pressure, large displacement, long cycle) have become very important. The total extent of the oil and gas pipeline network has been increasing, the pressure levels have been increasing, and the diameter of the pipelines has been expanding (the total extent has reached 16.9×10^4 km, with the highest pressure of 12 MPa, and the maximum diameter of the pipeline of 1.4 m). China's State Council's Work Safety Committee issued a "three-year action plan for special rectification of national work safety", which explicitly requires "strengthening the safety and security of oil and gas storage and expansion, focusing on the control of high-temperature and high-pressure, high-sulfur wells blowout out of control and the risk of hydrogen sulfide poisoning, and to strictly prevent the rush to progress, rush production capacity, and pressure on the cost of the accidents caused by. It has also strengthened technical research on safety of deep-sea oil and gas exploitation, and reinforced the risk control measures for marine oil safety in extreme weather". The General Office of the Central Committee of the Communist Party of China and the General Office of the State Council "on comprehensively strengthening the safety of production of hazardous chemicals", clearly pointed out that one must strengthen the oil and gas pipeline areas and other key aspects of safety control. Therefore, risk perception, assessment, and safety early warning are of great significance in capturing the risk factors in the oil and gas production process in a timely manner, eliminating the root causes of accidents and realizing system safety.

The complex system of oil and gas production is characterized by hierarchy, nonlinearity, openness, and fragility. As the saying goes, "A thousand miles of dike will collapse in an ant hole". The occurrence, development, aggravation, derivation, and secondary process of major catastrophic accidents are closely related to the characteristics of complex systems. The structure of oil and gas equipment is complex, and the operating conditions for oil and gas production are diverse. Different equipment has different spatial and temporal correlations with accident factors in different environments and working conditions, but their mechanisms are regular and can be followed. Therefore, the purpose of early-warning research is to avoid the occurrence of a strong "chain reaction" between the causal factors of complex system accidents and to realize the smallest preventive measures. On the one hand, it is necessary to suppress the triggering of accidents from the root cause; On the other hand, starting from the temporal and spatial correlation of accident causation factors and the process of accident propagation, we can reduce the degree of accident harm, enhance the safety resilience of the system, and reduce operational risks. Therefore, predicting accidents in advance, detecting possible accidents early, and transforming the emergency response management mode after accidents into a management mode of monitoring and preventing dangerous situations in advance; Transforming static security management methods into dynamic risk prediction and proactive maintenance methods; Transform dispersed single event handling methods into a systematic and composite management system.

This book is oriented toward the engineering needs of the safe operation of complex systems of oil and gas production and combines more than ten years of research results in dynamic risk assessment and early warning. This book includes the latest research and in-depth theories, methods, and techniques on how to utilize real-time process monitoring parameters, images, text, and line-of-sight tracking technology for early warning of risks in the whole process of oil and gas production. The book provides practical case applications for typical oil and gas production complex systems, such as petroleum refining systems, unconventional oil and gas fracturing systems, and offshore oil and gas extraction, which will help readers learn practical safety management.

The book consists of eight chapters, of which Chapters 2–5 and 7 have been written by Prof. Jinqiu Hu. Chapters 1 and 6 have been written by Academician Laibin Zhang. Chapter 8 has been written by Associate Professor Shengnan Wu. Doctoral students from China University of Petroleum (Beijing), Xi Ma, Qianlin Wang, Shuang Cai, Xin Zhang, Zhou Liuhui, and Chuanguang Chen, and master's students Yu Wang, Anqi Wang, Yan Yi, Bin Tian, Yaqin Cao, Jing Luo, Siyun Tian, Fang Guo, Siyang Li, Lijiang Zhang, Jinghua Hu, Xiyue Zhang, and Zhiqiang Wu compiled the research material and results during their study period, which laid the foundation for the successful completion of this book. We would like to express our heartfelt thanks to doctoral students Xinyi Li, Yiyue Chen, and Shangrui Xiao and master's students Tianyu Wang, Zihan Xu, Xinyi Chen, Ruoxin Liu, Linggen Meng, Xin Xu, and Xueni Li for their participation in the data collection and proofreading of the manuscript. On the occasion of the publication of this book, we would like to thank the National Natural Science Foundation of China (NSFC) for the project "New Disaster-Causing Mechanisms and Early Warning for Failure of Oil and Gas Intelligent Pipeline Systems under the Threat of Information Security (No. 52074323)", and External cooperation project for scientific research and technology development of CNPC Safety and Environmental Protection Technology Research Institute "Evolution mechanism of major risks in complex oil and gas drilling and production and safe intelligent operation and maintenance method (PetroChina: 2023DJ6508)".

Due to the limitations in our knowledge, there will inevitably be inaccuracies in the book. We sincerely hope that readers will make valuable comments and suggestions.

About the Authors

Jinqiu Hu, a female professor and doctoral supervisor, is recognized under the National Programme for the Promotion of Young People's Excellence project. Currently, the head of the Safety Engineering Department at China University of Petroleum (Beijing) and the director of a sub-laboratory under the Ministry of Emergency Management's Key Laboratory, extensive research has been conducted on safety and emergency support technology for complex oil and gas production systems, with significant achievements in risk assessment, early warning, and emergency decision-making related to deep earth, deepwater, unconventional oil and gas, and hydrogen energy production, storage, transportation, and utilization processes. Over the past five years, more than ten national and provincial-level projects have been led. More than 100 scientific papers have been published, with over 40 indexed by SCI, 4 academic monographs authored, and 2 textbooks co-authored. Over 30 invention patents and 15 software copyrights have been granted. Two Special Prizes and five First Prizes for technological progress have been awarded by provincial and ministerial-level organizations and industry associations, along with a Youth Science and Technology Award, a Technology Transfer Award, an Innovation Team Award, and an Excellent Work Award. Research achievements have been applied in more than 20 domestic enterprises, yielding significant economic and social benefits. Numerous honors have been received, including the title of Beijing's Model Teacher in Curriculum Ideology and Politics and the university's "Outstanding Young Teaching Backbone." Courses have been selected as Beijing's Model Courses in Curriculum Ideology and Politics, the university's first batch of "Quality Courses," and the college's flagship graduate courses. Two Teaching Achievement Awards have been received from the university.

Laibin Zhang, male, professor, doctoral supervisor, and academician of the Chinese Academy of Engineering at China University of Petroleum (Beijing), has been engaged in the creation of Chinese oil and gas safety science and engineering disciplines, as well as in research on oil well pipe damage detection, fault diagnosis of large-scale power units in oil and gas production, and risk assessment and early warning for oil and gas production systems. Multiple projects, including the "863 Program," key projects funded by the National Natural Science Foundation, and major enterprise projects, have been led. Four national scientific and technological awards have been received, including two second prizes for national technological inventions as the first author, one special prize for provincial and ministerial-level technological progress, and several first prizes. Numerous monographs and textbooks have been authored, over 300 papers have been published, and more than 50 patents have been granted. As the first author, two second prizes for national teaching achievements and two first prizes for provincial and ministerial-level teaching achievements have been awarded. Honors such as the Outstanding Member Award from the International Association of Petroleum Engineers, the Sun Yueqi Energy Award, and the IET-Fangzheng University President Award have been received, and special government allowances from the State Council have been granted. Technological achievements have been made in the areas of magnetic memory detection and diagnosis technology for oil well pipe damage based on the giant magnetoresistance effect, precise fault diagnosis and early warning technology for power units in oil and gas stations, and intelligent fault tracing and safety early warning technology for in-service refining equipment. These achievements have been recognized as internationally advanced, with some considered internationally leading.

Shengnan Wu, a female associate professor and doctoral supervisor at China University of Petroleum (Beijing), has long been engaged in research on complex oil and gas extraction, risk assessment of key safety equipment for oil and gas, early warning, reliability, and testing and maintenance. Honorary titles such as "Young Top Talent" from China University of Petroleum (Beijing) and "Outstanding Science and Technology Paper of Beijing Youth" have been awarded. Projects funded by the National Natural Science Foundation, Postdoctoral Science Foundation, and Youth Top-notch Talent Fund have been hosted, with participation as a key member in over 12 production and research projects for enterprises such as PetroChina and CNOOC. Research results have been published in more than 40 scientific papers domestically and internationally, and participation in compiling one international academic monograph has occurred. One first prize for provincial and ministerial-level scientific and technological progress and three other provincial and ministerial-level awards have been received, along with one invention patent application and four software copyrights. An invitation to serve as an editorial board member of the *International Journal of Reliability and Safety* was extended. Long-term scientific research cooperation and exchange visits with foreign universities, including the Norwegian University of Technology and the University of Stavanger, have been maintained.

Contents

Chapter 2 Early Intelligent Warning Based on Real-Time Monitoring Data of Process Parameters 29

Chapter 3 Early-Warning Technology for Hidden Dangers of Oil and Gas Extraction Equipment Based on Infrared Thermal Image Video Surveillance 133

Chapter 6 Typical Examples of Safety Alerts for Shale Gas Fracturing Systems 459

Chapter 1

Introduction

1.1 Risk Factors and Characteristics of Complex Systems for Oil and Gas Production

1.1.1 *Risk characterization of complex oil and gas extraction systems*

The oil and natural gas resources in China's deep stratum account for 30% and 60% of the respective total reserves, making it the main battlefield for oil and gas exploration and development. However, the exploitation of deep and complex oil and gas resources faces severe challenges from high-risk and complex factors such as complicated geological structures, high temperatures and high pressures, and great danger due to the formation of fluids. These factors cause the drilling and completion process to frequently encounter downhole complexities, making well control extremely difficult, wellbore integrity difficult to ensure, lengthening the operation cycle, increasing the development cost, increasing the problem of production wells with pressure in the annulus, and causing great difficulty in long-term comprehensive management. In fact, in the past 20 years, the vast majority of domestic blowout accidents have occurred from the complex oil and gas exploration and development production process. Although after years of research, great progress has been made in engineering safety design, well control technology, handling of complex situations and accidents, and emergency response technology and equipment, large gaps still exist in the realization of complex oil and gas resources through safe and efficient means. With the increasingly fierce international competition in the field of oil and gas supply, ensuring safe exploitation of complex oil and gas resources while constantly seeking ways to reduce costs and increase efficiency has become a major technical challenge in the field of domestic oil and gas production that needs to be solved urgently.

In terms of early warning of complex oil and gas accidents, most of the early warnings in oil drilling engineering include comprehensive on-site inspection and recording technology integrating geological observation and analysis, gas detection, drilling fluid parameter measurement, formation pressure prediction, and drilling engineering parameter measurement. The drilling operation process is monitored 24 h through recording technology, and the prediction and forecast of drilling abnormalities are realized through monitoring and quantitative calculation of drilling parameters. Fine pressure control drilling is a precise control method for bottomhole pressure. During the drilling process, low-density drilling fluid is used to monitor the annulus pressure in real time. After determining the downhole situation, fine control of bottomhole pressure is achieved by adjusting the back pressure at the wellhead. Determining the underground situation: Short section measurement of the annular pressure during drilling can be used to determine the underground situation by measuring the annular pressure of a certain section of the wellbore during the drilling process, combined with the inlet and outlet flow rates of the drilling fluid measured by the flowmeter. Adjusting wellhead back pressure: Attempt to use back pressure compensation devices and automatic throttling manifolds to achieve precise control of wellhead back pressure, thereby maintaining stable bottomhole pressure.

In terms of health assessment and prediction of key drilling facilities, China has systematically studied and summarized the corrosion failure characteristics and influencing factors of gas well pipes, revealing the variation laws of corrosion resistance of over 13Cr pipes with temperature, CO_2 partial pressure, Cl^- concentration and flow rate, acidification environment, completion fluid, loading stress, corrosion behavior and characteristics, and forming a set of corrosion resistance rules based on wellbore characteristics. In addition, corrosion integrity selection and evaluation technology based on the whole life cycle of the wellbore has been formed, and a physical tensile stress corrosion test system for simulating downhole service conditions of oil casing has been developed, which provides a basis for decision-making for the selection of tubing materials for the Tarim "Three Supers" gas field. Due to deep exploration and development, oil and gas wells are developing high temperatures, high pressure, and high sulfur content. The usage of high-grade steel and large thick wall tubing continues to increase; the complexity of the downhole environment leads to a decline in tubing integrity and service performance due to operational damage as well as synergistic effects with environmental media. The failure of tubing columns is still prone to occurring in greater frequency.

In view of the poor working conditions of wellbores in deep and complex formations, the complex engineering geological conditions of ultra-deep wells, the arduous challenges of emerging risks, accurate monitoring, and diagnosis, and early-warning technology and equipment for abnormal downhole conditions based on multi-source information, usage of supporting technology for safe and efficient drilling and completion of wells is a good solution, which can also fulfill the requirement of early prognosis of complicated downhole conditions such as leakage, overflow, and spillage, avoid the occurrence of accidents such as well blowouts, and realize safe and fast drilling.

1.1.2 *Unconventional oil and gas extraction system risk characteristics*

Most of the existing shale gas development blocks in China are located in fragile and sensitive geological locations, in mountainous and hilly areas, with poor road conditions and difficult well site layouts, facing challenges of small sites requiring continuous operation using large loads over a long time, which leads to the possibility of fire, leakage, and other accidents. Fracturing operation has been transformed into a "factory" mode, which puts forth stricter requirements for the safety of equipment and people and environmental protection. Large-scale hydraulic fracturing is a typical feature of shale (oil) gas extraction, which can significantly increase oil and gas production, but there are issues about the impact on the geological and ecological environment, and there are potential risks such as earthquakes, geological disasters, and environmental and ecological impacts. At present, the relationship between hydraulic fracturing and seismic or microseismical impacts has attracted a great deal of attention in the international arena, and research in this area is directly related to the safety and sustainable development of shale (oil) gas.

Well leakage is a complex problem in the process of unconventional oil and gas extraction, and the engineering risks of blowout, leakage, and collapse of the shale gas field are still are significant. The geological conditions in the shale gas field area in South Sichuan are complicated, with frequent well leakage and serious problems such as drilling equipment getting stuck and rotary guided tools falling down the wells. Shale gas wells are prone to spillage when drilling into natural fractures or to the operation being interfered with due to fracturing of neighboring wells, leading to safety risks. Large-scale segmental fracturing creates high-frequency alternating stress impacts on the wellbore, posing a serious challenge to the long-lasting seal integrity of the wellbore. Influenced by the differences in physical properties of the cemented casing, cement ring, and formation, the response state of force on the wellbore is different, and it is easy to fracture the cement ring body or form interfacial micro-annular gaps, which leads to the failure of interlayer sealing of the wellbore, and even the phenomenon of fluid from the formation running up to the wellhead to form the casing annulus under pressure, which affects the efficient and safe development of shale gas. For example, since the large-scale development of the Fuling shale gas field, the phenomenon of pressure buildup in the shale gas well casing annulus has become more common, which is a major hidden danger for safe production.

At the same time, abnormal downhole conditions such as sand plugging will inevitably occur during the actual construction process, leading to fracture failure, obstruction, or premature termination of construction. Usually, the cause of sand plugging can be considered from the perspective of oil production engineering as the structural damage of the oil formation due to loosening of the formation cement or excessive differential pressure, which results in a large amount of sand, and the sand content of the production fluid is so high that the sand is unable to be brought out and settles at the bottom of the wellbore. From the perspective of hydraulic fracturing, the phenomenon of sand settling in the near-wellbore zone or the tubing column of a wellbore is caused when the pressure of the required injected fluid exceeds the pressure limit of the injected wellbore tubing column, wellhead equipment, and pumping equipment. Thus, faster sand addition rates and higher pumping displacements are direct engineering factors that produce sand plugs. Sand plugging is also a direct cause of stuck wells and fracturing failures.

Compared with the commercial exploitation of unconventional oil and gas in developed countries in North America, China's research in unconventional oil and gas exploitation is relatively new. New processes, new technologies, new equipment, and large-scale localized development and application projects are constantly being promoted; however, the projects and their negative effects coexist, and various unpredictable and undesirable potential risk factors are present throughout the life cycle of the process, leading to a variety of emergencies and their secondary and derivative disasters. A dynamic safety risk evaluation and early-warning technology can provide reliable technical support for the systematic and technical guarantee for safe development of shale gas.

1.1.3 *Oil and gas pipeline system risk characteristics*

Oil and gas are the "lifeblood" of national energy, and the length of oil and gas pipelines in China is 13.6×10^4 km. In recent years, China's oil and gas pipeline projects have been steadily advancing, and the oil and gas pipeline network has been gradually improving, with the cooperative construction of a number of land-based oil and gas import corridors from China to Myanmar, Central Asia, Kazakhstan, and Russia. Due to the aging of pipeline material, manufacturing defects, third-party damage, natural disasters, poor operation conditions, and other factors, pipeline leakage, combustion, and explosion occasionally occur, damaging the ecological environment, resulting in casualties and causing huge economic losses. Based on statistics, pipeline safety accidents account for about 50% of all industrial accident, and serious accidents can lead to hundreds of casualties and direct losses of nearly one billion dollars.

In order to avoid pipeline accidents, the United States Congress approved the Act on Improving Pipeline Safety in 2001. The core content of the Act is to implement integrity management in high-consequence zones. Pipeline integrity management has gradually become an important means of preventing accidents and realizing pre-control in the global pipeline industry. It is a systematic management system with the goal of pipeline safety and continuous improvement, which involves design, construction, operation, monitoring, maintenance, replacement, quality control, and communication, and runs through the whole life cycle of the pipeline. Since 1995, China has carried out research on pipeline risks and pipeline safety.

In 2001, PetroChina took the lead in introducing pipeline integrity management and achieved fruitful results, covering pipeline storage and transportation facilities in various stages such as lines, stations, gas storage, and system platforms. In the field of lines, PetroChina set up technologies to ensure ontological safety and security, risk assessment and control, transport medium safety and security, rescue and repair, and emergency response security. In the field of station integrity management, technologies for testing and assessment of field station process facilities, diagnosis and assessment of compressor sets, quantitative risk assessment, safety level assessment, and facility integrity evaluation have been gradually formed. In the field of gas storage integrity management, technologies have been formed for risk control of underground gas storage and the construction and operation safety of gas storage. In the field of pipeline integrity, a pipeline emergency decision-making GIS system based on multi-source data has been set up, and an intelligent pipeline network has been built along the China–Russia Eastern Route.

In recent years, researchers in the pipeline industry have also been actively exploring the safe and intelligent construction of pipelines. An intelligent pipeline network is a product of a deep integration of the pipeline network and information technology based on key technologies such as big data, the Internet of Things, cloud computing, and artificial intelligence. It has the functional characteristics of comprehensive perception, automatic prediction, self-adaptation, self-feedback, self-learning, etc., and can realize the safe and efficient operation of a pipeline network. It features standardization, digitalization, visualization, automation, and intelligence in planning management, construction management, and operation management. However, at present, the construction of intelligent pipelines in China has not been fully completed and is in a small-scale digital stage. The large-scale database of the national pipeline network system has not yet been formed. The existing application cases of big data are relatively few, limited to traditional oil and gas pipeline risk analysis, pipeline inspection, and other aspects, lacking large-scale applications in oil and gas pipeline safety assurance and emergency decision support. At the same time, big data based oil and gas pipeline leak monitoring and early warning, disaster warning, and corrosion control management have not yet formed effective applications; The decision support platform for oil and gas pipelines based on big data has only completed the construction of the system architecture and has not yet achieved commercial application of decision support functions. How to improve the applicability and pertinence of the model, effectively apply it to pipeline operation management and assessment, and integrate the data and information system into one remain to be tackled.

On the other hand, while pipeline intelligence is being built, there is a lack of complete understanding and effective detection and warning methods for the new and complex security dangers of information security threats that include deliberate attacks and unintentional actions, leading to malfunctions or anomalies in the pipeline information space. Effective threats against oil and gas pipeline systems are usually initiated in the network domain through local or remote access, mimicking component failures while isolating the connection between the network and the physical system, thus leaving the physical process of oil and gas transportation uncontrolled. These will result in oil and gas transport process delays and power unit denial of service (DoS), which can interfere with the normal operation of the local pipeline network or unintended release of energy/materials. Information security threats have the potential to create more catastrophic consequences for the operation of oil and gas pipeline networks than the physical failure of pipeline entities. Cyber security threats will lead to cross-domain risks of cyber-physical failure in oil and gas pipelines. At the same time, complex and unknown cyber attack methods are constantly emerging. In this context, the security of oil and gas pipeline cyber physical systems is facing severe challenges. It is urgent to integrate and analyze the impact of cyber security threats on oil and gas pipelines from the information, physical, and social levels, reveal new disaster mechanisms, and establish early warning methods.

1.1.4 *Oil and gas processing and utilization system risk characteristics*

In oil and gas processing, to go from raw materials to the production of products requires many processes and complex processing units, multiple reactions and separations. The process conditions are harsh and the influencing factors are accompanied by extreme conditions such as high temperature, high pressure, low temperature, vacuum, large flow rate, and high speed. At the same time, the devices start and stop when the operating parameters change greatly or if the system is in an unstable operating state. At the same time, the frequent inflow and outflow of hazardous chemical media will increase the probability of accidents and the failure rate of equipment. Normal production of the process parameters is relatively stable, but in the long cycle of continuous operation involving process equipment, utility conditions, processing volume adjustment, personnel operating levels, instrumentation reliability, and many other factors, there are still some factors affecting production safety, which could cause an emergency shutdown of the device, leakage, and fire, and other accidents. In the process of maintenance and repair of these devices, improper operation, lack of safety measures, and the impact of environmental factors can also easily lead to major accidents. For example, in March 2005, a sudden explosion occurred in the startup process of BP's Texas refinery's atmospheric decompression unit, resulting in the deaths of 15 people, injuries to 180 people, and economic losses of more than 1.5 billion U.S. dollars. In May 1999, a Beijing Yanshan Branch refinery's tri-catalytic unit exploded in the startup stage of due to dirty oil in the pontoon. In January 2003, the Cangzhou refinery sulfur recovery unit exploded during a dismantling and repair process, causing one person to be seriously injured. In May 2007, a hydrogen sulfide poisoning accident occurred in a diesel oil hydrotreating and refining device at a refinery of the Urumqi Petrochemical Company during the process of shutting down the refining plant, resulting in the poisoning of five people.

Most of the safety accidents in oil and gas processing and utilization systems are caused by "changes" in the system, such as high liquid levels, excessive flow rate, degradation of pumping components, and pipeline cracks. These changes can be internal (also called "deviation") or external (also called "perturbation"). If these "variations" cause the system to operate outside of the safe range expected by the design, a system failure will occur. Due to the connection between media, information, and processes, there is interdependence, mutual constraint, and mutual influence between system units. A single equipment or process failure often triggers a chain reaction, starting from one failure and causing a series of failures, forming a fault chain. As the fault spreads and expands, it ultimately leads to accidents or disasters. At the same time, such failures spread from one geographical space to another, causing derivative accidents and bringing immeasurable social and environmental risks.

Due to the constantly changing working conditions of the oil and gas processing and utilization system, it is almost impossible to build a comprehensive fault sample database. Some faults lack historical event observation, and fault samples are scarce. The performance of oil and gas processing and utilization system equipment and the properties of raw materials often change, and there are interferences such as noise, frequent operations, and abnormal working conditions in the production process. Distributed control system (DCS) and other safety monitoring systems based on threshold alarm mechanisms are prone to a large number of false alarms, increasing the difficulty of on-site personnel's work and easily ignoring correct alarm information, resulting in the inability to handle abnormal working conditions in a timely manner. At the same time, there is also an alarm delay phenomenon when safety management personnel cannot find the abnormal conditions in time, delaying the best adjustment time.

Hazardous factors in the oil and gas processing and utilization system exist objectively, and these factors exist in various forms within the system and may be transformed into accidents under certain conditions. However, as long as the dangerous factors are noticed in time before the accident occurs, the accident is predicted, and measures are taken in advance to inhibit it, it is possible to prevent the dangerous factors from being transformed into accidents. Therefore, in view of the production and operation characteristics of oil and gas processing, it is of great social and safety significance to establish a theoretical system of accident early-warning, apply a method of system analysis, carefully analyze the human, material, and environmental factors in the system according to the roles of each unit of the constituent system, capture the dangerous factors in the system in time, and eliminate the root causes of accidents so as to realize the safety of the system.

1.2 Current Status of Early-Warning Technologies for Risks in Complex Systems of Oil and Gas Production

With the development of the modern petroleum industry, oil and gas equipment is growing in scale and transforming into a high-speed, automated, and intelligent industry, especially the gathering and transmission systems, booster station power units, various types of chemical equipment, transportation, treatment, and some other processes. The essence of the complex system of oil and gas production is that the subsystems with different are interrelated and interact with each other, constituting a large, open, and complex dynamic system with a unified structure and function.

The complex system of oil and gas production, on the one hand, has long-term operational viability (the operation period is generally more than 30 years). However, in the long operation process, accidents are inevitable; on the other hand, the complex system has the characteristics of hazard diffusion (due to the system's external environment). Moreover, once an accident occurs, it will inevitably lead to huge losses, so it is also a typical high-risk system. Its safety issues are different from other industries due to some different characteristics: (1) Most of the materials are flammable explosive, reactive, toxic, and corrosive. (2) The production equipment has a large scale and high integration, and the data presented in the production and operation process is highly nonlinear. (3) The system is composed of complex relationships and behaviors, as well as a high degree of correlation with its environment and strong coupling, resulting in the formation of system failure, propagation, and evolution of fault behavior.

In the complex system of oil and gas production, once any subsystem or component thereof fails, it often triggers a chain reaction, resulting in significant production losses and even leading to catastrophic safety accidents, for example, the BP Texas City refinery explosion, the PetroChina Jilin Petrochemical Company's "11 · 13" explosion, the Dalian pipeline explosion, and the Penglai 19-3 oil spill accident. The complex nonlinear interaction between the causal factors of these safety accidents is the main reason for the complexity of the safety accidents.

Since Holland proposed the theory of complex adaptive systems in the 1990s, the research of complexity science has been booming and is still in an ascendancy. Many scholars at home and abroad have introduced complexity science into the field of oil and gas safety science and technology and have carried out studies around different aspects, including the theory of large-scale accident causation, safety evaluation systems, intelligent diagnosis of complex equipment failure, and trend prediction. With the concept of predictive maintenance, the focus of oil and gas equipment inspection and maintenance has shifted from aftermath maintenance, preventive maintenance, etc., to the study of online prediction of when the equipment will fail and the type of faults, along with the implementation of targeted maintenance plans or emergency programs in advance to reduce unplanned shutdowns and even the occurrence of safety accidents.

Therefore, the combination of complexity science and predictive maintenance technology is the future development direction of complex system security warning research. Although there are still many challenges in this technology, it is still worth exploring and researching in depth. As the saying goes, "It is better to prevent the disease before the disease than to seek medical treatment after the disease". So, if we can move the time for accident warning forward and change the management mode that focuses on emergency response after accidents to a warning mode that focuses on monitoring dangerous states and strengthening pre-prevention, it is of great significance for ensuring the safe operation of complex oil and gas production systems and reducing or avoiding major catastrophic accidents.

1.2.1 Spatial and temporal correlation of accident causative factors in complex systems of oil and gas production

The complexity of the causes of accidents in oil and gas production is related both to the limitations of subjective cognitive abilities and the complexity of the accidents themselves, both of which are associated with a variety of complex nonlinear interactions between system components. The study of the "principle of systemic wholeness" and the "principle of systemic connectedness" points out that "interaction is the true ultimate cause of system failure". Exploring time-space correlations and hierarchical structures among factors, especially accident-causing factors, is the prerequisite and foundation for realizing early and accurate warning of systemic risks.

With the continuous improvement of equipment design and human awareness of prevention, the proportion of accidents induced by multi-factor coupling is on a rising trend. Analysis shows that 92% of global industrial accidents are caused by a combination of multiple factors, with an average of 4.39 risk factors per accident, and as many as 20. Multi-factor coupling-induced accidents are characterized by complex factor associations and strong concealment, and are prone to disastrous consequences.

The causative factors of accidents (such as changes in the external environment, degradation or failure of a certain component, poor coordination of the transmission system, etc.) will first affect a certain equipment, link or process within the system. Some subsystems have certain connections with the affected equipment, link or process, and the complex coupling relationship between subsystems leads to the diffusion and propagation of the impact of accident causative factors within the complex system, ultimately affecting the subsystems responsible for critical functions or directly threatening system safety (critical subsystems) within the system. The collapse (failure) of these subsystems will be manifested by the paralysis of certain functions. With the increase in the number of collapsed subsystems and an expansion in the hierarchy, it will eventually lead to the partial or total collapse of the entire complex system, i.e., the entire complex system cannot function normally, and in more serious cases, major security accidents will occur, which will bring about incalculable losses.

At the same time, the occurrence of accidents has a strong dynamic complexity, which is usually caused by the accumulation of accidental and coupled unsafe primitive events. The spatial and temporal correlation of accident causal factors is mainly reflected in the following aspects:

(1) *Adaptive and self-organizing*: When a component of a complex system fails, the spatiotemporal interaction between accident causing factors will have varying degrees of impact on other related components within the system. Simultaneous abnormalities in multiple components or abnormalities in a critical component may accelerate (or delay) the failure rate of the initial faulty component, resulting in complex coupling effects of system failures.

(2) *Uncertainty and randomness*: Random factors in complex systems (such as transient impact, environmental changes, and operational regulation) not only affect the organizational structure of the system but also affect its state and behavior, making the system unstable if coupled with the propagation path, the degree of impact, the form of performance, and a certain degree of uncertainty.

(3) *Emergence*: Interaction between subsystems can lead to the emergence of macro-integral properties that are significantly different from the behavior of individual subsystems. The faults coupling role of emergency is also reflected in the qualitative change: after the system is started, there is interaction between the main faults, which can self organize, self coordinate, self strengthen, and then propagate, diffuse, and develop, ultimately leading to a qualitative change in a single or small fault, forming a large-scale fault that leads to system collapse, defined as the emergence that has occurred.

(4) *Evolutionary*: The process of anticipation, adaptation, and self-organization of complex systems to external environments and states leads to the continuous evolution of system functions and structures. The eventual occurrence of a catastrophic accident in a system is a reflection of the stochastic evolution of a single fault itself and the interactive evolutionary process between multiple faults.

In addition, complex systems are characterized by hierarchy, non-linearity, openness, and fragility. As the saying goes, "A thousand miles of dike will collapse in an ant hole". The occurrence, development, aggravation, derivation, and secondary process of major catastrophic accidents are closely related to the characteristics of complex systems. The structure of oil and gas production systems is complex and the operating conditions are diverse. Although the causal factors of accidents in different environments for different equipment exhibit different spatiotemporal correlations, the evolution mechanism follows a regular pattern. Therefore, the purpose of early-warning safety research is to avoid the occurrence of such strong causal "chain reactions" between the accident-causing factors of the complex system and to prevent the smallest problems from occurring. On the one hand, starting from the root cause of the accident, intrinsic safety measures are adopted to suppress the occurrence of the root cause and risk of the accident. On the other hand, the degree of harm caused by the spatiotemporal correlation of accident factors in complex systems is reduced. By blocking the propagation and expansion of faults in complex systems, the occurrence of key component, subsystem and other critical links faults is restrained, enhancing the safety resilience of the system and reducing operational risks.

1.2.2   *Progress of basic research on early warning of complex systems in foreign countries*

According to the definition of the International Strategy for Disaster Reduction, early warning is an urgent warning of an impending disaster. In the field of system safety engineering, this refers to both an urgent warning of a possible impending accident and a delayed warning of secondary accidents that may arise from this accident after a certain period of time. Therefore, early warning is to measure the future state of the system and to make an anticipatory assessment of the development of future accidents, in order to find out in advance the possible problems of the future operation of the system and their causes, to forecast the temporal and spatial scope of the abnormal state and the degree of harm, and to put forward precautionary measures so as to avoid or minimize the possible losses.

Early warning includes early-warning analysis and early-warning control. Early-warning analysis is through signs-of-failure monitoring, identification, diagnosis, and evaluation and timely alarm technical activities. Early warning control is a management activity based on the analysis results of early warning, which seeks the precursor characteristics of faults or abnormal working conditions (such as abnormal trends), determines possible faults or abnormal working conditions, carries out targeted safety protection technical measures, and corrects, prevents, and controls the system. Therefore, early warning for complex systems is not only required for modern safety management but is also an effective measure to implement the policy of "safety first, prevention first".

In recent years, foreign scholars have made some outstanding progress in basic research and engineering applications in fields related to the safety warning of complex systems in oil and gas production.

1.2.2.1   *Study on the correlation law of accident causative factors and their fragility in complex adaptive systems*

The Laboratory for Intelligent Process Systems (LIPS) at Purdue University has been conducting research for nearly 25 years on risk identification, analysis, and management of complex adaptive systems (CAS). It explores the causal factors of systemic failures and their interconnections in complex systems through different fields and different accident forms, and studies the brittle characteristics of complex systems, as well as the commonalities and differentiations of accidents from a broader perspective, i.e., from the perspective of systems engineering, through which it is conducive to design and control these complex systems in the future.

LIPS points out that the challenge for the future is that the next generation of diagnostic and early-warning systems should be able to detect and monitor complex processes in real time, determine the degree of degradation of performance, predict potential failure scenarios, diagnose actual failures, and give recommendations or responses for corrective maintenance and control. The dynamics of complex processes and equipment need to be taken into account in the design of diagnostic and early-warning systems, and the difficulties center on the lack of effective sensing technologies, suitable analytical models, incomplete and uncertain data, heterogeneity of knowledge sources, and the significant expertise and effort required to develop and maintain these systems.

1.2.2.2 *Research on intelligent early-warning systems and predictive maintenance decision support technology*

The Center for Intelligent Maintenance Systems (IMS), a joint venture of the University of Cincinnati, the University of Michigan, and the Missouri University of Science and Technology, has established the Watchdog Intelligent Early Warning System and Decision Support Tool. This predictive technology anticipates downtime and identifies faults or failure modes that may lead to downtime. The system focuses on predicting when and why a system will fail and is currently working on predictive warning applications for large and complex industrial systems.

IMS conducts research on Decision Support Tools (DSTs) to provide the correct response that enterprise operators should execute when a machine is damaged or fails. When one or more pieces of equipment are likely to continue to fail in the future, DST can balance enterprise resources to reduce lost productivity due to downtime and facilitate optimization of equipment maintenance schedules to reduce downtime. IMS notes that in the area of complex system monitoring and safety warning technology, future research should focus on equipment performance degradation analysis and prediction, as well as safety warning-based predictive maintenance technology and engineering applications.

1.2.2.3 *Research on process safety uncertainty and risk resilience engineering for complex systems*

The Mary Kay O'Connor Process Safety Center (MKOPSC) is dedicated to research in risk management, consequence analysis, and risk resilience engineering in the area of process safety for complex systems. This includes complex system uncertainty issues, dynamic process operational risk analysis, and equipment degradation, prediction of inspection cycles, the impact of data fluctuations on prediction, and research on the ability of systems to recover from disturbances to a normal state. For example, MKOPSC uses Bayesian methods, LOPA, and fuzzy logic to process safety-related data for complex systems and to model and simulate causal relationships of causal factors.

MKOPSC states that in order to prevent catastrophic accidents, real-time safety decision-making needs to evolve with advances in complexity science, multi-perspective modeling, and hybrid intelligent systems. In the research process of safety early-warning systems, achieving a high level of warning and control requires focusing on (1) safety component functionality and (2) correct and safe interaction between components.

1.2.2.4 *Research on chemical equipment fault propagation law and intelligent process support system*

Based on a study of major chemical accidents that have occurred in Japan in recent years, Japanese scholar Kazuhiko Suzuki has proposed an Intelligent Operation Support System (IOSS) that can effectively prevent accidents from occurring. The complexity of equipment units, vessels, instrumentation, and control systems relating to chemical equipment makes it very difficult for operators to predict the impact of a failure and decide what corrective action to take. In addition, there is a need to implement a timely and rapid response, as any delay in responding to an abnormal event will result in the expansion and propagation of the fault. IOSS is able to calculate the impact of fault propagation during an abnormal event and suggest appropriate contingency measures to the operator, which will help the operator make quick safety decisions. IOSS is also able to use a simulator to predict the trend of the process variables during an abnormal condition. The status of all equipment is fed into the simulator to correctly calculate the predicted future values of process variables.

In contrast, China's safety early-warning research on equipment and process risks in the petroleum industry is in its infancy, but the safety early-warning technology has attracted extensive attention from petrochemical enterprises and has now become a hotspot for research. Combined with the characteristics of China's oil and gas equipment, production, and processing, the research should be focused on how to (1) find the first signs of accidents (i.e., eliminating implied faults); (2) predict the multiple paths of the propagation of hazards; (3) predict the trend of the future of a single or a combination of anomalous states; and (4) mitigate the failure of the combined, related, and other effects.

1.2.3 *New technology for safety early warning in domestic oil and gas production*

In recent years, the China University of Petroleum (Beijing) has carried out basic research on safety early warning of complex systems with the aim of improving the accuracy of identification of hidden dangers in complex systems, and has achieved some distinctive innovative results, which have solved some basic scientific and engineering problems in the field of safety early warning of complex systems. These include new methods for diagnosis and prediction of early implied faults, integrated multi-level correlation diagnosis-warning technology for coupled faults, and early-warning technology for accidents and events with multi-source information fusion and application research.

These research results can have two levels of significance: First, by applying such correlation diagnostic-warning mechanisms to specific oil and gas production complex systems to obtain hazard scenarios and safety pre-control measures, we can promote the research progress in the field of engineering where specific oil and gas production complex systems are located. Second, by studying the spatial and temporal correlation characteristics of accident-causing factors in as many specific oil and gas production complex systems as possible, this kind of correlation diagnosis-warning mechanism can be extended to as many complex systems in the oil and gas industry as possible, thus laying the foundation for finding a general theory describing the nonlinear interaction mechanism of compound faults in complex systems.

1.2.3.1 *A new approach to diagnosis and prediction of early implicit faults*

With the exception of a very small number of sudden failures, most of the failures that occur in complex systems are gradual failures, in which the functional indicators of the components gradually deteriorate until they fail, but continue to operate until they do. Gradual failures often induce nonlinear interactions between failures at an early stage and lead to the failure of more components or even the whole system with severe losses. In early implied fault diagnosis, there are two key issues that have been attracting much attention: First, the characteristic signals of early equipment fault prediction are very weak, often drowned by strong noise, with low signal-to-noise ratio, which is not easy to recognize and fix; second, most of the equipment is in the normal operation state most of the time in the actual operation, with fewer samples of anomalies and faults, and it is very difficult to accurately establish a discriminative model. After research, a new diagnosis and prediction method of early implied faults is proposed for oil and gas production equipment (Fig. 1.1),

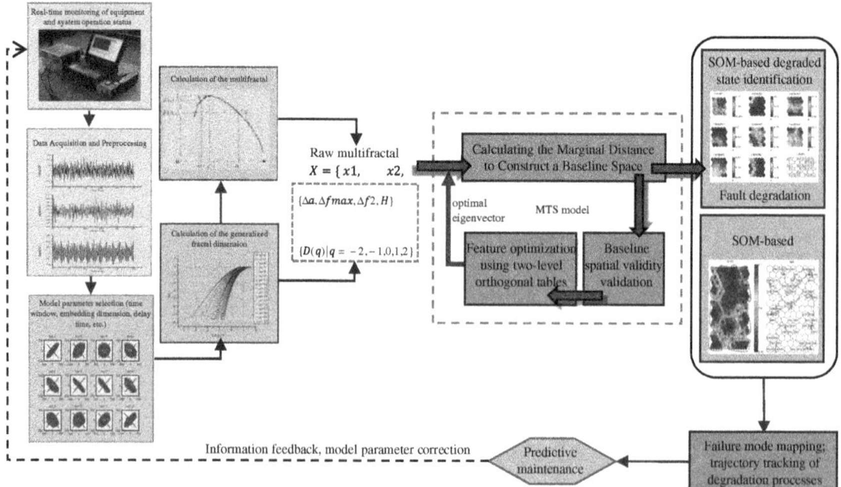

Fig. 1.1. A new approach to diagnosis and prediction of early implicit faults.

which can effectively capture the early signs of faults in the system and help control the development of faults. The significance of this research is as follows:

(1) For the first time, the quantitative diagnostic index of early equipment faults based on multiple fractal spectra and generalized dimensions is proposed from the multiple fractal nature of early faults, and the relationship between the local-scale characteristics and the overall characteristics of fault fractal phenomena is established.

(2) The traditional fault diagnosis accuracy is limited by the number of effective samples.

(3) The nonlinear relationship between the selection of multiple fractal feature parameters and diagnostic accuracy is revealed, and a statistical optimization method of diagnostic parameters is proposed by using a small number of abnormal samples, which greatly reduces the diagnostic process affected by the number of samples and subjective factors.

(4) Based on this method, a self-organized diagnosis and prediction method for the early degradation of key components of complex systems is further established, which reveals the multiple fractal features of the faults at each degradation stage of the components during the whole-life process. It draws the degradation state trajectories during the whole-life process based on the self-organized neural network, which is conducive to the predictive tracking of degradation states, and improves the accuracy of the fault prediction.

(5) A comprehensive diagnostic experimental platform for the early failure of indoor pipelines and power units has been established, which is capable of simulating different failure modes, failure locations, and failure degrees, and at the same time, a library of failure modes has been set up to serve the establishment and verification of the early-warning model.

1.2.3.2 *Multi-level correlation diagnosis-warning integration technique for coupled faults*

In the complex system of oil and gas production, most of the single-point failures have multiple propagation paths, and any small localized abnormal perturbation may propagate, diffuse, accumulate, and amplify through the failure causal chain, thus causing major safety accidents. In order to break through the assumptions of component failure independence required by traditional methods, a set of integrated multi-level correlation diagnosis-warning technologies are proposed for the randomness, complexity, and uncertainty of the causal chain of failures of complex systems (Fig. 1.2). These effectively solve the problem of accurately identifying the root causes of multi-component systems that generate multiple failures and accurately predicting the residual reliability life of the system under the conditions of component failures. The significance of this research result is as follows:

(1) For the first time in the domestic safety discipline, we have studied the inherent stochastic process and coupled interaction process of oil and gas equipment failure evolution and established a quantitative failure multiple propagation path model.
(2) A hybrid early-warning model integrating functional models, degradation models, behavioral models, event models, evaluation models, and predictive models based on the system's holistic principle and connectivity principle is proposed, which greatly expands the theory of multi-failure diagnosis and prediction of the system.
(3) On the basis of the hybrid warning model, an ant colony search and prediction algorithm is established to analyze the

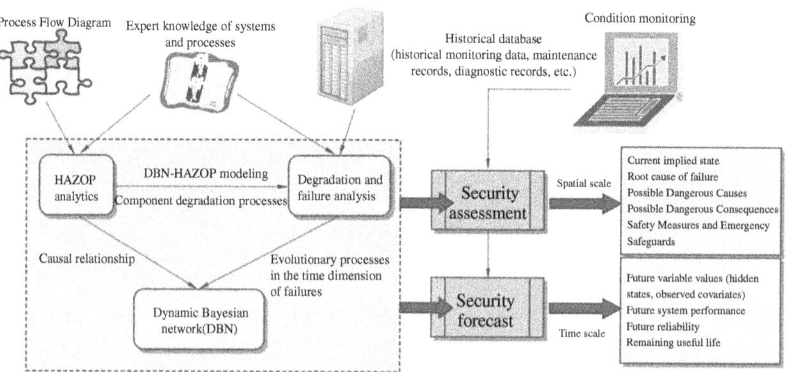

Fig. 1.2. Multi-level correlation diagnosis of coupled faults of oil and gas equipment and integrated technology of early warning.

uncertainty of multiple fault propagation paths, and the quantitative risk estimation value of each fault propagation path is determined by integrating two factors, namely, the risk fusion credibility and the severity.

(4) Successfully applied to oil and gas equipment complex systems such as large combustion and pressure unit systems in oil and gas production pressurization stations, large oil refining units, and oil and gas gathering and transportation joint stations in oil fields. Developed an expert system with independent intellectual property rights for hazardous source identification and HAZOP safety early-warning analysis, with the correct rate of fault trend prediction reaching more than 90%.

1.2.3.3 *Dynamic safety assessment and predictive maintenance technology for complex systems*

We have carried out research on dynamic safety assessment and predictive maintenance technology, and put forward an adaptive deployment mechanism for assessment indexes and their weights (Fig. 1.3). A series of comprehensive safety assessment models based on the three-dimensional perspective of time, a quantitative safety assessment model based on fuzzy information fusion, an intelligent risk classification assessment method for pipeline third-party damages, and a complex system predictive maintenance technology based on fault forward defense have been established, which can dynamically track the current safety status of the complex system of oil and gas equipment and foretell the future trend, which is helpful for suppressing any incipient faults.

1.2.3.4 *Establishing a theoretical system of cross-scale risk characterization and crisis early warning for process safety of oil and gas complex systems*

China's oil and gas production complex system (such as facilities, stations, reservoirs, and refineries) process complexity, harsh operating conditions, a high degree of influencing factors, high temperature, high pressure, low temperature, vacuum, large flow, high speed, and other extreme conditions. Any failure, to a large extent, affects the safe and smooth supply and utilization of oil and gas resources.

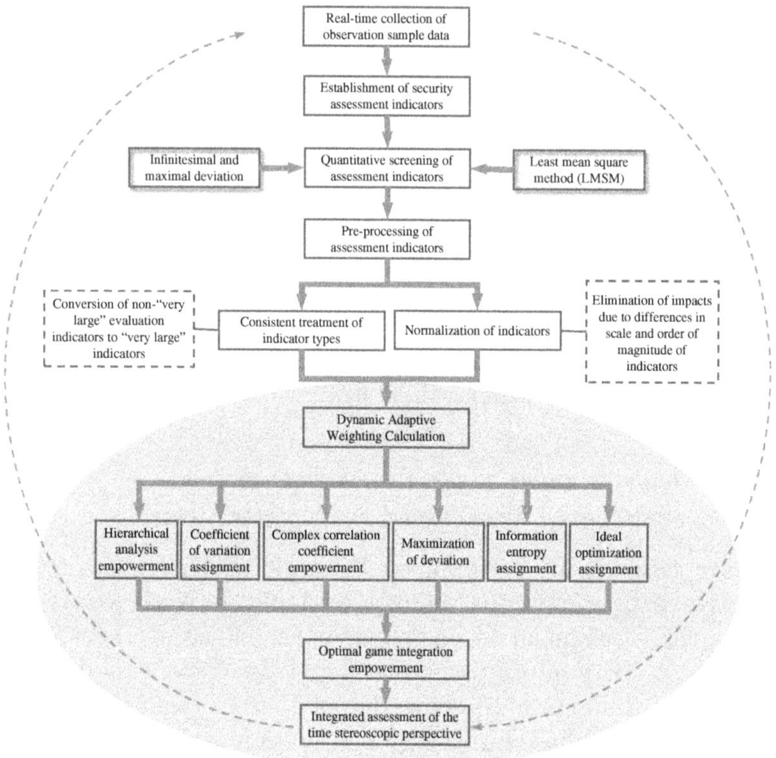

Fig. 1.3. Adaptive deployment mechanisms for dynamic security assessment indicators and their weights.

Oil and gas safety science and engineering also put forward higher requirements. The operational safety of complex systems of oil and gas production has become a social public safety issue.

The complex system of oil and gas production has a complicated structure, variable working conditions, hidden root causes, etc. Any risks have natural and social attributes, and the risk components have scale effects at different levels and different operation cycles. The problem of accurate monitoring, diagnosis, traceability, and early warning of faults under variable operating conditions requires a multi-scale and cross-scale approach. Based on the operational characteristics of complex oil and gas production systems, we have conducted theoretical research on process safety cross scale risk characterization and safety warning. With the main line of "disaster mechanism caused by multi-scale coupling effect of risk factors → extraction of multi-scale coupling characteristics of risk factors → integration of multi-source information of risk factors → fault monitoring based on multi-scale coupling characteristics of risk factors → cross scale intelligent warning and alarm optimization → promotion and application research", we have conducted in-depth research on basic scientific problems and method application practices. The key technologies mainly include the following: (1) adaptive analysis of accident scenarios and risk propagation paths based on the scale effect of risk factors; (2) multi-modal identification and intelligent monitoring of oil and gas production complex system failures; (3) fusion of multi-source heterogeneous risk information and risk scale deduction methods for oil and gas complex systems; (4) cross-scale intelligent warning and alarm optimization methods; (5) China's oil and gas production complex systems; and (6) complex systems for shale gas large-scale fracturing, refining and chemical production in China. Then, a cross-scale risk characterization and crisis warning theoretical system for process safety of complex systems of oil and gas production was formed (Fig. 1.4).

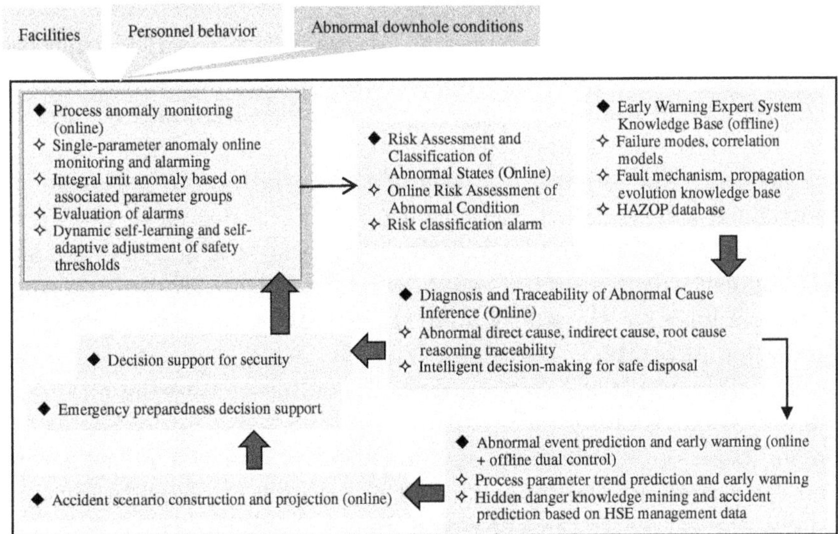

Fig. 1.4. Theoretical system of anomaly monitoring, risk tracing, assessment, and safety early warning for complex system of oil and gas production.

1.3 The Technical Challenge

Comprehensive domestic and foreign at this stage in the safety of complex systems early-warning basic research and engineering applications of the progress made and the problems that exist, summed up the phenomenon of "five more and five less": more research on component failures, less research on system failures; more qualitative research, less quantitative research; more research on a single failure, less research on complex failures; more research on static risk, less research on dynamic risk; more research on risk evaluation, less research on risk prediction. Dynamic risk research has been insufficient. For this reason, future safety warning research needs to realize the following four breakthroughs:

(1) Realize the breakthrough from qualitative research to quantitative research

Safety early-warning research work can usually be divided into five levels: first of all, monitoring the signs of failure or abnormal working conditions; second, identifying the type of failure; he third, diagnosing the root cause of the failure; fourth, evaluating the degree of damage; and fifth, predicting the future development trend of the failure to give an alarm in advance.

If the first to third levels are called qualitative research, then the fourth and fifth levels are quantitative research. Qualitative research is an important foundation for quantitative research, and the fourth and fifth levels of quantitative research are closely linked because the accurate and reliable analysis of the initial failure type and its root cause will lead to the identification of the degree of failure or degradation, failure trend prediction, and early-warning technology. The quantitative research of safety early warning is required so as to provide a basis for the safety analysis, reliability assessment, and life prediction of complex systems.

The first step is to carry out dynamic online monitoring and fault identification of abnormal states of equipment and processes; then, on the basis of the diagnosis of the causes of faults or degradation, one can carry out the evaluation of the degree of degradation; and finally, one can carry out the prediction of the failure trend, the prediction of the future impact on the other relevant components (or the system as a whole), and the prediction of the remaining safe life. It can be seen that the focus of future research on safety early warning for complex systems will shift from qualitative to quantitative research.

(2) Realize the breakthrough from single-fault early-warning research to compound-fault early-warning research

The early-warning research work of single fault relies on the signal processing method, which is usually easy to realize, but its accuracy is not high, the generalization ability is not strong, and the generality is poor, thus restricting its application in engineering. For the complex system of oil and gas production, since failure or accidents are often caused not by single faults but as a result of the coupling and association of a variety of failure factors, the early-warning analysis of the complex system from the perspective of a single failure will result in omissions or even false alarms.

The components or subsystems within complex systems interact, depend on, and constrain each other. The propagation of faults in complex systems involves simultaneous and cascading faults, and the fault characteristics are not simply superimposed, but present a state of mutual coupling. The generation of such coupling faults brings greater difficulties to safety warning work. Therefore, one should accurately identify the composite faults that lead to system anomalies and predict the trend of mutual influence between the composite faults, as well as predict the impact on the system.

(3) Realize the breakthrough from parts failure research to complex system failure research

Safety early-warning studies of equipment component failures are mainly focused on monitoring and diagnosing service failures and predicting trends in related parameters for key components or production units in the system, such as gears, bearings, rotors, pressure vessels, and process pipelines. However, it is the interaction of the various units within the complex system of oil and gas production that is the essential cause of accidents. Such early-warning analyses at the component level can often only diagnose induced failures, but cannot cure the system's hidden safety hazards. Due to the fact that this warning analysis method often ignores the interaction, propagation, and impact between system components and fault processes, only considers the impact of external environmental changes and operating conditions on the degradation process of components, and calculates the average service life based on the degradation process of components themselves, predicting the reliability of future remaining life. The reasoning and decision-making mechanisms are not very reasonable, and the results are often optimistic and unrealistic.

Therefore, future research should focus on viewing the oil and gas production system as a multi-level, nonlinear, and complex whole and establishing a multidimensional and multi-parameter complex system model. Then, the system should be studied in depth to explore the root causes of the complex system failures and to find out the primary failures, so as to eradicate the safety hazards.

(4) Realize the breakthrough from static safety evaluation research to dynamic risk prediction research

Safety early-warning analysis is different from fault diagnosis and evaluation in the traditional sense because the fundamental goal of early warning is to avoid the conditions under which the fault or abnormal working conditions arise, especially the conditions and characteristics at the time when the fault may be in the early stage of development, and to advance the preventive and control measures before the formation of a fault. Therefore, dynamically tracking the changes in physical properties and energy flow of input and output media in complex systems will be the focus of research on complex system safety warning. Due to the dynamic nature of the operation process, internal/external environmental factors, operation tasks, and degradation of units and components in complex systems in oil and gas production, future safety warnings need to study how to dynamically and automatically update the structure and parameters of prediction methods and models based on changes in internal and external environments, adjustments in operating conditions, and the degradation patterns of systems and components themselves. By understanding the relationship between the evolution process of failure or degradation and the early-warning signs, one can ensure the safety characteristics of the system in real time.

References

[1]   Zhang L, Hu J. Safety prognostic technology in complex petroleum engineering systems: Progress, challenges and emerging trends. *Petroleum Science*, 2013, 10(4): 486–493.

[2] Hu J, Zhang L, Liang W. Dynamic degradation observer for bearing fault by MTS-SOM System. *Mechanical Systems and Signal Processing*, 2013, 3(2): 385–400.

[3] Hu J, Zhang L, Liang W, *et al.* An adaptive online safety assessment method for mechanical system with pre-warning function. *Safety Science*, 2012a, 50(3): 385–399.

[4] Hu J, Zhang L, Liang W, *et al.* An integrated method for safety pre-warning of complex system. *Safety Science*, 2010, 48(5): 580–597.

[5] Hu J, Zhang L, Liang W, *et al.* An integrated safety prognosis model for complex system based on dynamic Bayesian network and ant colony algorithm. *Expert Systems with Applications*, 2011, 38(3): 1431–1446.

[6] Hu J, Zhang L, Liang W, *et al.* Intelligent risk assessment for pipeline third-party interference. *Journal of Pressure Vessel Technology, Transactions of the ASME*, 2012b, 134(1): 165–173.

[7] Hu J, Zhang L, Liang W, *et al.* Mechanical incipient fault detection based on multifractal and MTS method. *Petroleum Science*, 2009a, 6(2): 208–216.

[8] Hu J, Zhang L, Liang W, *et al.* Opportunistic predictive maintenance for complex multi-component systems based on DBN-HAZOP model. *Process Safety and Environmental Protection*, 2012c, 90(5): 376–388.

[9] Hu J, Zhang L, Liang W, *et al.* Quantitative HAZOP analysis for gas turbine compressor based on fuzzy information fusion. *Systems Engineering—Theory & Practice*, 2009c, 29(8): 153–159.

[10] Hu J, Zhang L, Liang W, *et al.* The application of integrated diagnosis database technology in safety management of oil pipeline and transferring pump units. *Journal of Loss Prevention in the Process Industries*, 2009b, 22(6): 1025–1033.

[11] Lee J, Deng CL, Tsan CH, *et al.* Product life cycle knowledge management using embedded Infotronics: Methodology, tools and case studies. *International Journal of Knowledge Engineering and Data Mining*, 2010, 1(1): 20–36.

[12] Lee J, Ni J, Djurdjanovic D, *et al.* Intelligent prognostics tools and e-maintenance. *Computers in Industry*, 2006, 57(6): 476–489.

[13] Markowski AS, Mannan MS, Bigoszewska A. Fuzzy logic for process safety analysis. *Journal of Loss Prevention in the Process Industries*, 2009, 22(6): 695–702.

[14] Markowski AS, Mannan MS, Bigoszewska A, *et al.* Uncertainty aspects in process safety analysis. *Journal of Loss Prevention in the Process Industries*, 2010, 23(3): 445–454.

[15] Mitchell SM, Mannan MS. Designing resilient engineered systems. *Chemical Engineering Progress*, 2006, 102(4): 39–45.

[16] Rajaraman S, Hahn J, Mannan MS. A methodology for fault detection, isolation, and identification for nonlinear processes with parametric uncertainties. *Industrial and Engineering Chemistry Research*, 2004, 43(21): 6774–6786.

[17] Suzuki K, Munesawa Y. How to prevent accidents in process industries? Recent accidents and safety activities in Japan. In: *The 4th World Conference of Safety of Oil and Gas Industry*, June 27–30, 2012, Seoul, Korea.

[18] Venkatasubramanian V. Prognostic and diagnostic monitoring of complex systems for product life cycle management: Challenges and opportunities. *Computers & Chemical Engineering*, 2005, 29(6): 1253–1263.

[19] Venkatasubramanian V. Systemic failures: Challenges and opportunities in risk management in complex systems. *AIChE Journal*, 2011, 57(1): 2–9.

[20] Yang X, Mannan MS. An uncertainty and sensitivity analysis of dynamic operational risk assessment model: A case study. *Journal of Loss Prevention in the Process Industries*, 2010a, 23(2): 300–307.

[21] Yang X, Mannan MS. The development and application of dynamic operational risk assessment in oil/gas and chemical process industry. *Reliability Engineering and System Safety*, 2010b, 95(7): 806–815.

Chapter 2

Early Intelligent Warning Based on Real-Time Monitoring Data of Process Parameters

2.1 An Adaptive Process Early-Warning Method Based on Trend Monitoring

With the expansion of the scale of the process industry and the improvement of automation levels, ensuring the safe operation of the process is a key issue to be solved in the production of enterprises. For some high-risk complex industrial processes, such as the petroleum refining process, the raw materials are mostly flammable and explosive hazardous materials, and most of the materials transported within the petroleum refining device are corrosive. If process failure occurs, it will cause huge economic losses, casualties, and environmental pollution. Such complex processes often have a large number of monitoring variables; however, the operator cannot take into account the trend of all variables. In some abnormal working conditions, there may be hundreds of process variable alarms (i.e., alarm flood phenomenon), which seriously interfere with the operator's judgment of the current process state and increase the operator's processing difficulty, making it very easy to introduce new risks during the emergency response and may even lead to the occurrence of derivative secondary accidents. Therefore, before a system failure occurs, if effective monitoring and timely warning of abnormal conditions can be carried out, and abnormal process information can be provided to the operator for early detection of abnormal conditions, more associated alarms can be avoided, and at the same time, the pre-control timeliness of the alarm system can be improved.

Trend analysis is a data analysis method that extracts process information in order to visualize the trend of process variables, and this type of method is widely used in process condition monitoring, fault detection, and diagnostic systems due to its computational simplicity. However, existing trend analysis methods generally lack adaptivity and are not applicable to large and complex systems with many process variables.

In complex oil and gas production processes, different process variables may produce different fluctuations and magnitudes of change. Influenced by process conditions, equipment status, external disturbances, and other factors, under different circumstances, different levels of interference noise may appear in each process monitoring signal. Process changes, signal transmission interruptions, and other factors may also cause step changes or instantaneous changes in process variables. Therefore, how to adaptively extract trend characteristics for different process variables is the key to realizing effective trend monitoring and early warning. In addition, in practice, sometimes there are multiple variables in a short period of time, which may lead to operators not being able to find important alarms in time, mishandling abnormal working conditions, and causing more serious accidents.

To this end, this section proposes an adaptive process early-warning method based on trend monitoring, which monitors process variables online through adaptive trend analysis and provides ultra-high or ultra-low warnings for process variables that exhibit non-smooth trends. When multiple variables all issue warnings within a short period of time, an adaptive weight calculation method is used to provide the operator with a reasonable order for processing the warning variables. Through online monitoring of process data trends and adaptive alarms for process variables, real-time tracking of process operation status is achieved to improve the predictive ability and timely warning control of the alarm system, in order to timely detect the characteristics of fault precursors, eliminate adverse factors that cause faults, and ensure safe and reliable operation of the process.

2.1.1 *Basic theory of trend analysis*

Trend analysis refers to the identification of time series data that exhibit the same behavior (e.g., rising, falling) over a period of time as a linear segment with a trend such as rising or falling, and the methodology mainly consists of two major parts, namely, segmentation of data segments and trend identification.

2.1.1.1 *Data fragmentation*

For given time series data, assume that the i-1st linear feature segment is estimated at moment t_s^{i-1} using least-squares fitting with slope p^{i-1}, the segment start moment is t_0^{i-1}, the vertical coordinate at moment t_0^{i-1} is x_0^{i-1}, and the output of the model at moment $t_j(t_j > t_s^{i-1})$ is the relationship shown in Eq. (2.1):

$$x(t_j) = p^{i-1}(t_j - t_0^{i-1}) + x_0^{i-1}. \tag{2.1}$$

The difference $e(t_j)$ between the measured value $x_m(t_j)$ and the model output value $x(t_j)$ at time t_j is shown in Eq. (2.2):

$$e(t_j) = x_m(t_j) - x(t_j). \tag{2.2}$$

The cumulative sum $cusum(t_j)$ of the differences is calculated from the moment t_s^{i-1} as shown in Eq. (2.3):

$$cusum(t_j) = cusum(t_{j-1}) + e(t_j) \sum_{k=s^{i-1}}^{j} e(t_k). \tag{2.3}$$

At each sampling moment t_j, the absolute value of the cumulative sum $|cusum(t_j)|$ is compared to two preset thresholds th_1 and th_2. th_1 is used to determine if the current linear model is appropriate; th_2 is used to determine when to make a new linear estimate. If $|cusum(t_j)| < th_1$, the current linear model is acceptable. If $th_2 \geq |cusum(t_j)| \geq th_1$, the moment $|cusum(t_j)|$ just reaches the threshold, th_1 is denoted as $t_{j_1}^i$, and the moment $|cusum(t_j)|$ reaches the threshold, th_2 is denoted as $t_{j_2}^i$. At the same time, the measured values $x_m(t_j)$ and $t_j(t_{j_1}^i \leq t_j \leq t_{j_2}^i)$ at the moments from $t_{j_1}^i$ to $t_{j_2}^i$ are recorded in the region of outliers, where the original linear model is no longer appropriate, at this time, based on the data in the region of outliers. A new linear segment i is fitted using the least-squares method (with no less than three fitted data), and the cumulative and $cusum(t_j)$ are reset to 0. The starting sampling moment of the new linear segment is $t_0^i = t_{j_1}^i$, the segment-fitting moment is $t_s^i = t_{j_2}^i$, and the segment-ending moment is $t_e^i = t_0^{i+1} - 1(t_e^i > t_{j_2}^i)$.

The output of this segmentation algorithm mainly includes the slope p^i, the segment start sampling moment t_0^i, the start moment t_0^i, longitudinal coordinate x_0^i, and the end moment of the segment t_e^i. The longitudinal coordinate x_e^i at the moment of t_e^i is shown in Eq. (2.4):

$$x_e^i = p^i(t_e^i - t_0^i) + x_0^i. \tag{2.4}$$

2.1.1.2 *Trend identification*

The segmented segments are categorized into seven types of trends: rising (a), falling (b), smooth (c), positive step (d), negative step (e), rising/falling change (f), and falling/rising change (g).

First, three parameters are defined: $Q^i(x_e^i - x_e^{i-1})$ represents the total upward (or downward) change of the model; discontinuous upward (or downward) values are represented by $Q_d^i(x_0^i - x_e^{i-1})$; and upward (or downward) values with the same slope are represented by $Q_s^i(x_e^i - x_0^i)$.

Based on the three parameters defined and the two preset thresholds t_{hc} and t_{hs}, trend identification is carried out through the decision tree shown in Fig. 2.1. It should be noted in particular that by continuous, we do not mean continuous in the strict mathematical sense, but rather in a qualitative range where the acceptable trend change is not significant. The threshold value t_{hc} is used to determine whether two segments are continuous or not, categorizing any trend change that exceeds this threshold as a discontinuous change. Threshold t_{hs} is used to determine whether the trend type is upward, downward, or flat.

2.1.2 *Adaptive process warning methods based on trend monitoring*

Based on the existing process industry alarm systems that have the problem of alarm flooding caused by untimely alarms, this section proposes an adaptive process warning method based on trend monitoring. Adaptive state monitoring of process variables is carried out by adaptively adjusting the relevant parameters of trend analysis, and ultra-high or ultra-low warnings are provided for process variables showing non-smooth trends. If multiple variables send warning messages within a short period of time, an adaptive weighting method is used to assign weights to each variable, providing the operator with a reasonable processing order for the warning variables. The main elements of the method are described in the following:

2.1.2.1   *Adaptive process trend monitoring method*

Firstly, the observed data of process variables are partitioned into linear segments by Eqs. (2.1)–(2.4), and combined with the trend identification method, the trend identification is performed based on the decision tree shown in Fig. 2.1 in order to get the change trend information of each process variable.

In the oil and gas production process, process conditions, equipment conditions, external disturbances, and other influencing factors may cause some process variables to undergo instantaneous sudden

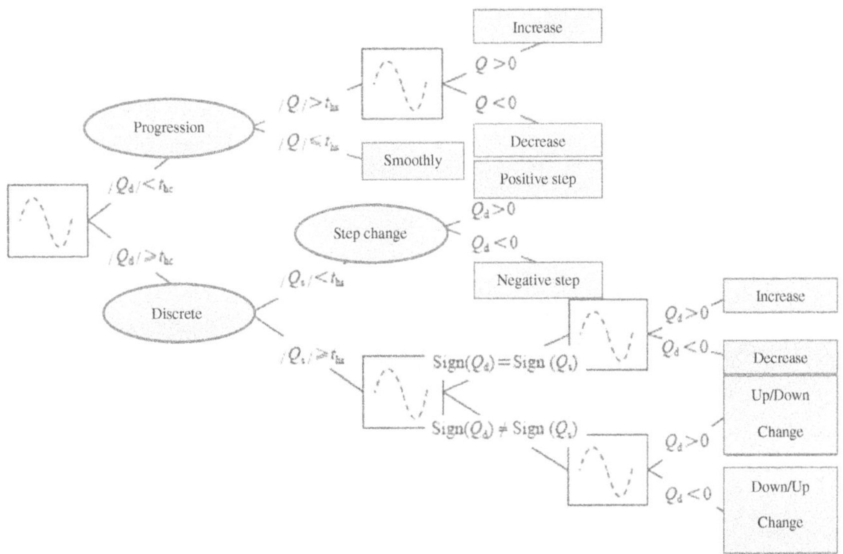

Fig. 2.1.   Trend recognition decision tree.

changes or step changes. Here, instantaneous sudden change refers to the sudden change of process variables whose duration is less than S (adjustment parameter) sampling period; step change refers to the sudden change of process variables whose duration is more than S (adjustment parameter) sampling period. If the instantaneous sudden change or step change occurs in a new linear estimation process, it will be difficult to obtain appropriate model parameters through least-squares estimation.

In order to obtain a reasonable trend of the process variable changes, a repetitive median filtering method is used to distinguish between instantaneous sudden changes and step changes before a new linear estimation as follows:

The derivative of a process variable measurement $x_m(t_j)$ is shown in Eq. (2.5):

$$x_d(t_j) = x_m(t_j) - x_m(t_{j-1}). \tag{2.5}$$

Let the size of the sliding time window be N. The median value of the derivative of the measured value of a variable, $x_d(t_j)$, within that time window is shown in Eq. (2.6):

$$M(t_j) = median_{i=j-N:j}[x_d(t_i)]. \tag{2.6}$$

The rate of change of the derivative of the measurement, $v_d(t_j)$, is shown in Eq. (2.7):

$$v_d(t_j) = median_{i=j-N:j}|x_d(t_i) - M(t_j)|. \tag{2.7}$$

For the derivative $x_d(t_{j+1})$ of the measured value of the process variable at the moment t_{j+1}, if $|x_d(t_{j+1}) - M(t_j)| \geq \omega v_d(t_j)$ (ω is the tuning parameter), it indicates that there is an instantaneous sudden change or step change at the moment t_{j+1}. For moment t_i ($i = j+1, j+2, \ldots, j+S$), if $|x_m(t_i) - x_m(t_j)| \geq \omega v_d(t_j)$, it is a step change, and vice versa, it is an instantaneous sudden change.

If an instantaneous mutation is detected, the mutation data are deleted and then linear estimation is performed; if a step change is detected, linear estimation is performed on the measurements recorded in the outlier zones from t_{j1} to t_j at the moments before the mutation occurs, and then the measurements recorded in the outlier zones from t_{j+1} to t_{j+S} at the moments after the mutation occurs, with a sampling period $S \geq 3$.

For the time window size N, it has to be short enough to obtain an accurate estimate of the rate of change, but it also needs to be long enough to ensure the smoothness of the measurement data. Therefore, a compromise between the desired smoothness assumption and the estimated accuracy is required. The tuning parameter ω can be adjusted on a case-by-case basis so that the detected change is slightly larger than the noise level.

2.1.2.2 *Adaptive parameter tuning methods*

If the four adjustment parameters th_1, th_2, t_{hc}, and t_{hs} are set to fixed values based on process knowledge and expert experience, the accuracy and applicability of the method for online application will be reduced. For this reason, an adaptive function is used to adaptively adjust the four parameters online at the fitting moment t_s^i for each segment i, taking into account the variation of the process measurement data.

The repeated median filtering method is also adopted, and the difference $e(t_j)$ between the measured value of the parameters $x_m(t_j)$ and the output value $x(t_j)$ of the linear estimation model at the time t_j can be calculated by Eq. (2.2), and the time window with the size of N_r, which is prior to the fitting moment t_s^i of each segment i, can be selected, and the median value of $e(t_j)$ within this time window is shown in Eq. (2.8):

$$M_e(t_s^i) = median_{j=s^i-N_r:s^i}[e(t_j)]. \tag{2.8}$$

The rate of change of $e(t_j)$ over the time window N_r, $v_e(t_s^i)$ is shown in Eq. (2.9):

$$v_e(t_s^i) = median_{j=s^i-N_r:s^i}|e(t_j) - M_e(t_s^i)|. \tag{2.9}$$

The choice of the median value, rather than being based on the mean and standard deviation, is intended to minimize the effect that transient abrupt changes have on the measured data. As long as the transient mutation duration is less than one-half of the time window N_r, the mutation will have no effect on the rate of change $v_e(t_s^i)$.

The values of t_{hc}, th_1, th_2, and t_{hs} are adaptively adjusted as shown in Eqs. (2.10)–(2.13) at each segment fitting moment t_s^i:

$$t_{hc}(t_s^i) = \lambda v_e(t_s^i), \tag{2.10}$$

$$th_1(t_s^i) = v_e(t_s^i)/2, \tag{2.11}$$

$$th_2(t_s^i) = \alpha t_{hs}(t_s^i) = \alpha \mu v_e(t_s^i), \tag{2.12}$$

$$t_{hs}(t_s^i) = \mu v_e(t_s^i). \tag{2.13}$$

In order to initialize $v_e(t_s^i)$, the initialization value v_{initial} of $v_e(t_s^i)$ is calculated based on the smooth process measurement data within a period of time N_{initial} as shown in Eq. (2.14):

$$v_{\text{initial}} = median_{i=j-N_{\text{initial}}:j}$$
$$\times \{|x_m(t_i) - median_{k=j-N_{\text{initial}}:j}[x_m(t_k)]|\}. \tag{2.14}$$

t_{hc} is used to determine whether a process variable has undergone a sudden change that is much larger than its noise level. In order to regulate t_{hc}, the parameter λ can be adjusted according to the sensitivity of the identification of the process trend, so as to detect small changes in the process trend up to larger changes. t_{hs} is used to determine the upward or downward trend of the process, and μ is the parameter used to regulate t_{hs}, so that different degrees of upward or downward changes of the process trend can be detected by adjusting the parameter μ. α is the parameter used to regulate th_2, indicating that it is used to detect trends with amplitudes of t_{hs}. α is a parameter used to regulate th_2, which represents the delay of the algorithm used to detect the trend change with the amplitude of t_{hs}, which is in essence a sampling period. In order to effectively identify smooth process trends, the value of th_1 is set below the noise level $v_e(t_s^i)$.

2.1.2.3 *Univariate warning methods*

If the new fitted segment i shows an increasing positive step or decreasing/increasing trend, for any moment $t_j(t_{j2}^i \leq t_j \leq t_e^i)$ between the moment t_{j2}^i and the end of the segment t_e^i, the difference between the measured value $x_m(t_j)$ at moment t_j and the high alarm threshold x_H of the process variable is calculated:

$$\delta_H(j) = x_H - x_m(t_j). \tag{2.15}$$

If $\delta_H(j) \leq 0.2(x_H - x_L)$, the variable is warned to be ultra-high in order to remind the operator to take effective measures in time to eliminate the abnormal working conditions.

If the new fitted segment shows a decreasing, negative step or rising/falling trend, for any moment $t_j(t^i_{j2} \leq t_j \leq t^i_e)$ between the moment t^i_{j2} and the end of the segment t^i_e, calculate the difference between the measured value $x_m(t_j)$ at the moment t_j and the process variable's low alarm threshold x_L value:

$$\delta_L(j) = x_m(t_j) - x_L. \tag{2.16}$$

If $\delta_L(j) \leq 0.2(x_H - x_L)$, the variable is given an ultra-low warning to alert the operator to take timely and effective measures to eliminate the abnormal working condition.

For process measurement data recorded in the outlier region (moments t^i_{j1} to t^i_{j2}), if $\delta_H(j)$ or $\delta_L(j)$ is less than $0.1(x_H - x_L)$ with respect to the measured value of the variable $x_m(t_j)$ at any moment $t_j(t^i_{j1} \leq t_j < t^i_{j2})$ in the intervening period, the process variable is subjected to a super-high or super-low alert.

2.1.2.4 *Multivariate adaptive early-warning methods*

If a number of process variables in a short period of time are issued warning signals, the following adaptive weight calculation method is used to calculate the weight of each variable, and based on the weight of each variable and the degree of deviation from the normal range of each variable at the time of the warning, the warning variables are prioritized and the priority order of the warning variables is given.

The proposed adaptive weight calculation method includes the following steps:

Step 1: Calculate the fluctuation weights.

If a process variable x^k $(k = 1, 2, \ldots, K)$, K is the number of warning variables) is warned at moment t_P, for x^k measured at n moments t_P before moment $x^k_m(t_{P-n+1}), x^k_m(t_{P-n+2}), \ldots, x^k_m(t_P)$, denote $\overline{x^k_m} = \frac{1}{n}\sum_{j=P-n+1}^{P} x^k_m(t_j)$ and $s_{x^k_m} = \{\frac{1}{n-1}\sum_{j=P-n+1}^{P}$

$[x_m^k(t_j - \overline{x_m^k})]^2\}^{\frac{1}{2}}$, and for each process variable x^k ($k = 1, 2, \ldots, K$), calculate its fluctuation weight as shown in Eq. (2.17):

$$w_{k1} = (S_{x_m^k}/\overline{x_m^k}) \bigg/ \left(\sum_{k=1}^{K} S_{x_m^k}/\overline{x_m^k}\right). \tag{2.17}$$

Fluctuation weights reflect the relative magnitude of change in each process variable. The higher the volatility weight, the greater the degree of volatility of the variable in the period prior to the occurrence of the warning. The fluctuation weight represents the fluctuation trend of a variable before the warning, and the larger the fluctuation weight, the greater the degree of fluctuation of the variable over time before the warning occurs.

Step 2: Calculate the information weights.

In complex modern industrial processes, there are often various nonlinear correlations between process variables. The Maximal Information Coefficient (MIC), which is based on Mutual Information (MI), can be used to measure the degree of nonlinear correlation between two variables and it performs well on high dimensional complex datasets. Literature studies have shown that MIC outperforms other methods in analyzing correlation problems.

In terms of information theory, MIC can be understood as the percentage of information in a process variable that can be explained by another variable. More intuitively, it is based on the idea that if there is a correlation between two process variables, by dividing the scatterplot consisting of the two variables x^k and x^l into a grid of i rows and j columns, a number of different gridding schemes can be obtained, and by identifying the gridding scheme with the largest mutual information value from these gridding schemes, the maximum mutual information value I_{ij} is obtained. Gradually increase the resolution i and j of the grid, calculate the maximum mutual information value I_{ij} at each resolution based on the number of points in the grid, and normalize these mutual information values [Eq. (2.18)] to ensure a fair comparison between grids of different resolutions.

$$I_{ij}^s = \frac{I_{ij}}{1b[\min(i, j)]}. \tag{2.18}$$

By comparing the maximum mutual information normalized values calculated at each resolution, the maximum of them is obtained as the MIC value for the two variables x^k and x^l as shown in Eq. (2.19):

$$MIC[x^k, x^l] = \max_{i \times j < B(n)} I_{ij}^s, \qquad (2.19)$$

where n is the number of process data samples and $B(n) = n^{0.6}$.

MIC represents the degree of correlation between two variables, and the larger the MIC value, the greater the correlation between the two variables. $MIC[x^k, x^l]$ represents the correlation between x^k and x^l, and the information weight of the warning variable x^k is shown in Eq. (2.20):

$$w_{k2} = \sum_{l=1}^{K(l \neq k)} MIC[x^k, x^l] \bigg/ \left(\sum_{k=1}^{K} \sum_{l=1}^{K(l \neq k)} MIC[x^k, x^l] \right). \qquad (2.20)$$

The information weight reflects the correlation level of each process variable with other early-warning variables, and the larger the information weight, the greater the correlation between the variable and other early-warning variables, and the greater the influence on the changes of other variables.

Step 3: Calculate the deviation weights.

If the process variable x^k ($k = 1, 2, \ldots, K$), K is the number of warning variables) is warned at moment t_P, the measured value of x^k at n moments prior to moment t_P is the vector $\mathbf{X_m^k} = [x_{\mathbf{m}}^k(t_{P-n+1}), x_{\mathbf{m}}^k(t_{P-n+2}), \ldots, x_{\mathbf{m}}^k(t_P)]^T$, so that the deviation and $d^k = \sum_{i=P-n+1}^{P} \sum_{j=P-n+1}^{P} |x_m^k(t_i) - x_m^k(t_j)|$ of the variable x^k, and then the deviation weight corresponding to the process variable x^k is as shown in Eq. (2.21):

$$w_{k3} = d^k \bigg/ \left(\sum_{k=1}^{K} d^k \right). \qquad (2.21)$$

The greater the sum of the deviations of a process variable, the greater the weight of the deviation of that variable, indicating that the measurement of that variable has changed over a greater span of time prior to the occurrence of the warning and should be emphasized.

Step 4: Calculate composite weights.

In order to combine the fluctuation weights, information weights, and deviation weights to obtain the combined weights for each process variable, a weight matrix $\boldsymbol{W}$ is required as shown in Eq. (2.22):

$$\boldsymbol{W} = \begin{bmatrix} w_{11} & w_{12} & w_{13} \\ w_{21} & w_{22} & w_{23} \\ \vdots & \vdots & \vdots \\ w_{K1} & w_{K2} & w_{K3} \end{bmatrix}, \tag{2.22}$$

where w_{k1}, w_{k2}, w_{k3} $(k = 1, 2, \ldots K)$, K is the number of early-warning variables, 1, 2, 3 denote the fluctuation weight, information weight, and deviation weight, respectively) denote the value of fluctuation weight, deviation weight, and information weight of each process variable, respectively.

The maximum values of the fluctuation weights, deviation weights, and information weights of each process variable (w_{M1}, w_{M2}, w_{M3}) are selected to construct the reference weight vector:

$$\boldsymbol{W_M} = [W_{M1}, W_{M2}, W_{M3}], \tag{2.23}$$

and calculate the distance between the weight vector of each process variable and the reference weight vector as shown in Eq. (2.24):

$$D_k = \sum_{r=1}^{3} |w_{Mr} - w_{kr}|, \quad k = 1, 2, \ldots, K. \tag{2.24}$$

Finally, the combined weight of each process variable w'_k [Eq. (2.25)] is obtained and normalized to obtain the normalized combined weight of each process variable [Eq. (2.26)]:

$$w'_k = 1/(1 + D_k), \quad k = 1, 2, \ldots, K, \tag{2.25}$$

$$w_k = w'_k \bigg/ \sum_{k=1}^{K} w'_k. \tag{2.26}$$

Step 5: Preprocess the process variables.

If the variable x^k undergoes a super-high warning at the moment t_P, it is preprocessed according to Eq. (2.27):

$$x_m^{*k}(t_P) = [x_m^k(t_P) - x_L^k]/(x_H^k - x_L^k), \qquad (2.27)$$

where $x_m^k(t_P)$ and $x_m^{*k}(t_P)$ are the measured values of variable x^k before and after preprocessing, respectively, at the moment t_P; x_H^k is the high alarm threshold of x^k; and x_L^k is the low alarm threshold of x^k.

If the process variable x^k undergoes an ultra-low warning at the moment t_P, it is preprocessed according to Eq. (2.28):

$$x_m^{*k}(t_P) = [x_H^k - x_m^k(t_P)]/(x_H^k - x_L^k). \qquad (2.28)$$

The value of $x_m^{*k}(t_P)$ ranges from 0 to 1. The closer $x_m^{*k}(t_P)$ is to 1, the greater the deviation of the variable from the normal range. Preprocessing the process variables can eliminate the influence of the difference in the order of magnitude and scale of each variable on the results of the early-warning ranking.

Step 6: Estimate the early-warning ranking of process variables.

For each warning variable x^k, $x^{*k} = w_k x_m^{*k}(t_P)$, x^{*k} is sorted in descending order to give the priority order of the warning variables, which provides a basis for the field operator to reserve more time to process the alarm information and avoid the alarm flooding phenomenon.

2.1.3   *Implementation steps*

The flow of the adaptive process warning method is shown in Fig. 2.2.
    The method mainly includes the following steps:

(1) For each process variable x, the initialization value v_{initial} of $v_e(t_s^i)$ is calculated by using the measured data $x_m(t_i)$ of the smooth process during the initial period of N_{initial} according to Eq. (2.14) to initialize the tuning parameters th_1, th_2, t_{hc}, and t_{hs}. Calculate the cumulative sum function $cusum(t_j)$ at different moments t_j according to Eqs. (2.1)–(2.3) and compare its absolute value with the two preset thresholds th1 and th2 to determine whether it

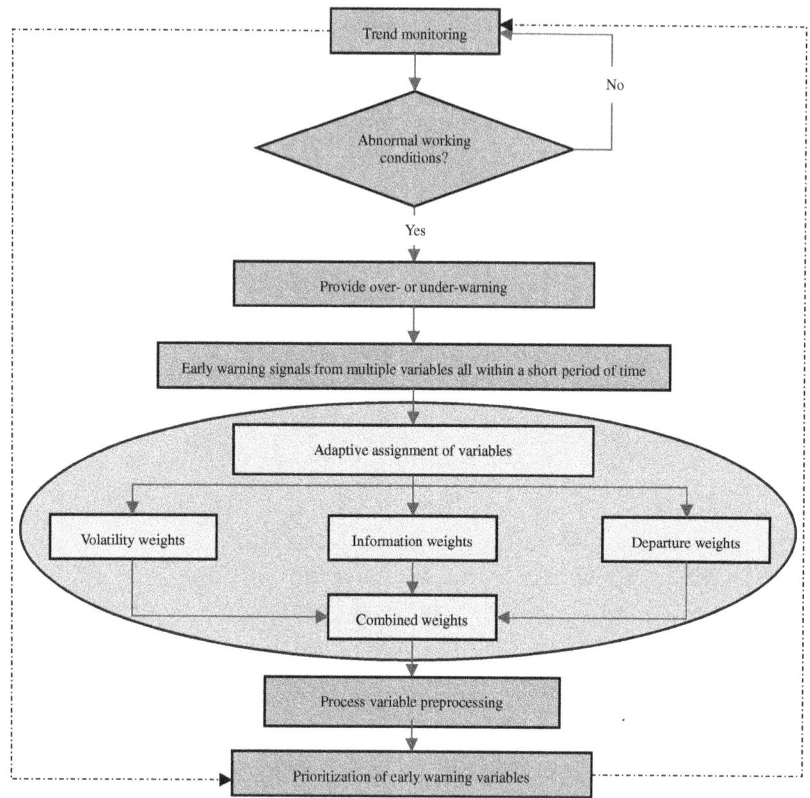

Fig. 2.2. Adaptive early-warning methodology process.

is necessary to split the linear feature segments and fit a new linear segment using a least-squares method (see 2.1.1). Before making a new linear estimate, determine whether there is an instantaneous mutation or step change in the process based on Eqs. (2.5)–(2.7). Trend identification of the linear segments is performed by means of the decision tree shown in Fig. 2.1, which determines whether the current trend of the process variable is smooth or not.

(2) If the current process variable shows a rising, positive step or falling/rising trend, calculate $\delta_H(j)$ according to Eq. (2.15) and determine whether to label the variable as ultra-high. If the current process variable shows a decreasing, negative step or rising/falling trend, calculate $\delta_L(j)$ according to Eq. (2.16) and

determine whether to label the variable as ultra-low, so as to remind the operator to take timely and effective measures to eliminate the abnormal working conditions.

For the process measurement data recorded in the outlier region (moments t^i_{j1} to t^i_{j2}), if $\delta_H(j)$ or $\delta_L(j)$ is less than $0.1(x_H - x_L)$ with respect to the measured value of the variable $x_m(t_j)$ at any moment $t_j(t^i_{j1} \leq t_j < t^i_{j2})$ in the intervening time, the process variable will be subjected to a super-high or super-low warning.

(3) If the warning signals are issued by multiple variables in a short period of time, the adaptive weighting method mentioned in 2.1.2 is used to calculate the weights of each warning variable x^k ($k = 1, 2, \ldots, K$, K is the number of warning variables). The fluctuation weight w_{k1} of each variable is calculated according to Eq. (2.17), the information weight w_{k2} of each variable is calculated according to Eqs. (2.18)–(2.20), and the deviation weight w_{k3} of each variable is calculated according to Eq. (2.21), and these weights are fused by Eqs. (2.22)–(2.26) to get the composite weight w_k of each variable.

(4) According to Eq. (2.27) or Eq. (2.28), the measured value $x^k_m(t_P)$ of the variable x^k, which has an ultra-high or ultra-low warning at moment t_P, is preprocessed to obtain $x^{*k}_m(t_P)$, which is used to indicate the degree of deviation of the variable from the normal range at the current moment. Finally, according to the comprehensive weight of each variable w_k and $x^{*k}_m(t_P)$, and the size of the product x^{*k}, the process variables are prioritized and the priority order of the warning variables is given, so as to reserve more time for the operators to deal with the alarm information and to avoid the flood of alarms that interferes with the operator's correct judgment of the abnormal working conditions, which may lead to serious consequences or accidents.

2.1.4 *Case study*

Take the oil refining process in the oil and gas production system as an example because its production raw materials are mostly flammable and explosive hazardous materials, and most of the materials transported in the petroleum refining device are corrosive, which makes it much more difficult to control production safety.

The petroleum refining process refers to the various processes used in the petroleum refining industry, such as crude oil pretreatment, atmospheric distillation, decompression distillation, catalytic cracking, and catalytic reforming processes. Here, two abnormal working conditions occurring in the atmospheric distillation and catalytic cracking processes of a refinery are used as examples to verify the effectiveness of the methods proposed in this section.

2.1.4.1 *Atmospheric tower flooding accident*

In the process of atmospheric distillation of crude oil, if the liquid from the atmospheric tower of an upward tower plate gradually accumulated to fill part of the tower section, so that the rising gas is blocked, a gas and liquid two-phase mass heat transfer process cannot be carried out normally, causing a flooding tower accident. The following is an example of a refinery atmospheric distillation unit flooding tower accident case study.

Prior to the accident, the distillation temperature of the atmospheric second line began to rise gradually, and in the following minutes, the level at the bottom of the atmospheric pressure tower dropped to 42%. After that, the liquid level at the bottom of the atmospheric pressure tower began to rise gradually, and the temperature of the second line of the atmospheric pressure tower began to rise rapidly from about 280°. At the same time, the temperature at the outlet of the decompression furnace began to rise continuously, and the temperature at the outlet of the decompression furnace was already significantly higher when the liquid level at the bottom of the tower rose to 57.9%, and the level indicator at the bottom of the tower was malfunctioning after that, which made it impossible for the field operators to obtain the level measurement value in a timely manner.

When the operator concludes that the liquid level display is out of order, immediately contact the instrumentation worker to report it and turn off the atmospheric tower bottom that is blowing steam. The operator confirms whether the bottom of the tower is full by observing the float, and judges whether an overflow accident has occurred. Then, the liquid level control at the bottom of the atmospheric tower is switched from automatic to manual, and the switching frequency is increased to increase the normal oil extraction at the bottom and reduce the amount of raw oil in the atmospheric furnace.

At this time, the operator needs to check the quality of the side line oil. Usually, one oil line is in a normal state, and the other two oil lines have black oil. The operator immediately closes the refining valves of the two commonly used oil lines, and then switches the refining tank to the secondary line. Because there is no dirty oil tank in the tank area, the on-site valve group of the two commonly used oil lines can only be disabled to prevent dirty oil from entering. Since the two commonly used oil lines in the whole process have been contaminated, when the flow meter is unloaded, the oil of the contaminated two commonly used oil lines is discharged to the upper part of the dirty oil tank through the leather hose, and then the oil is pumped into the tank area through the upper pump. Since the whole process of the Chang Er Line had been polluted by black oil, the black oil could only be completely replaced a few hours later when the device resumed normal operations.

(1) Trend analysis of process variables

The data used in the analysis are taken from the observed data of the distillation temperature of the normal second line, the liquid level at the bottom of the normal pressure tower, and the outlet temperature of the decompression furnace collected on-site during a period of time on the day of the accident, with the initial sampling moment recorded as 0, the data sampling interval as 20 s, and the length of the observed sample data as 300, as shown in Fig. 2.3.

The high and low alarm thresholds for each process variable are shown in Table 2.1. From Fig. 2.3, it can be seen that the distillation temperature ($X1$) of the normal second line starts to rise from 4680s until 5720s when the temperature rises to 324.6°, but from this moment onward, $X1$ has a short decline until 5820s when the temperature drops to 323.5°, and then the temperature continues to rise. The liquid level at the bottom of the atmospheric pressure tower ($X2$) starts to drop from 4860s to 42% in 4940s, and has been on an upward trend since then. The decompression furnace outlet temperature ($X3$) began to rise from 4960s, and has been on the rise since then.

The trend identification method in this section is used to identify the different trends of each process variable (as shown in the thick solid line in Fig. 2.3, with the tuning parameter λ in the respective adaptation parameter set to 3, α set to 60, N_r set to 60, and μ set to 5). The initialization value v_{initial} is calculated according to

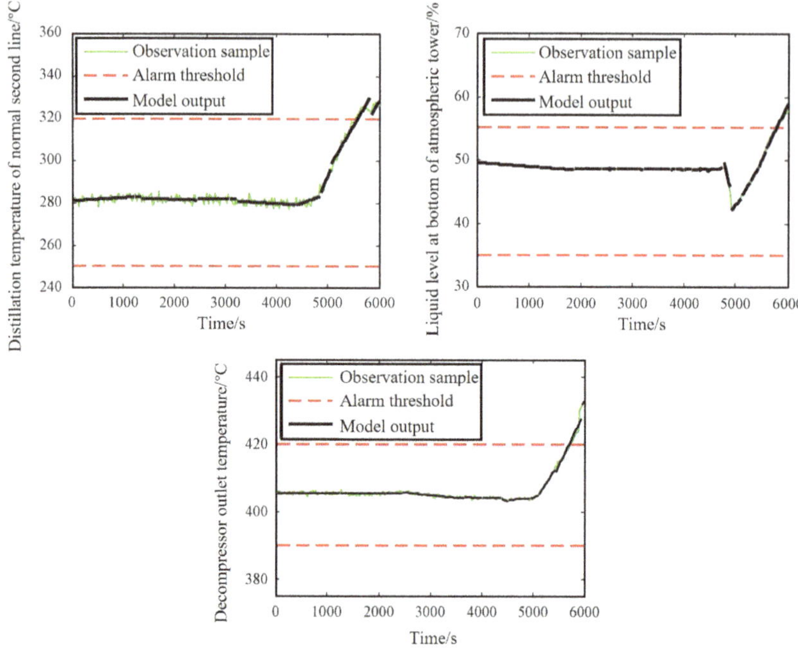

Fig. 2.3. Sample of observations and trend-fitting curves for each variable.

Table 2.1. High and low alarm thresholds for each process variable.

Variant	Variable description	High threshold	Low threshold	Unit
$X1$	Normal second-line distillation temperature	320	250	$^\circ$C
$X2$	Normal pressure tower bottom level	55	35	%
$X3$	Decompressor outlet temperature	420	390	$^\circ$C

Eq. (2.14) for $v_e(t_s^i)$ the first 60 samples of the data to initialize the values of the respective adaptation parameter. In order to evaluate the effectiveness of the method, the following three evaluation criteria are used:

(1) Calculate the mean value of the absolute difference between the measured value $x_m(t_j)$ and the output value $x(t_j)$ of the

fitted model for each variable in the time period, denoted as $E1 = \frac{1}{N_n} \sum_{j=1}^{N_n} |x_m(t_j) - x(t_j)|$, with N_n being the length of the sampling data in the time period and the size of the $E1$ value reflecting the accuracy of trend identification; the smaller the $E1$ value, the closer the identified trend value is to the actual measurement value.

(2) Calculate the ratio of the number of correctly recognized smooth trend data N_{ts} to the actual number of smooth trend data N_{as}, recorded as $E2 = N_{ts}/N_{as}$, with the size of the $E2$ value reflecting the ability of the method to identify smooth trends; the higher the $E2$ value, the better the performance of the method to capture smooth trends.

(3) Calculate the ratio of the number of correctly recognized upward/downward trend data N_{tv} to the actual number of upward/downward trend data N_{av}, denoted as $E3 = N_{tv}/N_{av}$, with the size of the $E3$ value reflecting the ability of the method to recognize upward/downward trends; the higher the $E3$ value, the better the method's performance in capturing upward/downward trends.

The $E1$, $E2$, and $E3$ values for each variable obtained using the methodology in this section are shown in Table 2.2, and it can be seen that the $E1$ values for each variable are small; as can be seen in Fig. 2.3, the identified trend values are very close to the actual measured values. The $E2$ and $E3$ values of the proposed method in Table 2.2 are higher than 90%, indicating that the method is highly accurate in identifying smooth and upward/declining trends. The trend identification analysis shows that $X1$ from 4400s to 4680s shows a steady trend but is identified as an upward trend, $X1$ from 5720s to 5820s shows a downward trend but is identified as an upward trend, and the trend of the rest of the sampled data of $X1$ is identified accurately. $X2$ observations from 4780s to 4840s show a steady trend, but are recognized as downward; $X2$ observations of the rest of the sampled data trends are identified accurately. $X3$ observations from 4520s to 4940s show a steady trend, but are recognized as upward; $X3$ observations of the rest of the sampled data trends are identified accurately.

At the same time, the Traditional Trend Extraction (TTE) method was used to identify the trend of each variable, the relevant parameters were adjusted by the repeated test method; the $E1$, $E2$,

Table 2.2. Value of each variable — $E1$, $E2$, and $E3$.

Variant	Methodologies	$E1$	$E2$	$E3$
$X1$	Methodology of this section	1.2116	95.0%	98.0%
	TTE	1.7435	90.0%	78.9%
$X2$	Methodology of this section	0.1006	98.9%	100%
	TTE	0.3973	98.0%	80.0%
$X3$	Methodology of this section	0.3687	92.7%	100%
	TTE	0.6254	89%	85.3%

Table 2.3. Difference between the variables $E1$, $E2$, and $E3$ obtained by the proposed method and the traditional TTE method.

Difference between the two methods for each parameter	$E1$	$E2$	$E3$
$X1$	0.5319	5.0%	19.1%
$X2$	0.2967	0.9%	20.0%
$X3$	0.2567	3.7%	14.7%
Mean value of the difference of each parameter	0.3618	3.2%	17.9%

and $E3$ values were calculated as shown in Table 2.2, and the differences between the $E1$, $E2$, and $E3$ values of each variable obtained by the two methods are shown in Table 2.3.

By comparing the $E1$, $E2$, and $E3$ values obtained by the two methods, Table 2.2 shows that the E1 values obtained by the method in this section are all smaller than those of the traditional TTE method, and for the three process variables mentioned earlier, the error between the fitted values and the true values obtained by the proposed trend analysis method is reduced by 0.3618 on average, which indicates that it is better than the TTE method in terms of the accuracy in identifying the trend. The E2 value obtained by the method in this section is only slightly higher than that of the traditional TTE method; for the three process variables mentioned earlier, the E2 value is increased by 3.2% on average, which indicates that the proposed method is still slightly better than the traditional TTE method in terms of its performance in capturing the smooth trend. The E3 value obtained by the method in this section is significantly higher than that of the traditional TTE method, with an average increase of 17.9%, indicating that the proposed method is significantly more accurate than the TTE method in recognizing the upward/downward trend. At the same time, because the proposed method can adaptively adjust the relevant parameters, it can adaptively recognize the trend of each process variable without repeating the parameter setting and has wider applicability.

Through adaptive trend recognition, the trend changes of each variable in the development process of the flooding tower accident can be recognized, which helps the operator grasp the dynamics of the device operation and discover abnormal working conditions in time.

(2) Multivariable adaptive early-warning analysis

Through trend identification analysis, the normal second-line distillation temperature ($X1$) was recorded in the outlier region from the 5080s onward. When recording to the 5400s, the measured value of $X1$ reaches 313.1°, and its high and low alarm thresholds are 320° and 250°, respectively. According to the formula (2.15), it can be obtained that $(320 - 313.1) < 0.1 \times (320 - 250)$, so $X1$ is labeled as ultra-high. The water level data of the bottom layer of the atmospheric tower ($X2$) were collected, and the data from 0 to 5579s were fitted into a new trend segment, which was confirmed to be an upward trend. The high and low alarm thresholds of $X2$ were determined to be 55% and 35%, respectively. The calculated water level measurement value of 5580s was 51.5%, and the difference with the high alarm threshold of $X2$ was $55 - 51.5 = 3.5$, which was less than $0.2 \times (55 - 35)$. Therefore, an ultra-high warning was issued for $X2$. The outlet temperature of the decompression furnace ($X3$) is recorded in the outlier region since 5440s, and when it is recorded in 5620s, the measured value of $X3$ reaches 417.3°, and its high and low alarm thresholds are 420° and 390°, respectively, and $(420 - 417.3) < 0.1 \times (420 - 390)$; therefore, $X3$ is labeled as ultra-high.

Comparing the proposed method with the traditional threshold alarm method, as shown in Fig. 2.4, the warning time of the proposed method for the distillation temperature of the normal second line ($X1$), the liquid level at the bottom of the normal-pressure tower ($X2$), and the outlet temperature of the decompression furnace ($X3$) are 2 min 20 s, 3 min 20 s, and 2mins in advance of the threshold alarm time, respectively, with an average advance time of 2min33s. More time can be reserved for field operators to process the alarm information, avoiding the generation of related alarms or even the phenomenon of alarm flooding.

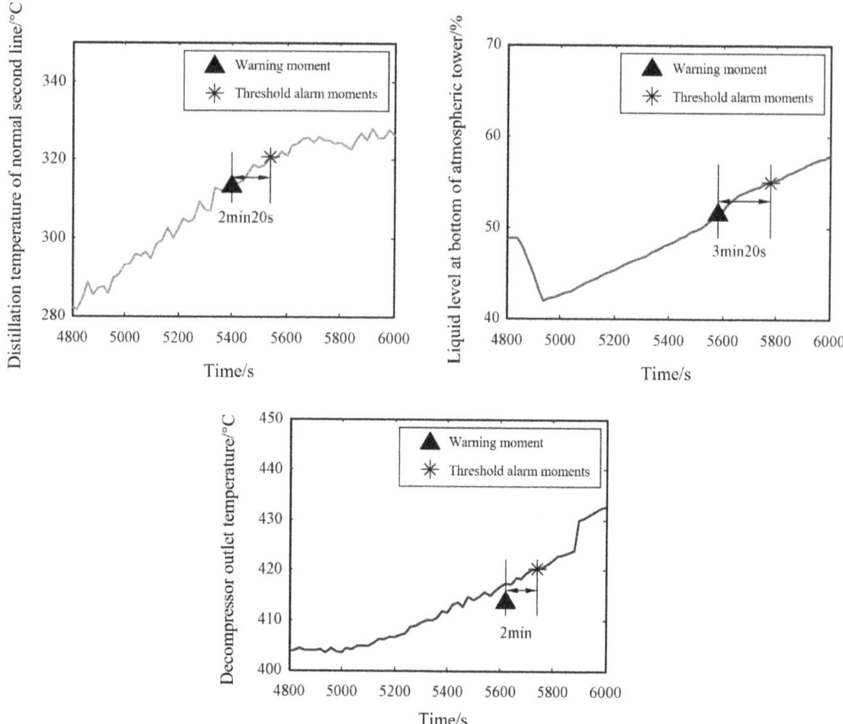

Fig. 2.4. Comparison of alarm time between the methods in this section and the traditional threshold alarm method.

Since all three process variables send out warning signals within a short period of time, the following analysis is carried out using the multivariate adaptive warning method proposed in this section, and the priority order of each warning variable is given.

Firstly, the 20 sampling data of each variable before the warning occurs are selected and subjected to maximum–minimum normalization, and the fluctuation weight and deviation weight of each variable are calculated according to Eqs. (2.17) and (2.21), respectively; based on 1,000 sets of historical data of each variable, the information weight of each variable is calculated by Eq. (2.20), as shown in Fig. 2.5.

As can be seen in Fig. 2.5, $X2$ has the largest fluctuation weight, indicating that the level at the bottom of the atmospheric column fluctuates the most in the period prior to the occurrence of

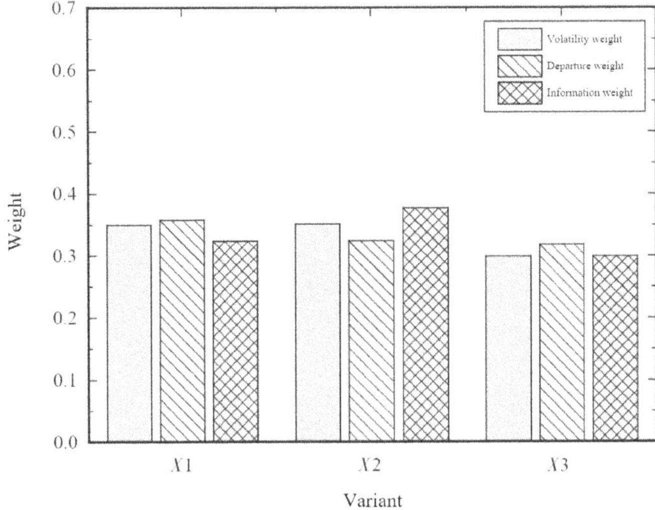

Fig. 2.5. Fluctuation weights, deviation weights, and information weights for each variable.

the warning. $X1$ has the largest deviation weight, indicating that the distillation temperature of the atmospheric second line has the largest span of change before the occurrence of the warning. $X2$ has the largest information weight, indicating that the level at the bottom of the atmospheric column is more correlated with the other warning parameters and has a greater influence on the changes in the other parameters.

The fluctuation weights, information weights, and deviation weights of each process variable are substituted into Eqs. (2.25) and (2.26), and the comprehensive weights of each variable can be obtained as shown in Table 2.4. According to Eq. (2.27), the measured value of each variable at the warning moment t_p, $x_m^k(t_P)$, is preprocessed, and the preprocessed measured value, $x_m^{*k}(t_P)$, is shown in Table 2.4, which eliminates the risk of the change in the temperature at the bottom of the atmospheric pressure column due to each variable. The effect of the difference in the order of magnitude and scale of the variables on the results of the early-warning ranking is eliminated. For each variable x^k, $x^{*k} = w_k x_m^{*k}(t_P)$, and x^{*k} is ranked from the largest to the smallest to prioritize the warning variables, as shown in Table 2.4.

Table 2.4. Composite weights, preprocessing values, and early-warning ranking for each variable.

Variant	$X1$	$X2$	$X3$
Combined weights w_k	0.342	0.349	0.309
$x_m^{*k}(t_P)$	0.901	0.825	0.910
x^{*k}	0.308	0.288	0.281
Early-warning priority	1	2	3

According to the warning sequence, the operator should prioritize the cause of the increase in the distillation temperature ($X1$) of the normal second line, and take timely measures to eliminate the abnormal working condition by reducing the pumping volume of the normal second line. Compared with $X1$ and $X3$, the excessively high liquid level ($X2$) at the bottom of the normal pressure tower is the most direct phenomenon before the flooding of the tower occurs. Because the liquid gradually accumulates and raises the liquid level at the bottom of the tower, the mass and heat transfer process of the gas and liquid phases cannot be carried out normally, and the product quality will be seriously affected, so the causes of the rise in the liquid level at the bottom of the tower can be considered later. The increase of decompression furnace outlet temperature ($X3$) is to some extent caused by the super-high level at the bottom of the tower, and measures can be taken subsequently to stabilize $X3$ in the normal operating range, by checking whether there are abnormal operating conditions in the atmospheric system and strictly controlling the product quality, to prevent the decompression system too much light components (such as low molecular weight hydrocarbons). To sum up, the warning priority of the method in this section is reasonable.

In fact, the flooding accident occurred mainly because the operator did not observe the changes in the operation and analyze the causes of the changes in a timely manner, delaying the best time to deal with the situation, until the furnace outlet temperature increased significantly before the abnormalities were found. If the operator considers the warning analysis results in this section and promptly considers the reasons for the ultra-high warning of X2, the failure of the tower bottom liquid level monitor can be discovered earlier and diagnosed. The liquid level control can be switched from automatic to manual, the position of the float can be confirmed in time, and appropriate adjustment measures can be taken to avoid tower flooding accidents.

(i) Trend analysis of process variables

The data used in the analysis were taken from the observed data of the top cycle pumping temperature of the fractionating tower, the top temperature of the fractionating tower, and the amount of crude gasoline and the amount of diesel oil pumped out collected on-site during a period of time on the day of the accident, with the initial sampling moment recorded as 0, the interval between data samples as 20s, and the length of the observed sample data as 300, as shown in Fig. 2.6 and the alarm thresholds for the high and low alarms of each variable as shown in Table 2.5. As can be seen from Fig. 2.6, the top circulating exhaust temperature ($X1$) of the distillation tower starts to rise from 4400s, and the temperature has risen to 138.8°C at 5000s. Then the heating rate accelerates, and the temperature has reached 155.6°C at 5660s, and the top circulating exhaust temperature will continue to rise. The temperature at the top of the fractional distillation column ($X2$) started to rise since 5080s, and has been on an upward trend since then. Crude gasoline volume ($X3$) started to increase since 5120s and has been increasing since then. Diesel oil extraction ($X4$) started to decrease since 4960s and has been decreasing since then.

Due to the interruption of data transmission, no data were collected at the 1200s, and the observed values of each variable were zero

Table 2.5.   High and low alarm thresholds for each variable.

Variant	Parameter descriptions	High thresholds	Low thresholds	Unit
X1	Fractionator top circulation temperature	160	130	°C
X2	Fractionator top temperature	130	100	°C
X3	Crude oil	170	140	t/h
X4	Diesel extraction	95	65	t/h

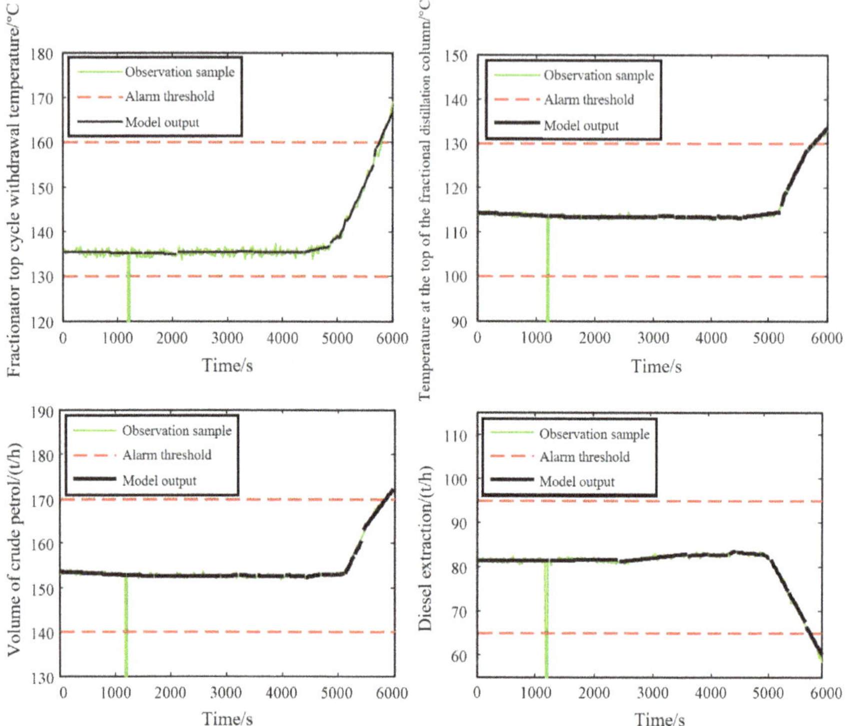

Fig. 2.6. Sample of observations and trend-fitting curves for each variable.

at that time, as shown in Fig. 2.6, which corresponds to the observation data at the 1200s of 0. In order to detect the step changes and transient mutations (see 2.1.2), ω was set to 5, the size of the sliding time window N was set to 60, and the maximum sampling time span S for transient mutations was set to 10.

The proposed trend identification method is used to identify the different trends of each process variable (Fig. 2.6, thick solid line part), and the adjustment parameter λ in the respective adaptation parameter is set to 3, α is set to 60, Nr is set to 60, and μ is set to 5. According to Eq. (2.14), the initialization value v_{initial} is computed for the first 60 sampling data of $v_e(t_s^i)$ to initialize the values of the respective adaptation parameters.

As can be seen in Fig. 2.6, the method in this section automatically filters the observation data at the 1200s and recognizes them as transient mutations. Due to the influence of process changes, equipment failures, signal transmission interruptions, and external disturbances in the refining process, which may cause step changes or instantaneous changes in the process variables, the proposed method can identify these changes in a timely manner to reduce the rate of false alarms and to avoid unnecessary interferences from affecting the operator's judgment of the process operating conditions.

In order to assess the effectiveness of the method, the $E1$, $E2$, and $E3$ values of each variable are calculated for the time period (Fig. 2.7). It can be seen that the $E1$ values of each variable obtained by the method in this section are small, and it can also be seen from Fig. 2.6 that the identified trend values are very close to the actual measured values. The $E2$ values of the variables in Fig. 2.7 are all above 80%, indicating that the accuracy of the method in identifying smooth trends is high, and the $E3$ values of the variables are all 100%, indicating that the method can accurately identify the upward/downward trends of the process variables.

Trend identification analysis shows that $X1$ has a smooth trend from 4400s to 4860s, but is identified as an upward trend, and the trend of the rest of the sampled data of $X1$ is identified accurately. $X2$ has a smooth trend from 3580s to 4100s, but is identified as a downward trend. $X2$ has a smooth trend from 4560s to 5060s, but is identified as an upward trend, and the trend of the rest of the sampled data of $X2$ is identified accurately. $X3$ has a smooth trend from 4420s to 4680s, but is identified as an upward trend, and the trend of the rest of the sampled data of $X1$ is identified accurately. $X4$ has a smooth trend from 4380s to 4500s, but is identified as

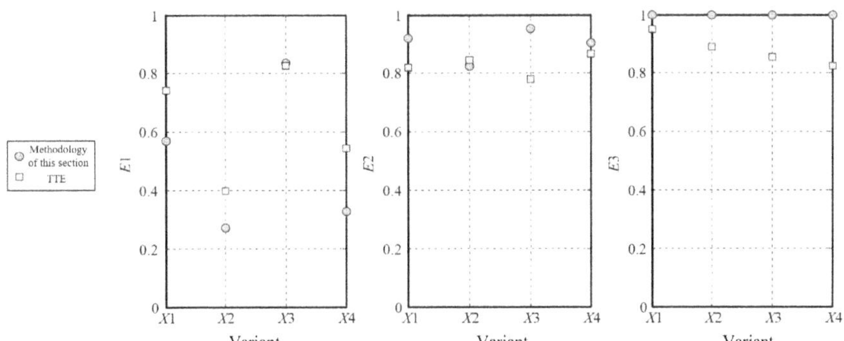

Fig. 2.7. The values of each variable $E1$, $E2$, and $E3$ for both methods.

an upward trend. $X4$ has a smooth trend from 4520s to 4940s, but is identified as a downward trend, and the trend of the rest of the sampled data of $X4$ is identified accurately.

At the same time, the traditional TTE method is used to identify the trend of each variable, and the relevant parameters are adjusted by the repeated test method, and their $E1$, $E2$, and $E3$ values are calculated as shown in Fig. 2.7. By comparing the sizes of the $E1$, $E2$ and $E3$ values obtained by the two methods, it can be seen that, except for the $E1$ value obtained by $X3$, which is slightly higher than that of the TTE method (0.0099), the $E1$ values of the other variables are significantly smaller than that of the traditional TTE method. For the above-mentioned four process variables, the fitted values obtained by the proposed trend analysis method have an average reduction of 0.1261 in the error between the fitted values and the true values, indicating that it is better than the TTE method in the accuracy of trend identification. Except for the $E2$ value obtained by $X2$ which is slightly lower by 2.2% than that of the TTE method, the $E2$ values of other variables are higher than the traditional TTE method, and the average $E2$ value is increased by 7.2%, which indicates that the method in this section is higher than the TTE method in the accuracy of recognizing the smooth trend. The $E3$ value of the method in this section is significantly higher than that of the TTE method, with an average increase of 12.1%, indicating that the proposed method is significantly more accurate than the TTE method in identifying upward/downward trends.

In conclusion, it seems that the adaptive trend identification method is better than the traditional TTE method in terms of the accuracy of trend identification of process variables. Through adaptive trend identification, the trend changes of each variable during the development of the fractional distillation tower rushing accident can be identified, which can provide a reference for operators to analyze the operation of the plant.

(ii) Multivariate adaptive warning analysis

Through trend identification analysis, a new trend segment was obtained by fitting the top recycle pumping temperature ($X1$) of the fractionator at 5660s, which was recognized as an upward trend. The high and low alarm thresholds of $X1$ are $160°$ and $130°$, respectively. According to Eq. (2.15), the difference between the measured value of $155.6°$ at 5660s and the high alarm threshold of $X1$ is $160 - 155.6 = 4.4$, which is less than $0.2 \times (160 - 130)$, so $X1$ is labeled as ultra-high. The temperature at the top of the fractionating tower ($X2$) is recorded in the region of outliers since the 5320s, and when it is recorded in the 5620s, the measured value of $X2$ reaches $127.0°$, and its high and low alarm thresholds are $130°$ and $100°$, respectively, and $(130 - 127.0)$ is less than $0.1 \times (130 - 100)$, so an ultra-high warning is issued for $X2$. The process data of crude oil gasoline quantity ($X3$) starts from 5440s, and then starts to record the outlier range of $X3$. At 5660s, the measured value of $X3$ reaches 167.7t/h, and its high and low alarm thresholds are 170t/h and 140t/h respectively, and $(170 - 167.7) < 0.1 \times (170 - 140)$, and $X3$ has an ultra-high warning. Diesel oil extraction ($X4$) is recorded in the outlier region since the 5440s, and when it is recorded to the 5700s, the measured value of $X4$ reaches 67.5t/h, and its high and low alarm thresholds are 95t/h and 65t/h, respectively, and $(67.5 - 65) < 0.1 \times (95 - 65)$; therefore, an ultra-low warning is issued for $X4$.

Comparing the proposed method with the traditional threshold alarm method, as shown in Fig. 2.8, the early-warning times of the method in this section for the top of the fractionating column cycle withdrawal temperature ($X1$), the top of the fractionating column temperature ($X2$), the crude gasoline volume ($X3$), and the diesel oil withdrawal volume ($X4$) are 2 min 40 s, 3 min, 2 min 40 s, and 2 min ahead of the threshold alarm time, respectively, with an average advance time of 2 min 35 s, thus reserving more time for the field operator to take preventive measures.

Since all four process variables send out warning signals within a short time, the following analysis is carried out by using the multivariate adaptive warning method proposed in this section to give the priority order of each warning variable.

Firstly, 20 sampling data of each variable before the warning occurs are selected, and the fluctuation weight and deviation weight of each variable are calculated according to Eqs. (2.17) and (2.21), respectively. Based on 1000 sets of historical data of each variable, the information weight of each variable is calculated according to Eq. (2.20), as shown in Fig. 2.9.

From Fig. 2.9, it can be seen that $X3$ has the largest fluctuation weight, indicating that the crude gasoline volume fluctuates the most in the period prior to the occurrence of the warning, $X2$ has

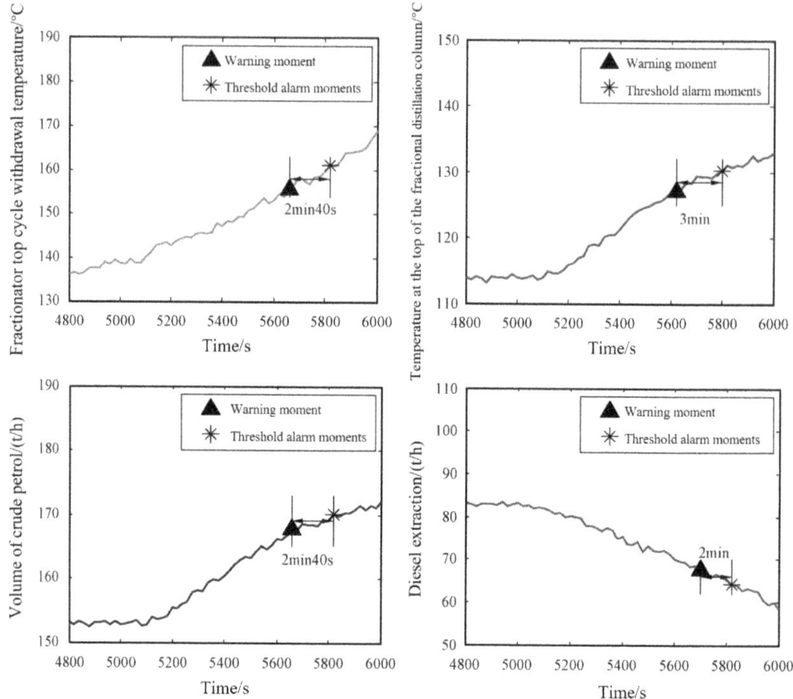

Fig. 2.8. Comparison of alarm time between the methods in this section and the traditional threshold alarm method.

the largest deviation weight, indicating that the temperature at the top of the fractionator has the largest change prior to the occurrence of the warning, and $X1$ has the largest information weight, indicating that the top of the fractionator recycle pumping temperature is more highly correlated with the other warning variables and has more impact on the changes of the other variables.

The fluctuation weights, information weights, and deviation weights of each process variable are substituted into Eqs. (2.25) and (2.26), and the comprehensive weights of each variable can be obtained, as shown in Table 2.6. According to Eqs. (2.27) and (2.28), the measured value $x_m^k(t_P)$ of each variable at the time of warning t_P is preprocessed, and its preprocessed measured value $x_m^{*k}(t_P)$ is shown in Table 2.6, and for each variable x^k, $x^{*k} = w_k x_m^{*k}(t_P)$, and x^{*k} is ranked in descending order to give a priority order of the early-warning variables, as shown in Table 2.6.

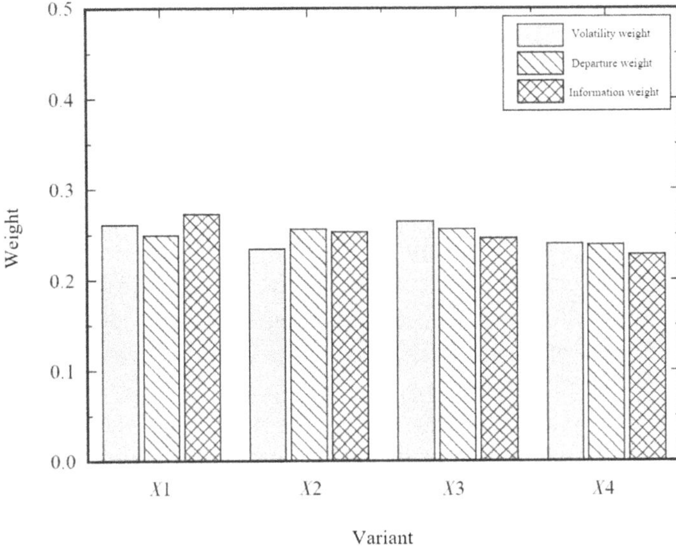

Fig. 2.9.   Volatility weights, deviation weights, and information weights for each variable.

Table 2.6. Composite weights, preprocessing values, and early-warning ranking for each variable.

Parameter number	$X1$	$X2$	$X3$	$X4$
Combined weights w_k	0.256	0.249	0.253	0.242
$x_m^{*k}(t_P)$	0.853	0.900	0.923	0.917
x^{*k}	0.218	0.224	0.234	0.222
Early-warning priority	4	2	1	3

Based on the sequence of warnings obtained, the operator should prioritize the cause of the excessively high warning for Crude Gasoline Volume ($X3$). As the amount of crude gasoline ($X3$) is increasing and the temperature at the top of the fractionator ($X2$) is too high, indicating that the problem may occur at the top of the fractionator, combined with the amount of diesel fuel extraction ($X4$) and other changes lagging behind the temperature at the top of the fractionator ($X1$), it can be deduced that the cause of the failure is the top of the fractionator followed by a few layers of the plate problems (such as plate salt and dirt), leading to the fractionator tower rushing accidents. When a punching tower occurs, the thermal balance of the entire fractionator is broken, thermal energy is displaced upward and the other middle temperature will also move up. When the flushing tower occurs, the amount of crude gasoline ($X3$) increases because the recombination fraction in the fractionator oil and gas is shifted upward, and the diesel fraction cannot be distilled out normally thereby entering the gasoline fraction, which affects the quality of the gasoline product.

In fact, after the occurrence of the flushing tower, it can be deduced that the flushing tower occurred due to the formation of ammonia salts at the top of the fractionating tower based on the flushing tower phenomenon and the analysis of the raw materials. The results of the early-warning analysis in this section are consistent with the conclusion, providing the operator with the basis for emergency disposal, so that the operator can carry out water washing treatment on the top of the tower plate in time to solve the problem of salt formation in the fractionating tower, avoiding the frequent occurrence of tower rushing accidents, and guaranteeing the safety of the process and the quality of the product.

(3) Analysis and summary

(i) Alarm flooding is one of the major problems in industrial alarm systems for complex processes of oil and gas production today. Due to the limitations of the alarm system design, it causes problems of poor system pre-control timeliness and repeated occurrence of associated alarms, which are very likely to lead to alarm flooding under some abnormal working conditions. For this reason, this section proposes an adaptive process early-warning method based on trend monitoring in view of the complexity, dynamics, and correlation characteristics of the complex process of oil and gas production, combined with difficult problems such as untimely alarms and lack of adaptability of the existing alarm management methods in practical applications. Through adaptive trend monitoring of process variables, the warning of non-smooth trend process variables is carried out and an adaptive weight calculation method is proposed for the multivariate warning problem to assign weights to each variable, which takes into account the degree and magnitude of fluctuation changes of each process variable in the period of time prior to the occurrence of the warning, as well as the level of correlation with other warning variables. Finally, based on the combined weights of the variables and the degree of deviation of the measured values from the normal range at the time of the warning, the operator is provided with a reasonable order of processing of the warning variables.

(ii) A case study was conducted to analyze normal-pressure tower flooding and fractional distillation tower rushing faults that occurred at a refinery. In the case of atmospheric tower flooding, compared with the traditional trend analysis method, the trend monitoring method in this section improves the accuracy of capturing smooth process trends by 3.2% on average and improves the accuracy of recognizing rising/declining trends by 17.9% on average; compared with the alarm method based on the threshold value, the warning method in this section advances the alarm time by 2 min 33 s on average. In the case of the flushing tower failure, compared with the traditional trend analysis method, the trend monitoring method adopted in this section improves the accuracy of capturing the trend of the smooth process by 7.2% on average and identifies the upward/declining trend by 12.1% on average; compared with the alarm method based on the threshold value, the proposed early-warning method advances the alarm time by 2 min 35 s on average. In addition, the multivariate adaptive warning mechanism further gives the priority order for multivariate warning, which provides a reference for operators to analyze the operation of the device and formulate a reasonable pre-control plan; this improves the timeliness and effectiveness of the alarm system, thus avoiding the generation of correlated alarms or even the phenomenon of alarm flooding.

2.2   A Multi-Step Forward Prediction Method for Shale Gas Fracturing Construction Curves

Abnormal working conditions such as sand plugging, pipe spikes, and sand sinking often occur during the shale gas fracturing process, which may lead to failure of the fracturing operation if timely measures are not taken. The change in the fracturing construction curve can reflect the abnormality of the system; therefore, the trend change of the fracturing construction curve is predicted to be of great significance for the pre-control of abnormal working conditions.

Existing prediction algorithms can be mainly classified into linear regression time series, pattern recognition prediction algorithms, and filtered prediction algorithms. Linear regression time series methods such as Auto Regressive (AR), Moving Average (MA), Auto Regressive and Moving Average (ARMA) and regression analysis are usually applicable to linear or weak nonlinear systems, but not to nonlinear systems containing non-Gaussian noise. In the face of complex refining systems, if the sample trend is not smooth or fluctuates greatly, it is easy to make the constructed time series model inaccurate, ultimately resulting in low prediction accuracy. Even after differentiation (i.e., ARIMA model), the prediction results are still not optimistic. Pattern recognition algorithms such as artificial neural networks and SVM can improve the prediction effect of one step forward or multiple steps, but a large number of historical samples are needed for training when constructing the network model. Filter prediction algorithms are difficult to achieve multi-step forward prediction and have poor real-time performance. For example, Kalman filtering, scalable Kalman filtering, unscented Kalman filtering, particle filtering, etc., although they have high prediction accuracy, require the construction of an initial state function. The commonly used method is to find an empirical function as the state transfer function, but the prediction accuracy cannot be guaranteed. Moreover, for many industrial control systems, the initial state function is difficult to determine. Therefore, filter prediction algorithms are difficult to use in practical applications.

Locally Weighted Linear Regression (LWLR) is a non-parametric regression algorithm for prediction. It considers a nonlinear sequence as multiple linear series, i.e., linear equations are established for each data point to be predicted by its mathematical characteristics. Each data point near the target area to be predicted is given a certain weight to measure the nearby data points' contribution to the multi-step prediction accuracy. Finally, the optimal linear equation of the data to be predicted can be constructed, which improves the forward multi-step prediction accuracy of the nonlinear series. However, even if k values with small fitting errors are used as parameters of the prediction model, there is no guarantee of the accuracy of the forward multi-step prediction.

The shale gas fracturing production system has nonlinear characteristics; it is difficult to find an empirical function that conforms to the change rule of the curve, and the prediction accuracy of the filtering algorithm cannot be guaranteed during the forward multi-step prediction process. To address these issues, a forward multi-step prediction method for fracturing construction curves based on the Locally Weighted Linear Regression (LWLR) algorithm was studied. The LWLR model parameters were optimized using the ARMA algorithm model (PF-ARMA) optimized by the particle filter (PF). This achieved accurate forward multi-step predictions, which can provide a theoretical basis for the prediction of abnormal shale gas fracturing conditions.

2.2.1 *Basic theory*

2.2.1.1 *LWLR model and modeling steps*

In order to the predict the behavior of most nonlinear systems, a relatively simple way is to model the input data which represents the system during a certain period of time and process them in a local modeling manner. The complexity of the traditional nonlinear model-based fitting method exhibits exponential growth with the increase of the dimension of input parameters, which leads to the training process requiring quite a long time and difficulty in meeting the prediction requirements. Therefore, if the output surface of the nonlinear system is assumed to be smooth enough, the system can be approximated by a series of local linear functions, while linear regression has been widely used for data analysis due to its good performance and low computation costs.

For a group of samples $(x_1, y_1), \ldots, (x_N, y_N)$, x_i is an independent variable, y_i is the dependent variable, $\{y_i, i, \ldots, N\}$ is assumed to be independent of each other and identically distributed. According to the relationship between independent variables and dependent variables, a non-parametric regression model as shown in Eq. (2.29) can be established:

$$y_i = f(x_i) + \varepsilon, \quad i = 1, 2, \ldots, N, \tag{2.29}$$

where $f(\cdot)$ is the non-parametric function to be solved; $\hat{f}(\cdot)$ is the estimate of $f(\cdot)$; ε is the system noise with mean value μ which is considered as zero and with variance considered as σ_ε^2.

LWLR is a non-parametric regression-based prediction method, the core principle of which is to find the sample data with the least similarity to the point to be verified $x = x_{\text{new}}$ among the

p historical sample data according to the minimum p nearest neighbor mechanism. And to find a suitable kernel function for this system from the commonly used kernel functions (Gaussian kernel function is used in this section) to calculate the value of the weight of this sample data, and, finally, according to the principle of weighted least-squares estimation, to Estimate the parameters of the local first-order linear function of the prediction point, that is, to obtain $\hat{f}(\cdot)$, bring in the value of the dependent variable, to get the estimated value of the prediction point $y_{\text{new}} = f(x_{\text{new}})$. LWLR can also be used for prediction modeling of multidimensional independent variables, if the number of dimensions is larger than the number of sample points, in order to prevent computational errors, can be introduced into the ridge regression, lasso method for optimization. In this section, the application object is a one-dimensional sample, and only the first-order linear model can be constructed.

Comparing with traditional parameter regression models (e.g., ARMA model), the LWLR model is not limited to a number of finite function equations, but based on the characteristics contained in the sample data with a certain degree of similarity to each prediction point instead, the estimation can be implemented, i.e., each prediction data point may correspond to different function equations.

The regression coefficients of $f(x_i)$ can be solved in Eq. (2.30):

$$\hat{\boldsymbol{w}} = (\boldsymbol{X}^T \boldsymbol{W} X)^{-1} \boldsymbol{X}^T \boldsymbol{W} y, \tag{2.30}$$

where $\hat{\boldsymbol{w}}$ is an estimated vector of coefficients; $\boldsymbol{X}$ is an independent variable matrix; $\boldsymbol{W}$ is a weight matrix given for each sample; y represents the dependent variable.

The basic steps of the traditional LWLR prediction process are as follows:

Step 1: initial condition setting.

Set x_{new} as a center (in the forward prediction, only the previous data before x_{new} are considered); suppose that there are a total of N samples and select the number of n ($n \leq N$) samples in historical data for modeling

Step 2: select the kernel function and calculate the weight of each sample.

The kernel function is used to calculate the weight of the modeled sample data. The types of kernel function mainly include

Tricube, Epanechnikov, and Gaussian kernel. However, some literatures pointed out that the choice of kernel function has little effect on the prediction results, so in this section, the Gaussian kernel function is selected to use (shown in Eq. (2.31)).

$$w(i,i) = \exp\left[\frac{|x^{(i)} - x|}{-2k^2}\right], \qquad (2.31)$$

where $w(i,i)$ denotes the weight of the sample point; $x^{(i)}$ denotes the horizontal coordinate of the ith sample point; k is the parameter controlling the weight of the weight, which is the only input parameter in the LWLR method.

Step 3: calculate the parameter matrix.

By substituting the samples $(x_1, y_1), \ldots, (x_N, y_N)$ and the weight matrix obtained by Eq. (2.31) into Eq. (2.30), the coefficients of the LWLR model are obtained, i.e., $\hat{f}(\cdot)$.

Step 4: calculate the predicted value $\hat{y}_{\text{new}}$.

By substituting the value of x_{new} into Eq. (2.29), the value of $\hat{y}_{\text{new}}$ is obtained, i.e., for each $\hat{y}_{\text{new}}$, $\hat{y}_{\text{new}} = \hat{f}(x_{\text{new}}) = \hat{a} + \hat{b}x_{\text{new}}$, where $\hat{a}$ and $\hat{b}$ are the calculated coefficients.

2.2.1.2 *ARMA forecasting methodology and modeling steps*

ARMA is one of the typical models used for time series trend forecasting, which was firstly proposed by Box-Jenkins, also known as Box-Jenkins method, but the traditional ARMA method is only applicable to the modeling of smooth series. Therefore, the ARIMA model for non-stationary series has been developed, which differentiates the non-stationary series to make them conform to the modeling requirements of ARMA. In this section, an ARIMA model is constructed based on the pressure sample data as a state function in particle filtering for prediction research.

The core principle of the Box-Jenkins method is to consider the time series (a sequence of data over time) as a stochastic sequence, in which the observed value at the nth moment in the sequence not only has a dependence on the previous $n-1$ observations, but also has a dependence on the perturbations that entered the system at the previous $n-1$ moments, according to which the relational function of the sequence is constructed, and the values of the next moment or the future period are predicted.

The general ARMA model is of the form ARMA (p, q), and the specific model expression is shown in Eq. (2.32):

$$x_t = \Phi_1 x_{t-1} + \Phi_2 x_{t-2} + \cdots + \Phi_p x_{t-p} - \theta_1 \varepsilon_{t-1}$$
$$- \theta_2 \varepsilon_{t-2} - \cdots - \theta_q \varepsilon_{t-q} + \varepsilon_t, \tag{2.32}$$

where when $q = 0$, the model is AR (p) model; and when $p = 0$, the model is MA (q) model. $\Phi_1, \ldots, \Phi_p$ and $\theta_1, \ldots, \theta_q$ are the autoregressive coefficients and moving average coefficients, respectively. $\{x_i\}$ denotes the zero-mean time series, and $\{x_t\}$ and $\{x_t\}$ denote the white noises which are independent of each other and obey the same distribution, where: $E(x_t) = 0$, $Var(x_t) = \sigma^2 > 0$.

The modeling steps for the ARIMA model are as follows:

Step 1: data acquisition.

First, obtain the time series of the variable to be analyzed and express them as $\{\lambda_1, \lambda_2, \ldots, \lambda_t\}$, then plot the original sequence and determine whether the sequence is stationary or not.

Step 2: judgment of the stationary property of the time series.

Perform autocorrelation analysis on the obtained data, i.e., calculate the autocorrelation coefficient $\hat{\rho}_k$ and partial correlation coefficient $\hat{\Phi}_{kk}$ according to Eqs. (2.33) and (2.34):

$$\hat{\rho}_k = \frac{\sum_{t=1}^{n-k}(x_t - \bar{x})(x_{i+k} - \bar{x})}{\sum_{t=1}^{n}(x_t - \bar{x})^2}, \tag{2.33}$$

$$\hat{\Phi}_{kk} = \begin{cases} \hat{\rho}_k & k = 1 \\ \frac{\hat{\rho}_k - \sum_{j=1}^{k-1}\hat{\Phi}_{k-1,j} \cdot \hat{\rho}_{k-j}}{1 - \sum_{j=1}^{k-1}\hat{\Phi}_{k-1,j} \cdot \hat{\rho}_j} & k = 2, 3 \ldots, \end{cases} \tag{2.34}$$

where $\Phi_{k,j} = \Phi_{k-1,j} - \Phi_{kk} \cdot \Phi_{k-1,k-j}$, $j = 1, 2, \ldots, k - 1$; $\bar{x}$ is the arithmetic mean of the sample data; k is expressed as the lag period, and n is the sample size of $\{(x_t - \bar{x})\}$.

Calculate the confidence interval $(-2/\sqrt{n}, 2/\sqrt{n},)$ of $\hat{\rho}_k$, and test the smoothness and randomness of $\{(x_t - \bar{x})\}$ according to the confidence interval. If $\hat{\rho}_k$ basically falls within $(-2/\sqrt{n}, 2/\sqrt{n},)$, then the sequence is random, and when $k > 3$, $\hat{\rho}_k$ falls into $(-2/\sqrt{n}, 2/\sqrt{n},)$

and gradually tends to 0, it can be initially judged that $\{(x_t - \overline{x})\}$ is smooth. Then the white noise test is performed, if the sequence is insignificant at the given significance level, then $\{(x_t - \overline{x})\}$ has non-white noise $\{(x_t - \overline{x})\}$ is determined to be a smooth sequence and can be used for ARMA modeling.

Step 3: difference operation.

If it is determined that the sample is a non-stationary sequence in Step 2, a difference operation is then performed. Generally, the sequence can become a stationary sequence after experiencing a first or second-order difference. If it is a stationary sequence, skip this step.

Step 4: model establishment and determination of its order.

On the basis of Step 2, the model orders as p and q are determined based on the Box-Jenkins method using the $\hat{\rho}_k$ and $\hat{\Phi}_{kk}$ of the sequence $\{(x_t - \overline{x})\}$. The judgment principle is shown in Table 2.7.

After preprocessing the modeling data, the model can be ranked, and the combination (p, q) is continuously trained according to the calculation results of the autocorrelation coefficients and partial correlation coefficients calculated in step 2 (the values of p and q determine the type of the model), where $0 \leq p \leq \sqrt{n}$, $0 \leq q \leq \sqrt{n}$. BIC is calculated and the combination (p, q) which corresponding to the smallest value of BIC is then chosen as the optimal order of the model.

Step 5: parameter estimation and residual test.

The commonly used method to evaluate the suitability of model parameters is the least square method. The applicability of the model needs to test to determine whether it has fully extracted the information of samples, i.e., the independence of ε_t should be tested.

Table 2.7.   Principle for ARIMA model recognition.

Autocorrelation coefficient graph (ACF)	Partial correlation coefficient graph (PACF)	Results of model recognition
q-order truncation	Tailing	MA (q)
Tailing	q-order truncation	AR (p)
Tailing	Tailing	ARMA (p, q)

Then the significance test of the parameters should be performed to eliminate the insignificant parameters. For significant variables, parameters can be re-evaluated and then an updated ARIMA model can be constructed.

2.2.1.3 *Particle filtering algorithm*

The Particle Filter algorithm (PF) is based on the principle of Sequential Monte Carlo and can be used to solve the problem of recursive Bayesian estimation. The PF algorithm relaxes the linear Gaussian constraints of the traditional filtering algorithms, such as KF, UKF and EKF, and provides an effective and reliable solution to the problem of trend prediction of nonlinear non-Gaussian curves.

The core idea of PF is to find a set of independent random samples in the state space of the system, and the posterior probability of the samples is approximated by the set of random samples, and the mean is calculated by replacing the sample mean with the sample mean, and finally, the minimum variance of the samples is estimated. In this case, this group of independent random samples is called "particles", and each particle has a right value, which can be expressed as $\{x_k^i, \omega_k^i,\ i = 1, 2, \ldots, n\}$. These particles with weights are passed in a certain way, and the weights of the particles are updated according to the Bayesian principle. When there are enough particles, the posterior probability distribution of the sample can be well approximated by these particles.

This set of independent random samples is then given weights that can be approximated as the state distribution probabilities of the system, see Eq. (2.35):

$$p(x_k|z_k) \approx \sum_{i=1}^{N} \omega_k^i \delta(x_k - x_x^i), \qquad (2.35)$$

where $p(x_k|z_k)$ is the state distribution of the system; $\delta(x_k - x_x^i)$ is the distribution of particles; ω_k^i denotes the weights of particles. According to the principle of Monte Carlo, $p(x_k|z_k)$ is sampled several times to find the mean value of the sample, see Eq. (2.36):

$$E(x_k) = \sum_{i=1}^{N} \omega_k^i x_k^i. \qquad (2.36)$$

Since the particles are usually not directly extracted from $p(x_k|z_k)$, the particles are sampled from $q(x_k|z_k)$, which is a relatively easy sample to collect and has a known probability density function, as shown in Eq. (2.37):

$$\omega_k^{(j)} \propto \frac{p(x_k|z_k)}{q(x_k|z_k)}. \tag{2.37}$$

The probability distribution of the particle is derived from the principles of probability theory, see Eq. (2.38):

$$p(x_k|z_k) \propto p(z_k|x_k)p(x_k|x_{k-1})p(x_{k-1}|z_{k-1}). \tag{2.38}$$

Bringing Eqs. (2.37) and (2.38) into Eq. (2.36) gives the structural frame of the PF, as shown in Eq. (2.39):

$$\omega_k^j \propto \omega_{k-1}^i \frac{p(z_k|x_k^i)p(x_k^i|x_{k-1}^i)}{q(x_k^i|x_{k-1}^i, z_k)}. \tag{2.39}$$

The PF algorithm is based on sequential importance sampling with the addition of a resampling technique (criterion). Sequential importance sampling is prone to degradation of particle diversity, i.e., as the number of sampling times increases, the large weighted particles tend to be concentrated among a few particles, which makes the a posteriori probability density obtained by the approximation not accurately reflecting the system state, and reduces the accuracy of the prediction results. Therefore, the resampling technique criterion is introduced, see Eq. (2.40); whether the algorithm needs to resample or not, it can be compared by calculating the number of effective particles N_{eff} with the threshold N_{th} (pre-set), when $N_{\text{eff}} \leq N_{\text{th}}$; the resampling algorithm will be used, otherwise the next step will be carried out.

$$N_{\text{eff}} = \frac{1}{\sum_{j=1}^N [\omega_k^{(i)}]^2}. \tag{2.40}$$

2.2.2 *A multi-step forward prediction method for shale gas fracturing construction curves*

For the fracturing construction curve, to address the problems of poor prediction effect of linear regression model and difficulty in selecting empirical function, a forward multi-step prediction method based on LWLR is carried out. In order to solve the problem of low prediction accuracy of traditional LWLR, the PF_ARMA model is combined to provide a model optimization basis for LWLR, and the range of parameter values of the LWLR model is determined, so as to achieve the accurate prediction of shale gas fracturing construction curves in multiple steps forward, and to analyze the optimal prediction step lengths after several attempts.

2.2.2.1 *Methodological step*

The specific implementation steps of the method for forward multi-step prediction of shale gas fracturing construction curves are as follows:

Step 1: select a cross-validation scheme.

Time series of fracturing pressure data $P = \{p_s | s = 1, 2, \ldots, m\}$ are selected from historical dataset, within which n samples are determined for test, while the previous $m - n$ samples for modeling.

Step 2: development of original LWLR model.

Calculate the Gaussian kernel weight matrix of all pressure data for modeling according to Eq. (2.31) and estimate the model regression coefficients of each modeling data according to Eq. (2.30), then the parameter matrix can be constructed.

Calculate the prediction value of each modeling data, and evaluate the prediction accuracy by test dataset according to its relative error δ (in shale gas fracturing practice, the acceptable level is $\delta < 5\%$). Considering k with different values, the fitting curves are presented, by which the values of k that can be considered as overfitting should be removed from candidate set.

Step 3: ARMA is used to construct the state function of particle filter.

For the selected modeling sequence $P = \{p_s | s = 1, 2, \ldots, m - n\}$, the data preprocessing and ARMA modeling are conducted according to the Step 4 and Step 5 in Section 2.2.1, where the type and order of the model are determined, and the estimation and verification of above parameters are implemented. The particle filtering algorithm was then substituted to fit the pressure modeling data, and the order and parameters of the ARMA were adjusted according to the Mean Absolute Percentage Error (MAPE) of the fitted results, so that the model could be used for particle filtering prediction.

Step 4: construct a model of particle filter prediction algorithm for pressure parameters.

According to the principle of the algorithm in 2.2.1, a particle filter optimization model for pressure parameters is constructed, in which the ARMA model derived in step 3 is used as the state function of the particle filter, and the optimized pressure PF_ARMA prediction model is obtained by taking the pressure modeling data as the observation values, and the test data are used to evaluate the prediction effect of the model.

Step 5: obtain the LWLR model optimization basis for the pressure parameters.

The model developed in step 4 was used to predict the pressure parameters in multiple steps forward, for forward steps of, e.g., $L = 10, 15, 20, 30$ (steps are per unit of time).

Step 6: forward multi-step prediction of pressure parameters based on LWLR.

With the LWLR model of the initially optimized pressure parameters obtained in step 2, a forward multi-step prediction is performed with the same step size as set in step 5.

Step 7: optimize the LWLR prediction model for the pressure parameters.

Based on the model optimization in step 5, calculate the MAPE of the predicted pressure values for the next steps in step 6, select the appropriate model parameters (k values), optimize the LWLR model built in step 2, and go back to step 6 to obtain the LWLR model with the optimal pressure parameters and the prediction results.

Step 8: evaluation of Pressure Prediction Results.

Calculate the MAPE of Step 7 with respect to the true value of the pressure, evaluate the prediction results, and analyze and summarize the results of the multi-step prediction step.

Step 9: comparison of results.

For the modeled data of the same pressure, the PF_ARMA and ARMA models were applied to predict the pressure in multiple steps

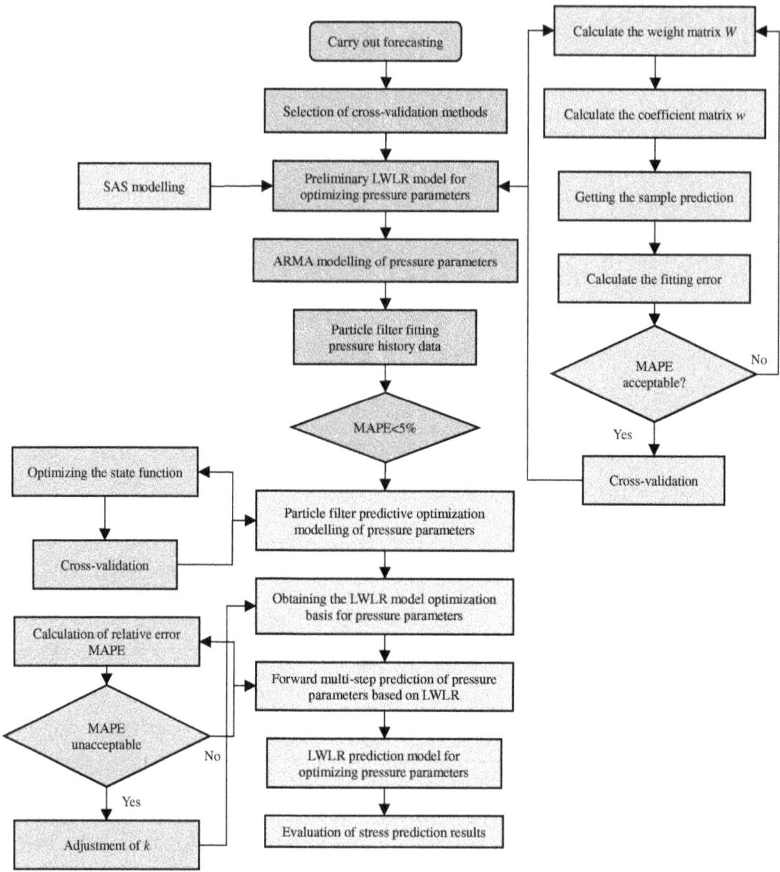

Fig. 2.10. Shale gas fracturing construction curve forward multi-step prediction method flow.

forward and compared with the pressure prediction results obtained in step 7.

2.2.2.2 *Methodological processes*

The flow of the shale gas fracturing construction curve forward multi-step prediction method is shown in Fig. 2.10.

2.2.3 *Case study*

In this section, the method of forward multi-step prediction of shale gas fracturing construction curves is validated using simulation data,

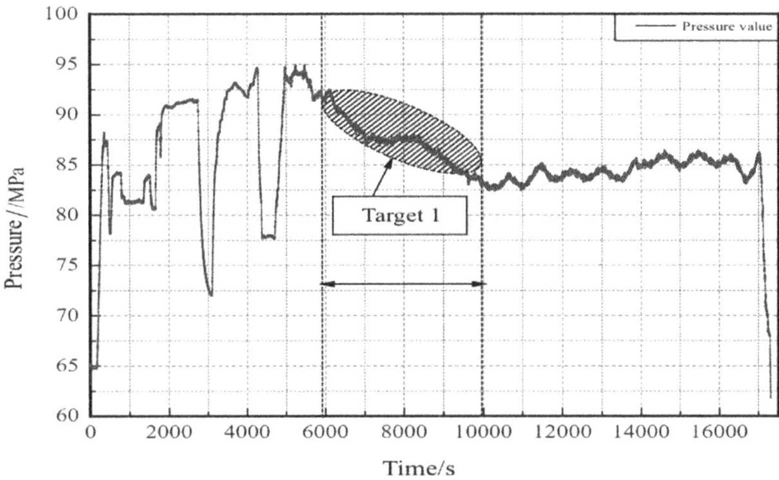

Fig. 2.11. Data curves of shale gas pressure changes in simulation experiments.

field data, and compared with the ARMA model, PF_ARMA model, and the traditional LWLR model, respectively.

2.2.3.1 *Comparison of simulation cases and results*

According to the simulation experiments and historical data on the simulation platform to calculate the shale gas fracture pressure change curve, as a prediction object, to carry out the shale gas fracture construction curve forward multi-step prediction method research, as shown in Fig. 2.11:

Step 1: select the cross-validation scheme. The fracturing construction curve of this process is shown in Fig. 2.11, and the data segment of "Prediction Object 1" is selected as the prediction object for method validation. Then, the pressure values from $T = 5340$ to 8580s are selected as the modeling data, in which one sample point is taken every 10s, with a total of 320 sample points, and the next 10 sample points are taken for validation.

Step 2: development of original LWLR model.

LWLR was applied to the modeled data and the k value was adjusted repeatedly, Fig. 2.12 shows the curve fitting at $k = 0.4, 1, 5, 10, 30, 50, 100, 200$, with 1 step in the figure representing 10s.

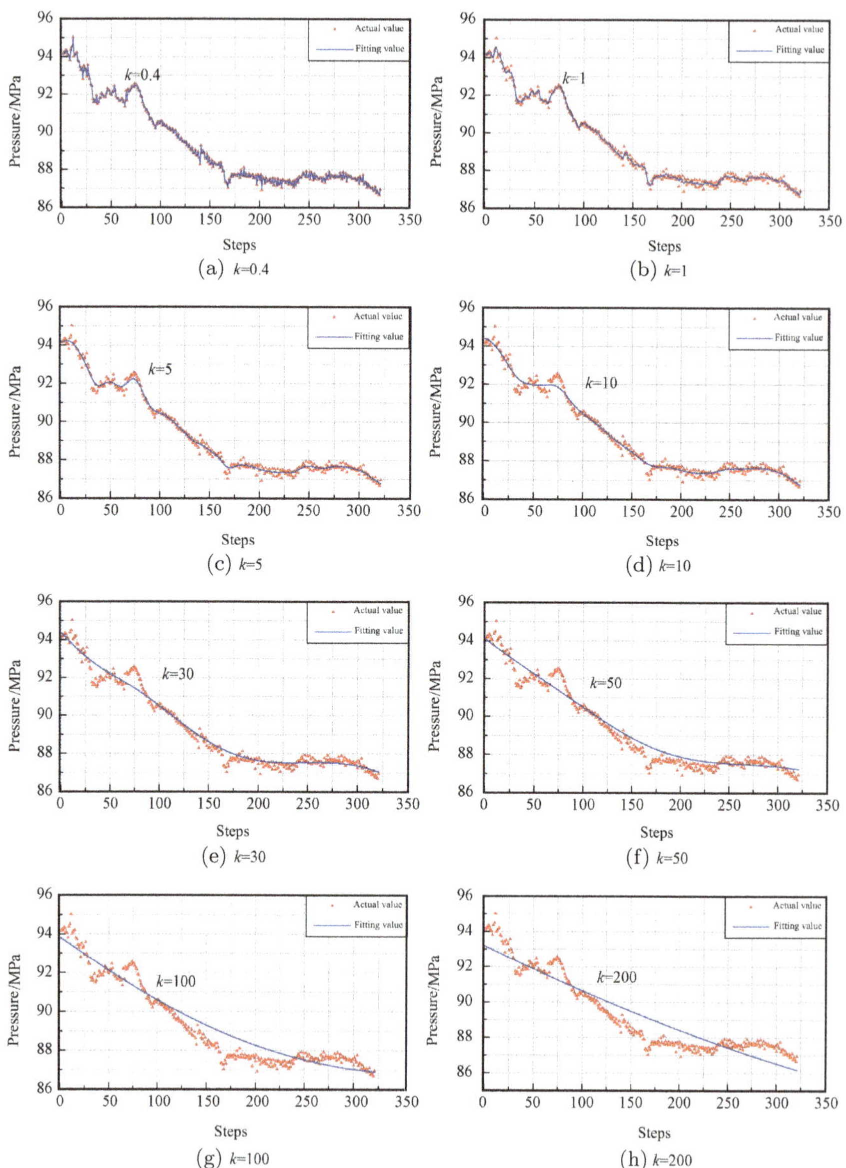

Fig. 2.12. Fitting of historical data with different k values.

From Fig. 2.12, it can be seen that the relative error of the fit gradually increases with the increase of k value, and the fitting effect is close to the standard linear regression when $k = 200$. Obviously, when $k = 0.4$, the fitted straight line is too close to the data points, incorporating too much noise, which leads to the phenomenon of overfitting, so the model determines that the parameters can be set as $k = 1, 5, 10, 30, 50, 100$ is more appropriate.

Prediction validation is carried out on the test dataset and the validation results are shown in Fig. 2.13, which shows that the MAPE is less than 5% in all cases, indicating that the model has good prediction performance and also has good prediction results at $k = 200$, where the prediction fails at $k = 0.4$, removing this k-value. Note: MAPE is within the acceptable range when $k = 1$.

Step 3: use ARMA to construct the state function of PF.

An ARMA model was built for the selected modeled series according to the ARMA modeling procedure in 2.2.2. The SAS analysis reveals that the series is non-stationary, and after 1st order differencing the series is stationary, and after particle filter fitting optimization, the model is determined as an AR (4) model, and the resulting model is shown in Eq. (2.41):

$$X_t = 3.548 \times \exp(-6) + 0.531X_{t-1} + 0.522X_{t-2}$$

$$+ 0.228X_{t-3} - 0.280X_{t-4}. \tag{2.41}$$

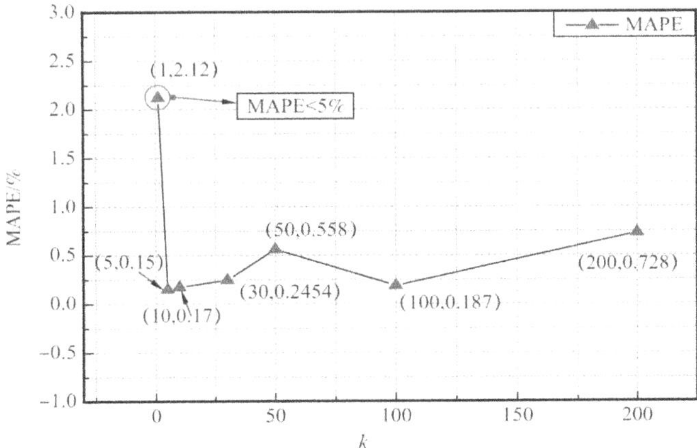

Fig. 2.13.   Prediction error of test data with different k values.

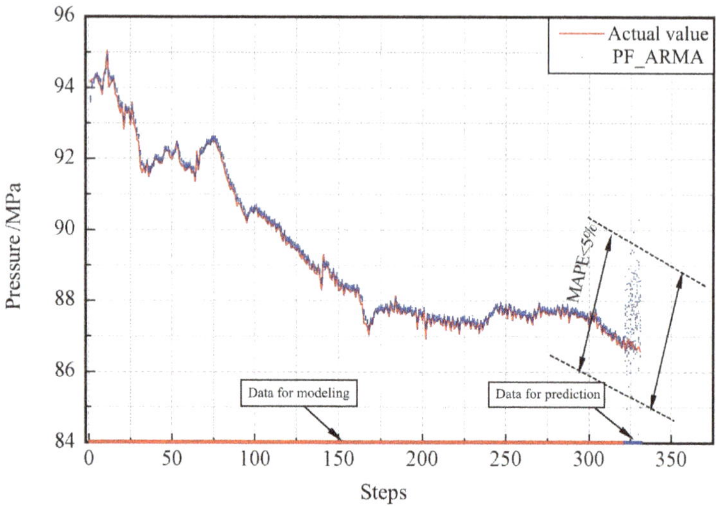

Fig. 2.14. Evaluation of prediction performance for PF_ARMA predictor.

Step 4: development of parallel PF_ARMA predictor.

Construct the particle filter prediction algorithm model for pressure parameters. The particle filtering algorithm is applied and Eq. (2.41) is used as the state equation, the number of particles is $N = 5000$, and it is run many times to optimize the ARMA model and test the prediction effect on the test dataset, and the results are shown in Fig. 2.14, and the real values fall in the prediction results of particle filtering, and the MAPE $<5\%$, which verifies the reliability of the model.

Step 5: obtain the baseline for LWLR model optimization.

The model built in step 4 is used for forward multi-step prediction, and the prediction analyses are carried out for the forward step lengths $L = 1$ to $L = 100$ (1 step means 10s), and the results of 20 predictions are shown in Fig. 2.15, where the particle filtering prediction results are the average of the results of the 20 runs at each moment in time.

From Fig. 2.15, it can be seen that most of the results predicted by the PF_ARMA model for the forward 30 steps (the prediction interval in Fig. 2.15) fall within the range of MAPE $<5\%$, and it can be seen that most of the absolute error is negative, i.e., it is slightly less than the true value. If the pressure change of the abnormal condition is

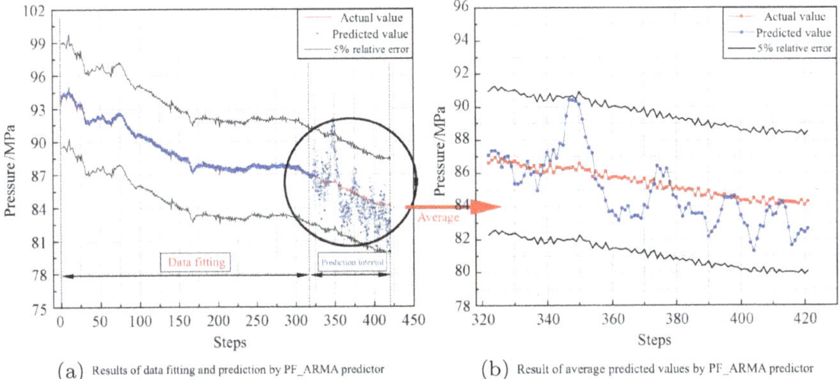

(a) Results of data fitting and prediction by PF_ARMA predictor   (b) Result of average predicted values by PF_ARMA predictor

Fig. 2.15. PF_ARMA predicts results 30 steps forward.

a sharp drop, the predicted value, if it is lower than the true value, can remind the operators to take precautions early to prevent the emergence of abnormal working conditions.

In addition, the significance of step 5 is to provide a factual basis for step 6, i.e., to judge the relative error of the predicted values of step 6 and step 5 100 steps forward, and to adjust the LWLR model (parameter k) constructed in step 2. According to the principle of LWLR, k, which is the best fit, may not be able to achieve a better prediction of the new data, and therefore, a factual basis is needed to prevent the value of k from deviating from the normal range.

Step 6: multi-step forward prediction based on the original LWLR model.

With the LWLR model obtained in step 2 ($k = 1$, 5, 10, 30, 50, 100, 200), the forward multi-step prediction ($L = 100$ steps) is performed and compared with the predicted value of PF_ARMA (1 step means 10s), and the results are shown in Fig. 2.16.

Several conclusions can be drawn from Fig. 2.16:

(1) When $k = 1$, $L > 36$, the prediction fails; when $k = 5$, $L > 33$, MAPE $< 5\%$; when $k = 10$, $L = 62$, MAPE $< 5\%$. When $k = 1, 5, 10$, the trend of the curve is gradually increasing, while the trend of the predicted value of PF_ARMA is gradually decreasing, so it is necessary to eliminate these three k values.

(2) The $k = 30$, 50, 100, and 200 predictions all show a decreasing trend, which is in line with the general trend of the PF_ARMA

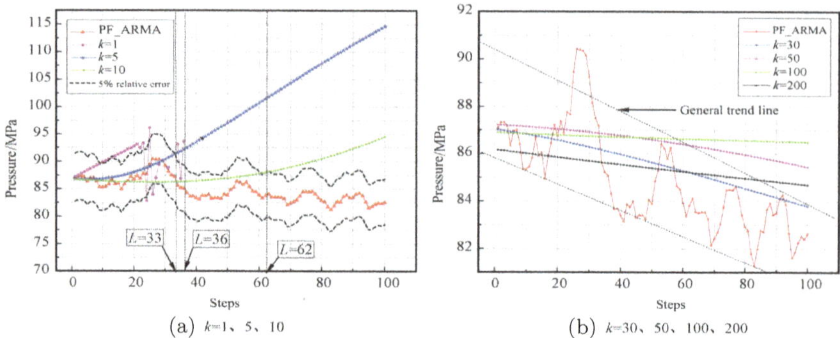

Fig. 2.16. Comparison of LWLR predictions with PF_ARMA predictions for different values of k.

predictions. Among them, $k = 30$ is the closest to the decrease of the PF_ARMA prediction, followed by $k = 200$. Although the result at $k = 30$ is better, it is still smaller than the change in the PF_ARMA prediction, so it can be further optimized.

Step 7: optimize the original LWLR based prediction model.

From Fig. 2.16, it can be concluded that $k = 1, 5, 10$ do not meet the requirements, although the fitting error is small in step 2, but it does not achieve a better prediction for the new data. The prediction effect of $k = 30$ is better than others, but still smaller than the change of PF_ARMA prediction value, which can be further optimized. The MAPE of $L = 100, 90, 80, 70, 60$, and 50 steps were used as the basis for adjustment, and the adjustment results are shown in Fig. 2.17.

The following conclusions can be drawn from Fig. 2.17:

(1) When the number of forward steps $L \geq 50$, the error is minimized at around $k = 25$, and when $k \geq 200$, the error is also relatively small and tends to stabilize, but it is still larger than the relative error at around $k = 25$, so the k value of the model can be set to 25.

(2) At $k = 76$, $L = 50$ to 100 is the worst prediction, and the error becomes progressively larger as L increases. In summary, after the optimization in step 7, the $k = 1, 5, 10, 30, 50, 100$ selected in step 2 is optimized to $k = 25$. Since the MAPE in this section is based on the predicted value of PF_ARMA, there is a certain

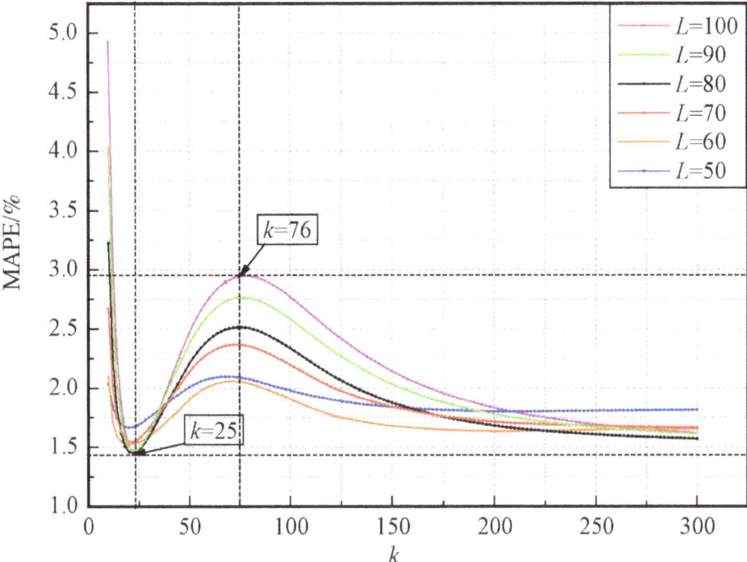

Fig. 2.17. MAPE of LWLR and PF_ARMA predictions for different values of k.

deviation in the optimal k value, but in practice, when the real value is not known, we can use the prediction result of PF_ARMA as the basis to adjust the k value to avoid the prediction result from deviating too much from the actual value and finally find the k-value with better prediction effect.

Step 8: evaluation of the final prediction results after optimization.

Calculate the predicted values when $k = 5, 10, 25, 30, 50, 100, 200$ as shown in Fig. 2.18 and calculate the MAPE by comparing it with the true value.

From Figs. 2.18 and 2.19, it can be seen that $k = 5$ and $k = 10$ have large errors and the predicted trends are not accurate. Although the MAPE of $k = 100$ and 200 is less than 2%, the trend is relatively flat and cannot describe the real trend of the curve. Although $k = 50$ is in line with the real curve, the predicted value is large, which is not favorable for safety warnings. The MAPE of $k = 25$ and $k = 30$ is smaller and can correctly describe the trend of the curve, so the prediction effect is better, and the MAPE of $k = 25$ is the smallest, which is closer to the real value.

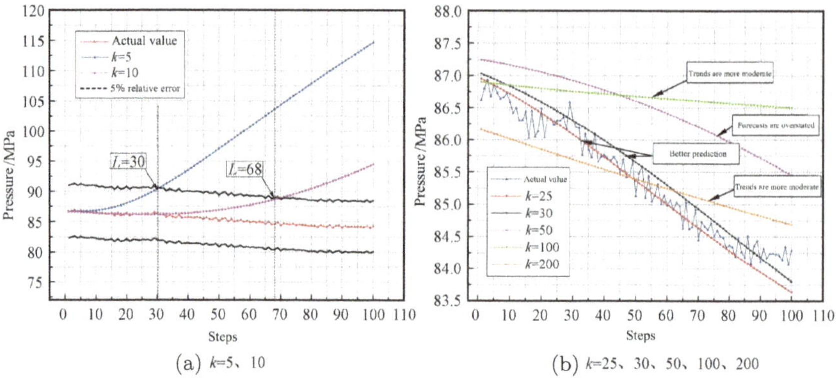

(a) k=5、10   (b) k=25、30、50、100、200

Fig. 2.18. Comparison of predicted and true values of LWLR for different values of k.

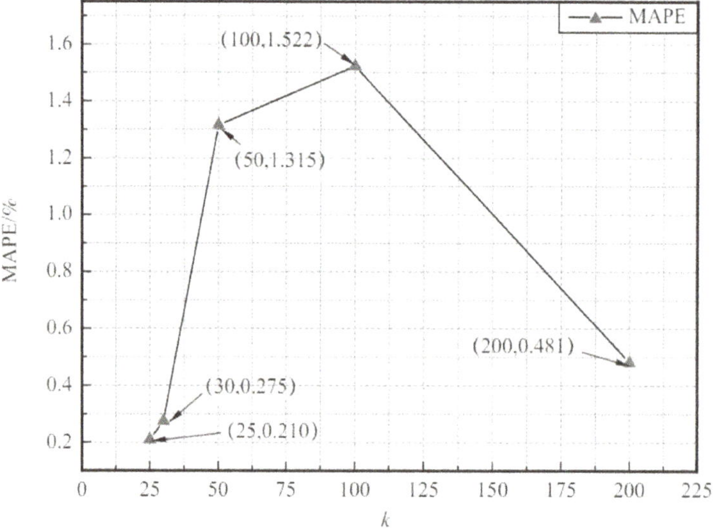

Fig. 2.19. True mean relative error of LWLR prediction for different k values.

Step 9: assessment of prediction results.

Figure 2.20 shows the comparison between the prediction results calculated by the three methods and the actual data. Some conclusions could be derived as follows:

(1) In terms of the overall trend of change, the three methods start to show changes from the decline around $L = 35$, the ARMA

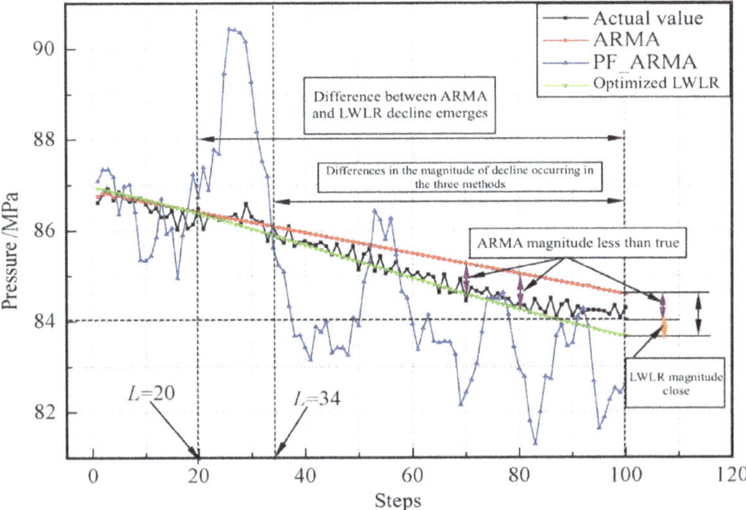

Fig. 2.20. Comparison of the results of the three forecasts.

prediction changes most gently and is smaller than the change of the true value, the PF_ARMA decreases the most and is slightly larger than the true value, and the optimized LWLR is the closest to the change of the true value.

(2) From the point of view of safety warning, the overall trend of PF_ARMA is close to the real value, and the absolute error of the prediction value is negative, while the absolute error of ARMA is positive. If fracturing abnormality occurs when the pressure is lower than a certain value, there is a possibility of false alarms if PF_ARMA is used for prediction, and the ARMA may have missed alarms. Therefore, PF_ARMA prediction is better.

(3) Comparing the traditional LWLR with the optimized LWLR, if the model established in step 2 is used, as shown in Figs. 2.18 and 2.19, it can be seen that the trend of the curve cannot be truly depicted by selecting $k = 1, 5, 50, 100, 200$, and therefore the optimized LWLR predicts more accurately than the traditional LWLR.

(4) In terms of MAPE from $L = 20$ to 100, the optimized LWLR improves the prediction accuracy by 0.25 percentage points over ARMA, 1.51 percentage points over PF_ARMA, and 4.24 percentage points over the conventional LWLR.

Table 2.8.　Summary of the results of the four methodological comparisons.

Methods	Overall performance	Early-warning accuracy	MAPE ($L = 20 - 100$)
Optimized LWLR	Most consistent with actual pressure changes	Correct warning is possible	Minimum
Original LWLR	Not in accordance with the actual change	False alarms may occur with missed alarms	4.24 percentage points more than the new methodology
PF_ARMA	Slightly less than actual pressure change	False alarms may occur	1.51 percentage points more than the new methodology
ARMA	Greater than actual pressure change	False alarms may occur	0.25 percentage points more than the new methodology
Effect Sort	Optimized LWLR > PF_ARMA > ARMA > LWLR		Optimized LWLR > ARMA > PF_ARMA > LWLR

In summary, as shown in Table 2.8, the optimized LWLR is better than the PF_ARMA, AMRA conventional LWLR in terms of overall trend, prediction accuracy, and safety warning; of which, in terms of safety warning, the PF_ARMA has better prediction results than ARMA.

2.2.3.2　*Comparison of field examples and results*

The pressure data of the 16th fracturing process of a gas well was selected as the prediction object to carry out the research on the forward multi-step prediction method of the shale gas fracturing construction curve.

Step 1: select the cross-validation scheme.

The dataset labeled "Target 2" in Fig. 2.21 was selected as the prediction object for method verification. The sample dataset $X = [1490, 1610]$ min was selected for modeling, while $X = [1610, 1615]$ min was used to verify the prediction effect.

Step 2: development of original LWLR model.

LWLR method was used to fit the modeling data, and the k value was adjusted repeatedly. Figure 2.22 showed the data fitting results when $k = 0.4$, $k = 0.7$, $k = 1$, $k = 4$, $k = 7$, and $k = 10$. It could be seen that as the value of k increases, the relative error of the data fitting gradually increases, but when $k = 0.4$ and 0.7, the fitted curves

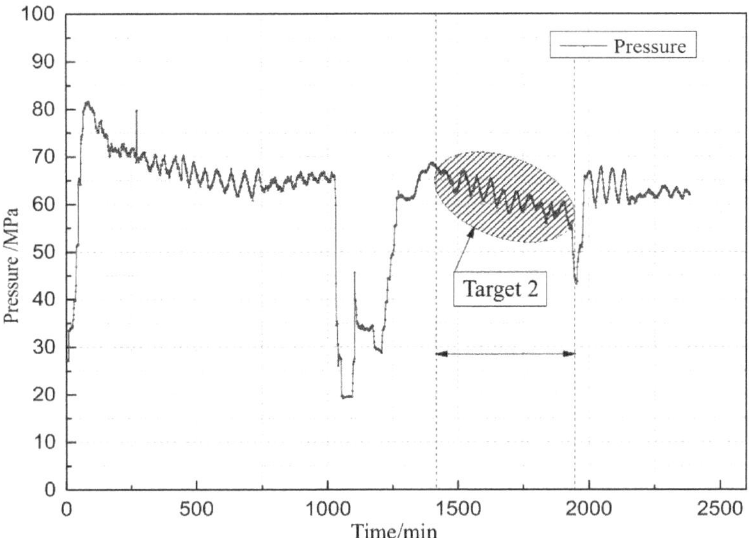

Fig. 2.21.   Field pressure change value.

could be considered as overfitting. Set $X = [1610, 1615]$ min for prediction verification, the verification results are shown in Fig. 2.23. The results showed MAPE $< 5\%$ indicating that the prediction performance of the model could meet the general requirements. The model parameters were set as $k = 1, 4, 7,$ and 10 as candidates for optimization.

Step 3: use ARMA (or ARIMA) to construct the state function of PF.

As mentioned in Section 2.2.1, the sequence $X = \{x_t | t = 1490, \ldots, 1610\}$ was checked and considered non-stationary. After the first-order difference operation, the innovation sequence was tested as stationary. Then ARMA model was used to fit data and adjust the order of the model according to fitting errors. The result was identified as the AR (2) which was shown in Eq. (2.42):

$$X_t = 0.728 + 0.987X_{t-1} + 0.001X_{t-2}. \qquad (2.42)$$

Step 4: development of parallel PF_ARMA predictor.

Equation (2.42) was used as the state function for particle filtering construction, where the number of particles was set as $N = 5000$.

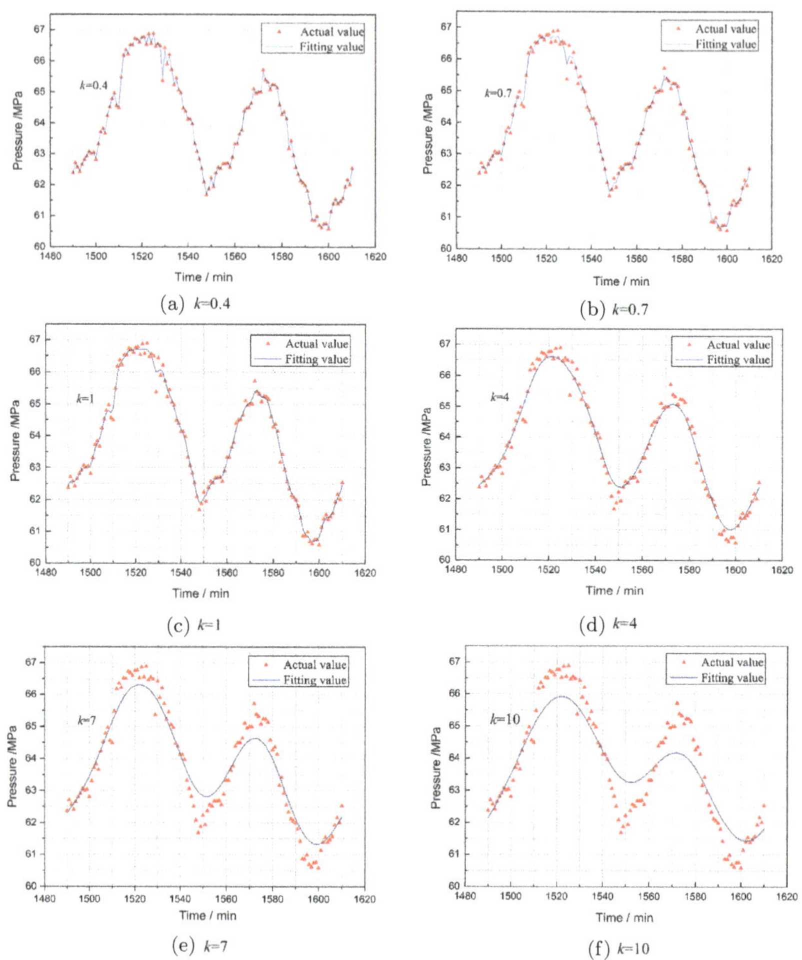

Fig. 2.22. Results of data fitting based on the original LWLR model with different k values.

The prediction effect for $X = [1610, 1615]$ min was assessed, the results of which were shown in Fig. 2.23 (the abscissa started from the modeling data). Figure 2.24 showed that the actual pressure data values all fell in the results of particle filtering and MAPE $< 5\%$, which indicated the reliability of the model was verified.

Step 5: obtain the baseline for LWLR model optimization.

Use the model built in Step 4 to make multi-step forward prediction with forward step length set from $L = 1$ to $L = 30$ (1 step

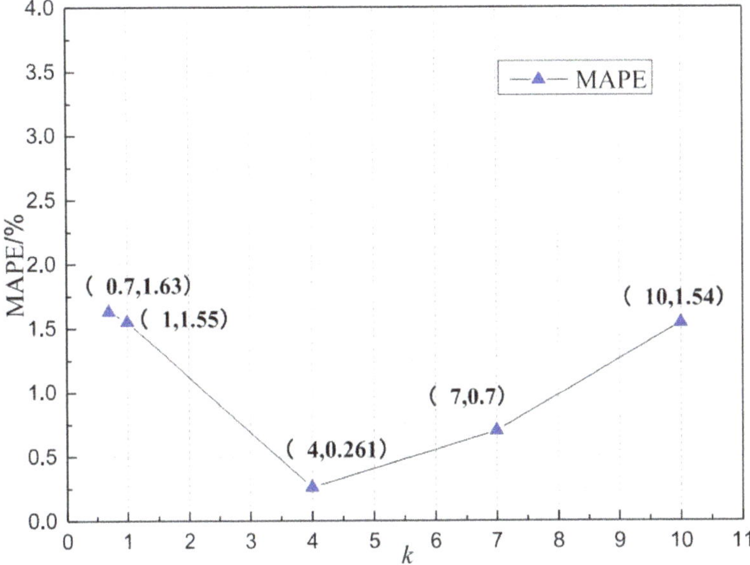

Fig. 2.23. MAPE of LWLR-based predicted data with different k values.

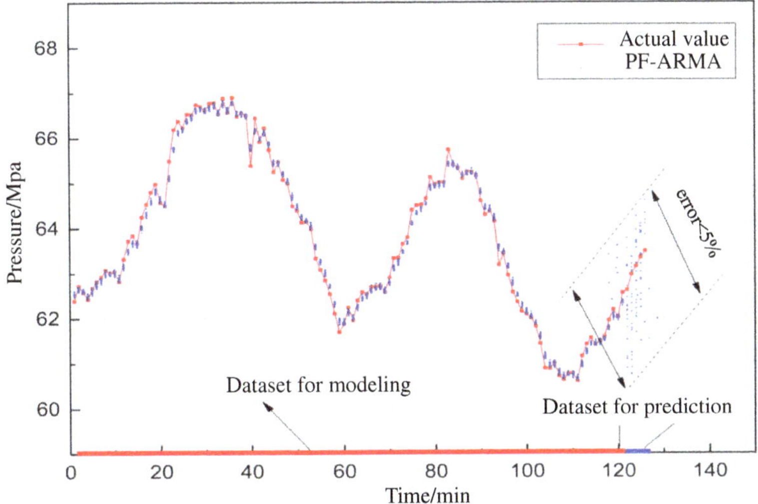

Fig. 2.24. Evaluation of prediction performance for PF_ARMA predictor.

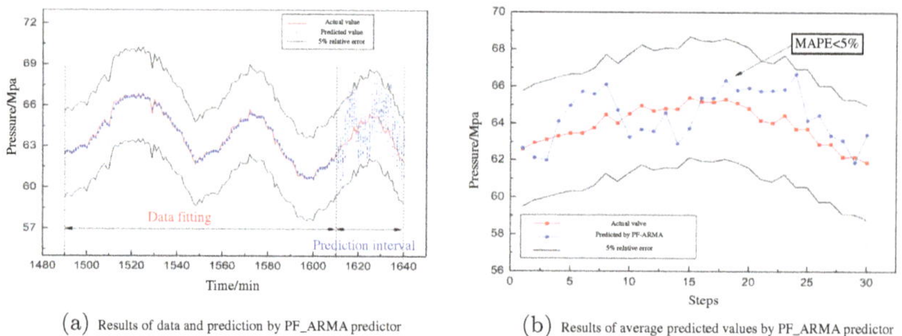

(a) Results of data and prediction by PF_ARMA predictor   (b) Results of average predicted values by PF_ARMA predictor

Fig. 2.25. Results of data fitting and 30-step forward prediction by PF_ARMA predictor.

represents $1\,\mathrm{min}$). The results are shown in Fig. 2.25, which presented the average value of 20 times running on each step. Most of the 30-step forward prediction values (prediction interval) predicted by the PF_ARMA model fell within the range MAPE $< 5\%$. It was also shown that most of the absolute errors were positive, i.e., slightly higher than the actual value, which was helpful in preventing the increased pressure data from deviating from the threshold and reporting in advance.

Step 6: multi-step forward prediction based on the original LWLR model.

Using the original LWLR model obtained in Step 2 ($k = 1, 4, 7, 10$), multi-step forward prediction (1 step represents $1\,\mathrm{min}$) was carried out, the results of which were shown in Fig. 2.17 (when $k = 1$, $L = 22, 28$, the calculation of weights were invalid, so the prediction process failed), compared with the predicted value of PF_ARMA parallel predictor. When $k = 1$, the relative errors of multi-step forward prediction by the LWLR model exceed the acceptable range of error, i.e., MAPE $> 5\%$. When $k = 4$, the predicted values of the first 22 steps fell within the acceptable range of error, i.e., MAPE $< 5\%$. When $k = 7$ and 10, most of the predicted values fell within the acceptable range of error (MAPE $< 5\%$). Therefore, $k = 1$ should be eliminated, and $k = 4, 7$, and 10 would be used for further optimization as candidates.

Step 7: optimize the original LWLR-based prediction model.

The results of the PF_ARMA predictor were used as a baseline for LWLR model optimization. The MAPE of predicted values between

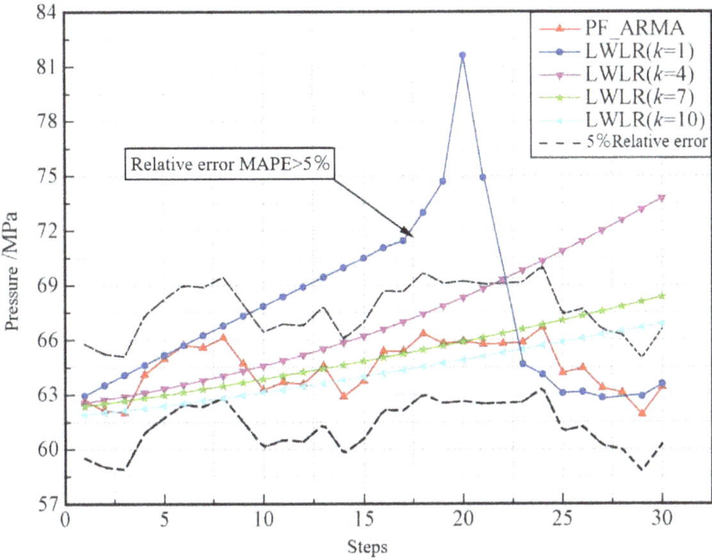

Fig. 2.26. Prediction result comparison of traditional LWLR and PF_ARMA.

the PF_ARMA predictor and the original LWLR model with different $k$values were calculated and shown in Fig. 2.26, from which each line indicated a certain setting of forward step lengths for prediction, such as $L = 10$, $L = 15$, $L = 20$, $L = 25$, and $L = 30$, respectively. Basically, considering the trend of each line, the value of MAPE gradually decreased firstly and then gradually increased again as abscissa k increased (Fig. 2.27). Within a certain acceptable range, the optimal value of k could be obtained according to the position of minimum MAPE. Table 2.9 shows the optimal value of k responding to each setting of forward step length, for example, when $L = 10$, the optimal value of k was 1.9; when $L = 15, 20, 25$, and 30, the optimal value of k was around 7.0.

Step 8: evaluation of the final prediction results after optimization.

It is shown in Figs. 2.28 and 2.29 that when $k = 7.7, 6.6, 7.3$, and 8.6, the MAPE of the first 25 steps is less than 5%, and the first 20 steps are less than 2%, which indicated the prediction effect was overall satisfactory. Therefore, the value of k could be determined according to the required prediction steps when applied in a practical shale gas fracturing system. However, Fig. 2.29 also showed that after

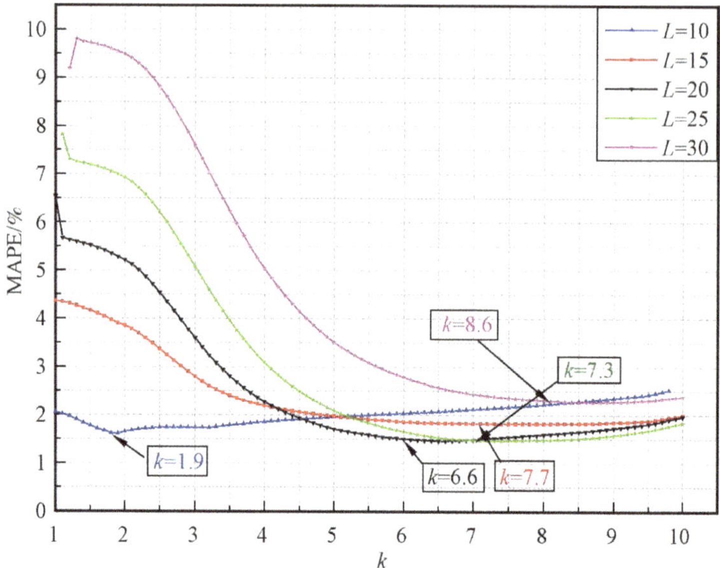

Fig. 2.27.   MAPE of LWLR-based predicted values with parameters k and L.

Table 2.9.   Optimal k value corresponding to L-step forward prediction.

L	Optimal k value	MAPE/%
10	1.9	1.612
15	7.7	1.824
20	6.6	1.514
25	7.3	1.465
30	8.6	2.296

20 steps, the actual pressure value gradually decreased while the predicted value was still increasing. Therefore, with 25 to 30 steps, and with the expectation that LWLR predictions will be less than optimal after 30 steps.

Step 9: assessment of prediction results.

Figure 2.30 shows the comparison between the prediction results calculated by the four methods and the actual data. Some conclusions could be derived as follows:

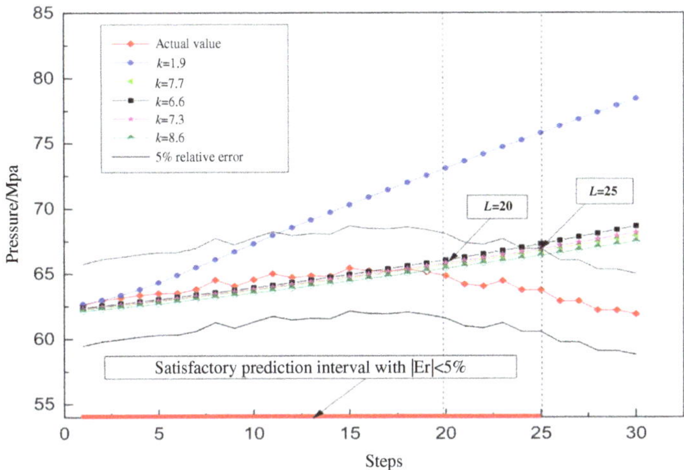

Fig. 2.28. Multi-step prediction results by LWLR with different k values.

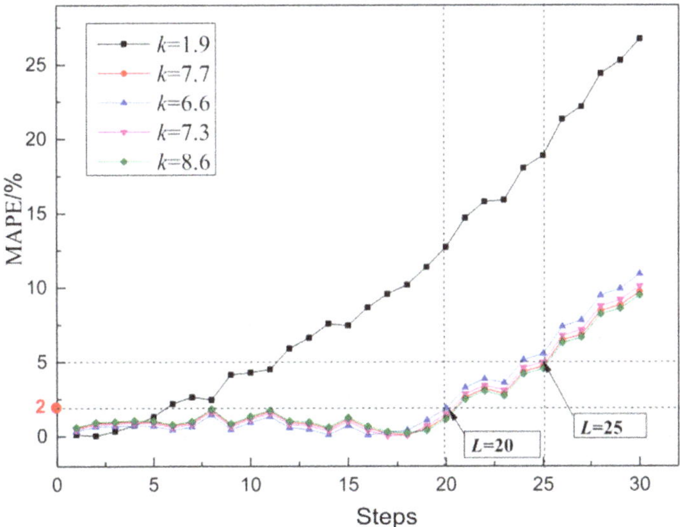

Fig. 2.29. The relationship between MAPE and the k values.

(1) From the perspective of the overall trend, step $L = 17$ could be considered as the turning point, beyond which the increasing trend turns to a decreasing trend. PF_ARMA could predict this changing situation of pressure data successfully, while

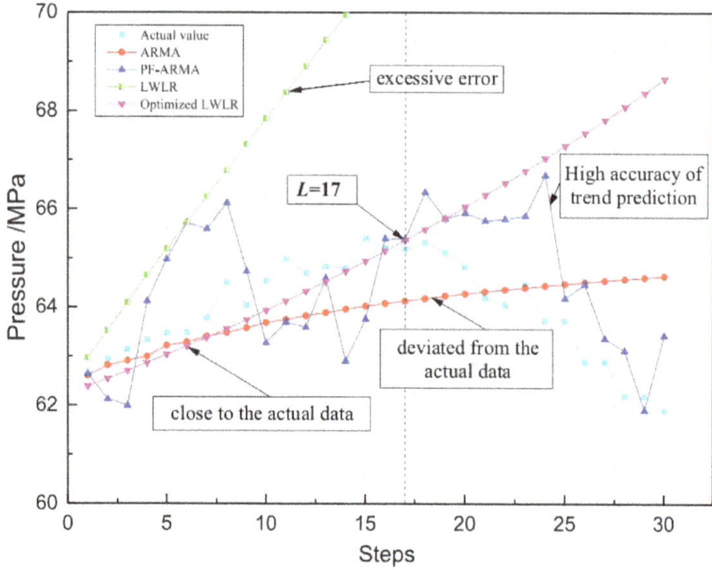

Fig. 2.30. Comparison of the prediction results by four models.

ARMA and optimized LWLR could only well predict the increasing trend.

(2) According to the prediction effect of the previous 17 steps, the optimized LWLR could predict the growth rate and growth trend more accurately, while the prediction results by PF_ARMA appeared sharp fluctuations, so the prediction accuracy of local trend by PF_ARMA was lower than the optimized LWLR. Prediction results by ARMA showed a slowly increasing trend, and the predicted growth rate was smaller than the actual situation, so the prediction accuracy of local data trend by ARMA was much lower than both of PF_ARMA and optimized LWLR model.

(3) Comparing the original LWLR with the optimized LWLR, if the LWLR model ($k = 1$) obtained in Step 2 was used for prediction, the error was rather unacceptable, and the prediction would also be invalid after only 5 steps forward, while the optimized LWLR prediction on the contrary had quite higher accuracy.

(4) According to the MAPE when the step length was set as 17, the optimized LWLR improved the accuracy by 0.4% over ARMA, 1.07% over PF_ARMA, and 4.20% over the traditional LWLR.

Table 2.10.　Application summary of the four methods.

Methods	Overall performance	Accuracy	MAPE ($L = 1 \sim 17$)
Optimized LWLR	Predicted data conforms to actual value within first 17 steps	Applicable for early warning	Least
Original LWLR	Excessive error	Missing alarms and false alarms may occur	+4.20% (compared to optimized LWLR)
PF_ARMA	Predicted trend follows actual data	Applicable for early warning	+1.07% (compared to Optimized LWLR)
ARMA	Predicted data is smaller than the actual value within first 17 steps	Missing alarms may occur	+0.4% (compared to Optimized LWLR)
Effect Sort	PF_ARMA > Optimized LWLR > ARMA > Original LWLR	Optimized LWLR = PF_ARMA > ARMA > Original LWLR	Optimized LWLR > ARMA > PF_ARMA > Original LWLR

In summary, as shown in Table 2.10, from the perspective of long-term trend the performance of PF_ARMA was overall better than the optimized LWLR. If the local details should be paid more attention to, the prediction accuracy of the optimized LWLR was much higher than PF_ARMA model, which attached much importance to the early warning of fracturing screen out. Because the screen-out scenarios usually occur in a short period of time, accurate prediction of local details could improve the accuracy of advanced alarming and reduce the false alarm rate.

2.2.3.3　*Analysis and summary*

(1) For the shale gas fracturing construction curve, the traditional linear regression model has a poor prediction effect, and the empirical function of the filtering algorithm is difficult to select. For this reason, a multi-step forward curve prediction method was developed on the basis of locally weighted linear regression, and the parameters of the LWLR model were optimized by combining the particle filtering algorithm and ARMA, so as to improve the prediction accuracy of the LWLR model.

(2) In the case study, the method was used for trend prediction of shale gas fracture pressure curves in simulation and in the field, respectively, to construct the LWLR model of the pressure parameters and optimize the model parameter k, which ultimately improves the accuracy of forward multistep prediction of the pressure parameters as well as the prediction accuracy of the overall trend.

(3) The case results show that the optimized LWLR model improves the prediction accuracy by an average of 2.28 percentage points over the traditional method, with an average improvement of 4.22 percentage points over the traditional LWLR and 0.33 percentage points over the ARMA method. Moreover, the average improvement in prediction accuracy is 1.29 percentage points compared to the PF_ARMA model. In addition, the optimized LWLR can more accurately describe the trend and magnitude of curve changes than the ARMA method.

2.3 Early Warning of Safety of Complex Systems Based on Condition Monitoring

2.3.1 *Grounded theory*

The methodology presented in this section is the combination of HAZOP analysis, degradation process modeling, dynamic Bayesian networks development, condition monitoring scheme, safety assessment steps, and at last safety prognosis steps. In this section, we term it "an integrated method", concerning how subsystems, components, humans, and environment interact with each other, based on the integration of system structure (hardware), process function, historical failure database, and condition monitoring data.

The nature of the abstraction made in the HAZOP analysis is dictated by the need to determine the system monitoring parameters, hidden state variables with corresponding state space, respectively, and interrelationship among subsystems, components, and environment which is essential for the safety per-warning, through brainstorming of experts taking advantage of historical database and maintenance records. The failure modes relating to each hidden state in HAZOP are further studied as physical mechanisms of deterioration in degradation process modeling. A dynamic Bayesian network (DBN) is developed to simultaneously integrate the structure of the system, hidden and observable variables, causal mechanisms (or interdependencies) provided by HAZOP analysis, and the dynamic deterioration process of each failure mode as a result of the degradation process modeling. The current value of each observable variable in DBN is obtained from condition monitoring based on PLC or SCADA systems relating to field sensors and then is stored in a database for further improvement of the proposed method. Then, we perform safety assessment studies using the inference mechanism of DBN with evidence provided by condition monitoring, in order to evaluate each hidden state and corresponding hazard reasons and consequences with quantified probabilities. Finally, the safety prognosis is conducted by use of the current observable and hidden states to predict the future degradation trends of components, providing a safety pre-warning alarm for proactive maintenance. The main steps of the HAZOP analysis method are as follows.

2.3.1.1 *HAZOP analysis*

The purpose of the HAZOP study is to uncover the causes and consequences of a facility's response to deviations from design intent or from normal operation, i.e., to reveal if the plant has sufficient control and safety features to ensure, that it can cope with expected deviations normally encountered during operation including startup, shutdown, and maintenance. The HAZOP study presented in our work is performed as a structured brainstorming exercise facilitated by a HAZOP study leader and exploits the collective experience of the participants. Specifically, HAZOP studies are used to investigate deviations from a norm: the normal operating conditions or the design intent, i.e., the goal of the system. This can be done by asking questions such as "What deviations can occur?" That is, "What guide words and process parameters are relevant?", "Why do they occur?" (causes), and "How are they revealed?" (consequences). In a HAZOP study, these questions are asked after dividing the whole system into its constituent parts, i.e., sections and nodes, shown in Fig. 2.31 The questions stated relate to the goal of the system while the process represents the means for achieving these goals. Several terms are considered as follows:

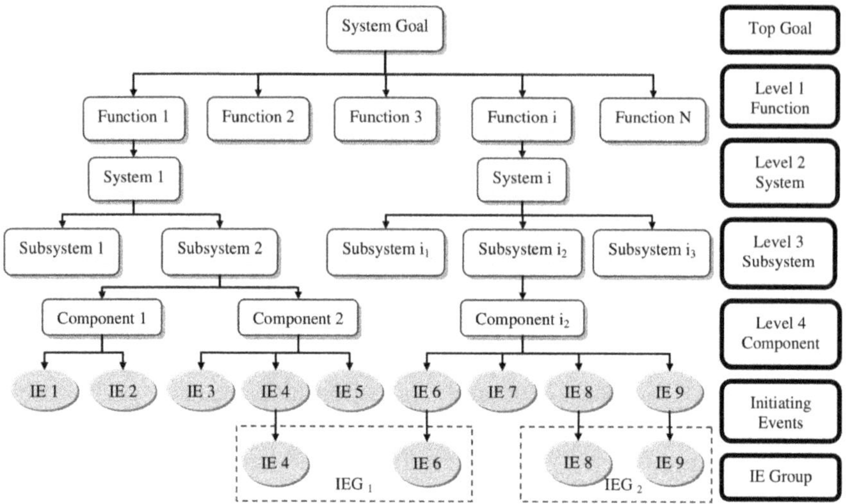

Fig. 2.31. Function system subsystem initiating event relationship.

- *Intentions*: defines how the part is expected to function.
- *Deviations*: departures from the design intention which are discovered by the systematic application of the guide words.
- *Causes*: the reasons why deviations might occur.
- *Consequences*: the results of the deviations.
- *Hazards*: consequences that can cause damage, injury, or loss.
- *Guide words*: simple words which are used to qualify the intention and hence deviations. The list of guide words includes but is not limited to: NO/NOT, MORE, LESS, AS WELL AS, PART OF, REVERSE.

The final results in this phase (shown in Table 2.12) of the proposed integrated method, relating to HAZOP study records presented here, shown in Table 2.11 taking air system in air system in gas turbine compressor system (GTCS) as an example, include but are not limited to:

(1) Failure modes (shown in Table 2.12a), which correspond to hidden states of components in a complex system, indicate components' safety condition and degradation process and are usually abstracted from the hazard reasons (possible cause) in HAZOP study records (Table 2.11). The state space of each failure mode

Table 2.11. HAZOP study records of the air system in GTCS.

Guide words	Deviation	Possible cause	Possible consequences	Safety-related action required
MORE OF	S1_1 Differential pressure of inlet filter (Kpad)	1. Dirty or blocked air filter	1. Air inlet pressure would get lower	1. Clean gas turbine system, or change air filter
	S1_3 Cabinet temperature (°C)	1. Malfunction in the fan on cabinet 2. Dirty air filter 3. Blocked air exhaust path 4. External fire accident	1. It would make the whole facility heat up, and impact the function of each component 2. The temperature of inlet air and the power consumption would increase 3. Corrosion process of internal metal material would be accelerated 4. It would cause fire	1. Change fan 2. Change inlet air filter 3. Check and clean air exhaust path
LESS OF	S1_5 Inlet differential pressure (KPa)	1. Air inlet path has worn or been fouled 1	1. The efficiency of the system would decrease 2. The barotropic state of the cabinet would be destroyed	1. Repair air inlet path
	S1_2 Cabinet pressure (KPa)	1. Partially opened cabinet doors, causing a mass of air leakage 2. Malfunction in the fan on cabinet 3. Dirty air inlet filter	1. If air contains ice, the components would be destroyed 2. The power loss would increase	1. Check and close the cabinet Door 2. Change fan 3. Change air filter
	S1_4 T1 temperature (°C)	1. Environment temperature is on the low side	1. If air contains ice, the components would be destroyed 2. The power loss would increase	1. Make air inlet temperature higher or append air heater

is determined by statistical analysis of maintenance records and failure data in a historical database.

(2) Observable variables (shown in Table 2.12b), the value of which can be obtained from condition monitoring of field sensors, indicate process safety condition and are directly abstracted from deviation in HAZOP study records (Table 2.11). The state space, safety region, and classification criteria are determined by facility specification and condition monitoring alarm standards (made by the company) with modification according to technology knowledge of field engineers.

(3) Exogenous factors, which indicate the influence of the environment (e.g., environment temperature), human factors (e.g., mistakes, operation adjustment), or external entities which relate to the system under consideration (e.g., inputs of the underlining system). Exogenous factors are usually abstracted from the hazard reasons and consequences in HAZOP study records, quantified by discrete states.

(4) Interrelationships (interactions) among degradation processes (hidden variables), environment (exogenous variables), and observable variables are developed from cause and consequence in HAZOP study records, considering system structure and process function. The interrelationship columns in Tables 2.12a and 2.12b only show the interactions within the air system in GTCS, but when it needs to analyze the whole GTCS, the interactions between each subsystem should be determined in detail.

Table 2.12a. Results of HAZOP analysis of the proposed integrated method — failure modes.

Subsystem	Failure modes	ID	State space	Interrelationship
Air system	1. Air filter degradation (AFD)	D1_1	{Normal, fouling, blocked}	{S1_1, S1_2, S1_5}
	2. Ventilation system degradation (VSD)	D1_2	{Normal, degraded, failure}	{S1_2, S1_3}

Table 2.12b. Results of HAZOP analysis of the proposed integrated method — observable variables.

Subsystem	Observable variables	ID	State space	State region	Classification criteria	Interrelationship
Air system	1. Differential pressure of inlet filter (Kpad)	S1_1	{Normal; on the high side; super-high}	<0.75	$\{<0.75; [0.75, 1.5); \geqq 1.5\}$	{D1_1}
	2. Cabinet temperature (KPa)	S1_2	{Normal; on the low side; Ultra low}	>0.062	$\{>0.062; (0.037, 0.062]; \leqq 0.037\}$	{ D1_1, D1_2}
	3. Cabinet pressure (°C)	S1_3	{Normal; on the high side}	<75	$\{<75; \geqq 75\}$	{D1_2}
	4. T1 Temperature (°C)	S1_4	{Normal; on the low side}	>5	$\{>5; \leqq 5\}$	{Exogenous factors}
	5. Inlet differential pressure (KPa)	S1_5	{Normal; on the high side}	<1	$\{<1; \geqq 1\}$	{D1_1}

The main outputs of the HAZOP analysis in the proposed integrated method, including variables with different attributes and explicit cause-and-effect relationships, will be used as the input to the dynamic Bayesian network development and degradation process modeling in the following phases.

2.3.1.2 *Degradation process modeling*

According to the failure modes with corresponding state spaces obtained by the HAZOP analysis in the integrated method, this stage consists of modeling all physical mechanisms of deterioration (e.g., fouling, wearing, cracking, corrosion).

Basically, the Markov process can be used to develop these degradation mechanisms, such as Markov chain with finite discrete states. Meanwhile, the transition probability distribution (or transition matrix) of each degradation process is calculated based on historical inspection and maintenance data (records) or determined by expert knowledge.

Considering that some components may continue to operate in a degraded mode following certain types of faults, the component may continue to perform its function but not at a specified operating level. For example, a lubricating oil pump that continues to work may experience a problem with some components. If it is important to distinguish the degraded state from that of a complete failure, then Markov analysis can be utilized if constant failure rates can be estimated from a historical event database.

Considering a Markov process $\{X(t), t \geq 0\}$ with state space $S = \{0, 1, 2, \ldots, r\}$ and stationary transition probabilities, the transition probabilities of the Markov process $P_{ij}(t) = \Pr(X(t) = j | X(0) = i)$ for all $i, j \in S$ may be arranged as a matrix:

$$
P(t) = \begin{pmatrix}
P_{00}(t) & P_{01}(t) & \cdots & P_{0r}(t) \\
P_{10}(t) & P_{11}(t) & \cdots & P_{1r}(t) \\
\vdots & \vdots & \ddots & \vdots \\
P_{r0}(t) & P_{r1}(t) & \cdots & P_{rr}(t)
\end{pmatrix}.
\tag{2.43}
$$

Three situations should be considered during the degradation process modeling based on historical data and expert judgment, which are (1) discrete degraded state with available maintenance records;

(2) continuous degraded state with available maintenance records; and (3) discrete or continuous state with expert judgment, when maintenance records are unavailable.

(1) Discrete degraded state with available maintenance records. Considering the mechanical seal in GTCS as an example, in the degradation process modeling, the degradation Markov process is determined, called the "mechanical seal degradation (MSD) process". The maintenance records of mechanical seals are shown in Table 2.13, from 1995 to 2009. When a leakage of the seal which is considered a failure is detected, a replacement takes place and the state gets back to normal; while when a fault is detected, a minimum repair takes place keeping the same hazard rate, so the state keeps in degraded. Through the statistical analysis, the number of days in a normal state is 3964 days, in a degraded state is 1009 days, changing to degraded from a normal state is 4 days, changing to leakage from a normal state is 5 days, and changing to leakage from the degraded state is 4 days. Therefore, the transition probabilities of the Markov processes of the above degradation mode are calculated, as shown in Table 2.14, based on historical data stored in databases. In the same way, other components with discrete degraded states, such as air filter degradation process (shown in Table 2.15), are determined in this phase.

(2) Continuous degraded state with available maintenance records. Considering bearing 1# in GTCS as an example, the maintenance records of bearing 1# in all GTCSs (1#–7#) are shown in Table 2.16. When a failure of bearing is detected, such as wearing, the replacement of the bearing is carried out immediately, which is considered as perfect repair, and the return to service state of the bearing is as good as new. Considering Weibull, Normal, and Exp. distribution which can be used to fit the distribution of failure data, the cumulative probability distribution, and probability density of time to failure of bearing 1# comparing with Weibull, Normal, and Exp. distribution are shown in Figs. 2.32 and 2.33, respectively, which demonstrate that the Weibull distribution can fit the failure data better than other distribution functions. The estimated parameters of Weibull distribution are calculated shown in Table 2.17. According to the hazard function of Weibull distribution which is $\lambda(t) = \frac{b}{a}t^{b-1}$, wherein the parameters a, b are shown in Table 2.17, the degradation

Table 2.13.   Maintenance records of mechanical seal.

Date failure discovered	Inspection state	Safety action	Return to service state	Time to failure/day
2009-05-25	Leakage	Replacement	Normal	64.00
2009-03-22	Fault	Repair	Degraded	280.00
2008-06-15	Fault	Repair	Degraded	26.00
2008-05-20	Leakage	Replacement	Normal	467.00
2007-02-08	Fault	Repair	Degraded	476.00
2005-10-20	Leakage	Replacement	Normal	135.00
2005-06-07	Fault	Repair	Degraded	260.00
2004-09-20	Leakage	Replacement	Normal	63.00
2004-07-19	Fault	Repair	Degraded	56.00
2004-05-24	Leakage	Replacement	Normal	168.00
2003-12-08	Leakage	Replacement	Normal	934.00
2001-05-18	Leakage	Replacement	Normal	790.00
1999-03-12	Leakage	Replacement	Normal	1202.00
1995-11-26 (last replace date 1995-10-05)	Leakage	Replacement	Normal	52.00

Table 2.14.   Mechanical seal degradation (MSD) process.

$\mathrm{MSD}_{k+1} =$	Normal	Degraded	Leakage (failure)
$\mathrm{MSD}_k = $ normal	—	0.0010	0.0013
$\mathrm{MSD}_k = $ degraded	0	—	0.0039
$\mathrm{MSD}_k = $ leakage (failure)	0	0	1

process model of bearing 1# is developed shown in Table 2.18, where the time variable k has the unit as a day.

(3) Discrete or continuous state with expert judgment, when maintenance records are insufficient. When the maintenance records are insufficient due to management problems or lack of failure history, the experiential model can be used utilizing expertise accumulated from operational experience, such as mean life and basic statistic records, and further adjusted by field engineer.

Considering an oil pump in GTCS, which experiences a useful life period (a nearly constant failure rate), an exponential distribution can be used according to technical experience. Practical statistic data

Table 2.15. Air filter degradation (AFD) process.

$\text{AFD}_{k+1} =$	Normal	Fouled	Blocked (failure)
$\text{AFD}_k = $ normal	—	0.0031	0.0010
$\text{AFD}_k = $ fouled	0	—	0.0022
$\text{AFD}_k = $ blocked (failure)	0	0	1

Table 2.16. Maintenance records of bearing 1#.

Return to service date	Date failure discovered	Safety action	Time to failure/day
2008-10-29	2009-05-22	Replacement	205
2004-07-26	2008-10-29	Replacement	1556
2002-06-13	2003-09-12	Replacement	456
2008-10-21	2009-06-12	Replacement	235
2003-01-18	2008-10-21	Replacement	2103
2002-05-14	2003-01-18	Replacement	249
2008-04-15	2009-03-24	Replacement	343
2005-01-27	2009-12-15	Replacement	1783
2003-12-09	2004-04-14	Replacement	127
2003-04-03	2003-12-19	Replacement	260
2002-09-22	2003-04-03	Replacement	193
1999-09-20	2003-06-03	Replacement	1352

Table 2.17. Weibull distribution estimate.

				Parameter a		Parameter b	
Log likelihood	Domain	Mean	Variance	Estimate	Std. err.	Estimate	Std. err.
−91.207	$0 < y < $ Inf	740.423	478,046	760.526	217.594	1.072	0.237

in the field station provides that during 100 days, only 1% of oil pumps have failures, so according to the exponential function $f(t) = \lambda e^{\lambda t}$, $R(t) = \lambda e^{\lambda t}$, $\lambda(t) = \lambda$, the failure rate k is then calculated as 1.005×10^{-4}. Therefore, the degradation process of the oil pump can be developed shown in Table 2.19.

In conclusion, the main outputs of this phase of the proposed integrated method, i.e., the degradation process modeling, will be used for dynamic nodes in the dynamic Bayesian network construction.

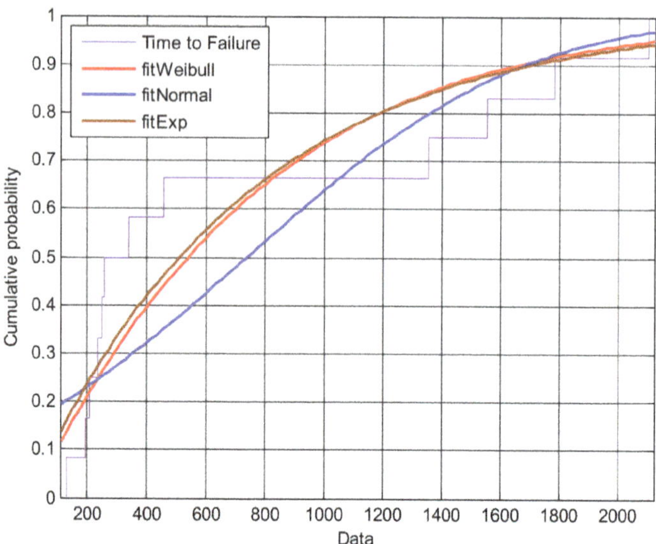

Fig. 2.32. Cumulative probability distribution of time to failure of bearing 1# comparing with Weibull, Normal, and Exp. Distribution.

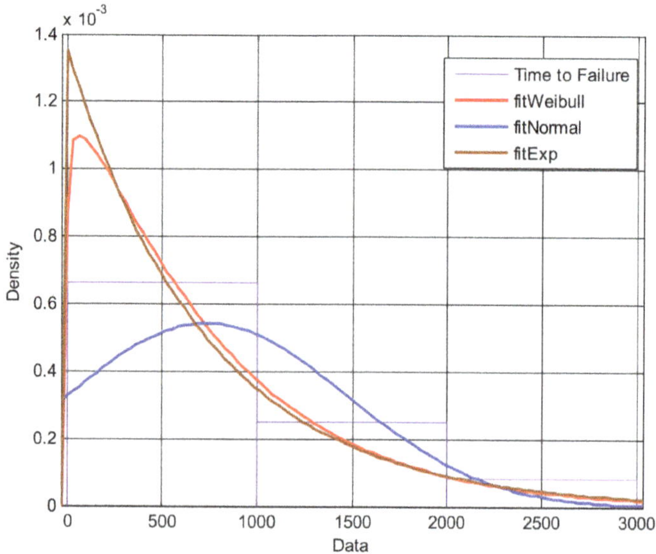

Fig. 2.33. Probability density of time to failure of bearing 1# comparing with Weibull, Normal, and Exp. Distribution.

Table 2.18.   Bearing 1# degradation process.

$X_{k+1}^{B_1}/X_k^{B_1}$	Normal	Failure
Normal	$1 - \lambda(k) = 1 - \frac{1.072}{760.526}k^{0.072}$	$\lambda(k) = \frac{1.072}{760.526}k^{0.072}$
Failure	0	1

Table 2.19. Oil pump degradation (OPD) process.

X_{k+1}/X_k	Normal	Failure
Normal	$1 - \lambda$	$\lambda = 1.005 \times 10^{-4}$
Failure	0	1

When new data (failure data from maintenance records or observable data from condition monitoring) is available, according to the updating (training) mechanism of the dynamic Bayesian network, the degradation process model can also be improved by the training of new data in order to ensure the accuracy of the final safety assessment and prognosis, which is considered as an advantage of this proposed method.

2.3.1.3 *DBN construction*

From the last two phases of the integrated method, the whole system has been systematically decomposed into variables with different attributes, state spaces, and interrelationships based on HAZOP analysis and degradation process modeling from both structural and functional system points of view. In this phase, the dynamic Bayesian network approach as a probability-based knowledge representation method is utilized to simultaneously integrate these elements (the variables and causal mechanisms or interdependencies), which is appropriate for the modeling of causal processes with uncertainty. As another advantage, DBN allows using different knowledge sources and integrating them, such as operative knowledge database (OKD), expert experience represented by logical rules, equations or subjective probabilities, and observable data from conditioning monitoring.

The main reasons for selecting the DBN approach in this phase are as follows:

(1) Because of the interdependence of coding variances of models, DBN can easily deal with the situation of insufficient or partially lost data;
(2) DBN can be used to learn the causal relationship, and hence obtain an understanding of the problem as well as predict the consequence of interference;
(3) As the semantics of a DBN have causal and probabilistic attributes, it is the ideal representative way to combine prior knowledge and data;
(4) DBN can be utilized to establish real-time dynamic data mining architecture, by which network structure is developed step by step from time series data, further making a contribution to a strong real-time decision-making system.

Theoretically, a Bayesian network is a directed acyclic graph (DAG) whose nodes represent random variables and links define probabilistic dependences between variables. These relationships are quantified by associating a conditional probability table with each node, given any possible configuration of values for its parents.

The static Bayesian network (BN) can be extended to a dynamic Bayesian network (DBN) model by introducing relevant temporal dependencies that capture the dynamic behaviors of the domain variables between representations of the static network at different times. Two types of dependencies can be distinguished in a dynamic Bayesian network: contemporaneous dependencies and non-contemporaneous dependencies. Contemporaneous dependencies refer to arcs between nodes that represent variables within the same time period. Non-contemporaneous dependencies refer to arcs between nodes that represent variables at different times. Murphy studied DBN in general terms and applied it to speech recognition.

In our work, DBN with discrete states is used in the third phase of the integrated method to model probability distributions over semi-infinite collections of random variables $\{Z_1, Z_2, \ldots\}$. The DBN is defined to be a pair $(B_1, B_\rightarrow)$, where B_1 is a BN which defines the prior $P(Z_1)$, and $B_\rightarrow$ is a two-slice temporal Bayesian net (2TBN) which defines $P(Z_t|Z_{t-1})$ by means of a DAG as follows:

$$P(Z_t|Z_{t-1}) = \prod_{t=1}^{N} P(Z_t^i|Pa(Z_t^i)), \tag{2.44}$$

where Z_t^i is the ith node at time t and $Pa(Z_t^i)$ are the parents of Z_t^i in the graph.

The nodes in the first slice of a 2TBN do not have any parameters associated with them, but each node in the second slice of the 2TBN has an associated conditional probability distribution (CPD) for continuous variables or conditional probability table (CPT) for discrete variables, which defines $P(Z_t^i|Pa(Z_t^i))$ for all $t > 1$. The parents of a node $Pa(Z_t^i)$ can either be in the same time slice or in the previous time slice.

The arcs between slices reflect the causal flow of time. If there is an arc from Z_{t-1}^i to Z_t^i, this node is called persistent. The arcs within a slice are arbitrary. Directed arcs within a slice represent "instantaneous" causation. In this section, the parameters of the CPTs used in this phase are assumed time-invariant, i.e., the model is time-homogeneous. Therefore, the semantics of a DBN can be defined by "unrolling" the 2TBN until there are T time-slices. The resulting joint distribution is then given by

$$P(Z_{1:T}) = \prod_{t=1}^{T} \prod_{i=1}^{N} P(Z_t^i Pa(Z_t^i)). \tag{2.45}$$

Practically, there are three procedures in this phase (Fig. 2.34):

(1) Identification of the DBN variables (nodes). Generally, all the variables in the results of the HAZOP analysis are associated with nodes in dynamic Bayesian networks. Specifically, each deterioration mechanism relating to failure modes further provided by degradation process modeling is associated with a hidden state variable in DBN as dynamic nodes, whereas other variables in the HAZOP study which can be observed by a condition monitoring system are associated with observable variables in DBN as static nodes, while exogenous factors in the HAZOP study are associated with hidden-state variables in DBN as static nodes.

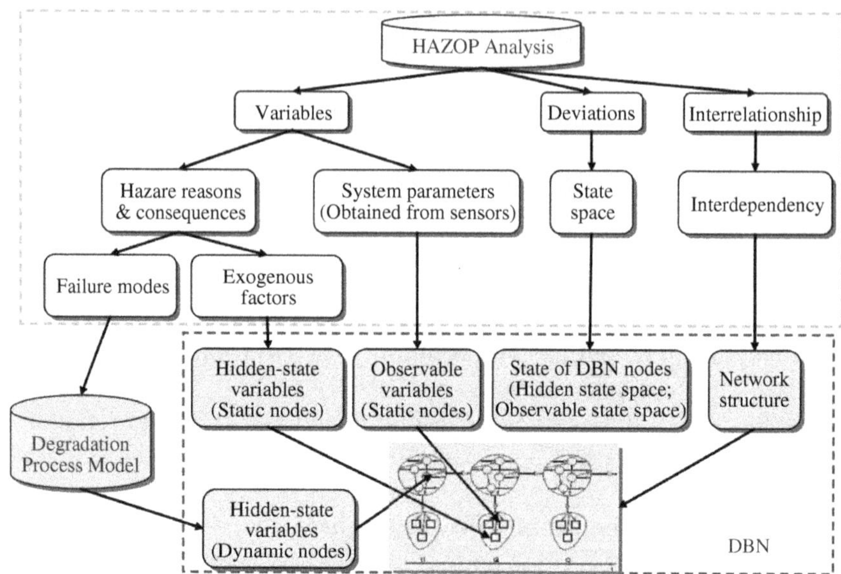

Fig. 2.34. Mechanism of DBN construction in the integrated method.

(2) Development of the network structure by creating the directed edges, which is indicative of a conditional dependency between the variables it links as a key step in this phase, based on the interrelationship in HAZOP analysis, where the directed edges are from possible causes to consequences with certain probability. Taking advantage of the network structure, "combinational explosion" in conventional assessment methods can be avoided, because the root cause can be easily located through the network structure with the largest occurrence probability, and a hazard propagation pathway (if it is possible) can be provided by DBN inference mechanism, instead of giving large numbers of hazard outcomes.

(3) Determination of DBN parameters, i.e., the definition of all the conditional probability tables (CPTs). CPTs can be determined by learning the parameters from the database, or depending on an expert's judgment.

- *Parameter learning based on historical data*: if there are sufficient samples in a database for every parameter in DBN, a sample statistic approach can be used for parameter learning; if there is partially observable or insufficient data, EM or gradient algorithm can be used for learning.

- *Parameter learning based on expert experience*: when the data are insufficient, another way to determine parameters (conditional probabilities) is according to expert experience first, then updating them by collecting condition monitoring data and maintenance records in the future periodically. As for conditional probabilities between static variables, the CPT can usually be determined by determinative functional equations as physical rules, such as the function of pressure, volume, and temperature.

In this way, the DBN model integrating the results of both HAZOP analysis and degradation process modeling describes the evolution of the states of some systems (subsystems, components) in terms of their joint probability distribution. At each instance in time t the states of the system depend only on the states at the previous $(t \rightarrow 1)$ and possibly the current time instance. Furthermore, the DBN model is a probability distribution function on the sequence of T hidden-state variables $X = \{x_0, \ldots, x_{T-1}\}$ and the sequence of T observables $Y = \{y_0, \ldots, y_{T-1}\}$ that has the following factorization, which satisfies the requirements for DBNs that state x_t depend only on state $x_t \rightarrow 1$:

$$\Pr(X, Y) = \prod_{t=1}^{T-1} \Pr(x_t | x_{t-1}) \cdot \prod_{t=0}^{T-1} \Pr(y_t | x_t) \cdot \Pr(x_0). \tag{2.46}$$

In order to further specify a DBN, the above-mentioned parameters as CPT can be classified as follows.

- State transition CPT $\Pr(x_{t+1} | x_t)$, which has already determined, i.e., degradation process modeling, and can be modified or improved in the DBN parameter learning process if it is needed.
- Observation CPT $\Pr(y_t | x_t)$, which is the other key step in this section, and will be illustrated in detail as follows, and
- Initial state probability distribution $\Pr(x_0)$.

Taking the system in GTCS as an example, the DBN is then developed by integrating the HAZOP and degradation models of the air system, and the DBN structure is shown in Fig. 2.35, according to

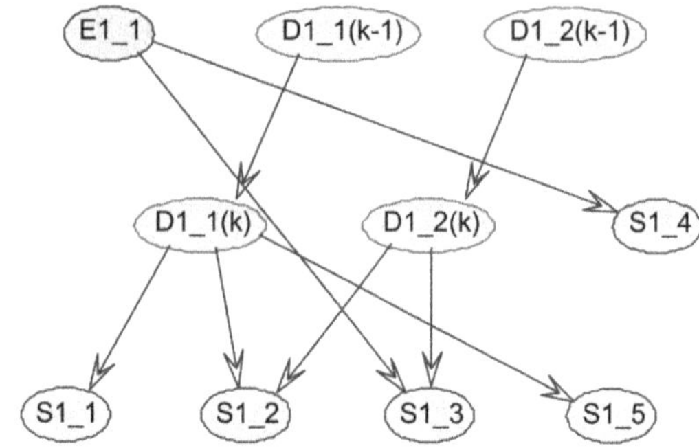

Fig. 2.35. Dynamic network structure of air system.

Table 2.20. Parameter learning modification with insufficient data and experience.

Observation CPT			Observation CPT_GUESS			Observation CPT_EST		
0.7490	0.1975	0.0535	0.50	0.30	0.20	0.7778	0.1978	0.0244
0.2635	0.3979	0.3386	0.30	0.60	0.10	0.1996	0.4442	0.3562
0.0664	0.3253	0.6083	0.20	0.50	0.30	0.0616	0.3744	0.5639

Table 2.21. CPT of $\Pr(S_{1\_2}|D_{1\_1}, D_{1\_2})$.

D_{1_1}	D_{1_2}	S_{1_2} = normal	On the low side	Ultra-low
Normal	Normal	0.9553	0.0447	0
Fouled	Normal	0.7557	0.2443	0
Blocked	Normal	0.5945	0.3539	0.0516
Normal	Degraded	0.2502	0.6063	0.1435
Fouled	Degraded	0.1527	0.7496	0.0977
Blocked	Degraded	0.1109	0.7453	0.1438
Normal	Failure	0.0563	0.7950	0.1487
Fouled	Failure	0	0.7016	0.2984
Blocked	Failure	0	0.5935	0.4065

which the CPTs of all the observations are shown in Tables 2.20–2.24 determined by historical databases and maintenance records.

Specifically, taking two nodes $(D_{1\_1} \rightarrow S_{1\_1})$ (i.e., air filter degradation as hidden-state variable and dynamic node — differential pressure of inlet filter as observable variable and static node) as a

Table 2.22. CPT of $\Pr(S_{1\_3}|E_{1\_1}, D_{1\_2})$.

E_{1_1}	D_{1_2}	$S_{1_3} = $ normal	On the low side
Normal	Normal	1	0
On the low side	Normal	1	0
On the high side	Normal	0.9819	0.0181
Normal	Degraded	0.7622	0.2378
On the low side	Degraded	0.8224	0.1776
On the high side	Degraded	0.7468	0.2532
Normal	Failure	0.4013	0.5987
On the low side	Failure	0.4511	0.5489
On the high side	Failure	0.3548	0.6452

Table 2.23. CPT of $\Pr(S_{1\_4}|\ E_{1\_1})$.

E_{1_1}	$S_{1_4} = $ normal	On the low side	On the high side
Normal	0.9787	0.0121	0.0092
On the low side	0.1536	0.8464	0
On the high side	0.1461	0	0.8539

Table 2.24. CPT of $\Pr(S_{1\_5}|D_{1\_1})$.

D_{1_1}	$S_{1_5} = $ normal	On the high side
Normal	0.8469	0.1531
Fouled	0.4464	0.5536
Blocked	0.2507	0.7493

simple DBN to illustrate the process of DBN parameter learning, by searching the historical inspection records, maintenance records and abnormal alarms of differential pressure in about 10 years (1997–2009), hundred samples are provided and shown in Fig. 2.36, wherein for S_{1_1}, state "1" indicates "normal" state, "2" indicates "On the high side", and "3" indicates "Super-high", while for D_{1_1}, state "1" indicates "Normal", "2" indicates "Fouled", and "3" indicates "Blocked". Each sample includes the state of S_{1_1} with the corresponding state of D_{1_1}. The Baum–Welch learning algorithm (Baum *et al.*, 1970) is used, and the results are shown in Fig. 2.37, wherein

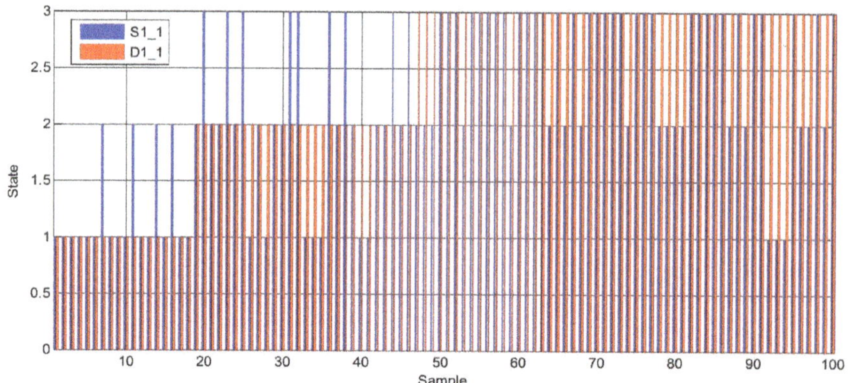

Fig. 2.36. ($D_{1\_1} \to S_{1\_1}$) historical data for parameter learning (S_{1_1}: "1"–"normal", "2"–"On the high side", "3"–"Super-high"; D_{1_1}: "1"–"Normal", "2"–"Fouled", "3"–"Blocked").

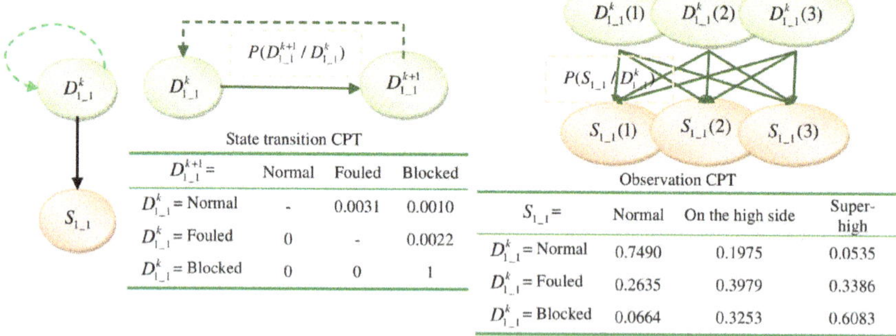

Fig. 2.37. Parameter learning of two nodes ($D_{1\_1} \to S_{1\_1}$) in air system.

the state transition CPT of D_{1_1} inherits the result of the degradation model in Table 2.15. Considering the situation of insufficient data, we take only 50 samples out of the original dataset, as a new insufficient dataset, and give an initial experiential Observation CPT named "Observation CPT_GUESS" shown in Table 2.20, while the "Observation CPT" column shows the standard parameters. With the Baum–Welch algorithm, the initial parameters are modified gradually by the available dataset, and finally the modified Observation CPT is estimated shown in Table 2.20 named "Observation CPT_EST", from which the estimated parameters from insufficient data as well as expert experience are quite close to the standard parameters. Then we use "Observation CPT_EST" as well as state transition CPT to estimate the hidden states (D_{1_1}) based on the original dataset (100 samples) given only observable variables (S_{1_1}). Only five estimated hidden states of original samples are wrong, and the correctness is 95%. Therefore, a dynamic Bayesian network integrating HAZOP study results and degradation process model can be developed by both historical data and technical experience in industrial applications as this case demonstrates, and as long as a condition monitoring system and maintenance action will provide updated data periodically, the DBN of the complex system can be modified online and close to the real world model finally. Tables 2.21–2.24 show the other observation CPTs of air systems which will be used in the following section, based on historical data with the assistance of experiential knowledge. The main results of DBN modeling of GTCS are shown in Appendix, in which Table 2A.1 records information of dynamic nodes in DBN, and Table 2A.2 records information of static nodes. The DBN model developed in this phase of the proposed method is the fundament of safety assessment and prognosis in the following steps.

2.3.2 Methods for safety early warning of complex systems based on condition monitoring

2.3.2.1 Condition monitoring framework

As mentioned above, observations need to be updated to modify the parameters of the DBN. A state monitoring framework is designed at this stage, which will provide further real-time evidence, i.e., the state of the observable variables, for the next stage of safety assessment and forecasting. The next stage of safety assessment and forecasting.

The online dynamic condition and process monitoring system is built on the SCADA system and obtains observable indicators by OPC technology with high acquisition frequency. Specifically, in the condition and process monitoring system, each observable variable value is obtained from field sensors based on local PLC or SCADA systems (Fig. 2.38). Figure 2.39 shows the application of condition monitoring with historical data display. Then, the variable value is mapped into probabilities of each state space defined in HAZOP analysis, which is explicated in detail in previous work. On the other

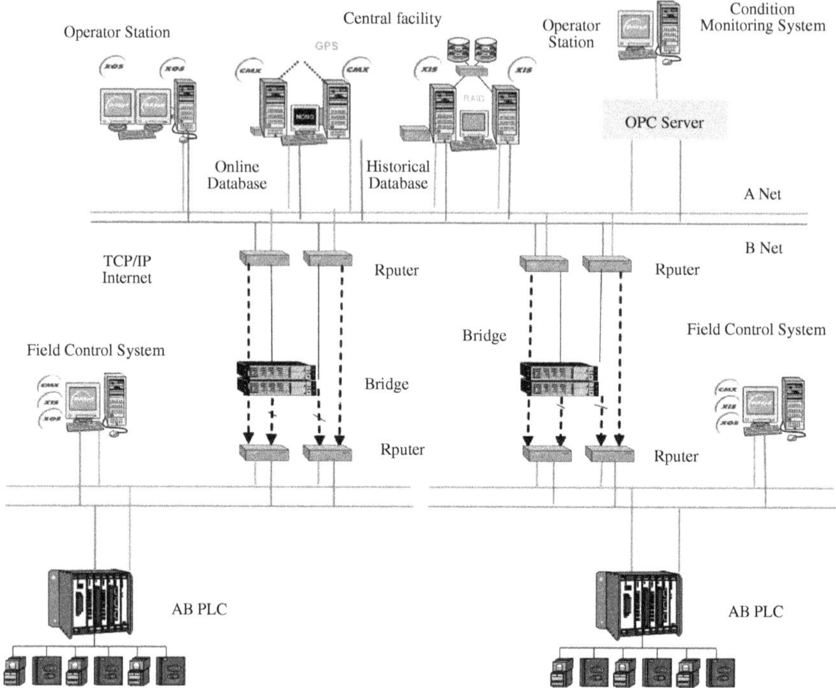

Fig. 2.38. Condition monitoring system with OPC technology based on SCADA system.

hand, the real-time data are stored in a safety database consisting of a real-time database, fault sample database, normal operating condition database, and unknown or potential fault database. It is developed to update the degradation and DBN models with proper parameters. Furthermore, the real-time monitoring values of indicators will further be propagated in the network providing hard or soft evidence for DBN inference, in order to calculate the posterior probabilities of each hidden state of the complex industrial system for safety assessment and prediction.

2.3.2.2 *Safety assessment steps*

In this phase, the basic aim is to determine the hidden states of the system defined in the DBN based on current observation data obtained by the condition monitoring as inference evidence, and then to locate possible hazard reasons and consequences, and decide on

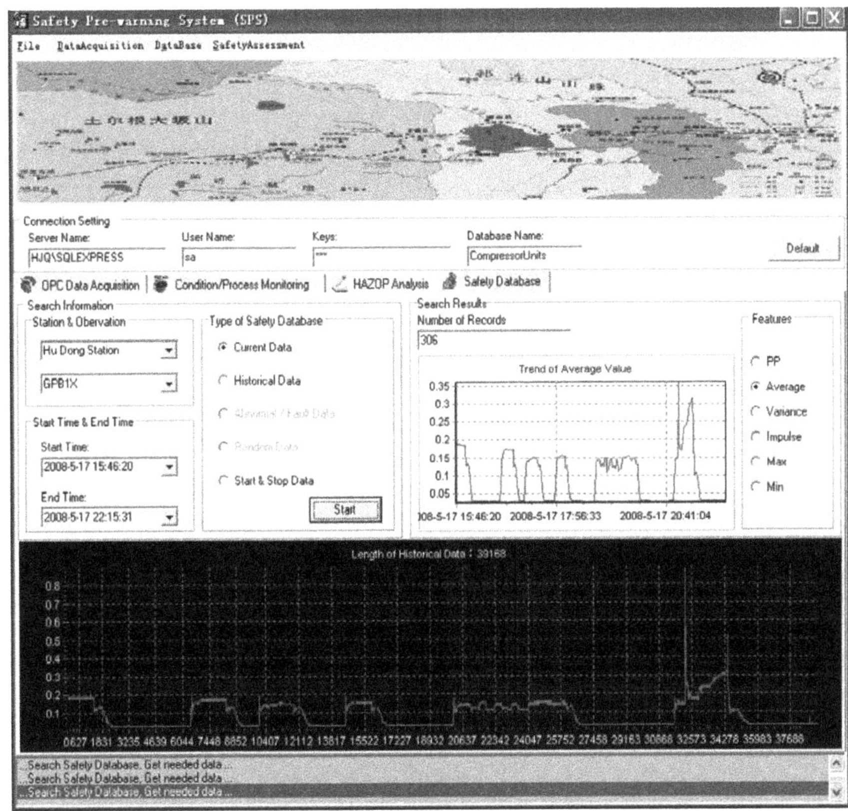

Fig. 2.39. Condition monitoring system and historical data.

proper safety measurements as well as maintenance plans. In other words, assessment with dynamic Bayesian networks consists of fixing the values of the observed variables and computing the posterior probabilities of the unobserved variables.

The problem of inference in assessment steps can be posed as the problem of finding $\Pr(X_0^T | Y_0^T)$, where Y_0^T denotes a finite set of T consecutive observations $Y_0^T = \{y_1, \ldots, y_T\}$ and X_0^T is the set of the corresponding hidden variables. The squares in Fig. 2.34 indicate that the distribution of x_t is to be estimated based on observations Y_0^T.

In this phase, the forwards–backwards (FB) algorithm is used for inference in assessment steps. The basic idea of the FB algorithm is to recursively compute $\alpha_t(i) = P(X_t = i | y_{1:t})$ in the forwards pass, to recursively compute $\beta_t(i) = P(y_{t+1:T} | X_t = i)$ in the backward pass, and then combine them to produce the final answer $\gamma_t(t) = P(X_t = i | y_{1:T})$:

$$P(X_t = i | y_{1:T}) = \frac{1}{P(y_{1:T})} P(y_{t+1:T} | X_t = i, y_{1:t}) P(X_t = i, y_{1:t})$$

$$= \frac{1}{P(y_{1:T})} P(y_{t+1:T} | X_t = i) P(X_t = i | y_{1:t}) \quad (2.47)$$

or

$$\gamma_t \propto \alpha_t \cdot *\beta_t. \quad (2.48)$$

Several other inference methods for a discrete state DBN inference can be used, e.g., unrolled junction tree and the frontier algorithm, which have advantages in more complicated situations. The assessment steps finally give the probability of each hidden state of the system, i.e., the degradation condition of each component, the probability of each potential exogenous hazard reason, as well as possible consequences. According to the results of the assessment model, a decision can be made to eliminate hazard origins, repair degraded entities, or even establish a proper contingency plan.

2.3.2.3 *Safety prognosis steps*

Based on the safety assessment results, the execution of the prognosis forecasts the future evolution of the system's degradation or reliability (temporal inference) in terms of hidden-state variables and also system performance in terms of corresponding observable variables, based on current observations from the condition monitoring, present hidden states from the safety assessment, and diachronic evolution mechanism from the DBN model.

The prognosis steps deal with predicting future observations which indicate system performance as well as future hidden states which indicate component reliability based on both the current observation data and hidden states obtained from the assessment model. A one-step prediction can be started as the following inference problem: $\Pr(x_{t+1} | Y_0^T)$ or $\Pr(y_{t+1} | Y_0^T)$:

$$\Pr = (x_{t+1} | Y_0^T) = \frac{\sum_{x_t} \Pr(x_{t+1} | x_t) \alpha_t(x_t)}{\sum_{x_t} \alpha_t(x_t)}. \quad (2.49)$$

Similarly,

$$\Pr = (y_{t+1}|Y_0^t) = \frac{\sum_{x_{t+1}} \alpha_{t+1}(x_{t+1})}{\sum_{x_t} \alpha_t(x_t)}. \qquad (2.50)$$

The result from the prediction model will be used as input for the decision-making procedure: if the future estimated situation is a state considered "safe" or "successful" to fulfill the goal of the system, no action should be planned. In a contradictory way, if the degradation degree can be the cause of risk or insufficient performance, an action must be started immediately (in safety case) or at a scheduled time (proactive maintenance).

The results of the safety prognosis include both future hidden and observable variable values, future reliability as well as the performance (degradation trend) of each entity in the system, remaining useful life with certain thresholds in respect of system safety requirements, and long-term proactive maintenance plans.

The results from both the safety assessment and prognosis play a fundamental role in safety pre-warnings in industrial safety management and are conducive to reducing hazard accidents and improving system performance with proactive maintenance.

2.3.3 *Case study*

2.3.3.1 *System description*

The proposed integrated method is applied to a real system to assess the safety state and predict the reliability and performance of a gas turbine compressor system, which is an important facility in long-distance gas pipeline systems. The information on the gas turbine compressor system is shown in Table 2.25. The gas turbine compressor system consists of seven main subsystems shown in Fig. 2.40 In this section, the former six subsystems are considered and analyzed in detail, while the centrifugal compressor system is relatively independent, and can be analyzed alone in the same way.

Table 2.25. Information of the system.

Gas turbine		Centrifugal compressor	
Company	SOLAR	Company	MAN TURBO
Type	Titan 130	Type	RVO50/04
NGP	11,220 rpm	Max. inlet flow	80,000 m^3/h
NPT	8856 rpm	First critical speed	4428 rpm
Power	14,000 Kw	Max. continuous speed	8856 rpm
Inlet airflow	83,000 Kg/h	Efficiency	770–84%
Heat efficiency	33.3%	Inlet pressure	4.5 MPa
Axial compressor stage	14	Exhaust pressure	6.4 MPa
Compression ratio	16	Pressure ratio	1.86
Variable inlet guide vanes	6 stages (adjustable)	Impeller	4 stages
Fuel nozzle	21	Max. temperature	193 (^{circ}C)
T5 °C	760 (°C)	Impeller diameter	500 mm

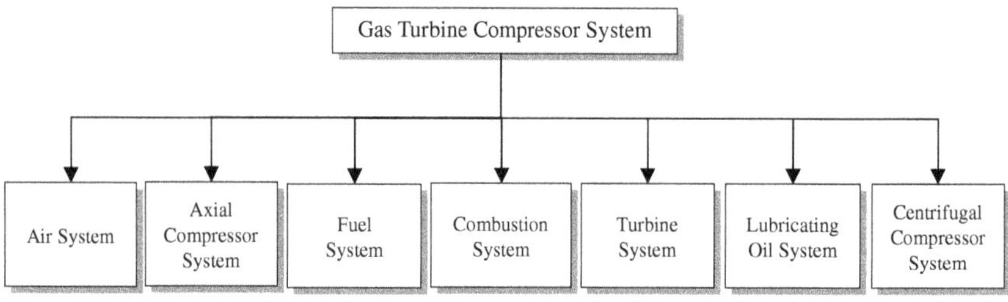

Fig. 2.40. Gas turbine compressor system and main subsystems.

2.3.3.2 *Results and discussion*

(1) The results of safety assessment

During the online monitoring of the gas turbine compressor system at the beginning of May 2008, the average lubricating oil temperature was 57 (°C) and the lubricating oil pressure was 0.195 MPa, all of which deviated from the normal value interval. Based on the principle of the safety assessment presented in Section 3.2.5, the analysis process of probability calculation based on the fuzzy method and Bayesian inference is as follows, and the meaning of node ID (e.g., S7_2) is explicated in Tables 2A.1 and 2A.2 in Appendix.

(i) The observable node deviations are fuzzily quantified as posterior probability. The normal value range of lubricating oil temperature at machine running state is [35, 55] (°C) (shown in Table 2A.2), while the measured lubricating oil temperature was 57 (°C), then the posterior probability of observable node

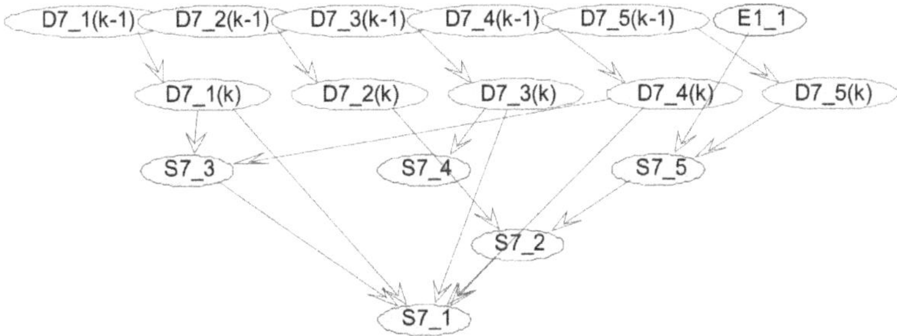

Fig. 2.41. Dynamic network structure of lubricating oil system.

S7_2 is calculated as $\{0, 0.1546, 0.7939, 0.0515\}$ according to the state space: {On the low side; Normal; On the high side; Super-high}. In the same way, the normal value range of lubricating oil pressure is [0.210, 0.449] MPa (shown in Table 2A.2), then the posterior probability of observable node S7_1 is calculated as $\{0.0593, 0.8536, 0.0871, 0\}$ according to the state space: {Ultra low; On the low side; Normal; On the high side}. The abnormal observations are in the lubricating oil system, whose dynamic network structure is shown in Fig. 2.41. The probabilities of normal state of D7_1, D7_2, D7_3, D7_4, D7_5, which can be considered as reliabilities of such entities, are then estimated as 0.3140, 0.6583, 0.7244, 0.5395, and 0.9016, respectively by inference in the safety assessment steps. The probabilities of the normal state of node D7_1 and node D7_4 have the lowest values, which means that lubricating oil tubes and oil pumps have the lowest reliabilities and need certain repairs.

(ii) The possible hazard causes and consequences of the lubricating oil system estimated by the safety assessment are shown in Tables 2.26 and 2.27, from which the most possible hazard cause is "Smaller diameter of oil inlet throttle" with the largest probability of 0.8035, and the most possible hazard consequence is "Oil film is difficult to sustain; bearing will be burnt; seal will be damaged or impeller will be destroyed" with the largest probability of 0.9132. Afterward, a safety engineer made an inspection of the gas turbine compressor system according to the hazard cause list and corresponding repair (or maintenance) advice. The manual examination result proved that the results given by the integrated safety pre-warning method proposed in the section fitted well with the actual situation, and the possible consequence

Table 2.26. Result of safety assessment steps of the integrated pre-warning method (possible hazard reasons with advised safety-related action).

Possible hazard reasons	Probability of occurrence	Safety-related action required
1. Smaller diameter of oil inlet throttle	0.8035	1. Checking the oil throttle orifice of bearing and increasing throttle diameter, making the flow of lubricating oil meet the operation requirement
2. Higher outlet water temperature	0.7142	2. Increasing the cooling circulating water
3. Fault in the main oil pump	0.6829	3. Switching pump for inspection and repair
4. Insufficient cooling water	0.6351	4. Increasing the cooling circulating water
5. Oil pipe rupture or oil leakage at conjunctions	0.5122	5. Inspection or replacement of pipe section
6. Bearing clearance is on the small side, causing severe friction and a mass of heat	0.4763	6. Inspecting and appropriately adjusting (enlarging) the bearing clearance to decrease the heat caused by friction
7. Obstruction in the oil filter or oil passageway, which causes oil pressure loss	0.4000	7. Disassembly checking and cleaning the lubricating oil filter
8. Deterioration of lubricant	0.3969	8. Releasing the gas from lubricating oil system or changing lubricating oil
9. Bearing fault	0.3616	9. Checking and repairing or replacing the bearing
10. Lubricating oil leakage at pipeline and valves	0.2561	10. Changing conjunction gasket at leak location, tightening connecting bolt, and making no leakage occurrence

Table 2.27. Result of safety assessment steps of the integrated pre-warning method (possible hazard consequences).

Possible hazard consequences	Probability of occurrence
1. Oil film is difficult to sustain; bearing will be burnt; seal will be damaged or impeller will be destroyed	0.9132
2. Shutdown caused by low oil pressure	0.6034
3. Carbonization of lubricating oil or decreasing lubrication performance	0.3886

lists as a pre-warning alarm are conducive for safety engineers to make contingency plans to avoid accidents in advance.

If the conventional HAZOP method is used in this situation for safety assessment, all the parallel possible hazard reasons will come up without any priority, so field engineers usually do not know which safety-related action should be taken first. If the observable deviations happen in more than one subsystem concerning more than 10 nodes, the conventional HAZOP method will result in large numbers of hazard causes and consequences, which usually makes field engineers lose their heads, and miss the best maintenance time, or even before engineers going through the huge hazard list, catastrophic happens. Therefore, the proposed method successfully overcomes this problem of the conventional HAZOP method. Field engineers only need to check the list by the probability of occurrence relating to each hazard cause, and take corresponding safety-related action advised by the method, and then the root cause will be found quickly. In this way, the efficiency of safety inspection is improved, and accidents or even catastrophes can be avoided as soon as possible.

(2) The results of safety prediction

According to the algorithm of the safety prognosis steps, future reliability trends of lubricating subsystems beginning with the present degraded states are predicted and shown in Fig. 2.42, which indicates that D7_1 (oil tube degradation) and D7_4 (oil pump degradation) with the least reliabilities will deteriorate gradually and their reliabilities will be under 0.5 in 30 days. Although D7_3 (oil filter degradation) has higher reliability at present, it will deteriorate severely, which will affect other components. The influence on the bearing degradation process is shown in Fig. 2.43, which indicates that in the situation of higher lubricating oil temperature and lower lubricating oil pressure, the degradation trends of bearings (solid lines in Fig. 2.43) are more severe than the natural degradation process (dash lines in Fig. 2.44). In other words, the degraded states of the lubricating system shorten the remaining useful life of bearings.

Considering the results of the safety assessment and prognosis, repair plans are made as follows:

- replacing the oil tube, which is considered perfect maintenance, so the state of D7_1 can be set as $[1\,0\,0\,0]$ after repair;
- cleaning the oil filter, which is considered unperfected maintenance, so the state of D7_3 can be set as $[0.9\,0.1\,0]$ after repair;
- repairing the oil pump, which is also considered unperfected maintenance, so the state of D7_4 can be set as $[0.85\,0.15\,0]$ after repair.

Figure 2.44 shows the future reliability trends of lubricating subsystems beginning with the states after repair. The reliabilities of the repaired subsystem are all above 0.5, which is acceptable in industrial practice, and except D7_3, the other four subsystems will run functionally well in the next four months. Due to the degradation mechanism of the oil filter, its reliability deteriorates more severely than other subsystems, so the oil filter should be cleaned (or replaced) every two or three months to ensure the whole system works normally.

(3) Discussion of predictive maintenance plan

The results coming from the integrated safety pre-warning method can be used as input for the decision-making procedure: when the reliability of the underlying system drops down to a certain threshold (or criteria), proper maintenance will be actioned. Take the air system as an example, whose dynamic network structure is shown in Fig. 2.35, and the corresponding parameters are shown in Tables 2.21–2.25, given the states of both entities (D1_1: Air Filter and D1_2: Ventilation System) are as good as new, the future degradation trends and

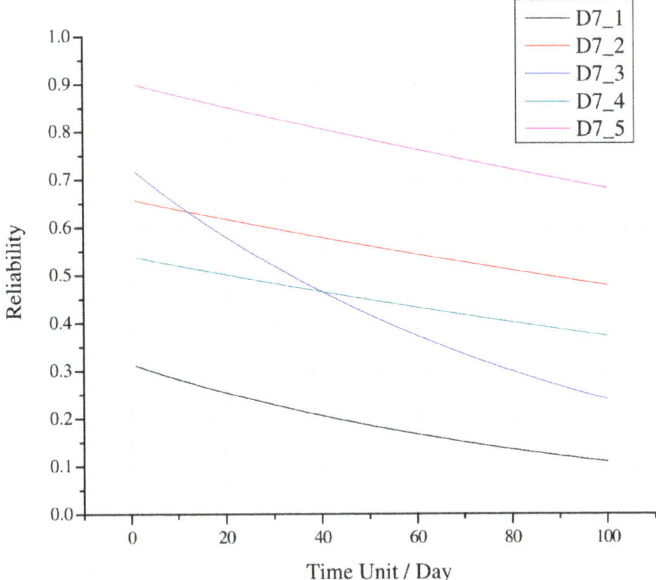

Fig. 2.42. Future reliability trends of degraded lubricating subsystem.

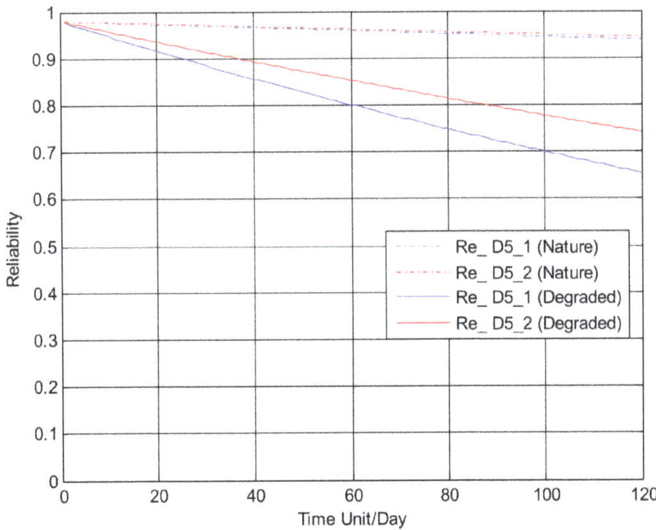

Fig. 2.43. Bearing degradation trends in the situation of normal and degraded lubricating system.

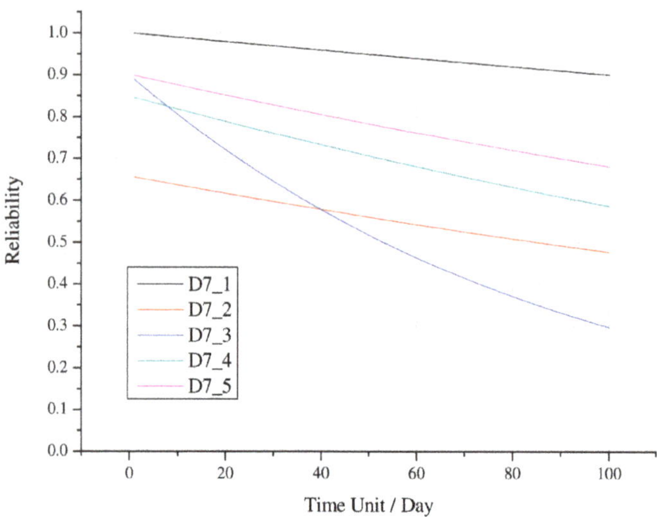

Fig. 2.44. Future reliability trends of repaired lubricating subsystem.

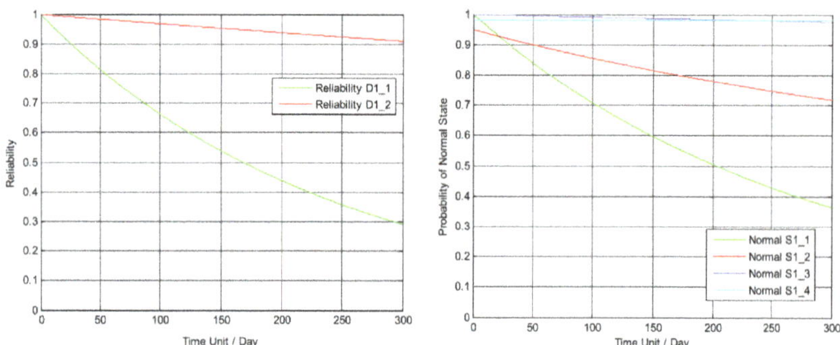

Fig. 2.45. Air system degradation and performance trends without maintenance.

performance trends in 300 days are shown in Fig. 2.46. The reliability of the air filter (D1_2) will drop down to 0.5 when the time unit is 168 (days), while the ventilation system (D1_1) will work well without much degradation. Analyzing the corresponding trend of observation nodes representing system performance (S1_1, S1_2,

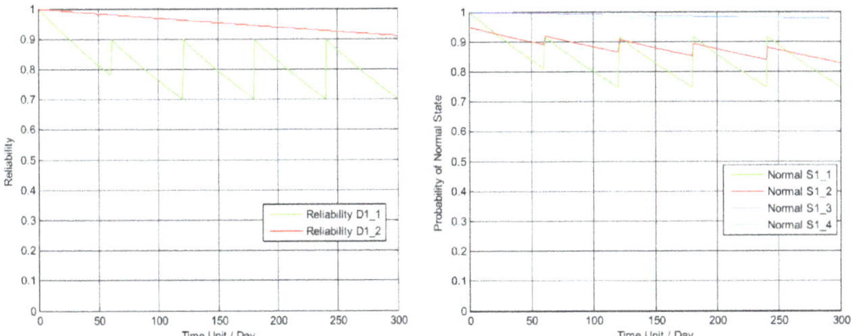

Fig. 2.46. Air system degradation and performance trends with predictive maintenance action.

S1_3, S1_4) in Fig. 2.45, the S1_1 (differential pressure of inlet filter) severely derivates from the normal state, which will have a negative impact on other interrelated subsystems, i.e., combustion system.

In this case, the maximum maintenance period is set as 168 days according to the 50% reliability criteria (pre-warning alarm), and the best maintenance period is between 60 and 90 days when the reliability drops down below 0.7. Figure 2.46 shows the degradation trends of the air system based on the 60-day maintenance period. The maintenance takes place in terms of cleaning the air filter, and the state space of the D1_1 which is set as [0.9 0.1 0] after repair is considered as an unperfected maintenance because although dusts and particles are removed after cleaning, the degradation of material and certain wearing (or corrosion) still exist, which cannot make the result of maintenance as good as new.

The overall reliabilities of the air system are all above 0.7 with such a maintenance plan, and the probabilities of the normal states of the observations are also all above 0.7, which indicates that the performance of the air system falls into the acceptable or safety area, hardly causing any hazards or having little impact on other interrelated subsystems. Although the maintenance of air filter is taken into consideration, there is still a derivation trend in S1_2, because S1_2 is also affected by the ventilation system (D1_1) which will deteriorate gradually in the long term.

Appendix

Table 2A.1. Information of dynamic nodes in DBN model.

Subsystem	Dynamic nodes	Node ID	State space	Parent nodes	Children nodes
1. Air system	1. Air filter degradation (AFD)	D1_1	{Normal, fouling, blocked}	$\emptyset$	{S1_1, S1_2, S1_5}
	2. Ventilation system degradation (VSD)	D1_2	{Normal, degraded, failure}	$\emptyset$	{S1_2, S1_3}
2. Axial compressor system	1. Compressor blade fouling degradation (CBFD)	D2_1	{Normal, fouling, failure}	{S1_3}	{S2_1, S2_2}
3. Fuel system	1. Fuel adjustor degradation (FAD)	D3_1	{Normal, degraded, failure}	$\emptyset$	{S3_1, S3_2}
	2. Fuel tube degradation (FTD)	D3_2	{Normal, slightly leakage, severe leakage}	$\emptyset$	{S3_1}
4. Combustion system	1. Hot path degradation (HPD) 2. Air cooling tube degradation (ACTD)	D4_1	{Normal, fouling, corrosion, thermal distortion, fouling & corrosion & distortion, failure}	{S1_3}	{S4_1}
		D4_2	{Normal, leakage, blocked, failure}	{S1_3}	{S4_1}
5. Turbine system_1 (gas turbine subsystem)	1. GT 1#2#3# bearing degradation (GTBD)	D5_1	{Normal, degraded, failure}	{S7_1, S7_2, S5_1, S5_3}	{S7_3, S5_3, S5_4}
	2. GT Thrust bearing degradation (GTTBD)	D5_2	{Normal, degraded, failure}	{S5_1, S5_2, S7_1, S7_2}	{S7_3, S5_5, S5_2}
	3. GT rotor degradation (GTRD)	D5_3	{Normal, unbalanced, unaligned, failure}	{S5_1}	{S5_3}
	4. GT blade degradation (GTBD)	D5_4	{Normal, thermal distortion, corrosion, failure}	{S4_1}	{S5_1}

6. Turbine System_2 (power turbine subsystem)	1. PT bearing degradation (PTBD)	D6_1	{Normal, degraded, failure}	{S7_1, S7_2, S6_3, S6_5}	{S7_3, S6_5, S6_6}
	2. PT thrust bearing degradation (PTTBD)	D6_2	{Normal, degraded, failure}	{S7_1, S7_2, S6_3, S6_4}	{S6_4, S7_3, S6_7}
	3. PT rotor degradation (PTRD)	D6_3	{Normal, unbalanced, unaligned, failure}	{S6_3}	{S6_5}
	4. PT blade degradation (PTBD)	D6_4	{Normal, thermal distortion, corrosion, failure}	{S4_1}	{S6_3}
7. Lubricating oil system	1. Oil tube degradation (OTD)	D7_1	{Normal, fouling, leakage, failure}	{S7_1}	{S7_1, S7_3}
	2. Oil cooler degradation (OCD)	D7_2	{Normal, degraded, failure}	∅	{S7_2}
	3. Oil filter degradation (OFD)	D7_3	{Normal, fouling, blocked}	{S7_4}	{S7_1, S7_4}
	4. Oil pump degradation (OPD)	D7_4	{Normal, degraded, failure}	∅	{S7_3, S7_1}
	5. Oil heater degradation (OHD)	D7_5	{Normal, degraded, failure}	∅	{S7_5}

Table 2A.2. Information of static nodes in DBN model.

Subsystem	Static nodes	Node ID	State space	Normal interval	Classification criteria	Parent nodes	Children nodes
1. Air system	1. Differential pressure of inlet filter (Kpad)	S1_1	{Normal; On the high side; Super-high}	<0.75	{<0.75; [0.75, 1.5); $\geq$ 1.5}	{D1_1}	{S2_1, S2_2}
	2. Cabinet temperature (KPa)	S1_2	{Normal; On the low side; Ultra low}	>0.06227	{>0.06227; (0.03736, 0.06227]; $\leq$ 0.03736}	{D1_1, D1_2}	$\emptyset$
	3. Cabinet pressure (°C)	S1_3	{Normal; On the high side}	< 75	{<75; $\geq$ 75}	{E1_1, D1_2}	{D2_1, D5_1, D4_1, D4_2, D6_1}
	4. T1 temperature (°C)	S1_4	{Normal; On the low side; On the high side}	>5	{>5; $\leq$ 5}	{E1_1}	{S6_2}
	5. Inlet differential pressure (KPa)	S1_5	{Normal; On the high side}	< 1	{<1; $\geq$ 1}	{D1_1}	{S2_2}
2. Axial compressor system	1. PCD (KPa)	S2_1	{Normal; On the low side}	>551	{>551; $\leq$ 551}	{S1_1, D2_1, S2_2}	{S4_1, S5_1, S6_2, S6_3}
	2. Serge (Bool)	S2_2	{Normal, Surge}	False	{False; True}	{D2_1, S1_1, S1_5}	{S2_1}
3. Fuel system	1. Fuel flow (K Nm³/h)	S3_1	{Normal; On the low side; On the high side}	[1900, 2400]	{[1900, 2400]; <1900; >2400}	{E2_3, D3_1, D3_2}	{S4_1}
	2. Fuel pressure (KPa) S	S3_2	{Normal; On the low side; On the high side}	[2000, 3450]	{[2000–3450]; <2000; > 3450}	{D3_1, E2_2}	$\emptyset$
	3. Fuel temperature (°C)	S3_3	{Normal; On the low side; On the high side; Super-high}	[38, 85]	{[38, 85]; < 38; (85, 96); $\geq$ 96}	{E2_1}	{S6_2}

4. Combustion system	1. T5 temperature (°C)	S4_1	{Normal; On the low side; On the high side} {S5_1, D5_4, S6_2, S6_3}	(500, 760)	{(500, 760); ≦ 500; ≧ 760}		{S2_1, S3_1, D4_1, D4_2}
5. Turbine system_1 (gas turbine subsystem)	1. NGP (%)	S5_1	{Normal; On the low side; On the high side; Super-high}	(80, 100) (100%= 11,220 rpm)	{(80, 100); ≦ 80; [100, 102.5); ≧ 102.5}	{S2_1, S4_1, D5_4}	{S5_2, S5_3, S6_1, D5_1, D5_2, D5_3}
	2. Axial displacement (mm)	S5_2	{Normal; On the high side; Super-high}	(−0.508, 0.102)	{(−0.508, 0.102); [0.102, 0.178) or (−0.584, −0.508]; ≧ 0.178 or ≦ -0.584}	{S5_1, D5_2}	{D5_2}
	3. Radial vibration (μm)	S5_3	{Normal; On the high side; Super-high}	<63.5	{<63.5; [63.5, 101.6);≧ 101.6}	{D5_1, D5_3, S5_1}	{D5_1}
	4. 1#2#3# bearing oil return temperature (°C)	S5_4	{Normal; On the high side; Super-high}	1# < 75 2# 3# <111	1#: {< 75; [75, 85); ≧ 85} 2#3#: {< 111; [111, 121); ≧ 121}	{D5_1, S7_2}	∅
	5. Thrust bearing oil return temperature (°C)	S5_5	{Normal; On the high side; Super-high}	<110	{< 110; [110, 121); ≧ 121}	{D5_2, S7_2}	∅

(Continued)

Table 2A.2. (*Continued*)

Subsystem	Static nodes	Node ID	State space	Normal interval	Classification criteria	Parent nodes	Children nodes
6. Turbine system_2 (power turbine subsystem)	1. Output power (KW)	S6_1	{Normal; On the low side; On the high side}	>5000	{> 5000; ≦ 5000}	{S6_3, S5_1}	∅
	2. Efficiency (%)	S6_2	{Normal; On the low side}	>0.24	{> 0.24; ≦0.24}	{S1_4, S2_1, S3_3, S4_1}	∅
	3. NPT (%)	S6_3	{Normal; On the low side; On the high side; Super-high}	(80, 93.7) (100%= 8856 rpm)	{(80, 93.7); ≦ 80; [93.7, 98.4); ≧ 98.4}	{S2_1, S4_1, D6_4}	{D6_1, D6_2, D6_3, S6_1, S6_4, S6_5}
	4. Axial displacement (mm)	S6_4	{Normal; On the high side; Super-high}	(-0.508, 0.102)	{(−0.508, 0.102); [0.102, 0.178) or (−0.584, −0.508]; ≧ 0.178 or ≦ −0.584}	{D6_2, S6_3}	{D6_2}
	5. Radial Vibration (μm)	S6_5	{Normal; On the high side; Super-high}	< 63.5	{<63.5; [63.5, 101.6); ≧ 101.6}	{D6_1, D6_3, S6_3}	{D6_1}
	6. 4#5# bearing oil return temperature (°C)	S6_6	{Normal; On the high side; Super-high}	< 75	{<75; [75, 85); ≧ 85}	{D6_1, S7_2}	∅
	7. Thrust bearing oil return temperature (°C)	S6_7	{Normal; On the high side; Super-high}	< 110	{<110; [110, 121); ≧ 121}	{D6_2, S7_2}	∅

7. Lubricating Oil System	1. Oil supply pressure (KPa)	S7_1	Normal; On the low side; Ultra low; On the high side} {D7_1, D5_1, D5_2, D6_1, D6_2}	(210, 449)	{(210, 449); (173, 210]; $\leq$ 173; $\geq$ 449}		{D7_1, D7_3, D7_4, S7_2, S7_3}
	2. Oil supply temperature (°C)	S7_2	{Normal; Om the low side; On the high side; Super-high}	(35, 55)	{(35, 55); $\leq$ 35; [55, 74); $\geq$ 74}	{D7_2, S7_5}	{S7_1, S5_5, S5_4, S6_6, S6_7, D6_1, D6_2, D5_1, D5_2}
	3. Oil level (cm)	S7_3	{Normal; On the low side; Ultra low; On the high side}	(48.3, 55.9)	{(48.3, 55.9); (40.6, 48.3]; $\leq$ 40.6; $\geq$ 55.9}	{D5_1, D5_2, D6_1, D6_2, D7_1, D7_4}	{S7_1}
	4. Differential pressure of oil filter (KPa)	S7_4	{Normal; On the high side}	< 207	{<207; $\geq$ 207}	{D7_3}	{D7_3}
	5. Oil tank temperature (°C)	S7_5	{Normal; On the low side; On the high side; Super-high}	(21, 68)	{(21, 68); $\leq$ 21; [68, 74); $\geq$ 74}	{D7_5, E1_1}	{S7_2}

(*Continued*)

Table 2A.2. (*Continued*)

Subsystem	Static nodes	Node ID	State space	Normal interval	Classification criteria	Parent nodes	Children nodes
8. Exogenous environment (or External System)	Environment temperature ($^\circ$C)	E1_1	Normal; On the low side; On the high side}	(0, 40)	$\{(0, 30); \leqq 0; \geqq 40\}$	$\emptyset$	{S7_5}
	Fuel supply temperature ($^\circ$C)	E2_1	{Normal; On the low side; On the high side}	Adjustable	According to manual setting based on supply and demand	$\emptyset$	{S3_3}
	Fuel supply pressure (KPa)	E2_2	{Normal; On the low side; On the high side}	Adjustable	According to manual setting based on supply and demand	$\emptyset$	{S3_2}
	Fuel supply flow (K Nm^3/h)	E2_3	{Normal; On the low side; On the high side}	Adjustable	According to manual setting based on supply and demand	$\emptyset$	{S3_1}

References

[1] Zhang L, Cai S, Hu J. An adaptive pre-warning method based on trend monitoring: Application to an oil refining process. *Measurement*, 2019, 139: 163–176.

[2] Tian S. Research on Shale Gas Fracturing Abnormal Condition Prediction and Cause Analysis Method. PhD Thesis, *China University of Petroleum*, Beijing, 2017. (In Chinese)

[3] Cai S. Adaptive Correlation Analysis and Prediction Method for Industrial Process Alarm based on Text Mining. PhD Thesis, *China University of Petroleum*, Beijing, 2019. (In Chinese)

[4] Hu J, Khan F, Zhang L, Tian S. Data-driven early warning model for screenout scenarios in shale gas fracturing operation. *Computers & Chemical Engineering*, 2020, 143: 107116.

[5] Luo J, Hu J. A study of adaptive composite-indicator alarm threshold optimization of chemical process parameters. *Petroleum Science Bulletin*, 2016, 3: 407–416. (In Chinese)

[6] Shu Y, Ming L, Cheng F, *et al.* Abnormal situation management: Challenges and opportunities in the big data era. *Computers & Chemical Engineering*, 2016, 91: 104–113.

[7] Kameswari US, Babu IR. Sensor data analysis and anomaly detection using predictive analytics for industrial process. In: *IEEE Workshop on Computational Intelligence: Theories, Applications and Future Directions*. IEEE, 2015: 1–8.

[8] Zheng B, Guo J. Analysis and strategy research on sand plug of gas well fracturing in YQ prospect area. *Complex Hydrocarbon Reservoirs*, 2010, 3(1): 70–72, 76. (In Chinese)

[9] Zhai H. Reason analysis and countermeasures of sand plug in shale gas fracturing. *Unconventional Oil & Gas*, 2015, 2(1): 66–70. (In Chinese)

[10] Yan H, DeChant CM, Moradkhani H. Improving soil moisture profile prediction with the particle filter-Markov chain Monte Carlo method. *IEEE Transactions on Geoscience and Remote Sensing*, 2015, 53(11): 6134–6147.

[11] Zhang P, He Z, Zhao J. Research and application of interpretation technique of net pressure fitting analysis for hydraulic fracturing. *Well Testing*, 2005(3): 8–10, 75. (In Chinese)

[12] Liang S, Zhang Y, Gao H, *et al.* Analysis of the reasons and countermeasures for sand plugging of high-resistance red layer fracturing in Dongpu Depression. *West-China Exploration Engineering*, 2010, 22(8): 72–75. (In Chinese)

[13] Charbonnier S, Portet F. A self-tuning adaptive trend extraction method for process monitoring and diagnosis. *Journal of Process Control*, 2012, 22: 1127–1138.

[14] Charbonnier S, Garcia-Beltan C, Cadet C, *et al.* Trend extraction and analysis for complex system monitoring and decision support. *Engineering Applications of Artificial Intelligence*, 2005, 18(1): 21–36.

[15] He J, Liu F. Improved local outlier factorization method for industrial process data anomaly detection. *Computers and Applied Chemistry*, 2013, 30(1): 53–56. (In Chinese)

[16] Ge Z, Song Z, Gao F. Review of recent research on data-based process monitoring. *Industrial & Engineering Chemistry Research*, 2013, 52(10): 3543–3562.

[17] Hu J, Zhang L, Ma L, *et al.* An integrated method for safety pre-warning of complex system. *Safety Science*, 2010, 48(5): 580–597.

Chapter 3

Early-Warning Technology for Hidden Dangers of Oil and Gas Extraction Equipment Based on Infrared Thermal Image Video Surveillance

3.1 Infrared Thermal Video Surveillance Technology

Oil and gas extraction equipment, such as shale gas fracturing equipment, involves a large amount of electrical energy, mechanical energy, and internal energy conversion. The local surface temperature can reflect the current operating state of the equipment. As the shale gas fracturing site equipment, environment, and other situations are more complex, it is not suitable to carry out close-range contact equipment monitoring and diagnosis work. Infrared thermal imaging monitoring is a non-contact equipment monitoring and diagnostic technology that analyzes and processes infrared thermal images and characterizes their parameters to indirectly grasp the value, distribution and changes of the surface temperature of the monitored equipment, and then understand the overall or local operating status of the equipment. It is suitable for equipment, systems or scenarios where the status parameters are strongly correlated with temperature and it is not suitable for close observation or deployment of monitoring instruments. In view of this, infrared thermal imaging monitoring technology has high applicability, feasibility, and broad application prospects in the field of equipment condition monitoring and fault diagnosis for oil and gas extraction equipment.

The core aspect of infrared thermal imaging technology is to convert the infrared radiation of an object into an image that can be recognized by the human eye. The process is to propagate the infrared radiation of an object to the infrared detector plane, focus and convert it into an electrical signal, amplify and digitize the electrical signal, and transmit it to the signal processor, which can be converted into an image to characterize the infrared thermal radiation of the object. Based on the infrared thermal imaging monitoring technology of oil and gas extraction equipment, the intensity and distribution of infrared radiation related to the surface temperature is transformed into an infrared thermal imaging map, and the differences or changes in the visual effect parameters through the image characterize the differences or changes in the temperature of the equipment as a whole, and based on the correspondence between the operating status of the equipment and the surface temperature, determine the current operating information or fault situation of the oil and gas extraction equipment.

For oil and gas extraction equipment, in using infrared thermal imaging technology for fault identification and early warning, and other industrial equipment infrared thermal imaging monitoring technology, the significant differences include the following:

(1) The oil and gas extraction monitoring site is completely open, and there is no warning area for the location of the infrared thermal imaging equipment and the field of view.

(2) The environment of the oil and gas extraction site is complex, and the randomness of the movement of personnel, vehicles, movable equipment, and even natural organisms is significant.

(3) The monitoring object is a large-scale oil and gas extraction equipment system, which consists of a variety of sub-modules with different performances, materials, structures, and functions.

(4) The monitoring range is larger, with more targets, and the thermal imaging information of each sub-module of the equipment system is equally important.

(5) The monitoring object's fault development speed is fast, so the visual presentation of its infrared thermal imaging information (including the original information and processing information) should be as quick as possible, intuitive, and easy to understand.

3.1.1 *Domestic and international technology status*

(1) Current status of research on anomaly data identification and cleaning for infrared thermography monitoring

Infrared thermography monitoring is a non-contact equipment monitoring and diagnostic technology based on the propagation, reception, and conversion of infrared radiation. The infrared radiation related to the surface temperature of the monitoring object is accurately, completely, and directly propagated to the infrared thermal imaging equipment and converted into image information that can be recognized and understood by the personnel, which is the key to realizing high-quality and high-efficiency infrared thermal imaging monitoring of the equipment. In terms of infrared thermal imaging monitoring of shale gas fracturing field equipment, compared with the infrared thermal imaging camera (or infrared thermal imaging camera unit), which has a fixed position, shooting angle, or running track, the movement status of personnel, vehicles, mobile equipment and even flying animals on the construction site is highly contingent and random. Therefore, they inevitably shuttle or stay in front of the camera, blocking the normal infrared radiation path of the monitored equipment and emitting infrared radiation to the camera that is significantly different from the infrared radiation of the monitored equipment. This phenomenon will seriously affect the data quality of infrared thermal imaging monitoring of fracturing equipment, weaken the credibility of the monitoring results, increase the rate of false alarms or missed alarms, and bring unnecessary troubles to the safety management of shale gas fracturing sites. Therefore, it is necessary to adopt technical means to accurately and efficiently determine the existence and movement of foreign objects in the lens field of view, and provide feedback on the abnormal monitoring data caused by them, so that the monitoring system can clean the abnormal infrared thermal imaging monitoring data in a timely manner, thus enhancing the credibility of the monitoring results and decreasing the rate of false alarms and leakage alarms.

(i) Anomaly data recognition based on out-of-bounds detection

Cross-border detection, i.e., the detection of unauthorized objects illegally entering a special area through the demarcation of the boundary, is commonly used in the field of security and access management based on video surveillance technology in important places, key areas, or guarded areas. In terms of the infrared thermal imager, its shooting field of view (or imaging field of view) can be regarded as a rectangular area with demarcated boundaries and special functions. The region is authorized to exist in the whole or part of the monitored equipment, any non-monitoring object of the personnel, and objects into the region are transboundary behavior. Objects crossing the boundary into the lens field of view will form an obstacle to the normal propagation of infrared radiation of the monitoring object, which in turn negatively affects the data quality of infrared thermal imaging monitoring. Therefore, accurately detecting the cross-border behavior of foreign objects is a key technology and an important prerequisite for cleaning abnormal data and improving data quality.

Boundary-crossing detection focuses on the relative position of moving objects and the delineated boundary, and thus moving target detection is an important part of boundary-crossing detection. Currently, the commonly used moving target detection methods are the frame difference method, background difference method, and optical flow method. In terms of optimization of algorithm robustness, computational efficiency, and complex background detection accuracy, researchers from Inner Mongolia University of Science and Technology proposed a motion target detection method for infrared sequence images based on local saliency and sparse representation, which has good robustness to external interference and dynamic background. Researchers at the National Institute of Technology, India, propose a motion target detection method based on vector graph modeling, which is suitable for target identification in video surveillance scenes under complex backgrounds. Researchers at the Economic Commission for Europe propose an adaptive motion target detection technique for complex scenes based on texture features, suitable for dynamic backgrounds and illumination changes. Researchers at the French National Institute of Scientific Research propose a method for detecting moving objects from two frames of continuous stereo images with strong robustness to global and local noise. Researchers at the National Penghu University of Science and Technology propose a method for detecting and tracking multiple moving objects from video sequences captured by a motion camera, which is applicable to cases where there is relative motion between the object and the camera. Researchers at Visvesvaraya Technological University in India and others propose a motion target detection algorithm based on improved background phase reduction and adaptive thresholding techniques, which performs well in terms of computing speed and image quality. Researchers from Zhejiang University propose a deep learning-based chunked scene analysis method, which solves the problems of low computational efficiency and poor parallelism in high-definition video processing. Researchers from Hohai University propose a spatio-temporal saliency model for infrared small motion target detection based on a three-dimensional difference of Gaussian filters, and its processing effect is better than the existing methods in a variety of complex background environments. Researchers from Nanjing University of Science and Technology propose an algorithm for infrared motion target detection and video surveillance security detection, which improves on frame difference nulling, error elimination, and noise suppression compared to the traditional W4 and frame difference methods. Researchers from Beijing University of Aeronautics and Astronautics propose a detection method based on optical flow estimation, which can obtain the complete boundary of a moving target. Researchers from the Institute of Automation, Chinese Academy of Sciences, and other researchers proposed a hierarchical motion target detection method based on spatio-temporal saliency, which solves the problems of background change, small target volume and poor processing real-time during aerial video processing.

As for transgression detection, the adaptive fast background modeling method proposed by researchers from the Guangdong Institute of Finance and other researchers is applicable to situations where there are few moving objects in the scene and their positions are scattered. Researchers from the University of Electronic Science and Technology and Sun Yat-sen University have designed two character-crossing detection systems based on fast background modeling, which meet the special needs of different places for intelligent surveillance character detection. Researchers from the National University of Defense Technology adopted the Hoff straight-line detection method that can accurately demarcate the position of the boundary line in the lens field of view, and other edge detection algorithms can also be applied to the situation where a special target is used as the set boundary. Researchers at the Northern Nationalities University designed a tripwire detection-based alert area intrusion and abnormal behavior recognition algorithm applicable to anti-theft surveillance. Researchers from Yanshan University designed a surveillance management system based on abnormal behavior detection for various situations such as wandering, crossing the boundary, lingering, and facial masking of illegal targets. Researchers from Shandong Normal University proposed an automatic pedestrian counting algorithm based on the three-frame difference method with center-of-mass matching, which has better processing speed and accuracy in the detection of dense crowds entering and leaving the target area. The researchers from the China Academy of Weights and Measures only detect a small range of neighborhoods of key points, avoiding the problem of large computing power caused by region traversal and target tracking. Researchers at the Northern Nationalities University designed an ARM-based out-of-bounds control system, which determines whether a target is out of bounds by calculating the position of the target's center of mass. Researchers from China Agricultural University designed a UAV electronic fence based on the ray detection algorithm, which can detect whether the UAV is out of bounds in real time and provide a timely warning.

(ii) Data cleaning

Data cleansing is a data preprocessing technique that identifies, removes, adds, and repairs noise, outliers, missing values, duplicate values, and other problems present in big data. Researchers at Northeastern University propose a data cleaning method based on Linked Data that can be used to solve the mixed problem of consistency and timeliness. Researchers from Xi'an University of Architecture and Technology propose a cleaning method based on clustering and neural networks, which makes up for the defects of low computational efficiency and poor accuracy of a single method, as well as poor results of outlier processing. Researchers from the Beijing University of Posts and Telecommunications propose a density-based method suitable for poor data quality, but its computational cost grows exponentially with the increase of data volume. Researchers at Donghua University designed an active learning-based data cleaning system that solves the problem of data that cannot be determined by the learning model through human participation, ensuring the accuracy of data cleaning. Researchers at Beijing University of Posts and Telecommunications proposed a regression interpolation-based approach for missing value cleaning and an approach based on the principle of up-to-date processing for duplicate records. Researchers from China Mobile Communications Group and other researchers proposed a cleaning method based on function dependency that can mine data segments with low credibility and feedback to the collection system, thus reducing the size of low-quality data from the source. Researchers from Huazhong University of Science and Technology and others propose an outlier detection method based on Euclidean distance for cleaning outliers in raw training data. Researchers at the National University of Singapore proposed a knowledge-based data cleaning method for eliminating duplicate information from databases, which is able to balance the needs of high checking completeness and high checking accuracy. Researchers at the University of Alberta and others have proposed an improved MT filter cleaning technique for online outlier detection and data cleaning when the process data model is unknown. Researchers at the University of Regina proposed a cleaning method for wastewater quality monitoring data based on wavelet multiresolution analysis, which can be used to reduce the noise caused by complex uncertainties, and then effectively remove anomalous data from water quality management systems.

(iii) Problems that exist

Taking into account the specificity and complexity of the environment, conditions, and possible events at the site of the oil and gas extraction system, the existing research on the identification of abnormal data and cleaning methods for infrared thermal imaging monitoring has the following problems:

(a) The cross-border detection method for video surveillance mainly focuses on modeling, localization, and tracking of moving targets, and takes the size, distribution, and change of the values of all pixel points within the field of view of the lens as the object of study, which is able to accurately detect the relative position and current morphology of the cross-border foreign objects at different moments from a spatial perspective. On the other hand, the identification and processing of abnormal monitoring data in the production site is mainly based on the determination and feedback of the moment of intrusion and removal of foreign objects, with the numerical value change of some pixel points at the boundary of the lens field of view as the object of study, and the positional information of the abnormal monitoring data is described from the perspective of time. That is, we only need to pay attention to the time period of the existence of the transgressing foreign object in the lens field of view, rather than the detailed motion status of the foreign object at each moment. Although the existing method can meet the technical requirements of foreign object recognition, its excessive processing unnecessarily increases the arithmetic complexity of the recognition process, which easily leads to the reduction of analysis efficiency and recognition accuracy.

(b) Typical faults of oil and gas extraction equipment develop at a fast pace, and the visual presentation of infrared thermal imaging information (including raw and processed information) should be fast, intuitive, and easy to understand as much as possible. When facing the batch processing of video or multi-frame time series images, the computational efficiency and cost of the existing methods cannot be guaranteed, and they cannot meet the needs of real-time processing and presentation of information for on-site monitoring. The higher the degree of optimization of the algorithm, the more processing steps, the larger the number of processing objects, and the worse the real-time performance.

(c) Differing from the cleaning and reconstruction of low-dimensional data anomalies in terms of data points, segments, rows (columns), etc., the identification and cleaning of infrared thermal imaging anomalous monitoring information is in terms of image frames or frame segments, which involves a large number of pixel point data, and the reconstruction is more difficult, and its main processing content is the marking, extraction, or deletion of anomalous data. Existing research on the repair and reconstruction of anomalous image information is yet to be improved.

(2) State of the art in infrared thermal imaging image enhancement research

The core of infrared thermal imaging technology is to transform the intensity and distribution of infrared radiation of an object into the form of an image that can be recognized and understood by the human eye, so the visual effect of the image is one of the key factors affecting the results of acquiring, analyzing, and evaluating the thermal image information. Image enhancement is an image processing technique that aims to highlight the overall or local characteristics of an image, amplify the characteristic differences between image elements, improve the visual effect of an image, and enhance the richness of information to meet the special needs of image analysis, and it can be applied to the optimization of the visual effect of infrared thermal imaging images and the enhancement of the level of visualization of thermal image information.

Image enhancement methods can be categorized into two types according to the different domains of action: spatial domain enhancement and frequency domain enhancement. Among them, spatial domain enhancement is a direct function of the pixels constituting the image, and commonly used methods include grayscale transformation, histogram equalization, image smoothing, and image sharpening.

(i) Grayscale transformation

As far as grayscale transformation is concerned, compared with the traditional methods, the multi-scale grayscale transformation method proposed by researchers from Hubei University is able to retain the detailed features of the original image and avoid the problem of grayscale saturation, but it is slightly insufficient in feature highlighting. The literature proposes an adaptive segmented linear grayscale transformation method, which is able to adjust the linear transformation segmentation points according to the image content, effectively suppressing the background gray-scale segments and stretching the object grayscale segments. However, it is based on the assumption that the dynamic range of the gray value of the infrared thermal imaging monitoring object is limited and the values are discrete, which is not applicable to the case of a large dynamic range of the gray value, average distribution of the values, or more complex composition of the gray value. The literature for sequential images proposed a grayscale transformation method based on the inverse histogram, according to the former frame of the image target information gray value distribution to guide the latter frame of the image gray value stretching, which can significantly optimize the visual effect of the target area. The S-curve grayscale transformation method proposed by researchers at SGGS Institute of Engineering and Technology, India, increases the grayscale gradient of the target region compared to the linear transformation method and highlights the edge effect of adjacent tissues.

(ii) Histogram equalization

To address some of the shortcomings of traditional histogram equalization methods, researchers at Dalian Maritime University summarized and evaluated the advantages and disadvantages of various improvement methods, including sub-histogram equalization methods, modified histogram equalization methods, histogram variational prescriptive methods, local histogram equalization methods, and transform domain-based equalization methods. Brightness Preserving Adaptive Sub-Histogram Equalization (BPASHE) to maintain image brightness and detail-enhanced histogram equalization based on edge information fusion are proposed in the literature. The former can avoid the reduction of the number of gray levels by equalizing the local peaks and keeping the ratio of the number of gray levels in each segment unchanged, which can help to improve the contrast while maintaining the brightness of the image. The latter combines the advantages of histogram equalization and Laplace filtering algorithms to adjust image brightness and enhance image details. Researchers at the University of Melbourne have proposed a recursively weighted multi-platform histogram equalization method that can more accurately maintain image brightness and increase contrast through a four-step process of histogram recursive segmentation, cropping, weighting, and equalization. Aiming at the problem of merging gray values and forming local peaks, which is easy to appear in the process of histogram equalization, researchers from Changjiang University and other researchers proposed an adaptive histogram shear function, which can effectively inhibit the merging of gray levels and solve the problem of missing information at the edges. Researchers from Xi'an Jiaotong University proposed a local histogram equalization contrast enhancement method based on adjacent block correction, which can significantly suppress the non-uniformity and over-enhancement effect of infrared images. In addition, for example, local entropy-weighted histogram equalization image enhancement algorithm based on particle swarm optimization, fuzzy histogram equalization image enhancement algorithm based on peak-limit separation, image enhancement algorithm based on iterative histogram equalization, and local contrast adaptive histogram equalization image enhancement algorithm are all improved solutions for a specific target to solve the problems of traditional histogram equalization.

(iii) Image smoothing

The main purpose of image smoothing is to reduce image noise, and commonly used algorithms include the mean filtering method and median filtering method. However, the smoothed image is prone to problems such as edge blurring and detail flooding. Researchers at Lanzhou University proposed a smoothing algorithm based on generalized random swimming, which can retain important features and edges as much as possible. Researchers from Hubei University of Technology proposed a smoothing algorithm based on improved Kalman filtering, which incorporates a variety of new information used to correct the prediction in order to improve the performance of the smoothing process. Researchers at CEREP, Tunisia, have proposed an improved adaptive intelligent median filter for removing impulse noise from grayscale images. Researchers at the Institute of Biological Sciences, Valencia, have proposed a color image processing model based on local graphic correlation, which can effectively differentiate between edge and non-edge regions of an image to take different smoothing approaches. Researchers at Tianjin Polytechnic University proposed an image smoothing algorithm based on an improved L_0-gradient minimization model, which can overcome the step effect and improve the smoothing performance and the ability to maintain edge information. The proposed image denoising algorithm based on the proposed normal distribution by researchers from Nanjing University of Information Engineering has superior performance in both edge retention and noise removal. In addition, the proposed algorithm based on the fusion model of gradient and curvature is able to retain more image information while keeping the edge features of the image well. The smoothing algorithm based on the double constraints of image pixel intensity and gradient proposed by researchers at Soochow University is suitable for image noise reduction with complex background noise and has higher computational efficiency than the bilateral filtering method. The edge-preserving smoothing algorithm based on extreme value constraints proposed by researchers from Harbin Institute of Technology can maintain the main edge information of the image, and also eliminate the irrelevant information such as secondary edges and noise, which has a better enhancement effect compared with the edge-preserving smoothing algorithm.

(iv) Image sharpening

The purpose of image sharpening is to enhance the edges or contours to improve the recognizability of blurred images, but it also enhances noise and other non-critical detail information. Researchers at Zhejiang University proposed a sharpening algorithm based on dynamic edge detection, which can effectively suppress edge noise and has low computational complexity. Researchers from the Second Artillery Engineering University considered the advantages and disadvantages of the Roberts operator and Laplacian operator and used a combination of the two to get a better enhancement effect. Researchers from Nanjing University of Information Science and Technology proposed a GPU-based Laplacian sharpening algorithm and a GPU-based Shared Memory Laplacian sharpening algorithm, which optimizes the operation efficiency of the traditional Laplacian algorithm. Researchers from the Guangdong University of Technology and other researchers proposed a double-layer image sharpening method based on parallel computing, which has good performance in terms of image visual effect and processing efficiency. Researchers from North Central University propose an image sharpening algorithm based on fuzzy logic and nonlinear modules, which is effective in noise suppression. Researchers at the Indian Institute of Technology, Vellore, India, propose a sharpening enhancement method based on Laplace Pyramid (LP) and Singular Value Decomposition (SVD), which is able to enhance the edge contrast of tiny tissues The ultrasound image sharpening enhancement technique based on the contour transform proposed by researchers at Mepco Schlenk Institute of Technology, India, allows the noise effect to be controlled by adjusting the parameters. Researchers at Xi'an Jiaotong University have proposed an image smoothing and sharpening model, in which the smoothing term is used to smooth the flat region, and the sharpening term is used to enhance the image edges, which has superior performance compared to the single-item model.

Frequency domain enhancement is an enhancement processing method in which the spectral components of the original image after the two-dimensional Fourier transform are processed separately, and then the desired image is generated by the Fourier inverse transform. The infrared image enhancement technology based on frequency compensation proposed by researchers from the Naval Engineering University compensates for the high-frequency detail information lost in the denoising process through an iterative algorithm, which is an optimization of the original image frequency domain hierarchical processing method. Researchers at the China Aerospace Science and Technology Industry Corporation 8511 Research Institute have proposed a frequency extrapolation-based detail enhancement algorithm for fuzzy infrared images, which makes the image clearer and enhances the details by adding higher frequency components generated by inverse prediction to the fuzzy image. Researchers from West Virginia University and others quantitatively analyzed the impact of the frequency filter enhancement algorithm on image quality, showing that it has more significant advantages in image quality improvement and detail enhancement. Researchers from Beijing University of Posts and Telecommunications proposed a frequency domain enhancement algorithm based on an improved Gabor filter, which achieved more satisfactory results in fingerprint image recognition. Researchers from Ningxia University proposed a hybrid filtering algorithm that combines the advantages of a high-pass filter and a low-pass filter to enhance the high-frequency and low-frequency parts of the image with different coefficients. Researchers from Anhui Xinhua College and Daqing Petroleum College proposed a spatial domain and frequency domain-based enhancement algorithm, which combines the advantages of the two methods and provides better results than a single method. Researchers from the Second Artillery Engineering University designed an enhancement algorithm for the effective frequency band of the infrared characteristics of the monitoring object, which improves the image contrast for more accurate identification of defects.

(v) Problems that exist

Taking into account the specificity and complexity of the equipment attributes, monitoring objects, and monitoring needs at oil and gas extraction sites, the existing research on image enhancement processing for infrared thermography monitoring has the following problems:

(a) The objects of interest of the existing enhancement methods should have common (or similar) image parameter features, and the features will be significantly different from other non-focused image elements after enhancement processing. For example, the infrared thermal imaging monitoring objects in oil and gas extraction sites, such as shale gas fracturing, are mostly large-scale equipment systems, consisting of a variety of sub-modules with different performances, materials, structures, and functions, which are difficult to become the objects of interest for one image enhancement process at the same time.

(b) The infrared thermal imaging monitoring range of shale gas fracturing field equipment is large, with more targets, and the thermal imaging information of each sub-module of the equipment system is equally important. The feature highlighting, difference enhancement, and visual effect improvement of existing methods for a certain target can easily lead to the blurring of features, weakening of differences, and image quality degradation of other targets. If a unified image enhancement standard (or rule) is applied to multiple sub-modules of the monitoring object, it is prone to lead to problems such as reduced local contrast and reduced level of visualization of detailed information.

3.1.2 *Overview of early-warning technologies based on infrared thermal video surveillance*

As can be seen from the foregoing, infrared thermography monitoring of oil and gas extraction field equipment represented by shale gas fracturing is characterized by open sites, complex environments, more and wider monitoring objects, and higher requirements for real-time and accuracy of processing results. Compared with other fields, scenes, or equipment applications, infrared thermal imaging monitoring of oil and gas extraction field equipment is prone to the following types of problems:

(1) Foreign objects (personnel, vehicles, movable equipment, natural organisms, etc.) abnormally pass in front of the lens of the infrared thermal imager, impede the normal propagation of

infrared radiation of the monitoring object, and emit their own infrared radiation that is significantly different from that of the monitoring object to the infrared thermal imager, contaminate the monitoring data, interfering with the results of the monitoring, and even triggering false alarms or missed alarms.

(2) The monitoring object of each sub-module attribute differences, if the unified "attribute — image parameters" transformation rules, will lead to a reduction in the local contrast, and the level of visualization of detailed information to reduce the problem.

(3) When the image processing algorithm faces the batch processing of video or multi-frame time series images, the computational efficiency and cost cannot be guaranteed, and it cannot meet the needs of real-time processing and presentation of information for on-site monitoring. The higher the degree of optimization of the algorithm, the more processing steps, the larger the number of processing objects, and the worse the real-time performance.

Aiming at the above problems, taking the early warning of hidden dangers of shale gas fracturing field equipment as an example, the following three key technologies are proposed in this section:

(1) Field equipment infrared thermography monitoring anomaly data identification and cleaning method based on transboundary detection.

In view of the problem that foreign objects passing in front of the lens (intruding into or moving out of the field of view of the lens of the infrared thermal imager) during the infrared thermal imaging monitoring process of the field equipment of shale gas fracturing have an impact on the monitoring results and cause false alarms, this section proposes a method of identifying and cleaning abnormal data during the infrared thermal imaging monitoring process of the shale gas fracturing equipment based on the trans-boundary detection. By setting fixed position vectors in the transformed data matrix of the infrared thermal imaging map and observing and analyzing the changes of its elemental values frame by frame, we can efficiently and accurately identify the time when the foreign object breaks into or moves out of the lens field of view. Then feedback on the number of acquisition frames corresponding to the existence of foreign objects in the lens field of view, delete the relevant data segments, and retain and splice the normal infrared thermal imaging monitoring data segments without the existence of foreign objects, in order to improve the accuracy of the infrared thermal imaging monitoring results of shale gas fracturing field equipment, and reduce the rate of false alarms.

(2) An image enhancement method for infrared thermography monitoring of shale gas fracturing field equipment based on the optimization of gray value distribution.

Aiming at the problems of reduced local contrast and reduced visualization level of detailed information when multiple subsystems, components, or regions with large differences in temperature attributes are transformed into imaging based on a uniform functional relationship in the process of infrared thermography monitoring of shale gas fracturing field equipment, this section proposes a method of infrared thermography monitoring image enhancement for shale gas fracturing field equipment based on the optimization of grayscale value distribution. The image segmentation technique is applied to construct the segmentation matrix of each subsystem, component, or region of the monitored equipment, which is used to extract the relevant information of the corresponding subsystem, component, or region in each frame of the infrared thermal imaging image, and filter out the irrelevant information of other parts. Specific and independent functional relations for transforming imaging are set for the extracted single (or several) subsystems, components, or regions with similar attributes so that they can optimally allocate imaging parameters according to the characteristics of their own temperature distributions and variations, in order to improve the visualization of the equipment monitoring information and to enhance the recognizability of the differences in or variations in the surface temperature of the equipment based on the visual differences in the images.

(3) Early accident monitoring and identification method of water leakage at the output end of shale gas fracture pump based on RGB value distribution statistics of infrared thermography.

For the monitoring and identification of early water leakage accidents at the output end of shale gas fracking pumps, this section proposes a method for identifying and determining abnormal temperature changes based on the RGB value distribution statistics

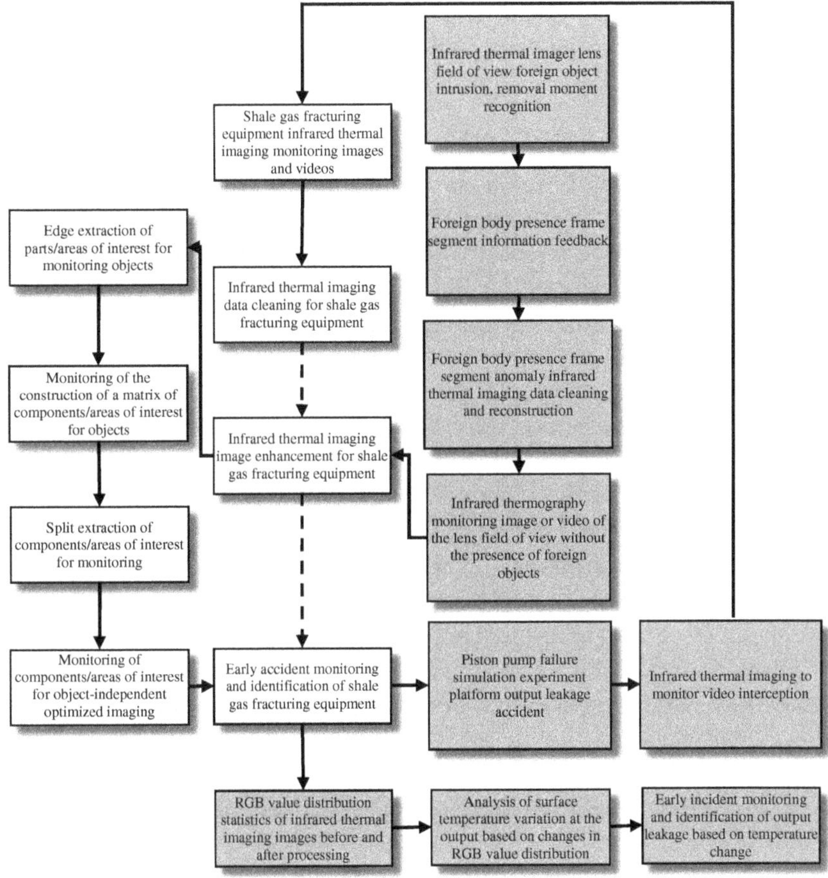

Fig. 3.1. Methodological and technical framework for this section.

of infrared thermography. By doing statistical analysis and feature extraction on the RGB value distribution of the infrared thermography map of the monitoring area, the temperature change caused by the accident is transformed into image information or other forms of information that can be recognized by the human eye, and the visual discriminability of small differences in parameters is optimized so as to facilitate the observation, recognition, and determination by on-site monitors.

The methodological and technical framework of this section is shown in Fig. 3.1.

3.2   Shale Gas Fracturing Equipment Infrared Thermal Imaging Monitoring Anomaly Data Identification and Cleaning

In view of the characteristics of infrared thermal imaging monitoring of the production equipment at the shale gas fracturing site, compared with the infrared thermal imaging camera (or infrared thermal imaging camera unit) which has a relatively fixed erection position, shooting angle, or running track, the movement state of the personnel, vehicles, mobile equipment and even flying animals at the construction site is extremely accidental and random, and inevitably shuttles through or stays in front of the lens of the infrared thermal imaging camera. They not only block the normal infrared radiation path of the monitored equipment but also emit significantly different infrared radiation to the thermal imaging camera. This phenomenon will seriously affect the infrared thermal imaging monitoring results of the fracturing equipment, increase the monitoring false alarm rate or leakage alarm rate, and bring unnecessary trouble to the safety management of the shale gas fracturing site. Therefore, it is necessary to adopt anomaly data identification and cleaning technology with strong real-time performance and high accuracy. Timely detection and identification of foreign objects passing in front of the lens of the infrared thermal imager, determining the characteristic moments of its entry and exit from the lens field of view, feeding back the number of acquisition frames corresponding to the existence of foreign objects in the lens field of view and deleting the relevant data fragments, reconstructing the normal monitoring data without the existence of foreign objects, and laying a good foundation for the processing and analysis of subsequent infrared thermal imaging monitoring data.

3.2.1   *Principles of infrared thermal imaging anomaly data recognition and cleaning methods based on out-of-bounds detection*

(1) Infrared thermal imaging to monitor foreign object intrusion during processes

During infrared thermography monitoring, the infrared radiation from each location of the monitored equipment propagates to the infrared thermal imager and reaches the infrared detector plane through the lens. The neatly arranged temperature measurement points on the infrared detector each record the intensity of the received infrared radiation, forming a specific radiation intensity distribution that corresponds exactly to the distribution of the infrared radiation intensity at each location of the equipment being monitored. The infrared radiation received at each temperature measurement point is converted into a computer-processable digital signal according to uniform rules, which is then converted into the image parameters required for imaging. All the temperature measurement point radiation intensity corresponds to the image parameters of the collection of presentation that is in the display to form the human eye can recognize, and understand the infrared thermal imaging map.

The analysis and processing of infrared thermography monitoring data is concerned with the temperature data matrix and the image parameter matrix formed in the above process. The temperature data matrix is an objective representation of the surface temperature of the monitored object, while the image parameter matrix is a subjective representation that visualizes the temperature distribution.

If the set of infrared radiation intensity values of the temperature measurement points on the infrared detector is defined as data matrix $A_{m \times n}$ (the relative position of each element in the matrix corresponds to the relative position of each temperature measurement point in the temperature measurement point array, and the value of each element is equal to the infrared radiation intensity of the corresponding temperature measurement point), the set of temperature values of each pixel in the infrared thermal imaging map is defined as data matrix $B_{m \times n}$, and the set of image parameter values of each pixel in the infrared thermal imaging map is defined as data matrix $C_{m \times n}$, then the values of the elements of the matrix $A_{m \times n}$, $B_{m \times n}$, $C_{m \times n}$ follow exactly the same functional relationship, that is, the temperature measurement point on the infrared radiation intensity values, the surface temperature of the monitored equipment at each location and the corresponding pixel image parameter values of the conversion between the relationship is shown in Eq. (3.1):

$$\begin{matrix} b_{ij} = \xi(a_{ij}) & i = 1, 2, \ldots m & j = 1, 2, \ldots, n \\ c_{ij} = \theta(a_{ij}) & i = 1, 2, \ldots m & j = 1, 2, \ldots, n \end{matrix}, \qquad (3.1)$$

where a_{ij} is the element in the infrared radiation intensity data matrix $\boldsymbol{A}_{m \times n}$; b_{ij} is the element in the temperature data matrix $\boldsymbol{B}_{m \times n}$; c_{ij} is the element in the image parameter data matrix $\boldsymbol{C}_{m \times n}$; i, j denote the number of rows and columns of the matrix elements, respectively; $\xi(\cdot)$ denotes the function relationship between the infrared radiation intensity and temperature; $\theta(\cdot)$ denotes the function relationship between the infrared radiation intensity and the image parameters.

As far as the research content of this section is concerned, the focus is on using the temperature matrix as the transformed data matrix of the infrared thermal imaging monitoring video, in order to obtain a more concise and intuitive exposition and results (using the image parameter matrix as the transformed data matrix, the principle, operation, analysis, and conclusions of the methods described in this section are exactly the same).

Take any frame of infrared thermal imaging map $Image_p$ and its transformed data matrix $\boldsymbol{I}_p$ as the object of study, if the infrared thermal imager has a total of $m \times n$ temperature measurement points, then each frame should contain $m \times n$ pixel points, which can be transformed into $m \times n$ of the data matrix $\boldsymbol{I}_{m \times n}$, see Eq. (3.2):

$$
\boldsymbol{I}_p =
\begin{bmatrix}
i_{11} & i_{12} & \cdots & i_{1n} \\
i_{21} & \ddots & & i_{2n} \\
\vdots & & \ddots & \vdots \\
i_{m1} & i_{m2} & \cdots & i_{mn}
\end{bmatrix},
\tag{3.2}
$$

where $\boldsymbol{I}_p$ denotes the pth data matrix generated from the transformation of the pth frame of the infrared thermogram; i is the value of the matrix element.

Whereas an infrared thermography monitoring video is essentially a sequential arrangement of a number of infrared thermographic images (frames). The sequential arrangement of still images is displayed continuously at a specific frame rate to form a visually dynamic video. The transformed data matrix of the infrared thermal imaging monitoring video can be regarded as the sequential arrangement of the transformed data matrix of a number of infrared thermal imaging images (frames), see Eq. (3.3):

$$\begin{cases} Video = \{Image_1, Image_2, \ldots, Image_p, \ldots, Image_q\} \\ \quad p = 1, 2, \ldots, q \\ Video \to \boldsymbol{V} \\ Image_p \to \boldsymbol{I}_p \ p = 1, 2, \ldots, q \\ \boldsymbol{V} = \{\boldsymbol{I}_1, \boldsymbol{I}_2, \ldots, \boldsymbol{I}_p, \ldots, \boldsymbol{I}_q\}, \end{cases} \tag{3.3}$$

where *Video* means infrared thermal imaging monitoring video; *Image$_p$* means the pth frame of infrared thermal imaging map; p is the number of frames; $\boldsymbol{V}$ is the transformed data matrix of infrared thermal imaging monitoring video; $\boldsymbol{I}_p$ is the transformed data matrix of the pth frame of infrared thermal imaging map.

That is, the infrared thermography monitoring video *Video* is a collection of infrared thermogram *Image$_p$*, and the *Image$_p$* is arranged in frame order. The infrared thermal imaging monitoring video transformation data matrix $\boldsymbol{V}$ is a collection of single-frame infrared thermal imaging map transformation data matrices $\boldsymbol{I}_p$, and $\boldsymbol{I}_p$ is arranged in frame order.

If a foreign object intrudes into the camera's field of view during the monitoring process, the foreign object will block the infrared radiation emitted by the monitored equipment and will itself emit infrared radiation to the thermal imaging camera. In view of the fact that most of the monitored equipment at the construction site is metal system with uneven distribution of infrared radiation intensity [e.g., the motor, transmission device, power end, hydraulic end of the fracturing pump system, etc., the difference in the surface temperature of the equipment after a period of time of work is large, and the corresponding difference in the intensity of the infrared radiation is also large, so that the distribution of the intensity of the infrared radiation of the equipment as a whole is seriously uneven], and the foreign object intruded into the scene is mostly a person walking around the scene (or a part of the body of the person). A part of the body), infrared radiation intensity is more uniform mobile equipment (such as cars, carts, and their goods), accidentally flew over the lens in front of flying animals (such as birds, bats), etc., the intensity of its infrared radiation, the distribution of the form of a large difference compared with the equipment being monitored.

Therefore, when a foreign object breaks into the lens field of view, the infrared radiation received on the infrared detector intensity, distribution form, etc. Compared with the conventional situation, there will be a large difference; this difference can lead to abnormal signal reception, conversion, and transmission in the infrared detector plane, resulting in a decrease in the reliability of infrared thermal images, which in turn affects and biases the automatic recognition and processing of machine learning.

In most cases, foreign objects are in motion (appearing, moving, disappearing, and other behaviors) in the lens field of view of the infrared thermal imager, and can be viewed as motion (or flow) of an anomalous object in the temperature field of view of the lens. In the field of fluid mechanics, the Lagrangian and Eulerian methods have been proposed to describe the motion of fluids. Among them, Euler's method does not directly track the motion process of the mass point in the fluid, but takes the flow field as the object of study, and comprehensively analyzes and calculates the state of motion of the fluid by studying the change of a certain parameter with time at a fixed position in the flow field. It should be noted that the foreign matter does not affect the objective existence of the temperature field of the monitored equipment, but its presence affects the normal propagation, reception, and transformation of the local infrared radiation of the monitored equipment, making the temperature field presented in the form of images available for the monitoring personnel to observe and understand the temperature field abnormal, that is, it affects the temperature field of the visualization of the monitored equipment.

Based on this idea, in the field of kinematics, in order to study the motion state of an object, a fixed position (point, line, surface, body, etc., in space) can be calibrated in the selected reference system, when the object continuously passes through the calibrated fixed position, a certain parameter describing the motion of the position will change over time, and the analysis of this change can be indirectly obtained by the relevant information on the motion state of the object.

(2) Foreign object recognition based on anomalous changes in special position vectors

The motion of a foreign body in the field of view of the lens of an infrared thermal imager can be regarded as a kinematic problem of studying the motion state of a foreign body using the field of view of the lens as a reference system. From the idea of Euler's method mentioned above, at least one fixed position in the reference system, i.e., the lens field of view, should be selected as the object of study. Given that the reference system itself is a two-dimensional plane, the selected fixed position can be a point, a line, or a surface. In order to improve the efficiency of the analysis and processing, and at the same time comprehensively reflect the motion of the movement of the object, it is appropriate to choose a smaller number of dimensions, but a larger capacity of the parameters of the "line" as a fixed position of the setup form. That is, in the lens field of view marked at least one imaginary stationary "line", when the object moves through this special "line", it will cause changes in the motion parameters on the line. Through the study of the fixed position of the "line" on the change of motion parameters with time, you can deduce the object's state of motion.

The single-frame infrared thermogram acquired under normal conditions can be transformed into a data matrix $\boldsymbol{X}_{m \times n}$, and the single-frame infrared thermogram acquired during the intrusion of foreign objects can be transformed into a data matrix $\boldsymbol{X}'_{m \times n}$. The difference between the two matrices can be used to construct the difference matrix Δ, which is shown in Eq. (3.4):

$$\Delta = \boldsymbol{X}'_{m \times n} - \boldsymbol{X}_{m \times n}$$

$$\boldsymbol{X}_{m \times n} = \begin{bmatrix} x_{11} & x_{12} & \cdots & x_{1n} \\ x_{21} & \ddots & & x_{2n} \\ \vdots & & \ddots & \vdots \\ x_{m1} & x_{m2} & \cdots & x_{mn} \end{bmatrix} \quad \boldsymbol{X}'_{m \times n} = \begin{bmatrix} x'_{11} & x'_{12} & \cdots & x'_{1n} \\ x'_{21} & \ddots & & x'_{2n} \\ \vdots & & \ddots & \vdots \\ x'_{m1} & x'_{m2} & \cdots & x'_{mn} \end{bmatrix}.$$

$$\tag{3.4}$$

Δ is a non-zero matrix. The set of locations of all non-zero elements in matrix Δ can be used to characterize the morphology of the foreign objects in the infrared thermogram, and the construction of a specific statistic for matrix Δ can be used to characterize the difference between the transformation matrix $\boldsymbol{X}_{m \times n}$ of the infrared thermogram in the conventional case and the transformation matrix $\boldsymbol{X}'_{m \times n}$ of the infrared thermogram in the case of the intrusion of a foreign object, so as to identify the formation of the anomalous infrared thermogram signals in case of intrusion of a foreign object into the field of view of the lens.

As can be seen from the foregoing, the essence of infrared thermal imaging monitoring video is a multi-frame static infrared thermal imaging image of the frame sequence arrangement, the foreign body intrusion video captured foreign body "movement" can be split into a number of consecutive frames of the image of the foreign body appeared in a different location (the nature of the movement that the object position changes over time). Each frame is transformed into a data matrix X'_p (p is the number of frames), where X'_p is not exactly the same for different frame order p. The frame order p is the number of frames, and X'_p is the number of frames. Comparing the difference matrix X_p generated by transforming each frame in the conventional case, the difference matrix Δ is not exactly the same as the frame ordinal number, i.e., it varies with time.

As far as the difference matrix Δ is concerned, its core content is the non-zero elements in the matrix, and the set of locations of all non-zero elements can be used to characterize the morphology of the foreign objects in the infrared thermogram. In order to facilitate the subsequent description, the set of all the attributes (e.g., numerical attributes, positional attributes, structural attributes, etc.) of all the non-zero elements in each difference matrix Δ is defined as Φ, which is a set of data containing some specific elements in the matrix and keeping the relative positions of the elements unchanged, with an irregular planar pattern (the planar pattern of Φ is the discretized projection pattern of the foreign object in the corresponding plane), and the process of generating the Φ is as illustrated in Fig. 3.2. In each frame of the foreign body, the position and size of the foreign body are different, and the position, shape and number of elements of Φ in the difference matrix Δ of the data matrix of each frame will also be different.

If the difference matrix Δ of the image data matrices of the frames is arranged in frame order and displayed in the screen at a certain frequency, Φ will show a dynamic effect, characterizing the changes in the appearance of the object projected in the plane of the lens and its state of motion, which is similar to the principle of formation of the video. In this case, the problem of describing and studying the motion of an object in an infrared thermal imaging video is transformed into the problem of describing and studying the changes in the position and shape of Φ in the difference matrix Δ.

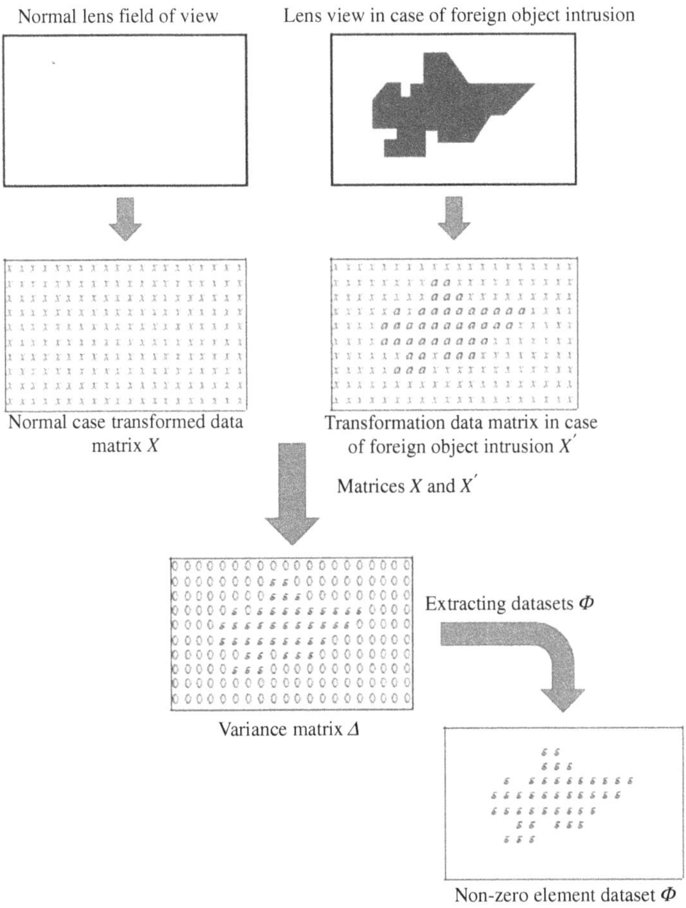

Normal lens field of view

Lens view in case of foreign object intrusion

Normal case transformed data matrix X

Transformation data matrix in case of foreign object intrusion X'

Matrices X and X'

Extracting datasets Φ

Variance matrix Δ

Non-zero element dataset Φ

Fig. 3.2. The set of non-zero element data Φ and its generation process.

In the presence of foreign objects in the lens field of view, multiple difference matrices Δ obtained from the transformation and processing of consecutive infrared thermograms are studied, and the difference matrices are named according to the number of frame sequences. In each difference matrix Δ, there is a dataset Φ, which is named according to the frame order. Then, with the change of time, the difference matrices are displayed in order from smallest to largest frame order, and the position and shape of Φ in each matrix will be changed.

The same column (row) is selected as the "fixed position" in each of the different matrices. Normally, this selected column (row) is a zero vector (if there are no foreign objects present, the difference matrix Δ is a zero matrix, and therefore all rows and columns are zero vectors), and is defined as Θ. When a foreign body is present and moving, Φ appears in the difference matrix and changes its position and morphology over time.

At frame p, the movement of Φ to the selected column (row) changes the value of some or all elements of the vector Θ (from zero to non-zero), and the number of changed elements is related to the morphological scale of the part of Φ that touches the "fixed position" at the current moment (i.e., the length of the part of Φ that intersects with Θ at the current moment).

At frame $p + 1$, a change in the position, shape, or value of an element of Φ causes a new change in the value of some or all of the elements of the vector Θ, e.g., a change in the value of some of the elements from a non-zero value to a zero value, a change in the value of an element from a zero to a non-zero value, a change in the value of a previously non-zero value to a non-zero value, and so on. This is because the part of Φ in contact with the "fixed position" at frame $p + 1$ is different from that at frame p, both in scale and in value.

By analogy, the contact part of Φ with the "fixed position" in the difference matrix Δ is different in each frame, and the change of the selected vector Θ will be different in each frame. By analyzing these differences together, we can realize the state description of moving objects at the level of the difference matrix based on the idea of Euler's method.

Given that the content of the infrared thermal imaging monitoring video layer and the difference matrix layer is a layer-by-layer derivation, one-to-one correspondence, and the analytical ideas and analytical methods are exactly the same, the results of the motion state description of the object based on the idea of Euler's method at the level of the difference matrix can be equated to the results of the description at the level of the infrared thermal imaging monitoring video. That is, at the level of infrared thermography video, when the temperature distribution of the monitored object in the lens field of view and its imaging results are transformed into the data matrix, the projection of the foreign object on the monitoring plane of the infrared thermography camera also appears in the transformed data matrix in the form of a special dataset Φ'.

Similar to Φ, Φ' is also a set of data that contains some specific elements in the matrix and keeps the relative positions of the elements unchanged, with irregular planar morphology, whose planar morphology is the projection morphology of the foreign body on the corresponding plane. The relationship between the dataset Φ' formed on the transformed data matrix and the dataset Φ formed on the difference matrix at the same moment is shown in Eq. (3.5):

$$\Phi = \Phi' - \Phi_{\text{norm}}, \tag{3.5}$$

where Φ_{norm} is the set of partial data in the transformed data matrix of the infrared thermal imaging map when there is no foreign object present, and its data elements reflect the temperature distribution of the monitored equipment in the part obscured by the foreign object, and its contour and morphology are the same as that of the foreign object's projection in the detector plane.

A column or a row in the transformed data matrix is selected as a "special position", defined as Θ'. If the selected Θ' on the transformed data matrix has the same number as the selected Θ row (column) on the corresponding difference matrix, the relationship is shown in Eq. (3.6):

$$\Theta = \Theta' - \Theta_{\text{norm}}, \tag{3.6}$$

where Θ_{norm} is a row (column) vector in the transformed data matrix of the infrared thermogram when no foreign objects are present. The values of its elements are exactly the same as Θ' when the foreign object has not passed through a special location.

When a foreign object appears and moves in the lens field of view, the dataset Φ' on the transformed data matrix appears and moves accordingly. When passing through the selected special position, the dataset Φ' intersects with the special position vector Θ', and in the intersection part, the data of the elements of the dataset Φ' will overwrite the data of the elements of Θ' at the corresponding position, so that the values of some or all of the elements of Θ' will be changed, and the number of changed elements is related to the morphological scale of the part of Φ' that is in contact with the "fixed position" at the current moment (i.e., it is related to the length dimension of the part of Φ' that intersects with Θ' at the current moment).

In this case, the part of the dataset Φ' formed by the foreign object in the multi-frame transformed data matrix that is in contact with the "fixed position" will be different from time to time, and the change of the selected special position vector Θ' in each frame will also be different from time to time. By comprehensively analyzing these differences, the state description of moving objects at the level of infrared thermal imaging video signals based on the idea of Euler's method can be realized.

For the problem of foreign object intrusion in the process of infrared thermal imaging monitoring, the premise of accurately identifying foreign objects is to shoot them into the lens field of view and capture the moment when they move in and out. Therefore, the selected "fixed position" should be the boundary part of the lens field of view, including the left boundary, right boundary, and upper boundary, the lower border. In the transformed data matrix, it is described as column 1, column n, row 1, and row m of the matrix.

(3) Foreign body detection and tracking based on morphological differences in numerical curves

Infrared thermography monitoring data is stored in the system in the form of a data matrix $\boldsymbol{X_p}$. A single frame of an infrared thermography image has a total of $m \times n$ pixels, so the data matrix of the pth frame is shown in Eq. (3.7):

$$\boldsymbol{X_p} = \begin{bmatrix} x_{11} & x_{12} & \cdots & x_{1n} \\ x_{21} & \ddots & & x_{2n} \\ \vdots & & \ddots & \vdots \\ x_{m1} & x_{m2} & \cdots & x_{mn} \end{bmatrix}, \tag{3.7}$$

where p denotes the frame order of the infrared thermogram; m denotes the number of pixel rows; n denotes the number of pixel columns.

Row 1 of the data matrix corresponds to the selected upper boundary special location, row m corresponds to the selected lower boundary special location, column 1 corresponds to the selected left boundary special location, column n corresponds to the selected right boundary special location, and the other rows and columns correspond to each selected special location.

The monitoring video captured by the thermal imager (t frames in total) is arranged in frame order, i.e., from the first frame, the second frame to the last frame, then its data matrix V is shown in Eq. (3.8):

$$V = \begin{bmatrix} X_1 \\ X_2 \\ \vdots \\ X_p \\ \vdots \\ X_t \end{bmatrix} \quad \text{or } V = \begin{pmatrix} X_1 & X_2 & \cdots & X_p & \cdots & X_t \end{pmatrix}, \qquad (3.8)$$

where X_p is the transformed data matrix of the pth frame of the infrared thermogram, $p = 1, 2, \ldots, t$.

Select the same data rows/columns in the transformed data matrix of each thermal image frame (e.g., the nth column or the mth row of the transformed data matrix of each frame), there are t data rows/columns vectors in total, and the data vector S is composed according to the sequence of the frames, see Eq. (3.9). The construction process is shown in Fig. 3.3:

$$S = \begin{bmatrix} X_1^n \\ X_2^n \\ \vdots \\ X_p^n \\ \vdots \\ X_t^n \end{bmatrix} \quad \text{或 } S' = \begin{pmatrix} X_1^m & X_2^m & \cdots & X_p^m & \cdots & X_t^m \end{pmatrix}, \qquad (3.9)$$

where X_p^n is the nth column of the pth frame and X_p^m is the mth row of the pth frame. That is, S has $m \times t$ data elements and S' has $n \times t$ data elements.

Or select the corresponding row/column (e.g., the nth column of the data matrix V, or the mth row of the data matrix V) of the data matrix V, and name it as vector S. The construction process is schematically shown in Fig. 3.4.

The numerical curve of the data vector S (or S') is an approximate periodic image with period m (or n). The image (line) of each cycle is called a periodic numerical curve fragment of the numerical curve of the vector S. The images of each cycle are similar in

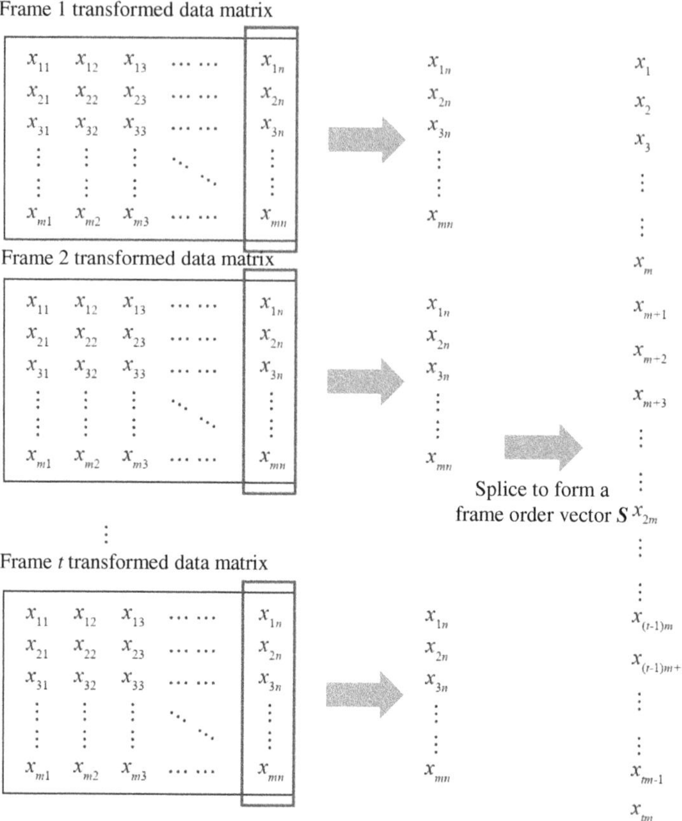

Fig. 3.3. Frame sequence data vector S generation method I.

shape, with small differences reflecting slow changes (or even constancy) in temperature values at the same location in the image over time.

When a foreign object intrudes at a certain point in time, the kth frame of the thermal imaging camera captures it, i.e., the foreign object passes through a special position at the boundary of the kth frame. When passing through the special location, the infrared radiation of the foreign object is clearly different from that of the object being monitored, which causes an abnormal change in the numerical curve of the data vector corresponding to the special location. The number of abnormally changed data points reflects the vertical/horizontal dimensions of the foreign object at that moment

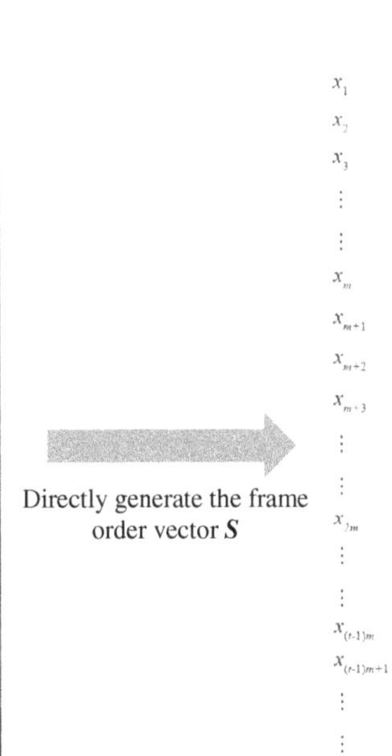

Fig. 3.4. Frame sequence data vector S generation method II.

(e.g., if a square with side length a enters the field of view of the camera from the right side of the image and passes through the special position at the right boundary, it will cause an abnormal change of y data points in the nth column of data, and y is related to the size of a). This change is shown in the numerical curve of S as a change in the periodicity of the image, and the shape of the numerical curve in the kth cycle is clearly different from that in the k-1st cycle.

If a foreign object is still present in the lens field of view at the next moment (frame), the portion of the $k + 1$st frame image in which the foreign object enters the lens field of view may change in the following three ways:

(1) The portion of the foreign body that enters the lens field of view increases (or decreases). The increased (or decreased) portion passes through the special position at the corresponding moment of the $k + 1$st frame, which causes an abnormal change in the

numerical curve of the data vector of the special position, and the number of abnormally changed data points reflects the longitudinal/horizontal dimensions of the foreign body at that moment, i.e., the longitudinal/horizontal dimensions of the portion of the foreign body that newly enters (exits) the field of view of the lens. This is reflected in the numerical curve of S by a further change in the periodicity of the image, with the shape of the numerical curve in the $k+1$st cycle being significantly different from that in the $k-1$st cycle.

(2) The foreign body is translated perpendicular to the direction of entry in the lens field of view, i.e., the area of the part of the foreign body in the lens field of view remains unchanged and the position is changed. The changed part in the kth frame image is translated perpendicular to the special position in the $k+1$st frame image, so that the numerical curve of the data vector of the special position changes abnormally, and the position of the abnormally changed data point reflects the position of the foreign body at the current moment. In the numerical curve of S, the change position of the curve shape in the $k+1$st cycle is different from that in the kth cycle, but the shape of the changed part is the same as that in the kth cycle. The $k+1$st cycle is clearly different from the $k-1$st cycle.

(3) The position of the foreign body in the lens field of view remains unchanged and the temperature changes. That is, at the corresponding moment of the $k+1$st image frame, the dataset Φ' formed by the foreign object has the same intersection elements with the special position vector Θ', but the values within the elements are different. In the numerical curve of S, it is manifested that the change position of the curve shape in the $k+1$st cycle is the same as that in the kth cycle, but the magnitude of the change is different, and the shape of the data curve in the $k+1$st cycle is obviously different from that in the $k-1$st cycle.

By analogy, as the foreign body moves in and out of the field of view, moves within the field of view, or changes within the field of view, the shape of the numerical curve of S will change accordingly. Until the foreign body moves out of the lens field of view through a special position at a certain boundary. When the foreign object moves out of the lens, it again triggers an abnormal change in the data curve of the moving out special position, the form of which is similar to the process of foreign object entry.

In summary, the movement and change of foreign objects in the lens field of view will lead to abnormal changes in the selected rows/columns of the image data matrix of a certain frame, and by analyzing the starting and ending times of these changes and the number of data points, the motion state of the object can be inferred.

It is known that the numerical curve of the data vector S can be divided into t-periodic numerical curve segments, and adjacent periodical numerical curve segments are similar in shape, with small differences reflecting slow changes (or even invariance) in the temperature values at the same location in the image over time.

Normally, the temperature of the monitored equipment changes with some continuity and at a slow rate. The shape of the numerical curve of the data vector S is characterized by a high degree of similarity of the numerical curve segments of adjacent cycles. In contrast, when a foreign object intrudes into the field of view of the camera, the infrared detector receives infrared radiation from the foreign object itself that is significantly different from that of the monitored object. In view of the fact that the infrared camera cannot recognize whether the received infrared radiation originates from the monitored object or not, it can only faithfully convert it into temperature parameters, which is manifested in the temperature change as a jump in the temperature value at a very fast rate, and in the numerical curve shape of the data vector S, which is manifested in the abnormally low degree of similarity of neighboring cycles of numerical curves segments.

In order to measure the similarity of neighboring segments of periodic numerical curves, the Euclidean distance d is introduced as an evaluation metric. The Euclidean distance d represents the true distance between two variables and can also be used to measure the degree of similarity between two variables, see Eq. (3.10):

$$d(X, Y) = \sqrt{\sum_{i=1}^{n} (X_i - Y_i)^2}, \tag{3.10}$$

where $d(X, Y)$ denotes the Euclidean distance between variables X and Y; X_i denotes the ith dimension of the n-dimensional variable X; Y_i denotes the ith dimension of the nth dimension variable Y.

The larger the Euclidean distance between neighboring cycle value curves, the less similar they are, and the more likely it is that an abnormal change will occur at that location. Conversely, the smaller the Euclidean distance between neighboring cycle value curves, the higher the degree of similarity between the two, and the lower the likelihood of an anomalous change at that location.

(4) Feedback and excision of anomalous data fragments.

Morphological similarity analysis of segments of numerical curves of each cycle based on the Euclidean distance $d(X, Y)$ enables to identify and determine the time (number of frames) at which an abnormal change in temperature occurs, i.e., the time (number of frames) at which a foreign body passes through a special location. In terms of the boundary special position, if it is recognized that the shape of the numerical curve segment corresponding to the ith frame has changed abnormally, it can be judged that the foreign object intrudes into the field of view of the lens at the current frame; if it is recognized that the abnormal change of the shape of the numerical curve segment corresponding to the jth frame has disappeared, it can be judged that the foreign object has moved out of the field of view of the lens at the current frame. In the infrared thermal imaging maps of all frames between ith frame and jth frame, there are foreign objects present, and the thermal image information reflected therein cannot accurately describe the current state of the monitored device.

After identifying the number of frames with foreign objects, it will be fed back to the monitoring system, cut out the data information of the relevant segments, and then splice and integrate the remaining segments, that is to say, to complete the cleaning of abnormal data of infrared thermal imaging monitoring. For example, if it is recognized that there is a foreign object in the infrared thermal imaging map between the ith frame and the jth frame, the frame sequence information from the ith frame to the jth frame will be fed back to the monitoring system, the infrared thermal imaging map from the ith frame to the jth frame will be resected and the relevant information will be removed, and the i-1th frame will be spliced with the j+1th frame, and a new infrared thermal imaging monitoring video will be formed. In view of the research content of this section is aimed at the monitoring system false alarms triggered by foreign objects accidentally intruding into and moving out of the lens field of view, then usually the speed of the foreign object through the front of the lens is faster, the existence of the lens field of view of a shorter period of time, cleaning the information related to its monitoring process and splicing of data segments of the process of parameter change has less impact on the continuity of the process. If the foreign matter stays in front of the lens field of view for a long time, the data segment of cleaning is longer, which seriously affects the monitoring process and the continuity of the parameter change process of the splicing data segment, the monitoring data of this group has no reference value, so it is not considered in this section of the research content.

3.2.2 *Implementation steps*

Step 1: construct the infrared thermography monitoring video transformation data matrix.

The infrared thermal imaging monitoring video is converted into a video data matrix V, see Eq. (3.3). The video data matrix V can be regarded as a number of frames (frames) of a single-frame infrared thermographic image converted to the data matrix I in sequence. Let a single frame of thermal imaging map have $m \times n$ pixels, then its transformation generates $m \times n$ data matrix $I_{m \times n}$.

If the acquisition frequency of the thermal imaging camera is αHz (i.e., frame rate αfps) and the acquisition duration δs, the total number of frames for generating the thermal imaging monitoring video is $\alpha \times \delta$ frames, i.e., the video consists of $\alpha \times \delta$ single-frame infrared thermal imaging maps. It is known that each infrared thermal imaging frame can be transformed into a corresponding image transformation data matrix I. The data matrix V generated from the transformation of this infrared thermal imaging monitoring video should contain $\alpha \times \delta$ sequential image transformation data matrices I. The data matrix V should contain $\alpha \times \delta$ sequential image transformation data matrices I.

(1) If the image transformation data matrix I of adjacent frames is in the form of end-row and first-row docking, then the resulting video transformation data matrix $V_{A \times n}$ of $A \times n$ is generated, where $A = \alpha \times \delta \times m$.

(2) If the adjacent frame image transformation data matrix $\boldsymbol{I}$ is in the form of end-column and first-column docking, the resulting video transformation data matrix $\boldsymbol{V_{m \times B}}$ is $m \times B$, where $B = \alpha \times \delta \times n$.

Step 2: select the special position and the corresponding transformed data vector.

The foreign object passes in front of the lens of the infrared thermal imager, which is equivalent to the foreign object as a whole or localized infrared thermal imaging to break into or move out of the lens field of view. As far as the lens field of view is concerned, no matter what shape, size, movement speed, movement trajectory of foreign objects into or out of the field of view will inevitably pass through the field of view range of the upper, lower, left and right boundaries of at least one of the boundaries. Regardless of the shape, size, speed, and trajectory of the foreign object in the field of view, it will inevitably pass through the selected special location in the field of view (the special location can be a point, a line, a surface). Selecting a representative special location, we can determine whether a foreign object has passed through based on the abnormal changes in the relevant parameters at the special location. The special positions to be selected for foreign matter intrusion or removal judgment under normal circumstances include the upper border, lower border, left border and right border. Other special positions can be set as required.

From the correspondence between the single-frame infrared thermal image and the image transformation data matrix, it can be seen that the "special position" in the selected image corresponds to a certain data vector(s) in the image transformation data matrix, i.e., the special position can be transformed into the form of a data vector. In other words, the special positions can be transformed into data vectors. The object of the method described in this section is the selected special positions and their corresponding data vectors.

For example, if the left boundary is regarded as a special position, the left boundary of the lens field of view corresponds to the first column of the image transformation data matrix $\boldsymbol{I_{m \times n}}$; if the upper boundary is regarded as a special position, the upper boundary of the lens field of view corresponds to the first row of the image transformation data matrix $\boldsymbol{I_{m \times n}}$.

Step 3: construct the frame order vector of the special position transformation data vector.

After selecting the special positions of a single frame of infrared thermal imaging map and transforming them into corresponding data vectors, the exact same special positions in other frames of infrared thermal imaging maps of the same group of infrared thermal imaging monitoring videos are selected and transformed into corresponding data vectors, generating t special positions and t special position transformed data vectors. The t special positions and t special position transformed data vectors are then arranged and spliced according to the number of frames to construct a complete frame sequence vector S of the special position transformed data vectors, which is shown in Fig. 3.3.

Alternatively, a frame order vector S may be constructed directly in the infrared thermal imaging monitoring video transformation data matrix V based on the relative position information of the selected special location in a single frame of the infrared thermal imaging image, wherein the frame order vector S is a long vector generated by combining the transformation data vectors of the same special location in each frame of the infrared thermal imaging image in the form of a stitching of the frame order numbers. If the nth column of the transformation data matrix of a single frame is selected as the transformation data vector of the special position, the nth column selected in the transformation data matrix V of the infrared thermal imaging monitoring video is the required frame order vector S. The construction process is shown in Fig. 3.4.

Step 4: generate a graph of the values of the frame order vector S.

Generate a numerical curve for the frame order vector S, where the horizontal coordinate is the pixel point order and the vertical coordinate is the temperature value.

If the frame order vector S is taken from $V_{A \times n}$ (where $A = \alpha \times \delta \times m$), then its numerical profile is composed of t numerical profile segments, each containing m data (corresponding to m rows of the transformed data vectors at the special positions of each frame).

If the frame order vector S is taken from $V_{m \times B}$ (where $B = \alpha \times \delta \times n$), then its numerical profile is composed of t numerical profile segments, each of which contains n data (corresponding to the

n columns of the transformed data vectors for each frame's special position).

Step 5: numerical curve morphology similarity analysis.

Select any segment of the numerical curve of the frame order vector S as the reference benchmark, and do the Euclidean distance-based similarity analysis [Eq. (3.10)] on other numerical curve segments frame by frame, i.e., calculate the Euclidean distance between the reference segment and the numerical curve segments of other frames. By calculating the Euclidean distance between any two segments of the numerical curve, the t-order Euclidean distance similarity matrix is generated, and the abnormal data segments are determined based on the set similarity threshold.

Step 6: second special location analysis verification.

Boundary-crossing detection based on the special position of the boundary can accurately identify the transient behavior of foreign objects entering and exiting the lens field of view, but it is unable to determine whether their presence or movement tendency constitutes an actual interference with the subject part of the monitored object in the lens field of view (e.g., the foreign object only stays at the boundary position of the lens field of view, and does not continue to move towards the center area), thus, the second special position needs to be selected to analyze and validate the foreign object recognition results based on boundary special position crossing detection.

The second special position in the image and its transformed data matrix near the boundary to be verified is selected, and the transformed data vector corresponding to the second special position is constructed, the implementation steps of which are described in step 2.

The foreign object recognition based on the second special position boundary crossing detection is described in steps 2 to 5. Comparing the results of the foreign object recognition based on the boundary position and the verification analysis results based on the second special position, if the frame or frame segment is judged to be abnormal by the boundary position transgression detection is again judged to be abnormal by the second special position transgression detection, it can be confirmed to be an abnormal frame segment of the monitoring data. On the contrary, it means that although the foreign object touches or passes the boundary of the lens field of view, it does not form actual interference with the monitoring subject, so it may not be cleaned.

Step 7: feedback and clearing of anomalous data information.

Feedback the information about the frame sequence of the numerical curve segments in the Euclidean distance similarity matrix that are higher than the set threshold to the monitoring system. According to the feedback, the system automatically searches for the abnormal frames of the infrared thermal imaging map and its transformed data matrix, and deletes the abnormal information of the relevant segments. After the removal is completed, the remaining segments of the infrared thermal imaging map and its transformed data matrix are spliced and combined in the original order to generate a new infrared thermal imaging monitoring video without the presence of foreign objects.

3.2.3   *Case study*

An infrared thermogram of the experimental reciprocating piston pump system during normal operation was acquired as shown in Fig. 3.5. Each frame consists of 288×384 pixels, and the acquisition frequency is 5 Hz, i.e., the frame rate is 5fps.

During the acquisition process, the infrared thermal imager captured a pedestrian passing in front of the lens at the 24th frame, i.e., the foreign object entered the lens field of view; at the 30th frame, it captured the pedestrian basically completing the passing action, i.e., the foreign object moved out of the lens field of view. Then the infrared thermographic images from the 24th frame to the 30th frame cannot accurately reflect the local or overall temperature change of the monitored equipment because of the presence of foreign objects (pedestrians).

As the pedestrian passes in front of the camera, various parts of the body block the localized infrared radiation from the monitored equipment and emit infrared radiation from the human body to the thermal imaging camera. The obvious difference in the intensity and distribution of infrared radiation between the human body and the monitored equipment will lead to an abnormal increase or decrease in the local parameters of the collected infrared thermal imaging signal, which will not accurately reflect the changes in the localized

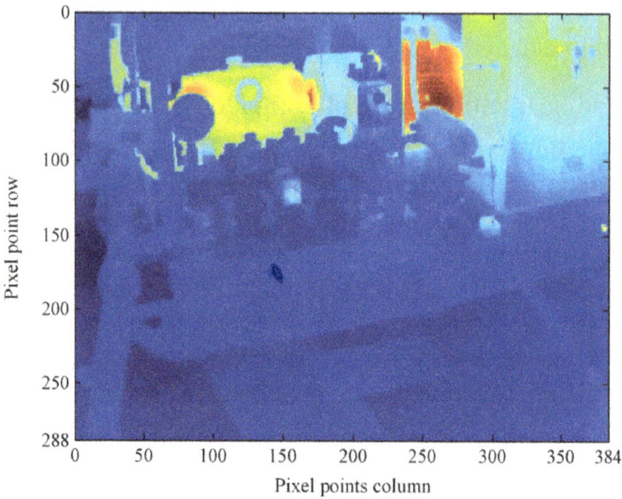

Fig. 3.5. Infrared thermography of the monitored object under normal conditions.

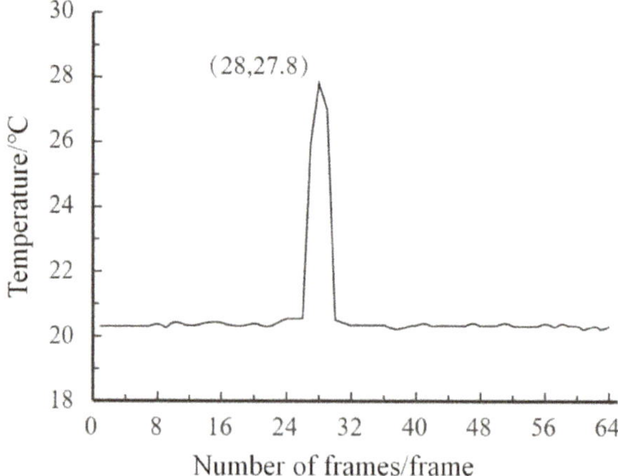

Fig. 3.6. Temperature change of pump head body measurement point during pedestrian passage.

or overall situation of the monitored equipment, and will even trigger a false alarm of the infrared thermal imaging monitoring system.

The anomalous changes in the temperature values of various parts of the equipment monitored based on infrared thermography during the acquisition process are shown in Figs. 3.6–3.11.

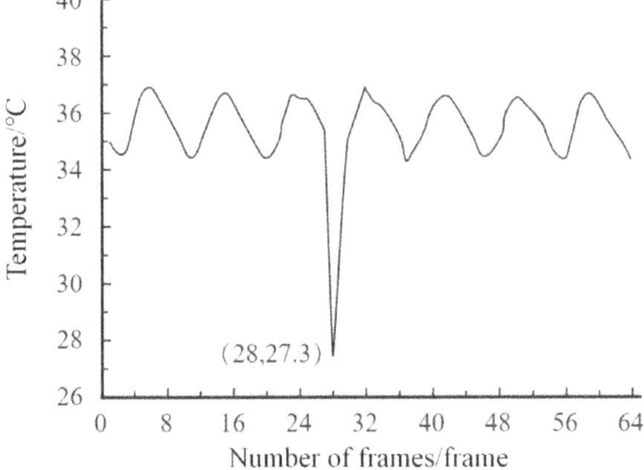

Fig. 3.7. Temperature change of drive shaft measurement point during pedestrian pass.

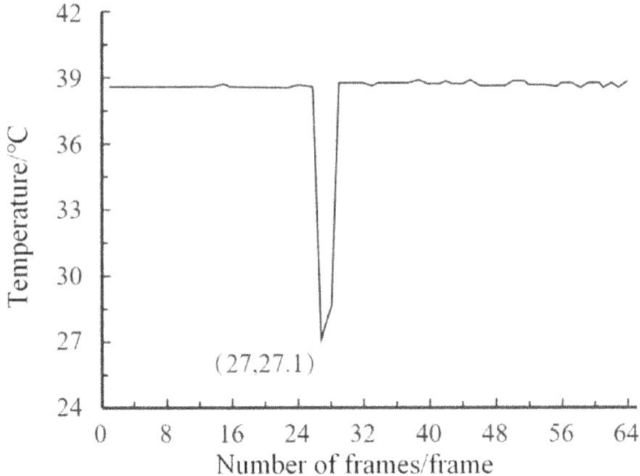

Fig. 3.8. Temperature change of the motor measuring point during pedestrian passage.

Figure 3.6 shows that the average temperature of the pump head is 20.3°C, and the temperature increases to 26.0°C, 27.8°C, and 26.9°C in the 27th, 28th, and 29th frames, respectively, because of the pedestrians passing by. As shown in Fig. 3.7, the average temperature of the drive shaft measurement point is 35.6°C, and the temperature decreases to 27.3°C and 33.1°C in the 28th and 29th frames, respectively, because of the pedestrian passing. From Fig. 3.8, it can be seen that the average temperature of the motor measurement point

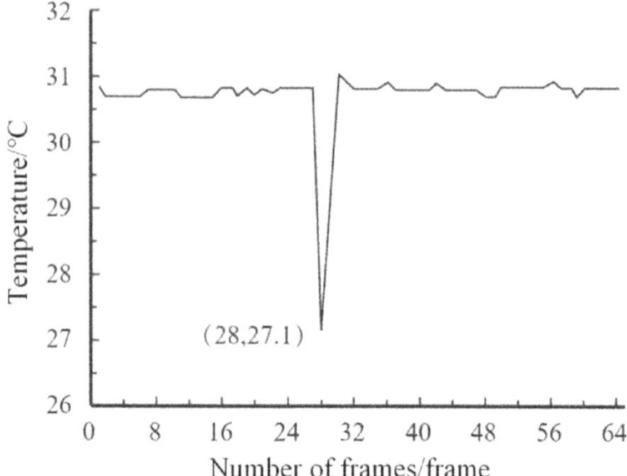

Fig. 3.9. Temperature change at the power end of the measurement point during pedestrian passage.

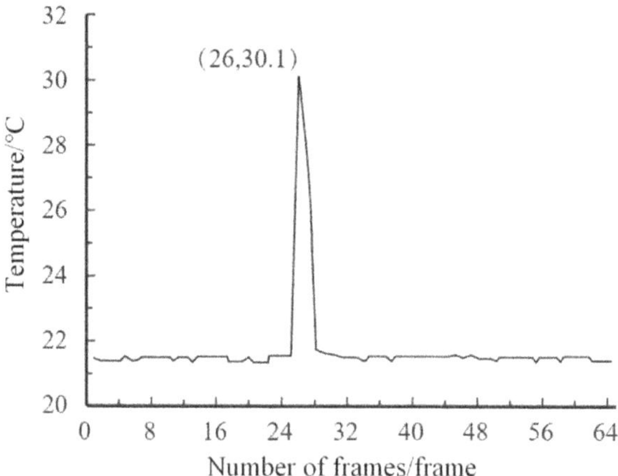

Fig. 3.10. Temperature change of the measurement point at the output during pedestrian passage.

is 38.7°C, and the temperature is reduced to 27.1°C and 28.7°C in the 27th and 28th frames, respectively, because of the pedestrians passing by. From Fig. 3.9, it can be seen that the average temperature of the measurement point of the power end is 30.8°C, because the pedestrian passes through, the temperature decreases to 27.1°C and 28.7°C in the 28th and 29th frames, respectively. From Fig. 3.10, it can be seen that the average temperature at the output point is

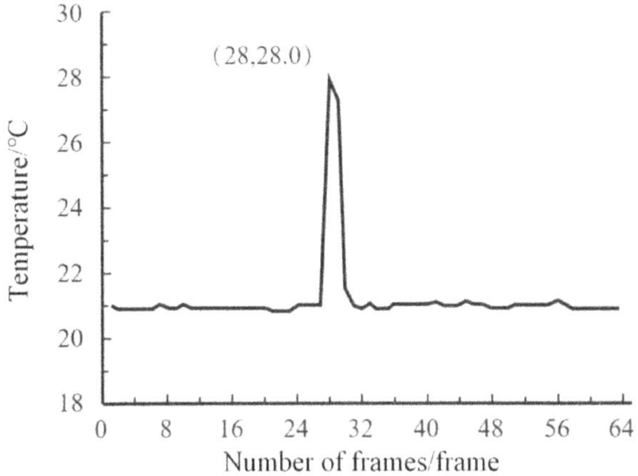

Fig. 3.11. Temperature change of the measurement point at the input during pedestrian passage.

$21.5°$C, and the temperature increases to $30.1°$C and 27.9℃ at frames 26 and 27, respectively, because of the pedestrians passing by. From Fig. 3.11, the average temperature of the measurement point at the input is $21.0°$C, and the temperature increases to $28.0°$C and $27.3°$C at frames 28 and 29, respectively, due to the passage of pedestrians.

In order to eliminate the negative impact of the accidental passage of pedestrians (foreign object intrusion) on the results of infrared thermal imaging monitoring, to ensure the accuracy of the information reflected in the infrared thermal imaging map, and to reduce the rate of false alarms in the monitoring system, it is necessary to clean the collected infrared thermal imaging monitoring data.

(1) Infrared thermography foreign object migration moment detection

(i) Constructing a video conversion data matrix for infrared thermal imaging monitoring

The infrared thermal imaging video is converted frame by frame into a 288×384 data matrix $\boldsymbol{X_p}$, $p = 1, 2, \ldots, 63, 64$, and is arranged in frame order to form a data matrix $\boldsymbol{V}$, which is a large $18,432 \times 384$ matrix, as described in Eq. (3.11):

$$\boldsymbol{V}_{\mathbf{18432 \times 384}} = \begin{pmatrix} X_1 & X_2 & X_3 & \cdots & X_p & \cdots & X_{64} \end{pmatrix}^{T}. \qquad (3.11)$$

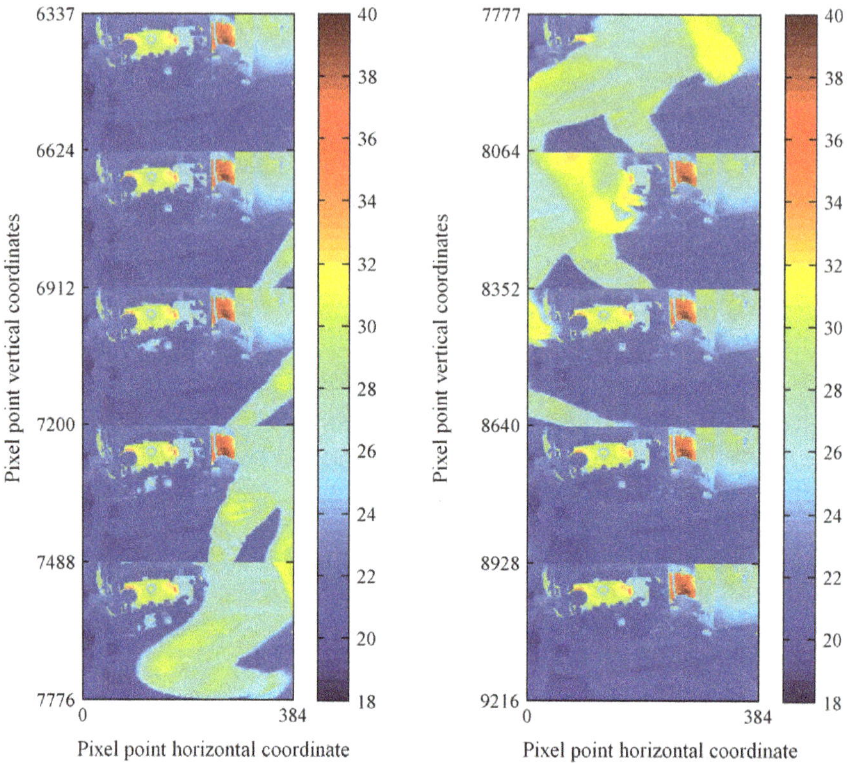

Fig. 3.12. Frame 23 to frame 32 continuous image capture segments.

The data matrix V is imaged to form a contiguous image, i.e., the frames are arranged in frame order, as shown in Fig. 3.12 (the illustration shows an extract of the contiguous image from frame 23 to frame 32).

(ii) Constructing special positional transformation data vectors

As can be seen from the case scenario, the foreign object enters the field of view from the right boundary of the lens, then the 384th column vector S_{384} corresponding to the right boundary should be selected from the data matrix V. Then S_{384} can be regarded as a long vector composed of 64 column vectors of 288×1 sequentially spliced together (the construction process is shown in Fig. 3.3), or it can be regarded as a column vector of $18,432 \times 1$ (the construction process is shown in Fig. 3.4).

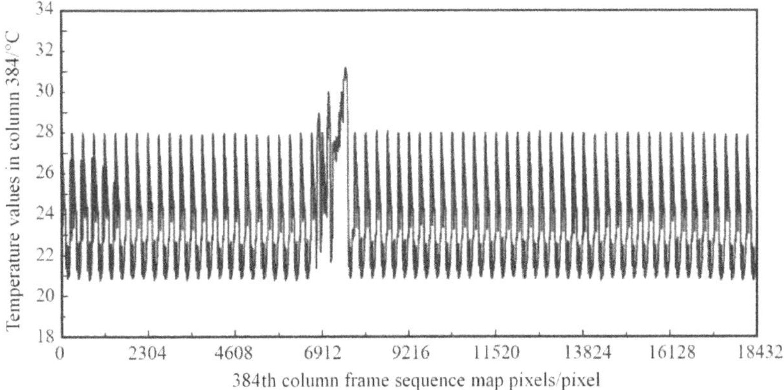

Fig. 3.13. Column 384 vector pixel point temperature value change.

(iii) Generate a graph of the values of the frame sequence vector $\boldsymbol{S}$

The variation of the temperature value of each element of the column vector $\boldsymbol{S}_{384}$ with the pixel point order is shown in Fig. 3.13. Under normal conditions, the image of the temperature values of each element of the column vector $\boldsymbol{S}_{384}$ can be approximated as a periodic curve with a period of 288 pixels. When a pedestrian passes by, the shape of the curve of each element temperature value changes abnormally, which means that the temperature value of the corresponding segment pixel point changes abnormally.

As shown in Fig. 3.13, the starting pixel point of the abnormal change of the curve shape when the pedestrian passes is 6851, and the conversion of 288 pixels as a period shows that the abnormal change occurs in the 24th period, i.e., the 24th frame of the image.

The ending pixel point of the anomalous change in the curve pattern is 7617, and the conversion of 288 pixel points as a period shows that the anomalous change ends in the 27th period, i.e., the 27th image frame.

The 24th, 25th, 26th, and 27th cycles of the column vector S384 are selected, and the variation of the temperature values of its elements with the number of pixel points is shown in Fig. 3.14. In addition, the 23rd (the cycle before the anomaly), 28th (the cycle after the anomaly) and randomly selected 1st, 20th, 40th, and 60th cycles of S384 are used as reference comparisons, and the variations of the

     Intelligent Early Warning of Risks in Complex Systems

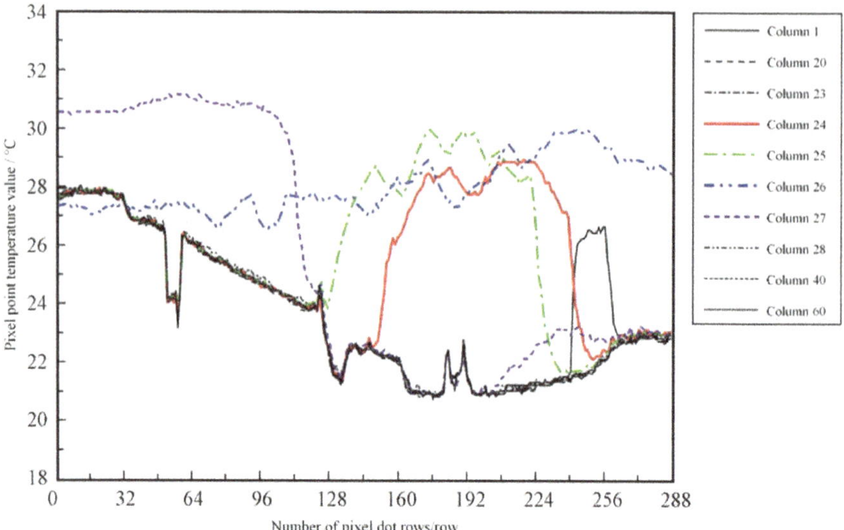

Fig. 3.14. Column 384 comparison of the variation curves of pixel point temperature values for selected cycles of the vector.

temperature values of each element with the number of pixel points are shown in Fig. 3.14.

From Fig. 3.14, it can be seen that the 24th, 25th, 26th, and 27th cycle value curve segments have obvious deviations from the 1st, 20th, 23rd, 28th, 40th, and 60th cycles, which indicates that anomalies have occurred in the 24th, 25th, 26th, and 27th frames of the infrared thermography video corresponding to the 24th, 25th, 26th, and 27th cycle value curve segments.

(iv) Numerical curve shape similarity analysis

Intercepting the numerical curve segments of any cycle in S_{384} as a reference benchmark, and calculating the Euclidean distance between it and the numerical curve segments of other cycles, the similarity of the morphology of the numerical curve segments of each cycle can be known. The larger the similarity, the smoother the temperature change of the corresponding position of the infrared thermal imaging signal in the neighboring frames. The smaller the similarity, the greater the change in the temperature value of the corresponding position of the infrared thermal imaging signal in the neighboring frame, which may be related to the intrusion of foreign objects.

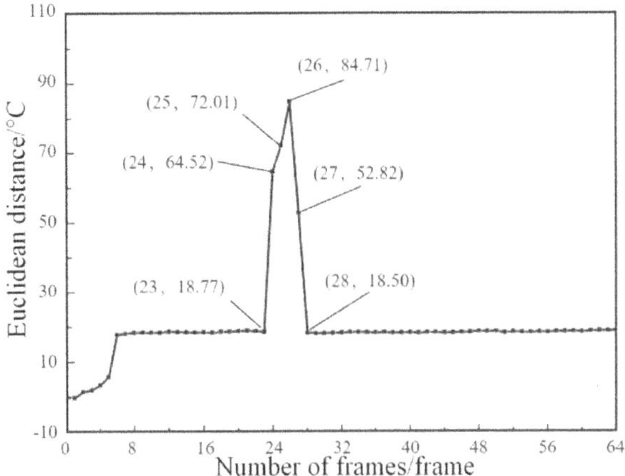

Fig. 3.15. Cycle 1 numerical curve segment Euclidean distance.

If the Euclidean distance is calculated between the segment of the 1st cycle numerical curve (pixel points 0 to 288, corresponding to column 384 of the data matrix of the 1st frame) and the segment of the 2nd to 64th cycle numerical curves, the results are shown in Fig. 3.15.

If we take the 64th cycle numerical curve segment (pixel points 18145~18432, corresponding to the 384th column of the 64th frame data matrix) as the base, we calculate the Euclidean distance between it and the 1st~63rd cycle numerical curve segments respectively, and the result is shown in Fig. 3.16.

From Fig. 3.15, it can be seen that the Euclidean distance between the numerical curve segments of the 1st cycle and the 23rd cycle is 18.77°C, and the Euclidean distance with the 24th cycle is 64.52°C, which indicates that the numerical curve segments of the 24th cycle have an abnormal change; The Euclidean distance between the numerical curve segments of cycle 1 and cycle 28 is 18.50°C, and the Euclidean distance with cycle 27 is 52.82°C, which indicates that the numerical curve segments of cycle 27 still have anomalous variations, but the anomalies of the numerical curve segments of cycle 28 disappeared and returned to normal. Then the 24th, 25th, 26th, and 27th cycles of the numerical curve segments corresponding to the 24th, 25th, 26th, and 27th frames of the infrared thermograms may have the presence of foreign objects.

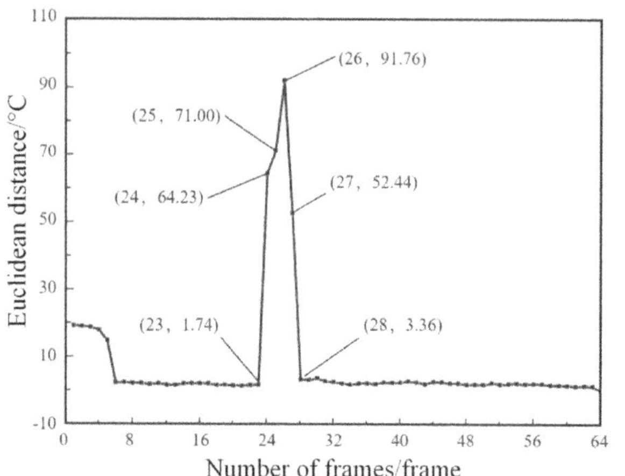

Fig. 3.16. Cycle 64 numerical curve fragment Euclidean distance.

From Fig. 3.16, it can be seen that the Euclidean distance between the segment of the numerical curve of the 64th cycle and the 23rd cycle is 1.74°C, and the Euclidean distance with the 24th cycle is 64.23°C, which indicates that the segment of the numerical curve of the 24th cycle has an abnormal change; The Euclidean distance between the segment of the numerical curve of cycle 64 and cycle 28 is 3.36°C, and the Euclidean distance with cycle 27 is 52.44°C, which means that the segment of the numerical curve of cycle 27 still has an abnormal change, but the segment of the numerical curve of cycle 28 has disappeared abnormally and returned to normal. Then, there may be foreign objects in the 24th, 25th, 26th, and 27th infrared thermograms corresponding to the 24th, 25th, 26th, and 27th cycles of the numerical curve segments.

In addition, as shown in Figs. 3.15 and 3.16, the Euclidean distances between the segments of the numerical curves of the 1st to 5th cycles and the reference frame have more obvious differences compared to the other cycles, and it is not possible to determine whether the anomalies exist in the 1st to 5th cycles or the 6th to 64th cycles. This phenomenon shows that when a segment of the numerical curve of a certain cycle is used as a reference standard to measure the similarity of the segments of the numerical curves of each cycle and to determine the location of the anomalous changes, there is a tendency to have ambiguities or paradoxes due to the existence of anomalies in the reference standard itself. This requires that there is absolutely no foreign matter in the infrared thermal imaging map in the frame where the period value curve segment as the reference frame is located, i.e., it is necessary to select and judge the selected period value curve segment of the reference frame.

In order to solve the above problems and improve the applicability and accuracy of the identification and determination method, it can be improved by calculating the Euclidean distance between any two segments of periodic numerical curves and generating the Euclidean distance matrix.

Calculate the Euclidean distance between any two segments of periodic numerical curves in S_{384}, and get the 64×64 Euclidean distance matrix $Eu_{64 \times 64}$, as shown in Fig. 3.17. Based on the Euclidean distance matrix $Eu_{64 \times 64}$, the three-dimensional surface and plane contour maps are plotted, as shown in Figs. 3.18 and 3.19, respectively. Calculate the average value of each column (row) element of the Euclidean distance matrix $Eu_{64 \times 64}$, as shown in Fig. 3.20.

From Figs. 3.17–3.20, it can be seen that the Euclidean distance of the numerical curve segments in the 24th cycle shows a

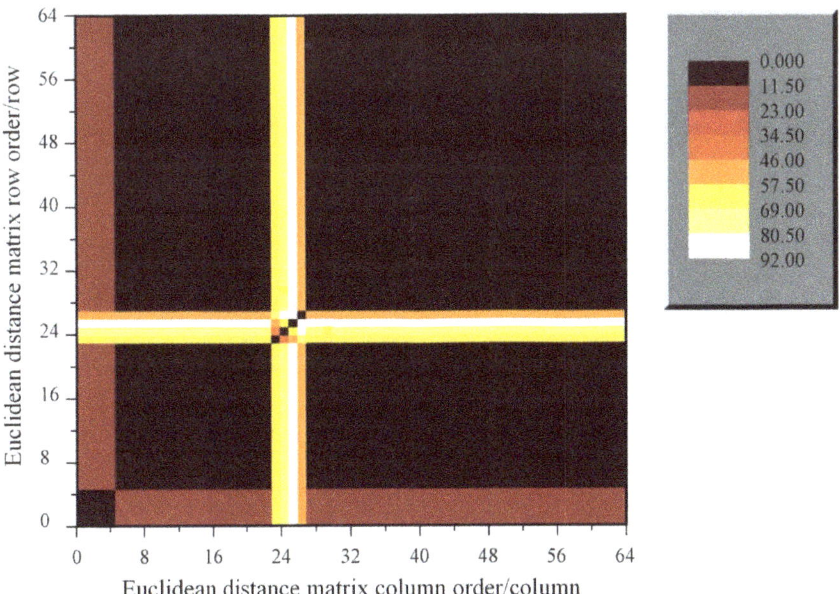

Fig. 3.17. Column 384 numerical curve fragment Euclidean distance matrix.

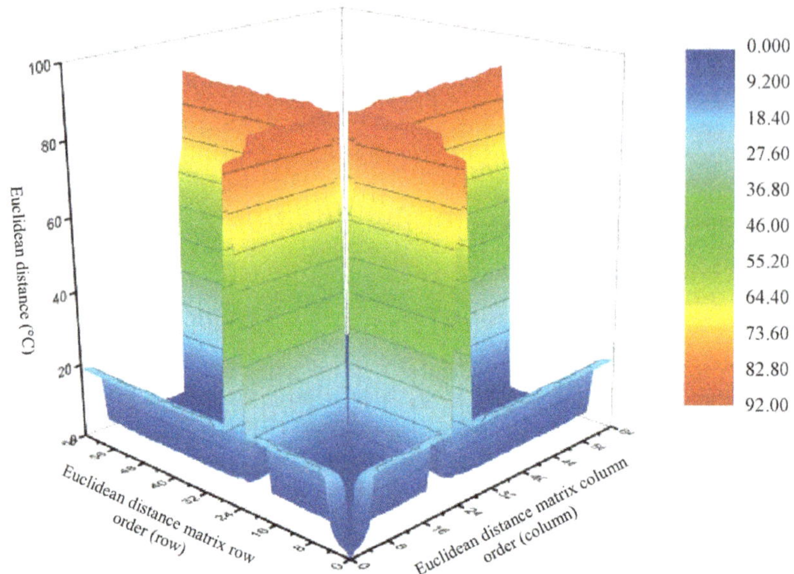

Fig. 3.18. Column 384 numerical curve fragment Euclidean distance matrix 3D surface plots.

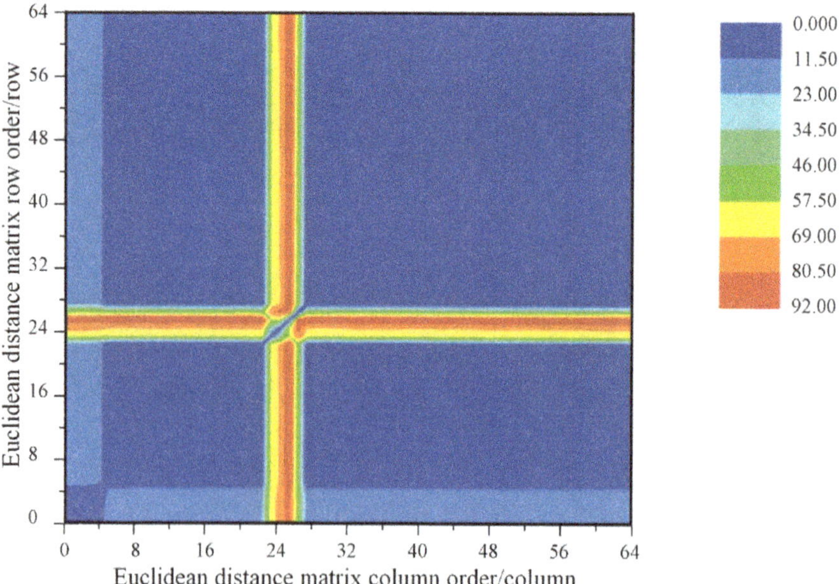

Fig. 3.19. Column 384 numerical curve fragment Euclidean distance matrix plane contour map.

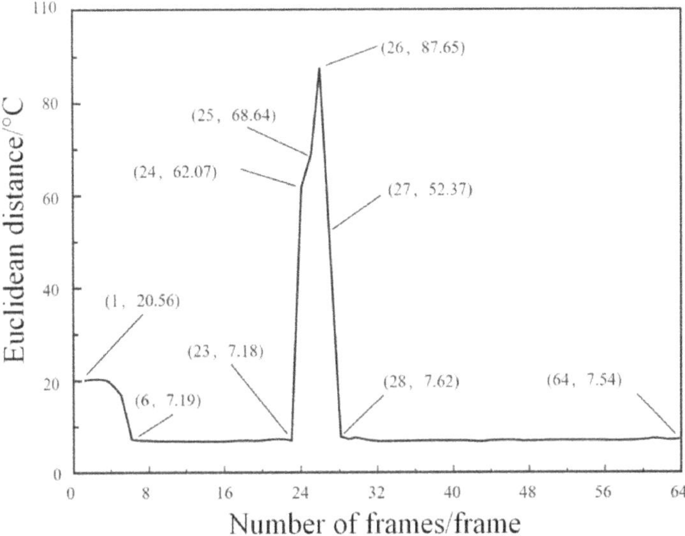

Fig. 3.20. Column 384 average Euclidean distance of segments of numerical curves for each cycle.

significant increase compared with that of the 23rd cycle (the average value increases from $7.18°C$ to $62.07°C$), and the degree of similarity decreases greatly; the Euclidean distance of the numerical curve segments in the 26th cycle is the largest (the average value is $87.65°C$), which indicates that the right border of the current frame image corresponds to the largest vector anomaly change; The Euclidean distance of the numerical curve segments in the 28th cycle decreased significantly compared with that in the 27th cycle (the mean value decreased from $52.37°C$ to $7.62°C$), and the degree of similarity increased greatly. The Euclidean distance has returned to normal in the 28th cycle, which indicates that there may be a foreign object in the 24th to 27th infrared thermograms corresponding to the 24th to 27th segments of the numerical curves in the 24th to 27th cycles. The Euclidean distances between the segments of the numerical curves of the 1st to 5th cycles and the segments of the numerical curves of the other cycles are large, i.e., the degree of similarity is low, which indicates that there may also be a foreign body in the 1st to 5th frames of the infrared thermograms corresponding to the numerical curves' segments of the 1st to 5th cycles.

(v) Second special position analysis validation

The boundary crossing detection based on the special position of the boundary can accurately identify the instantaneous behavior of the foreign object entering the lens field of view, but it is not possible to determine whether its movement trend will form an actual interference with the main part of the monitoring object in the lens field of view, and thus it is necessary to select the second special position to analyze and validate the results of the recognition of the intrusion of the foreign object based on the special position of the boundary.

Select the 375th column vector S_{375} corresponding to the second special position near the right boundary in the data matrix V. Then, S_{375} can be regarded as a long vector consisting of 64 column vectors of 288×1 sequentially spliced together (the construction process is shown in Fig. 3.3), or can be regarded as a column vector of 18432×1 (the construction process is shown in Fig. 3.4). Calculate the Euclidean distance between any two segments of periodic numerical curves for S_{375}, and obtain a 64×64 Euclidean distance matrix, as shown in Fig. 3.21. Based on the Euclidean distance matrix, a three-dimensional surface map and a planar contour map are plotted, as shown in Figs. 3.22 and 3.23, respectively. Calculate the average value of each column (row) of the Euclidean distance matrix, as shown in Fig. 3.24.

From Figs. 3.21–3.24, it can be seen that the Euclidean distance of the numerical curve segments in the 24th cycle shows a significant increase compared with that of the 23rd cycle (the average value increases from $5.81°C$ to $64.09°C$), and the degree of similarity decreases greatly; the Euclidean distance of the numerical curve segments in the 26th cycle is the largest (the average value is $92.52°C$), which indicates that the right border of the current frame image corresponds to the largest vector anomaly change; The Euclidean distance of the numerical curve segments in the 28th cycle decreased significantly compared with that in the 27th cycle (the mean value decreased from $41.94°C$ to $6.65°C$), and the degree of similarity increased greatly. The Euclidean distance has returned to normal in the 28th cycle, which indicates that there may be foreign objects in the 24th to 27th infrared thermograms corresponding to the 24th to 27th segments of the numerical curves in the 24th to 27th cycles.

The average Euclidean distance of the numerical curve segments of the 1st cycle is $12.72°C$, and the average Euclidean distances of

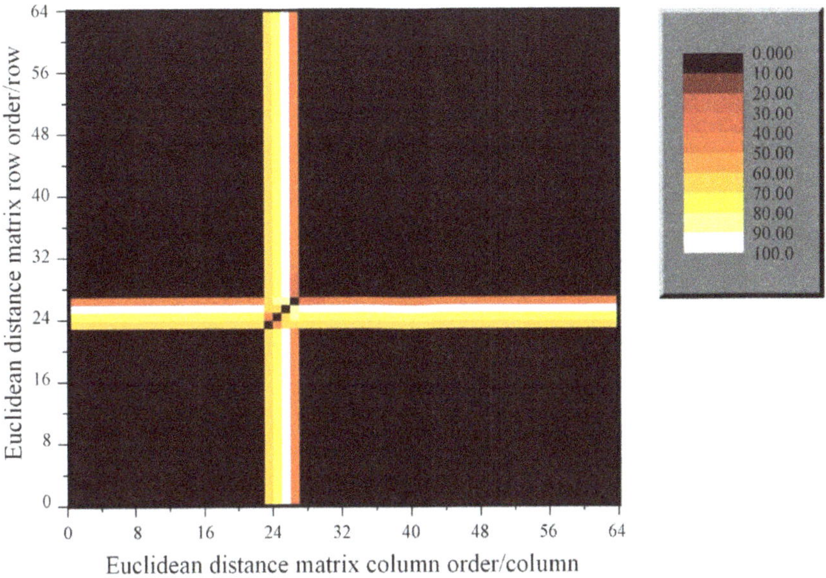

Fig. 3.21. Column 375 numerical curve fragment Euclidean distance matrix.

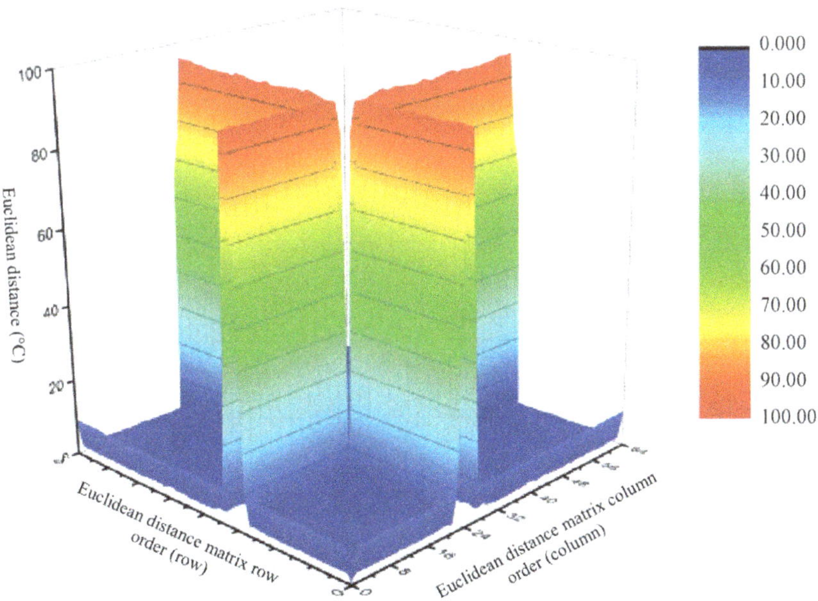

Fig. 3.22. Column 375 numerical curve fragment Euclidean distance matrix 3D surface map.

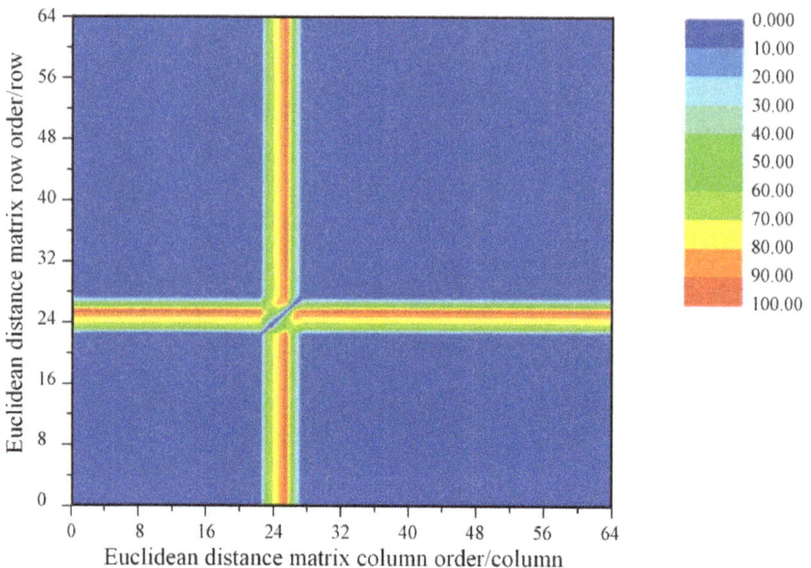

Fig. 3.23. Column 375 numerical curve fragment Euclidean distance matrix plane contour map.

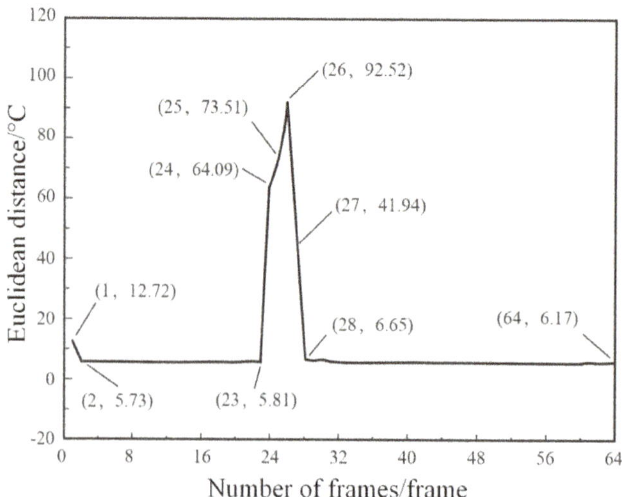

Fig. 3.24. Column 375 average Euclidean distance of segments of numerical curves for each cycle.

the numerical curve segments of the 2nd to 5th cycles are 5.73°C, 5.97°C, 5.93°C, and 5.80°C, respectively, which are similar to the average Euclidean distances of the curve segments of the majority of the cycles (except for the 1st and 24th to 27th cycles), which indicates that in the images of the 1st to 5th frames corresponding to the 1st to 5th cycles, although there is a foreign body present at the boundary position, it does not continue to move towards the center of the lens field of view, i.e., it does not constitute an actual interference with the monitoring subject and may not be cleaned.

(vi) Anomaly data cleansing and reconstruction

If data cleaning is performed for the right border, i.e., there is a foreign object on the right border of the 24th to 27th frame of the infrared thermogram. The frame sequence information is fed back to the monitoring system, and the data information in the video transformation data matrix $V_{18432\times384}$ from row 6625 (row 1 of the image transformation matrix of frame 24) to row 7776 (row 288 of the image transformation matrix of frame 27) is deleted, and the original row 6624 is spliced with the vector of the original row 7777 to form a new video transformation data matrix $V_{17280\times384}$. After cleaning, the numerical curves of the 384th column corresponding to the special position of the right boundary are shown in Fig. 3.25. The Euclidean distance matrix of the segments of each numerical curve is shown in Fig. 3.26.

After completing the cleaning, the average value of the Euclidean distance matrix is 4.12°C, which is 65.32% lower than the average value of the Euclidean distance of the original data which is 11.88°C, indicating that the similarity of each segment of the frame order vector numerical curves corresponding to the special position of the right boundary is improved after the data cleaning.

After monitoring the video splicing reconstruction, the average Euclidean distances of the original 23rd and 28th cycles of the right boundary numerical curves are 3.04°C and 3.59°C, respectively, with a standard deviation of 0.28°C, which is 98.98% lower than that of the standard deviation of the average Euclidean distances of the original 23rd and 24th cycles of the original uncleaned video with a standard deviation of 27.45°C; Compared with the standard deviation of 22.38°C between the mean Euclidean distances of the original 27th and 28th cycles, the standard deviation is reduced by 98.75%, which

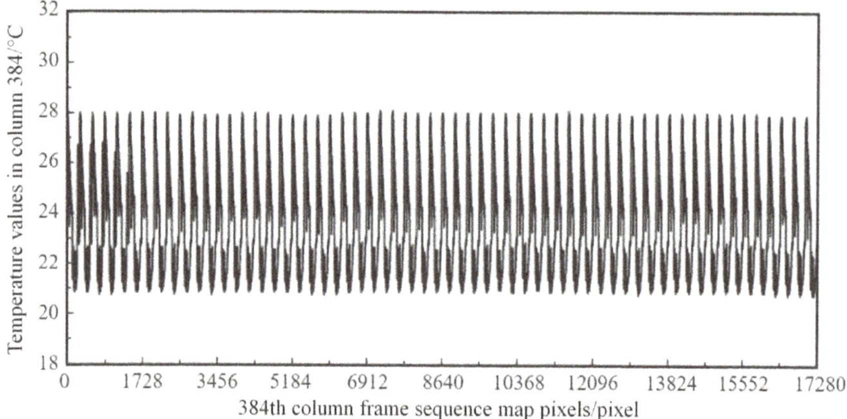

Fig. 3.25. Change in temperature values of pixel points in column 384 vector after data cleaning.

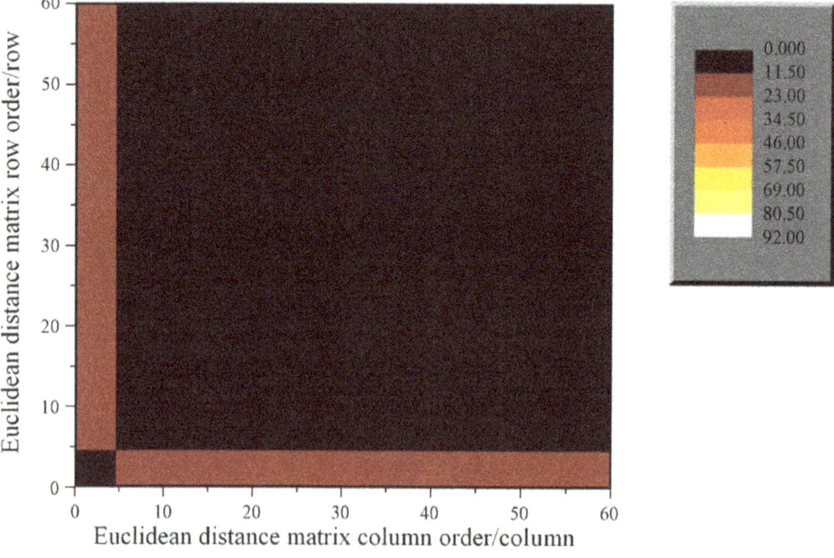

Fig. 3.26. Euclidean distance matrix of segments of numerical curves in column 384 after data cleaning.

indicates that the difference of temperature values at each point of the neighboring frames after the cleaning and reconstruction process has been greatly reduced, i.e., the change of temperature values of the neighboring frames tends to be continuous and smooth, and there is no more anomalous object interfering with the monitoring process.

(2) Infrared thermography foreign body removal moment detection

(i) Constructing a video conversion data matrix for infrared thermal imaging monitoring

The infrared thermal imaging video is converted frame by frame into a 288×384 data matrix $\boldsymbol{X_p}$, $p = 1, 2, \ldots, 63, 64$, and is arranged in frame order to form a data matrix $\boldsymbol{V}$, which is a large $18{,}432 \times 384$ matrix, as described in Eq. (3.12):

$$\boldsymbol{V}_{\mathbf{18432 \times 384}} = \left(X_1\ X_2\ X_3\ \cdots\ X_p\ \cdots\ X_{64} \right)^T. \qquad (3.12)$$

The data matrix $\boldsymbol{V}$ is imaged to form a contiguous image, i.e., the frames are arranged in frame order, as shown in Fig. 3.12 (the illustration shows an extract of the contiguous image from frame 23 to frame 32).

(ii) Constructing special positional transformation data vectors

As can be seen from the case scenario, the foreign object moves out of the field of view from the left boundary of the lens, then the first column vector $\boldsymbol{S_1}$ corresponding to the left boundary should be selected from the data matrix $\boldsymbol{V}$. Then $\boldsymbol{S_1}$ can be regarded as a long vector composed of 64 column vectors of 288×1 spliced together sequentially (the construction process is shown in Fig. 3.3), or it can be regarded as a column vector of $18{,}432 \times 1$ (the construction process is shown in Fig. 3.4).

(iii) Generate a graph of the values of the frame sequence vector $\boldsymbol{S}$

The variation of the temperature value of each element of column vector $\boldsymbol{S_1}$ with the pixel point order is shown in Fig. 3.27. Under normal conditions, the image of the temperature values of each element in column vector $\boldsymbol{S_1}$ can be approximated as a periodic curve with a period of 288 pixels. When a pedestrian passes by, the shape of the curve of each element's temperature value changes abnormally, i.e., the temperature value of the pixel point in the corresponding segment changes abnormally.

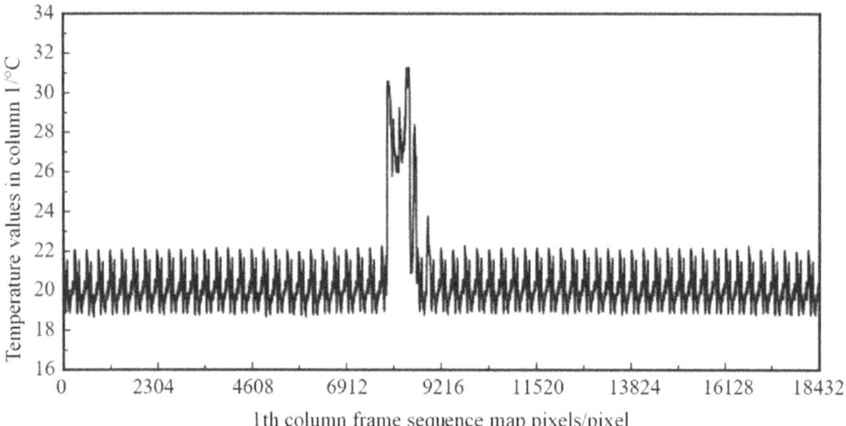

Fig. 3.27. Change in temperature value of pixel point in column 1 vector.

As shown in Fig. 3.27, the starting pixel point of the abnormal change of the curve shape when a pedestrian passes is 7913 and converted into 288 pixels, the abnormal change occurs in the 28th cycle, i.e., in the 28th frame of the image.

The ending pixel point of the anomalous change in the curve pattern is 8912, and the conversion of 288 pixel points to a period shows that the anomalous change ends in the 31st period, i.e., the 31st frame of the image.

The 28th, 29th, 30th, and 31st cycles of the column vector S_1 are selected, and the variation of the temperature values of its elements with the pixel point order is shown in Fig. 3.28. In addition, the 27th (the cycle before the anomalous change appears), 32nd (the cycle after the anomalous change ends) cycles, and the randomly selected 1st, 20th, 40th, and 60th cycles are selected as the reference comparisons, and the changes of their respective elemental temperature values with the pixel point ordinates are shown in Fig. 3.28.

From Fig. 3.28, it can be seen that there are significant deviations in the numerical curve segments of the 28th, 29th, 30th, and 31st cycles compared to the 1st, 20th, 27th, 32nd, 40th, and 60th cycles, which indicates that anomalies have occurred in the 28th, 29th, 30th, and 31st frames of the infrared thermography video corresponding to the numerical curve segments of the 28th, 29th, 30th, and 31st cycles.

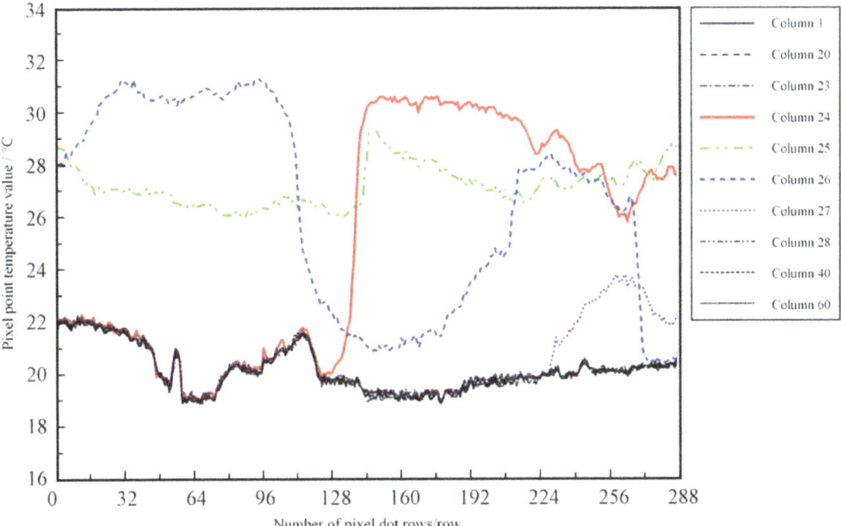

Fig. 3.28. Comparison of the variation curves of pixel temperature values for selected cycles of column 1 vectors.

(iv) Numerical curve shape similarity analysis

The similarity of the morphology of the numerical curve segments of each cycle can be known by intercepting the numerical curve segments of any cycle in S_1 as a reference frame and calculating the Euclidean distance between them and the numerical curve segments of other cycles. The larger the similarity, the smoother the temperature change of the corresponding position of the infrared thermal imaging signal in the neighboring frames. The smaller the similarity, the greater the change in the temperature value of the corresponding position of the infrared thermal imaging signal in the neighboring frame, which may be related to the intrusion of foreign objects.

If the Euclidean distance between the 1st cycle numerical curve segment (pixel points 0 to 288, corresponding to column 1 of the 1st frame data matrix) and the 2nd to 64th cycle numerical curve segments are calculated, the results are shown in Fig. 3.29. If the Euclidean distance between the 64th cycle segment ($18145 \sim 18432$ pixels, corresponding to column 1 of the 64th frame data matrix) and the 1st to 63rd cycle segment is calculated based on the 64th cycle segment, the result is shown in Fig. 3.30.

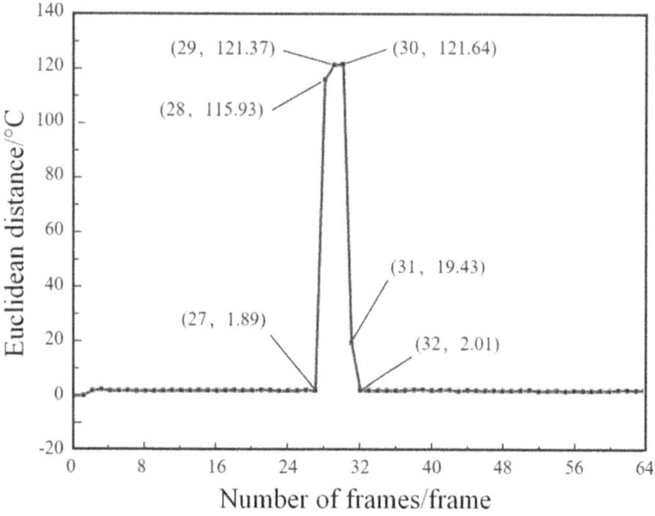

Fig. 3.29. Cycle 1 numerical curve segment Euclidean distance.

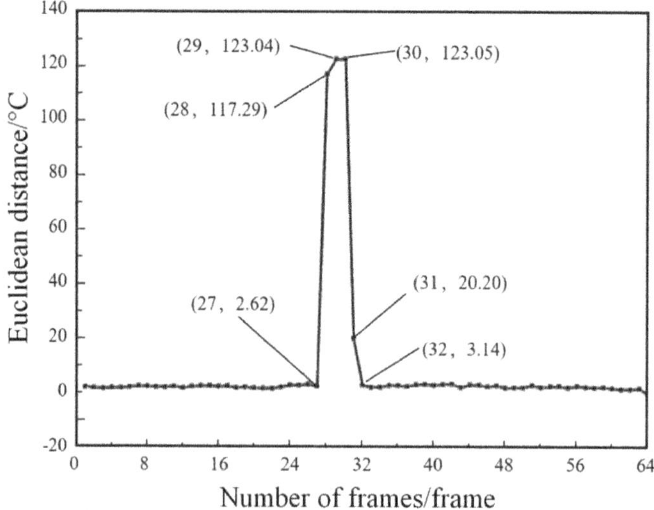

Fig. 3.30. Cycle 64 numerical curve segment Euclidean distance.

From Fig. 3.29, it can be seen that the Euclidean distance between the numerical curve segments of the 1st cycle and the 27th cycle is 1.89°C, and the Euclidean distance with the 28th cycle is 115.93°C, which indicates that the numerical curve segments of the 28th cycle have an abnormal change; The Euclidean distance between the numerical curve segments of cycle 1 and cycle 32 is 2.01°C, and the Euclidean distance with cycle 31 is 19.43°C, which means that the numerical curve segments of cycle 31 still have abnormal changes, but the numerical curve segments of cycle 32 disappear abnormally and return to normal. Then, there may be foreign objects in the 28th, 29th, 30th, and 31st infrared thermograms corresponding to the 28th, 29th, 30th, and 31st cycles of the numerical curve segments.

From Fig. 3.30, it can be seen that the Euclidean distance of cycle 64 from the numerical curve segments of cycle 27 is 2.62°C, and the Euclidean distance from cycle 28 is 117.29°C, which indicates that there is an anomalous change in numerical curve segments of cycle 28. The Euclidean distance between the segment of the numerical curve of cycle 64 and cycle 32 is 3.14°C, and the Euclidean distance with cycle 31 is 20.20°C, which means that the segment of the numerical curve of cycle 31 still has an abnormal change, but the segment of the numerical curve of cycle 32 has disappeared abnormally and returned to normal. Then, there may be foreign objects in the 28th, 29th, 30th, and 31st infrared thermograms corresponding to the 28th, 29th, 30th, and 31st cycles of the numerical curve segments.

Calculate the Euclidean distance between any two segments of periodic numerical curves in S_1, and get the 64×64 Euclidean distance matrix $\boldsymbol{Eu}_{64 \times 64}$, as shown in Fig. 3.31. Based on the Euclidean distance matrix $\boldsymbol{Eu}_{64 \times 64}$, the three-dimensional surface and plane contour maps are plotted, as shown in Figs. 3.32 and 3.33, respectively. Calculate the average value of each column (row) element of the Euclidean distance matrix $\boldsymbol{Eu}_{64 \times 64}$, as shown in Fig. 3.34.

From Figs. 3.31–3.34, it can be seen that the Euclidean distance of the numerical curve segments in the 28th cycle increases significantly compared with that in the 27th cycle (the average value increases from 7.73°C to 113.61°C), and the degree of similarity decreases significantly; the Euclidean distance of the numerical curve segments in the 30th cycle is the largest (the average value is 118.96°C), which indicates that the magnitude of the anomalous change of the corresponding vector at the left boundary of the current frame image is the largest. The Euclidean distance of the numerical curve segments in cycle 31 showed a significant decrease compared with cycle 30 (the mean value decreased from 118.96°C to 23.54°C), and the

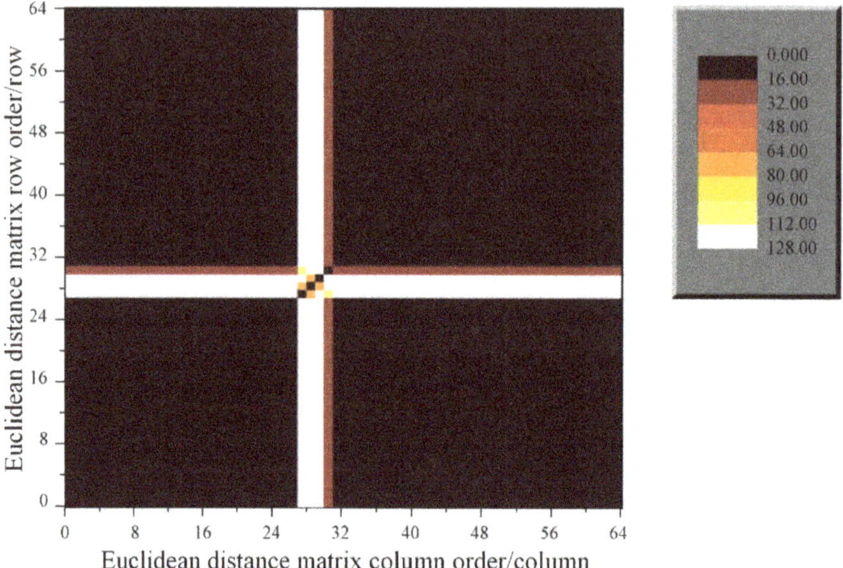

Fig. 3.31. Euclidean distance matrix for numerical curve segments in column 1.

degree of similarity increased dramatically. The Euclidean distance returned to normal in the 32nd cycle, indicating that there was a foreign body in the 28th–30th infrared thermograms corresponding to the numerical curve segments in the 28th–30th cycles. The average Euclidean distance between the numerical curve segments of the 31st cycle and the numerical curve segments of the other cycles is larger (23.54 °C), i.e., the degree of similarity is lower, which indicates that there may be a foreign body in the 31st frame of the infrared thermogram corresponding to the numerical curve segments of the 31st cycle as well.

(v) Second special position analysis validation

The boundary crossing detection based on the special position of the boundary can accurately identify the instantaneous behavior of the foreign object moving out of the lens field of view, but it is not possible to determine whether its movement tendency will form an actual interference with the main part of the monitoring object in the lens field of view, and thus it is necessary to select the second special position to analyze and verify the results of the recognition of

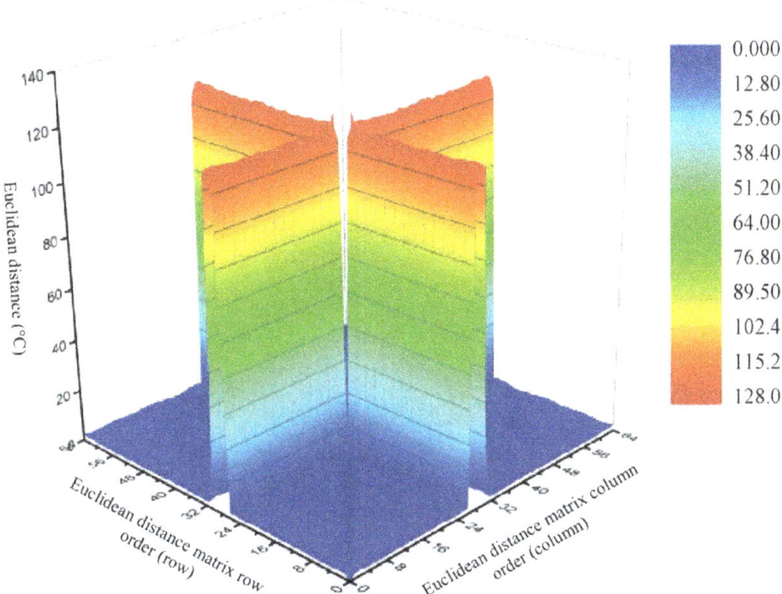

Fig. 3.32. Column 1 numerical curve fragment Euclidean distance matrix 3D surface plots.

the foreign object moving out of the special position of the boundary based on the special position of the boundary.

Select the 10th column vector S_{10} corresponding to the second special position near the left boundary in the data matrix V, then S_{10} can be regarded as a long vector consisting of 64 column vectors of 288×1 sequentially spliced together (the construction process is shown in Fig. 3.3), or can be regarded as a column vector of 18432×1 (the construction process is shown in Fig. 3.4). Calculate the Euclidean distance between any two segments of the periodic numerical curves for S_{10}, and obtain a 64×64 Euclidean distance matrix, as shown in Fig. 3.35. Based on the Euclidean distance matrix, a three-dimensional surface map and a planar contour map are plotted, as shown in Figs. 3.36 and 3.37, respectively. Calculate the average value of each column (row) of the Euclidean distance matrix, as shown in Fig. 3.38.

From Figs. 3.35–3.38, it can be seen that the Euclidean distances of the segments of the numerical curves in the 28th cycle showed a significant increase compared to the 27th cycle (the mean value

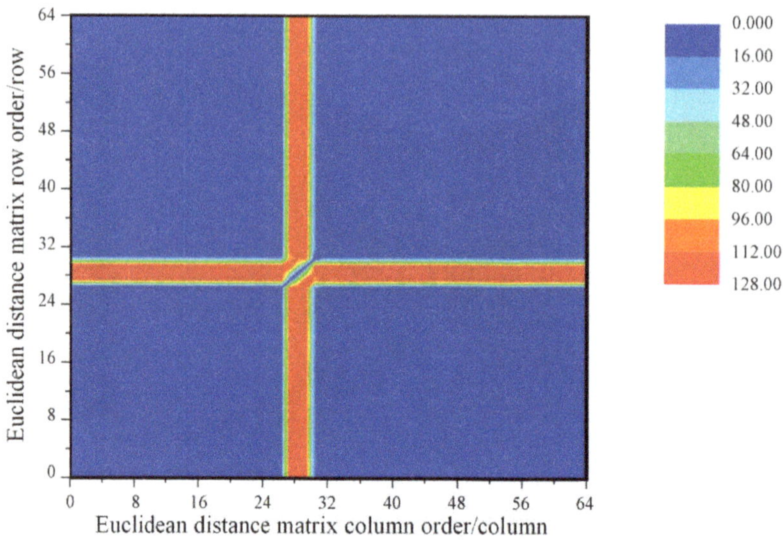

Fig. 3.33.   Column 1 numerical curve fragment Euclidean distance matrix plane contour map.

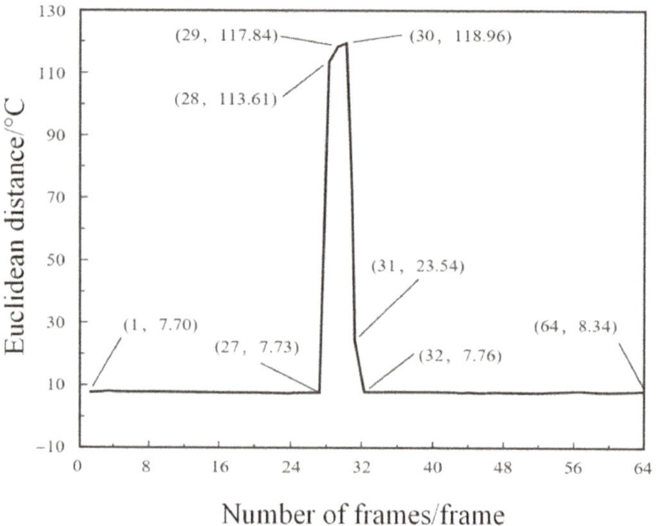

Fig. 3.34.   The average Euclidean distance of the segments of the numerical curves of each cycle in column 1.

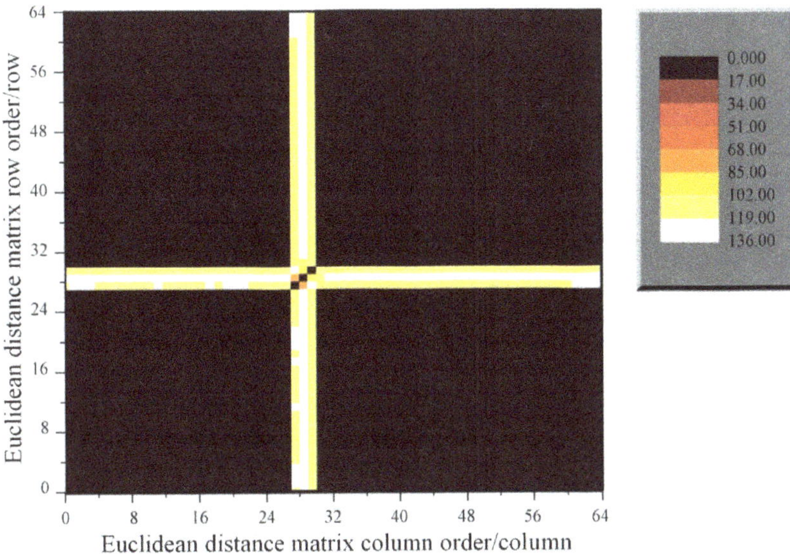

Fig. 3.35. Euclidean distance matrix for numerical curve segments in column 10.

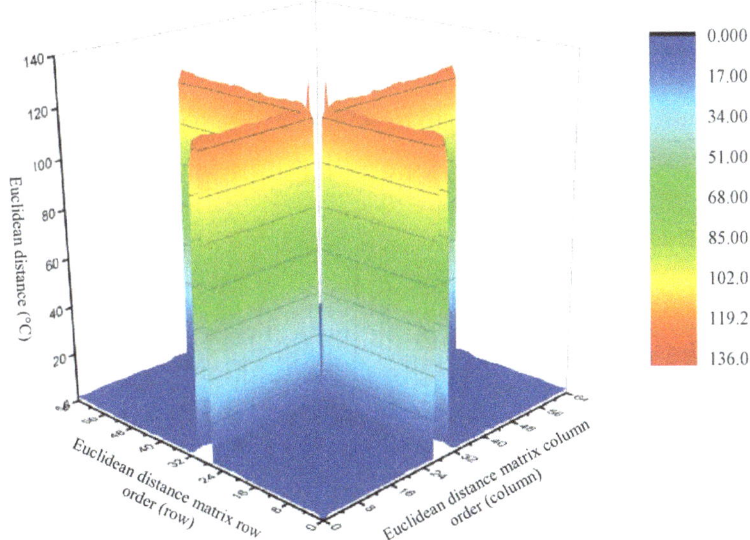

Fig. 3.36. Column 10 numerical curve fragment Euclidean distance matrix 3D surface plots.

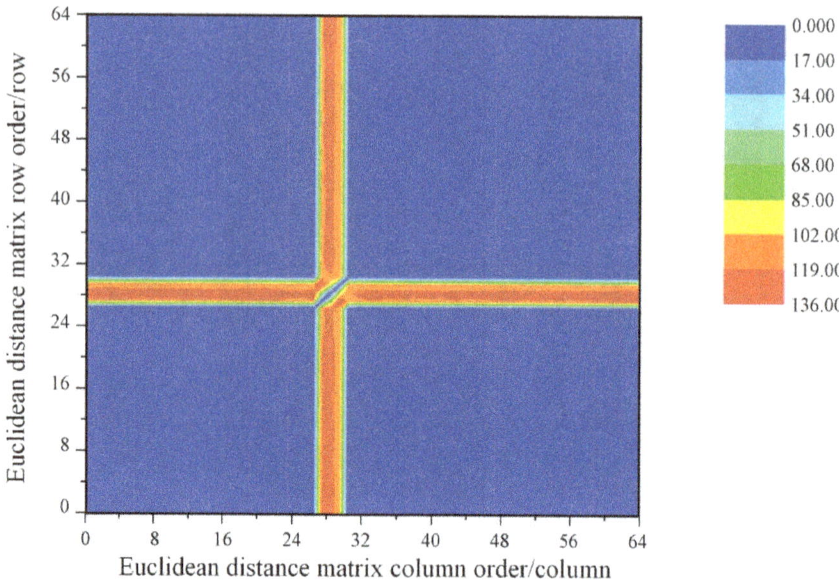

Fig. 3.37. Column 10 numerical curve fragment Euclidean distance matrix plane contour map.

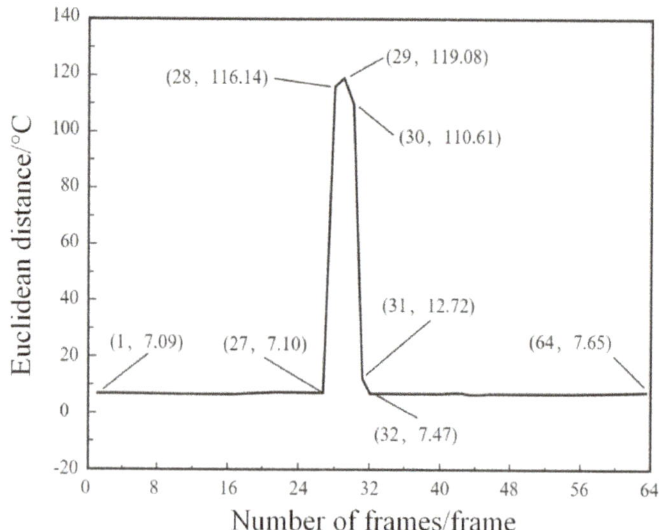

Fig. 3.38. The average Euclidean distance of the segments of the numerical curves for each period in column 10.

increased from $7.10°C$ to $116.14°C$), and the degree of similarity was greatly reduced. The Euclidean distance of the numerical curve segments in the 29th cycle is the largest (mean value $119.08°C$), indicating that the magnitude of the anomalous change of the corresponding vectors in the 10th column of the current frame image is the largest; the Euclidean distances of the numerical curve segments in the 31st cycle show a significant decrease compared with that of the 30th cycle (the mean value is lowered from $110.61°C$ to $12.72°C$), and the degree of similarity is greatly increased. The Euclidean distance of the numerical curve segments in the 31st cycle has returned to the normal level, indicating that there may be foreign objects in the 28th to 30th infrared thermograms corresponding to the numerical curve segments in the 28th to 30th cycles.

The average Euclidean distance of the numerical curve segments in the 31st cycle is $12.72°C$, which is obviously smaller than the average Euclidean distance of the numerical curve segments in the 31st cycle of the 1st column of the 31st cycle of the $23.54°C$, which means that in the 31st frame of the 31st cycle of the corresponding image, although there is a foreign body in the left border position, it has already had a tendency to move out of the center of the field of view of the lens, i.e., it does not constitute a real interference with the subject of the monitoring, so it can be excluded from the cleaning process.

(vi) Anomaly data cleansing and reconstruction

If data cleaning is performed on the left border, it is determined that there is a foreign object on the left border of the 28th to 30th frame of the infrared thermogram. Feedback the frame sequence information to the monitoring system, delete the data information of the video transformation data matrix $V_{18432 \times 384}$ from row 7777 (row 1 of the image transformation matrix of frame 28) to row 8640 (row 288 of the image transformation matrix of frame 30), and splice the vectors of row 7776 and row 8641 to form a new video transformation data matrix $V_{17568 \times 384}$. The cleaned data vector curves corresponding to the special positions of the left boundary in column 1 are shown in Fig. 3.39. The Euclidean distance matrix of the segments of the numerical curves is shown in Fig. 3.40.

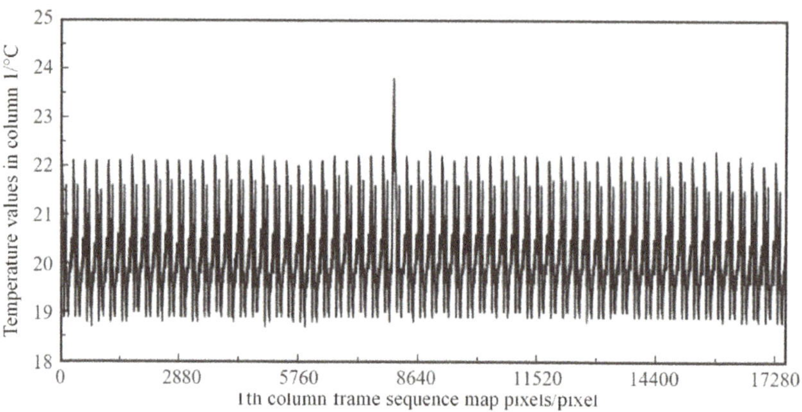

Fig. 3.39. Change in temperature value of pixel point in column 1 vector after cleaning.

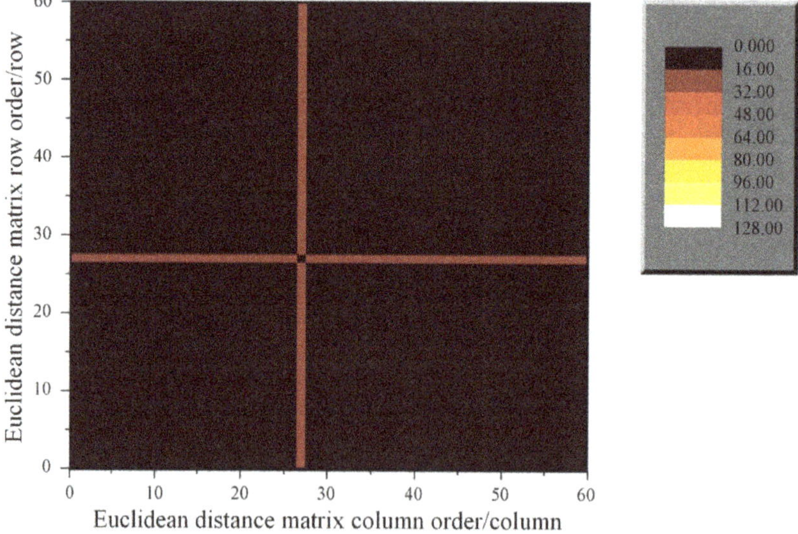

Fig. 3.40. Euclidean distance matrix for each segment of the cleaned column 1 data vector.

After completing the cleaning, the average value of the Euclidean distance matrix is $2.59°C$, which is 80.32% lower than the average value of the Euclidean distance of the original data, $13.16°C$. This indicates that the similarity of each segment of the frame order vector numerical curves corresponding to the special position of the left boundary has been improved after the data cleaning.

After monitoring the video splicing reconstruction, the average Euclidean distances of the original 27th and 31st cycles of the left boundary numerical curves are $2.23°C$ and $19.18°C$, respectively, and the standard deviation of the two is $8.48°C$, which is 83.98% lower than the standard deviation of the average Euclidean distances of the original 27th and 28th cycles of the original without the cleaning process, which is $52.94°C$; Compared with the standard deviation of $47.71°C$ between the average Euclidean distances of the original 30th and 31st cycles, the standard deviation is reduced by 82.23%, which indicates that the difference of temperature values at each point of the neighboring frames after the cleaning and reconstruction process has been greatly reduced, i.e., the change of temperature values of the neighboring frames tends to be continuous and smooth, and there is no more anomalous object interfering with the monitoring process.

(3) Identification of anomalous data and evaluation of cleaning results in infrared thermography monitoring

(i) Recognition accuracy analysis and evaluation

According to the detection results of foreign objects passing through the left and right borders, it can be determined that there are foreign objects in the infrared thermography from the 24th frame to the 30th frame. Among them:

(a) Right border crossing detection determines the presence of a foreign object in the 24th to 27th frame of the infrared thermogram. From the infrared thermal imaging monitoring video of the selected case (Fig. 3.12), it can be seen that in the 24th frame, the foreign object starts to move into the lens field of view from the right side, and at the 27th frame, the foreign object body passes through the right boundary of the lens field of view completely, and completes the moving-in action. In other words, there is a foreign object at the right border of the image in frames 24 to 27, and the process of moving in has a large impact on the observation of the main part of the object in the field of view of the lens, so it must be cleaned. It is proved that the method is correct in recognizing the moment of foreign object intrusion.

(b) Left border crossing detection determines the presence of a foreign object in the 28th to 30th frame of the infrared thermogram. From the infrared thermal imaging monitoring video of the selected case (Fig. 3.12), it can be seen that at the 28th frame, the foreign object starts to move out of the lens field of view from the left side, and at the 30th frame, the foreign object body passes through the left boundary of the lens field of view completely and completes the moving out the action. In other words, there is a foreign object at the left border of the image in frames 28 to 30, and the process of moving it out has a large impact on the observation of the main part of the object in the field of view of the lens, so it must be cleaned. It is proved that the method is correct in recognizing the moment of moving out of the foreign object.

(ii) Evaluation of recognition processing algorithm analysis

Compared with the traditional motion target detection method, the system needs to operate and analyze a total of 6,967,296 pixel points $(288 \times 384 \times 63 = 6,967,296)$ globally, the mentioned anomalous data identification method based on boundary-crossing detection only takes the boundary position as the research object, which is a total of 85,760 pixel points $(288 \times 2 + 384 \times 2 - 4 = 1340, 1340 \times 64 = 85760)$, with a reduction of 98.77% of the operation amount.

Let there be a total of t frames in the infrared thermal imaging monitoring video, then the traditional motion target detection method needs to operate and analyze a total of 110592 $(t - 1)$ pixel points globally. The proposed anomaly data recognition method based on boundary crossing detection only takes $1340t$ pixels at the boundary location as the object of study, and the computation reduction rate can be up to 98.79% in Eq. (3.13):

$$\frac{110592(t - 1) - 1340t}{110592(t - 1)} \times 100\% = \frac{109252(t - 1) - 1340}{110592(t - 1)} \times 100\%$$

$$= 98.79\% - \frac{1.21\%}{t - 1}. \tag{3.13}$$

(iii) Analysis and evaluation of the effectiveness of data cleaning and reconstruction

Feedback the frame sequence information to the monitoring system, delete the data information of the video transformation data matrix $V_{18432 \times 384}$ from row 6625 (row 1 of the 24th frame image transformation matrix) to row 8640 (row 288 of the 30th frame image transformation matrix), and splice the original row 6624 with the vector of the original row 8641 to form the new video transformation data matrix $V_{16416 \times 384}$. In terms of cleaning and reconfiguration effects:

(a) At the moment of foreign body intrusion, the average Euclidean distance of the 384th column corresponding to the right boundary of the lens field of view is 11.88°C. After the cleaning process, the average Euclidean distance is reduced to 4.12°C, which is 65.32% less than that before the process.

At the moment of foreign body removal, the mean Euclidean distance of the 1st column of data corresponding to the left boundary of the lens field of view is 13.16°C. After the cleaning process, the average Euclidean distance is reduced to 2.59°C, which is 80.32% less than that before the process.

It shows that the similarity of each segment of the frame order vector numerical curves corresponding to the special positions of the left and right boundaries is significantly improved after data cleaning.

(b) The standard deviation of the mean Euclidean distance between the original Cycle 23 and Cycle 28 (0.28°C) is 98.98% lower than that between the mean Euclidean distance between the original Cycle 23 and Cycle 24 without cleaning (27.45°C), and 98.75% lower than that between the mean Euclidean distance between the original Cycle 27 and Cycle 28 (22.38°C).

The standard deviation of the mean Euclidean distance between the original 27th and 31st cycle of the reconstructed data left boundary numerical curves (8.48°C) is reduced by 82.98% compared to the standard deviation between the mean Euclidean distances of the original 27th and 28th cycles without cleaning (52.94°C), and reduced by 82.23% compared to the standard deviation between the mean Euclidean distances of the original 30th and 31st cycles (47.71°C).

It means that the difference of temperature values at each point of adjacent frames after cleaning and reconstruction process appears to be greatly reduced, i.e., the change of temperature values of adjacent frames tends to be continuous and smooth, and there is no longer any abnormal object interfering with the monitoring process. The reconstructed data can maintain the continuity of the temperature change of the monitoring object based on the infrared thermal imaging observation, avoiding the problems of data distortion and loss of key information caused by excessive cleaning.

(4) Analysis and summary

In this section, to address the problem of foreign objects passing in front of the lens (intruding into or moving out of the field of view of the lens of the infrared thermal imager) during the infrared thermal imaging monitoring process of the shale gas fracking field equipment, which affects the monitoring results and triggers false alarms, we propose a method for recognizing and cleaning abnormal data during the infrared thermal imaging monitoring process of shale gas fracking equipment based on the transgressing boundary detection. By recognizing and feeding back the number of acquisition frames corresponding to the period of time when foreign objects exist in the lens field of view, deleting the relevant data segments, and retaining and splicing the normal infrared thermal imaging monitoring data segments without the presence of foreign objects, we can improve the accuracy of the infrared thermal imaging monitoring results of the shale gas fracturing equipment at the site and reduce the rate of false alarms.

(i) Compared with the conventional boundary-crossing detection for modeling, tracking, and detecting the two-dimensional characteristics of objects, the described method adopts the discriminative index based on the similarity of the numerical curves of the time-series one-dimensional vectors, and transforms the two-dimensional problem-oriented to the region into the one-dimensional problem-oriented to the boundary, and the reduction of the computation rate can reach up to 98.79%.

(ii) For a set case scenario, the said method can accurately recognize the video start frame (frame 24) in which the foreign object intrudes into the lens field of view and the video end frame (frame 30) in which the foreign object moves out of the lens field of view during the monitoring process of the case scenario, and feedback the relevant information about the anomalous monitoring data of the 24th to 30th frames.

(iii) After cleaning and reconstruction, the average Euclidean distance of column 384 corresponding to the right boundary of the lens field of view is reduced by 65.32% compared to the preprocessing period, and the average Euclidean distance of column 1 corresponding to the left boundary of the lens field of view is reduced by 80.32% compared to the preprocessing period, which indicates that the described method can significantly improve the similarity of the numerical curves of the boundary vectors of the images of neighboring frames.

(iv) The standard deviation of the average Euclidean distance of the reconstructed data at the spliced position of the numerical curve at the right boundary is reduced by more than 98% compared with that before processing, and the standard deviation of the average Euclidean distance at the spliced position of the numerical curve at the left boundary is reduced by more than 82% compared with that before processing, which indicates that the temperature values of the neighboring frames tend to be changed continuously and steadily after cleaning and reconstructing, and there is no more abnormal object interfering with the monitoring process. The reconstructed data can maintain the continuity of the temperature change of the monitoring object based on the infrared thermal imaging observation, avoiding the problems of data distortion and loss of key information caused by excessive cleaning.

3.3 Infrared Thermal Imaging Monitoring Image Enhancement for Shale Gas Fracturing Equipment

Most of the shale gas fracturing equipment is a large-scale system composed of several functional modules, each of which has different shapes, structures, materials, executive functions, energy conversion forms, etc., and the surface temperature and infrared radiation changes in the process of operation have large differences in the form, amplitude, and rate. Imaging processing, different functional module temperature values or infrared radiation intensity values, and a limited number of image parameter elements of the corresponding relationship are not identical, that is, the monitoring process of each functional module infrared radiation intensity (and equipment surface temperature) and the function of the image parameters of different relationships. If all the subsystems, components, or regions of the monitored equipment within the field of view of the lens set a uniform functional relationship to do imaging processing, prone to imaging rules interfering with each other, the local contrast is reduced, and the level of detail information visualization is reduced and other problems. Such as temperature or infrared radiation intensity changes in large amplitude, fast rate, or monotonous module, its imaging function relationship will form a greater impact on the overall imaging function relationship of the system, and even play a dominant role; While the temperature or infrared radiation intensity change amplitude is small, slow rate, strong volatility of the module, its imaging function relationship of the system as a whole imaging function relationship of the influence of the smaller, had to be in accordance with the other modules of the imaging function relationship of the intensity of infrared radiation and image parameters of the transformation, resulting in monitoring information visualization of the degree of reduction in the sensitivity of response to the parameters of the small differences in the diminution of the sensitivity.

Aiming at the problems of reduced local contrast and reduced visualization level of detailed information when multiple subsystems, components, or regions with large differences in temperature attributes are transformed into imaging based on a uniform functional relationship in the process of infrared thermography monitoring of shale gas fracturing field equipment, this section proposes a method of infrared thermography monitoring image enhancement for shale gas fracturing field equipment based on the optimization of grayscale value distribution. The image segmentation technique is applied to construct the segmentation matrix of each sub-system, part, or region of the monitored equipment, which is used to extract the relevant information of the corresponding sub-systems, parts, or regions in each frame of the infrared thermal imaging map, and filter out the irrelevant information of other parts. Specific and independent functional relations for transforming imaging are set for the extracted single (or several) subsystems, components, or regions with similar attributes so that they can optimally allocate imaging parameters according to the characteristics of their own temperature distributions and variations, in order to improve the visualization of the equipment monitoring information, and to enhance the recognizability of the differences in or variations in the surface temperature of the equipment based on the visual differences in the images.

3.3.1 *Principles of infrared thermal imaging image enhancement method based on gray value distribution optimization*

(1) Optimization of gray value distribution of pixel points in infrared thermograms

An infrared thermogram is a visual radiation intensity distribution based on a specific functional relationship between the intensity of infrared radiation and image parameters, the presentation of which is determined by the color values assigned to each pixel point and its relative position. The number and relative position of the pixels in the map correspond exactly to the number and relative position of the temperature measurement points in the plane of the infrared detector, and the color value of each pixel is a specific function of the intensity of the infrared radiation received at the corresponding temperature measurement point, as shown in Eq. (3.14):

$$N_{ij} = f(Ie_{ij}), \tag{3.14}$$

where N_{ij} denotes the color value of the pixel point in the ith row and jth column of the pixel array; Ie_{ij} is the value of the infrared radiation intensity received by the infrared detector in the ith row and jth column of the flat temperature measurement point array; and $f(\cdot)$ denotes the transformed function relationship between the color value and the intensity of the infrared radiation.

Assuming that the attenuation of infrared radiation intensity caused by differences in distance and angle from each region of the monitored equipment to the infrared detector plane can be ignored, then there exists a specific functional relationship between the intensity of infrared radiation Ie_{ij} received at each temperature measurement point of the infrared detector plane and the corresponding position of the monitored equipment's local surface temperature T_{local}, as shown in Eq. (3.15):

$$Ie_{ij} = g(T_{\mathrm{local}}), \tag{3.15}$$

where $g(\cdot)$ is the transformation function between the intensity of infrared radiation Ie_{ij} at each temperature measurement point and the surface temperature T_{local} at the corresponding location of the monitored equipment.

By converting the infrared thermal imaging map into a grayscale image to take advantage of its single-channel, one-dimensional variation to reduce the cost of computing and improve the processing efficiency, then the presentation of the image will be determined by the grayscale value assigned to each pixel point and its relative position. Correspondingly, N_{ij} in Eq. (3.14) will denote the grayscale value of the pixel point in the ith row and jth column of the pixel array, and $f(\cdot)$ denotes the transformation function relationship between the grayscale value and the intensity of infrared radiation.

The grayscale value is the identification value of the grayscale level of the image. The grayscale image is divided into 256 gray levels between "black" and "white" according to the difference in the proportion of black and white, which are used to represent the brightness of each pixel. Each gray level corresponds to one of the values from 0 to 255. 0 corresponds to black and 255 corresponds to white. From the imaging characteristics of the infrared thermal imager, it can be seen that the higher the temperature of the monitored equipment, the stronger the infrared radiation, the higher the brightness of the generated image, which is interpreted by the grayscale image, i.e., the larger the gray value; on the contrary, the lower the temperature of the monitored equipment, the weaker the infrared radiation, the lower the brightness of the generated image, which is interpreted by the grayscale image, i.e., the smaller the gray value.

Most of the application objects of infrared thermal imaging monitoring technology are the whole or part of the equipment system containing multiple functional modules. Each functional module of the form, material, executive function, contact media, distribution location, electrical and thermal properties and other parameters is different, and thus the same process conditions of the surface temperature of the functional modules of the distribution of the law, the law of change and the corresponding distribution of infrared radiation intensity of the law, the law of change will be different.

From Eq. (3.14), Eq. (3.15) can be seen, infrared thermal imaging map converted grayscale image of a pixel in the grayscale value of the corresponding area of the equipment being monitored (part) of the surface temperature. Then the temperature distribution of each functional module and each region (part) of the monitored equipment can be transformed into the gray value distribution of each pixel point in the grayscale image, as shown in Eq. (3.16):

$$N_{ij} = f(Ie_{ij}) = f[g(T_{\text{local}})], \qquad (3.16)$$

where N_{ij} denotes the gray value of the pixel point in the ith row and jth column of the pixel array; Ie_{ij} denotes the value of the infrared radiation intensity received by the infrared detector in the ith row and jth column of the planar array of temperature measurement points; $f(\cdot)$ denotes the transformed function relationship between the gray value and the intensity of the infrared radiation; T_{local} denotes the surface temperature value of the corresponding position of each temperature measurement point of the monitored equipment; $g(\cdot)$ denotes the transformed function relationship between the intensity of infrared radiation Ie_{ij} of each temperature measurement point and the surface temperature T_{local} of the corresponding position of the monitored equipment.

During the operation of the equipment, each functional module involves different values, directions, and forms of energy conversion, heat transfer, and medium transport, so the size, direction, and form of temperature change over time in each functional module and each region (part) are not exactly the same (e.g., when the motor of the fracturing plunger pump is running, some electric energy is converted into internal energy, and thus the temperature rises to a larger magnitude and at a faster rate; The part of the pump head body constantly has the flowing liquid to take away the heat, thus the temperature change is smaller and the rate of change is slower). It is known that there is a certain functional relationship between the gray value of each pixel in the converted gray image of the infrared thermal imaging map and the surface temperature of the corresponding area (part) of the monitored equipment, so the temperature difference or change of the monitored equipment's functional modules and areas (parts) can also be converted into the gray value of each pixel in the gray image of the difference or change.

Among them, the "difference" in surface temperature can be manifested either as the difference in temperature values of specific points on the monitored equipment belonging to different regions, systems, and parts at the same time, i.e., mainly reflecting the spatiality of the temperature difference, or as the difference in temperature values of specific points on the same region, the same system, or the same part of the monitored equipment at different times, i.e., mainly reflecting the spatiality of the temperature difference. The difference, that is, mainly reflects the temporal temperature difference. Reflected in the infrared thermal imaging map is manifested in the image color and brightness differences in space and time.

To sum up, the relationship between the distribution and change rule of the surface temperature of each functional module and each area (part) of the monitored equipment and the distribution and change rule of the gray value of each pixel in the transformed grayscale image of the infrared thermal imaging map is shown in Eq. (3.17):

$$\begin{cases} C = \{T \mid T_{\min} \leq T \leq T_{\max}\} \\ N_g = \{n_g \mid n_{g\min} \leq n_g \leq n_{g\max}\} \\ n_g = f[g(T)] \\ n_{g\min} = f[g(T_{\min})] \\ n_{g\max} = f[g(T_{\max})], \end{cases} \tag{3.17}$$

where C is the temperature range of the radiation plane of the monitored equipment; T is the temperature value at a location on the radiation plane of the monitored equipment at a given moment; $T_{\min}$ and $T_{\max}$ represent the minimum and maximum temperature values of the radiation plane of the monitored equipment, respectively; N_g is the range of gray values of the pixel points in the infrared thermogram; n_g is the gray value on a pixel point corresponding to a position in the radiation plane of the monitored device at a given moment; $n_{g\min}$ and $n_{g\max}$ denote the minimum and maximum values of the gray value of each pixel point in the infrared thermogram, respectively; $f(\cdot)$ denotes the pixel point gray value as a function of the intensity of infrared radiation at the temperature measurement point; $g(\cdot)$ represents the intensity of infrared radiation at the temperature measurement point as a function of the surface temperature at the corresponding location of the monitored equipment.

In each frame of the infrared thermal imaging map generated at different moments, the value range of the pixel color identification value is the same and fixed. Taking the grayscale value as an example, the grayscale value of each frame of the infrared thermal imaging map is in the range of 0 to 255, and each area (part) of the monitored equipment occupies a specific grayscale value (or range of grayscale values) according to its temperature (or temperature range), where the highest temperature value corresponds to the maximum grayscale value of 255, and the lowest temperature value corresponds to the minimum grayscale value of 0, as shown in Eq. (3.18):

$$\begin{aligned}
n_g &= f[g(T)], \quad T_{\min} < T < T_{\max} \\
n_{g\,\min} &= f[g(T_{\min})] = 0 \\
n_{g\,\max} &= f[g(T_{\max})] = 225.
\end{aligned} \tag{3.18}$$

When all the values of the surface temperature of the monitored equipment exist with their corresponding gray values, the difference in the surface temperature of the monitored equipment is manifested as the difference in the brightness effect of the infrared thermal imaging map. If we do not consider the limit of the human eye's ability to perceive light and darkness, theoretically, the 256 gray levels can represent the difference or change of 256 temperature gradations at most. In other words, when the minimum and maximum values of the surface temperature of the monitored object are less than or equal to 256 grades, each grade will have its own brightness characteristics that are different from the other grades; when the minimum and maximum values of the surface temperature of the monitored object are greater than 256 grades, there will be a number of grades that are indistinguishable from each other in terms of brightness. Therefore, in order to optimize the efficiency and quality of the human eye in recognizing the differences in image color and brightness, and then quickly and accurately identify the differences or changes in the surface temperature of the monitored device, the distinguishability of the temperature differences of the monitored device should be improved in the color dimension. That is, to enhance the difference in the corresponding grayscale values (color values) of adjacent temperature divisions so that small differences in temperature values are converted into the largest possible differences in grayscale values (color values).

(1) Evaluation of image enhancement effect based on gray value metrics

(i) Average gray value temperature ratio

From Eq. (3.17), it can be seen that any temperature value in the temperature range C of the surface of the monitored equipment has a corresponding gray value in the range N_g after the functional transformation, and the difference ΔT between any two temperature values can be transformed into the difference Δn_g of the corresponding gray value. In order to improve the color discriminability of temperature differences, the difference in grayscale values per unit of temperature difference should be as large as possible, i.e., the ratio of Δn_g to ΔT should be as large as possible, and there is a metric for measuring the color discriminability of temperature differences — the temperature ratio of the average grayscale values k, which is shown in Eq. (3.19):

$$k = \frac{\Delta n_g}{\Delta T} = \frac{f[g(T')] - f[g(T)]}{T' - T},\tag{3.19}$$

where k is the temperature ratio of the average gray value; Δn_g denotes the amount of difference or change in the gray value; ΔT denotes the amount of difference or change in the temperature value; T' and T are the two unequal temperature values in the radiation plane of the monitored equipment; $f(\cdot)$ denotes the function relationship between the gray value of the pixel and the intensity of the infrared radiation at the measurement point; $g(\cdot)$ denotes the function relationship between the intensity of the infrared radiation at the measurement point and the surface temperature of the corresponding position of the monitored equipment.

From Eq. (3.19), it can be seen that if the k value is larger, it means that the difference in grayscale value corresponding to the unit temperature difference is larger, reflecting that the color attribute difference in the imaging process can be recognized by the human eye is larger, and it is easier for the monitoring personnel to identify and determine; on the contrary, if the k value is smaller, it means that the difference in grayscale value corresponding to the unit temperature difference is smaller, reflecting that the color attribute difference in the imaging process is smaller (or even unrecognizable).

Further, according to the different evaluation objects, the average gray value temperature ratio can be divided into two categories: the overall average gray value temperature ratio k_{global} and the local average gray value temperature ratio k_{local}, wherein the overall average gray value temperature ratio k_{global} is used for evaluating the color distinguishability of the temperature difference of the infrared thermal imaging map as a whole, see Eq. (3.20); k_{local} is used to evaluate the local temperature difference color distinguishability of the infrared thermal imaging map, as shown in Eq. (3.21):

$$k_{\text{global}} = \frac{\Delta n_g}{\Delta T} = \frac{n_{g\max} - n_{g\min}}{T_{\max} - T_{\min}} = \frac{f[g(T_{\max})] - f[g(T_{\min})]}{T_{\max} - T_{\min}}, \quad (3.20)$$

where k_{global} is the temperature ratio of the overall average gray value; Δn_g is the change of the gray value of the pixel; ΔT is the change of the temperature value T; $n_{g\min}$ and $n_{g\max}$ denote the minimum and maximum gray value of the pixel in the infrared thermal imaging map, respectively; $T_{\min}$ and $T_{\max}$ denote the minimum and maximum temperature of the radiation plane of the monitored equipment, respectively; $f(\cdot)$ denotes the function of the gray value of the pixel to the intensity of the infrared radiation of the temperature measurement point; $g(\cdot)$ denotes the function of the intensity of infrared radiation of the temperature measurement point to the surface temperature of the corresponding position of the monitored equipment.

$$\begin{cases} k_{\text{local}} = \dfrac{\Delta n_g}{\Delta T'} = \dfrac{f[g(T'_a)] - f[g(T'_b)]}{T'_a - T'_b} \\[2mm] C_{\text{local}} = \{T' \mid T'_{\min} < T' < T'_{\max}\} \\[2mm] C_{\text{local}} \subseteq C \\[2mm] T'_a, T'_b \in C_{\text{local}}, \quad T'_a \neq T'_b, \end{cases} \qquad (3.21)$$

where k_{local} is the temperature ratio of the local average gray value; Δn_g is the difference or change of the gray value of the pixel point; T' is the temperature value of a position in the local area of the radiation plane of the monitored equipment at a certain moment; $\Delta T'$ is the change of the temperature value T'; $f(\cdot)$ denotes the function relationship between the gray value of the pixel point and the intensity of the infrared radiation at the measurement point; $g(\cdot)$ denotes the function relationship between the intensity of the infrared radiation at the measurement point and the corresponding surface temperature of the monitored equipment; C_{local} is the range of temperature values in the local area of the radiation plane of the monitored equipment; $T'_{\min}$ and $T'_{\max}$ denote the minimum and maximum temperature value in the radiation plane of the monitored equipment, respectively; C is the range of temperature values in the radiation plane of the monitored equipment; T'_a and T'_b denote the temperature values at position a and position b, respectively, in the local region of the radiation plane of the monitored equipment.

For the overall average gray value temperature ratio k_{global}, the range of gray values for the same imaging system is the same and fixed (0 to 255). When the number of elements in the grayscale value range is greater than or equal to the number of elements in the temperature range of the surface of the monitored device, each temperature value element will theoretically be assigned at least one grayscale value that is distinct from the grayscale values corresponding to the other temperature value elements; When the number of elements in the grayscale value range is smaller than the number of elements in the surface temperature range of the monitored equipment, there is bound to be the phenomenon that multiple temperature-valued elements share a single grayscale value, and at this time, the difference in temperature will not be able to be transformed into a difference in color that can be recognized by the human eye. That is, the numerator in Eq. (3.20) is a constant quantity, and the size of the k_{global} value depends on the size of the denominator, which is related to the range of the surface temperature of the monitored equipment, the larger the range, the smaller the k_{global} value; the smaller the range, the larger the k_{global} value.

The local average gray value temperature ratio, on the other hand, is concerned with the color discriminability of temperature differences in a region (corresponding to a component or area of the monitored device) in an infrared thermogram. Assuming that a region W is arbitrarily delineated in an infrared thermogram, the set N_W consisting of the gray values of all the pixels that make up W is a subset of the range N of gray values of the imaging system, and the minimum value of the elements in the set N_W is $n'_{g\,\text{min}}$, and the maximum value of the elements in the set N_W is $n'_{g\,\text{max}}$. The range of surface temperatures of specific components or areas of the monitored device corresponding to region C_W, is a subset of the range of overall surface temperatures of the monitored device C, and the minimum value of an element in the set is T'_{min} and the maximum value is T'_{max}. The minimum value of T'_{min} and the maximum value of T'_{max} in the set are T'_{min} and T'_{max} respectively, which corresponds to $n'_{g\,\text{min}}$ and $n'_{g\,\text{max}}$ by the specific functional relationship transformed in Eq. (3.22):

$$\begin{cases} C_W = \{T'(x, y) \mid T'_{\min} < T' < T'_{\max}, (x, y) \in W\} \\ N_W = \{n'_g(x, y) \mid n'_g(x, y) = f[g(T')], (x, y) \in W, \\ \quad T' \in C_W\} \\ C_W \subseteq C, \quad N_W \subseteq N_g \\ n'_{g\,\min} = f[g(T'_{\min})] \\ n'_{g\,\max} = f[g(T'_{\max})], \end{cases} \tag{3.22}$$

where C_W is the set of temperature values at each position in any region W of the infrared thermogram; T' is the temperature value at a location within any region W of the infrared thermogram; $T'_{\min}$ and $T'_{\max}$ denote the minimum and maximum temperature values in any region W of the infrared thermogram, respectively; W is an arbitrarily delineated region in the infrared thermogram; x and y denote the values of the horizontal and vertical coordinates of the pixel points in any region W of the infrared thermogram, respectively; N_W is the set of gray values of each pixel point in any region W of the infrared thermogram; n'_g is the gray value of a pixel point in any region W of the infrared thermogram; $f(\cdot)$ denotes the pixel point gray value as a function of the intensity of infrared radiation at the temperature measurement point; $g(\cdot)$ represents the intensity of infrared radiation at the temperature measurement point as a function of the surface temperature at the corresponding location of the monitored equipment; C is the range of temperature values in the radiation plane of the monitored equipment; N_g is the range of gray values for each pixel of the infrared thermogram; $n'_{g\,\min}$ and $n'_{g\,\max}$ denote the minimum and maximum values of gray values in any region W of the IR thermogram, respectively.

The size of the k_{local} value depends both on the size of the numerator (i.e., the size of the set of gray values C_W to which the region W is "assigned") and on the size of the denominator (i.e., the size of the range of values of the local temperature of the monitored equipment corresponding to the region W). In order to make the k_{local} value as large as possible, it is necessary to increase the number of elements in the set of gray values C_W corresponding to the region W, and to reduce the range of values of the local temperature of the monitored equipment corresponding to the region W by technical means.

(2) Gray value span

Grayscale value span, i.e., the range of values between the minimum and maximum grayscale values in a grayscale image, is used to characterize the overall brightness and darkness of the image and the visualization level of detailed information.

If it is known that in the range of $0 \sim 255$, 0 corresponds to black and 255 corresponds to white, then the closer the gray value is to the lower part of the range, the grayer the corresponding pixel or image is, and the closer the gray value is to the higher part of the range, the brighter the corresponding pixel or image is. When converted into a pseudo-color image, the closer the grayscale value is to the lower value area, the closer the corresponding pixel or image's dominant color is to blue; the closer the grayscale value is to the higher value area, the closer the corresponding pixel or image's dominant color is to red.

If the maximum gray value of the image is in the low-value area, then all the gray values must be in the low-value area, resulting in the overall visual effect of the image being dark, corresponding to the main color of the pseudo-color image tends to be close to the blue, and cannot make full use of the other hues to characterize the temperature information of the device. Similarly, if the minimum value of the gray value of the image is in the high-value area, all of its gray value must be in the high-value area, resulting in the overall visual effect of the image being bright, corresponding to the pseudo-color image of the main color tends to be close to the red, and cannot make full use of other hues characterize the temperature information of the device. Therefore, the enhancement process of the image should make the minimum and maximum values of the grayscale values as much as possible be located in different areas on the grayscale value axis, in other words, make the concentrated area of the grayscale values of the image as close as possible to the center of the range of values, in order to enhance the color (visual) legibility of the temperature difference.

As far as the scale property of gray value span is concerned, the larger the gray value span is, the more available gray value levels in the current image, i.e., the larger the theoretical gray value difference corresponding to the unit temperature difference is, and the higher the level of image visualization of the temperature difference or change of the device is. On the contrary, the smaller the gray value span, the less the available gray value class of the image, the smaller the theoretical gray value difference corresponding to the unit temperature difference, and the lower the image visualization level of the equipment temperature difference or change. Therefore, image enhancement processing should extend the gray value span of the image as much as possible to improve the visualization level of image detail information.

(3) Main gray level displacement

The main gray level, i.e., the number of gray values or a more concentrated range of gray values at all pixel points of an image, is used to characterize the overall visual effect of the current image. When the main gray level is shifted on the gray value axis, it means that the size of the gray value of most pixels or the distribution of the gray value of all pixels in the image has changed, and the overall visual effect is bound to be different. The larger the shift of the main gray level, the larger the change of the gray value and the overall distribution of the pixels in the image caused by the temperature change at different moments, and the larger the corresponding visual difference of the image, which is easier to be recognized and understood by the monitors. On the contrary, the smaller the shift of the main gray level is, the smaller the change of the gray value and the overall distribution of each pixel of the image caused by the temperature change at different moments is, the smaller the corresponding visual difference of the image is, and the more unfavorable it is for the monitor to identify and understand. As far as the infrared thermal imaging monitoring video is concerned, the infrared thermal imaging images intercepted at different moments (frames) indirectly characterize the operation status of the monitoring equipment at different stages. In order to highlight the differences in the visual effect of the monitoring images corresponding to changes in the operating state, the image enhancement process should increase the displacement of the main gray level of the image at different moments as much as possible, so as to improve the visual distinguishability of the differences in the operating state of the equipment at different stages.

3.3.2 *Implementation steps*

Step 1: construct the infrared thermal imaging map transformation data matrix.

The infrared thermal imaging map of the monitoring object is transformed into a data matrix $\boldsymbol{I_{m\times n}}$, i.e., the pixel point array of the image layer $m \times n$ is transformed into the matrix of numerical elements $\boldsymbol{I_{m\times n}}$ of the data matrix layer $m \times n$. The relative position (x, y) of each pixel point in the pixel point array corresponds one to one with the relative position (i, j) of each element in the data matrix. Then, the geometric element descriptions such as points, lines, surfaces, etc., in the infrared thermal imaging map can all be replaced with the forms of elements, vectors, and sub-matrices in the data matrix.

Step 2: construct the segmentation matrix of each part/region of the monitoring object.

Image segmentation is done on the infrared thermal imaging map of the monitoring object to extract the edges of the parts/regions of interest, and the relative positions of the pixel points on the edge line are labeled in the image data matrix layer. That is, the external contour of the part/area of interest is outlined on the transformed data matrix of the infrared thermogram by transforming the coordinates of the image pixel points to the positions of the matrix elements. The contour line and its internal elements are defined with a value of 1, and the external elements of the contour line are defined with a value of 0. The segmentation matrix $\boldsymbol{F_{m\times n}}$ of the part/region of interest is constructed, and its properties are shown in Eq. (3.23):

$$\boldsymbol{F_{m\times n}} = \{\omega_{ij}\}, \quad 1 \leq i \leq m, \ 1 \leq j \leq n$$

$$\omega_{ij} = \begin{cases} 1 & (x, y) \in \Lambda \\ 0 & (x, y) \notin \Lambda, \end{cases} \tag{3.23}$$

where ω_{ij} is the data element in the segmentation matrix $\boldsymbol{F_{m\times n}}$; x, y denote the horizontal and vertical coordinates of the pixel point corresponding to each data element in the segmentation matrix $\boldsymbol{F_{m\times n}}$ in the pixel array of the image layer, respectively, and numerically $x = j$, $y = i$; Λ denotes the external contour of the part/region of interest and the internal region.

Various segmentation matrices can be constructed for different parts/regions of interest. For example, the pump head body segmentation matrix $\boldsymbol{H}$, the power end segmentation matrix $\boldsymbol{D}$, the motor segmentation matrix $\boldsymbol{E}$, the transmission box segmentation matrix $\boldsymbol{Tr}$, the input/output end segmentation matrix $\boldsymbol{O}$, and so on.

Step 3: do image segmentation for each part/region of the monitoring object.

The segmentation matrix $\boldsymbol{F}_{m \times n}$ of each part/area is Hadamard product with the transformed data matrix $\boldsymbol{I}_{m \times n}$ of the infrared thermal imaging map respectively, and the extraction data matrix $\boldsymbol{I'}_{m \times n}$ of each part/area is constructed, see Eq. (3.24):

$$\boldsymbol{I'}_{m \times n} = \boldsymbol{F}_{m \times n} * \boldsymbol{I}_{m \times n}, \tag{3.24}$$

where $\boldsymbol{I'}_{m \times n}$ is the extraction data matrix of each component/region; $\boldsymbol{F}_{m \times n}$ is the segmentation matrix of each component/region; $\boldsymbol{I}_{m \times n}$ is the transformation data matrix of the infrared thermal imaging map; $*$ is the matrix Hadamard product operator.

The Hadamard product rule is shown in Eq. (3.25):

$$\begin{aligned} \boldsymbol{A} = \{a_{ij}\}_{m \times n}, \quad \boldsymbol{B} = \{b_{ij}\}_{m \times n} \\ \boldsymbol{A} * \boldsymbol{B} = \{a_{ij} \cdot b_{ij}\}_{m \times n}, \end{aligned} \tag{3.25}$$

where $\boldsymbol{A}$ and $\boldsymbol{B}$ denote any two $m \times n$ data matrices; a_{ij} and b_{ij} are elements in matrices $\boldsymbol{A}$ and $\boldsymbol{B}$, respectively.

Step 4: independent imaging of each part/area of the monitoring object.

The extraction matrix $\boldsymbol{I'}_{m \times n}$ of each part/area is imaged. As can be seen from Steps 2 and 3, all the elements in the extraction matrix $\boldsymbol{I'}_{m \times n}$ except the parts/areas of interest have a value of 0, which cannot be processed during imaging. Then the imaging rule is shown in Eq. (3.26):

$$n_g = f_F[g_F(T_F)], \quad T_F \in C_F, \tag{3.26}$$

where n_g is the pixel point grayscale value; $f_F(\cdot)$ denotes the functional relationship between the grayscale value of the segmented extracted part/region and the infrared radiation intensity; $g_F(\cdot)$ denotes the functional relationship between the infrared radiation intensity of the segmented extracted part/region and the surface temperature value; T_F is the surface temperature of the concerned part/region after segmentation and extraction; and C_F is the range of the surface temperature of the concerned part/region after segmentation and extraction.

Step 5: comparison of visual effects before and after monitoring object enhancement processing.

Comparatively analyze the visual effect of the infrared thermal imaging map of the concerned parts/areas before and after enhancement processing based on the average gray value temperature ratio, gray value span, and main gray level displacement indexes, and quantitatively evaluate the optimization effect of the described method on the distribution of gray values of different parts/areas of the image.

3.3.3 *Case study*

In this section, the infrared thermographic images corresponding to the startup of the piston pump failure simulation experimental platform and after running for a period of time are used as examples to illustrate the proposed method.

The color images when the platform is at the initial moment $(t = 0\,\text{s})$ and during operation are shown in Figs. 3.41 and 3.42, and the grayscale images are shown in Figs. 3.43 and 3.44, respectively.

(1) Analysis of the effect of image enhancement in the pump head body region

(i) Constructing the pump head body region splitting matrix

The region Area_H where the pump head body part is located in the infrared thermal imaging map is extracted by image segmentation, and the pump head body segmentation matrix $\boldsymbol{H}$ is constructed, see Eq. (3.27). The imaging effect of the pump head body segmentation matrix $\boldsymbol{H}$ is shown in Fig. 3.45:

$$\boldsymbol{H} = \{\omega_{ij}\}, \quad 1 \leq i \leq 288, \ 1 \leq j \leq 384$$

$$\omega_{ij} = \begin{cases} 1 & (x,y) \in \text{Area}_H \\ 0 & (x,y) \notin \text{Area}_H, \end{cases} \tag{3.27}$$

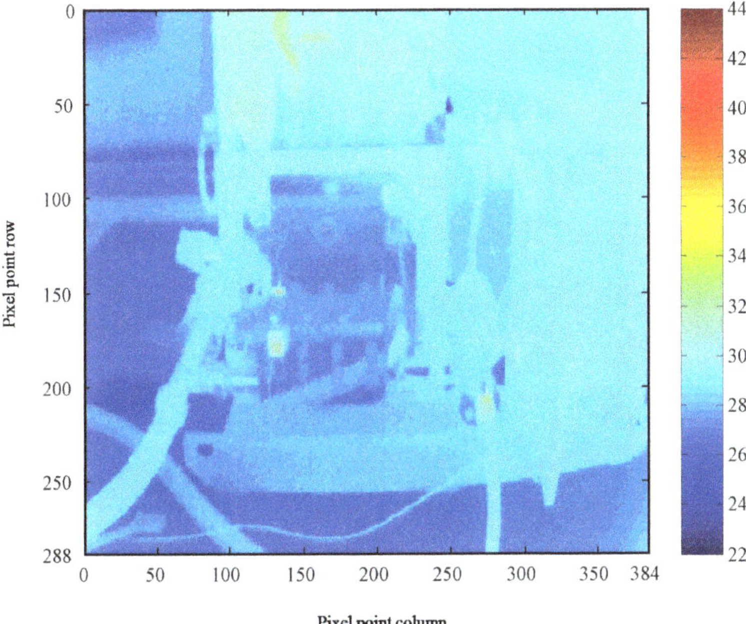

Fig. 3.41. Infrared thermogram of the experimental setup at the initial moment.

where ω_{ij} is the data element in the segmentation matrix $\boldsymbol{H}$; x and y denote the horizontal and vertical coordinates of the pixel corresponding to each data element in the segmentation matrix $\boldsymbol{H}$ in the pixel array of the image layer, and numerically $x = j, y = i$; Area$_H$ denotes the external contour and internal area of the pump body part.

(ii) Constructing the pump head body region segmentation extraction matrix and image processing

The infrared thermography images formed during the infrared thermography monitoring process are transformed into a data matrix $\boldsymbol{I}_i$ $(i = 1, 2, \ldots, t)$, which is respectively hashed with the pump head segmentation matrix $\boldsymbol{H}$. That is to say, in the transformed data matrix of the infrared thermography images of the frames, the data information of the region where the pump head is located is extracted, and the data information of the other regions of the matrix is filtered out, so as to construct an extracted data matrix $\boldsymbol{I}_i$.

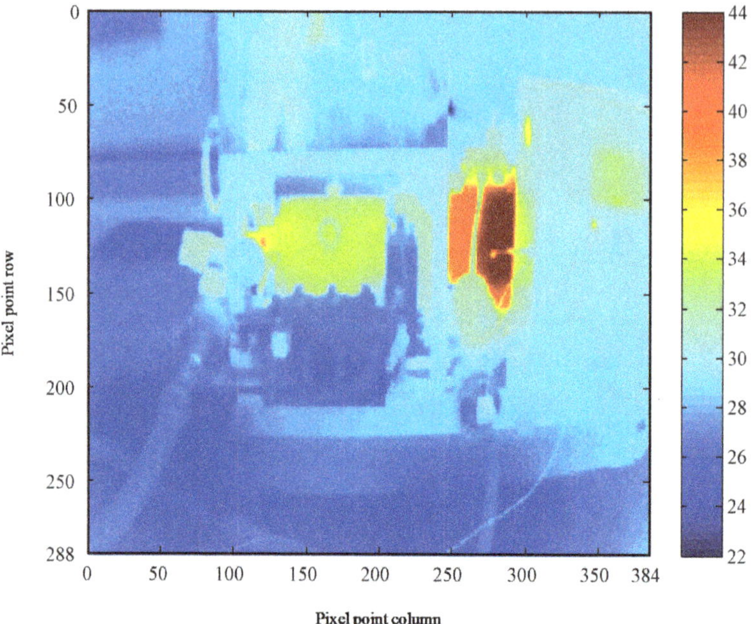

Fig. 3.42.   Infrared thermal imaging of the experimental device during operation.

The extracted data matrix $\boldsymbol{I}_i$ is imaged to obtain the infrared thermal imaging map of the pump head part of each frame. The initial moment and the infrared thermal image of the pump head after running for a period of time are shown in Figs. 3.46 and 3.47, respectively.

(iii) Comparative analysis of the average gray value temperature ratio before and after the enhancement process

The color distinguishability of the temperature difference in the infrared thermal imaging map of the pump head part is evaluated: before the segmentation and extraction treatment, the temperature change range of the monitored equipment as a whole is from 22.1 to 45.0°C, and the average gray value temperature ratio is $k = 11.18$; after the segmentation and extraction treatment, the temperature change range of the pump head part is from 24.8 to 28.6°C, and the average gray value temperature ratio is 67.37. The average number

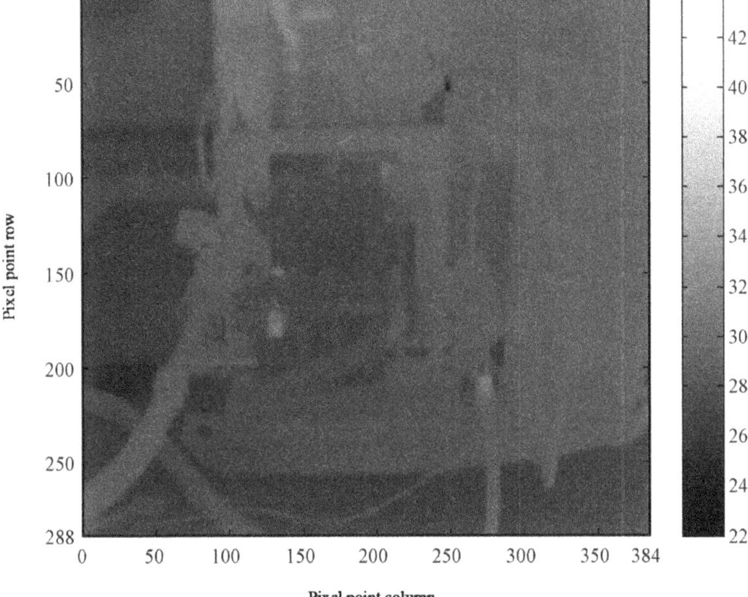

Fig. 3.43. Grayscale image of the experimental setup at the initial moment.

of gray value levels that can be assigned per degree Celsius after the extraction treatment is 6.03 times of that before the treatment.

(iv) Comparative analysis of the span of gray values before and after enhancement treatment

Before the segmentation and extraction process, the functional relationship (or conversion rule) between the temperature values of each area/part of the monitored equipment and the gray level of the corresponding pixel in the infrared thermal imaging map is exactly the same. The conversion rule is shown in Eq. (3.28):

$$n_g = f[g(T)], \quad T \in C, \tag{3.28}$$

where n_g is the gray value level; $f(\cdot)$ indicates the gray value level of the monitored equipment as a function of the intensity of infrared radiation; $g(\cdot)$ indicates the intensity of infrared radiation of the monitored equipment as a function of the surface temperature value; T is the value of the surface temperature of the radiation plane of the

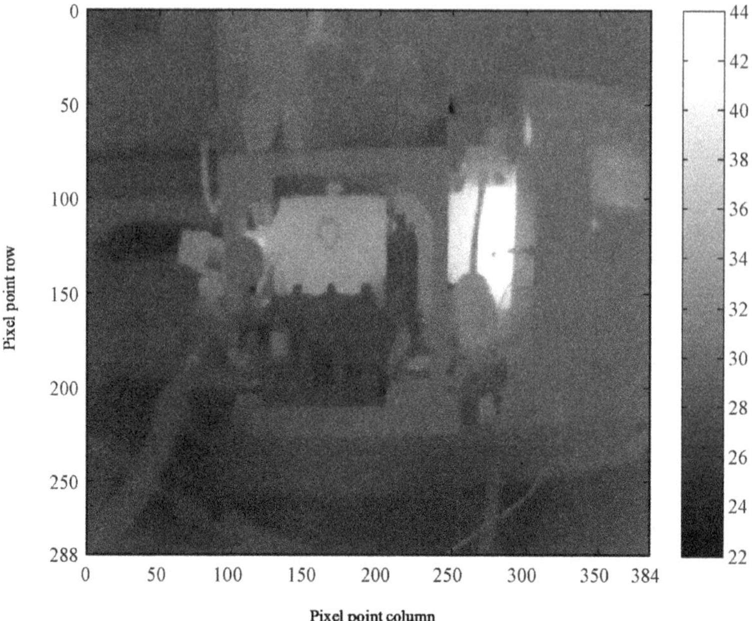

Fig. 3.44.   Grayscale image of the experimental setup during operation.

monitored equipment; C is the range of the surface temperature value of the radiation plane of the monitored equipment.

After completing the transformation, the gray value level of the pixels corresponding to the pump head part is extracted from the gray value level of all pixels corresponding to the monitored equipment as a whole, and the distribution intervals are counted to generate a bar chart.

At the initial moment of equipment operation, there are 3445 pixel points in the region where the pump head body part is located in the infrared thermal imaging map. The minimum gray value is 52.11, the maximum gray value is 73.17, and the arithmetic mean is 60.77; the gray value spans from 50 to 75. After running for a period of time, among the pixel points in the area of the pump head body, the minimum grayscale value is 31.04, the maximum grayscale value is 54.33, and the arithmetic mean value is 45.38; the grayscale value spans the interval of 30 to 55.

After the segmentation and extraction process, the functional relationship (or conversion rule) between the temperature value of

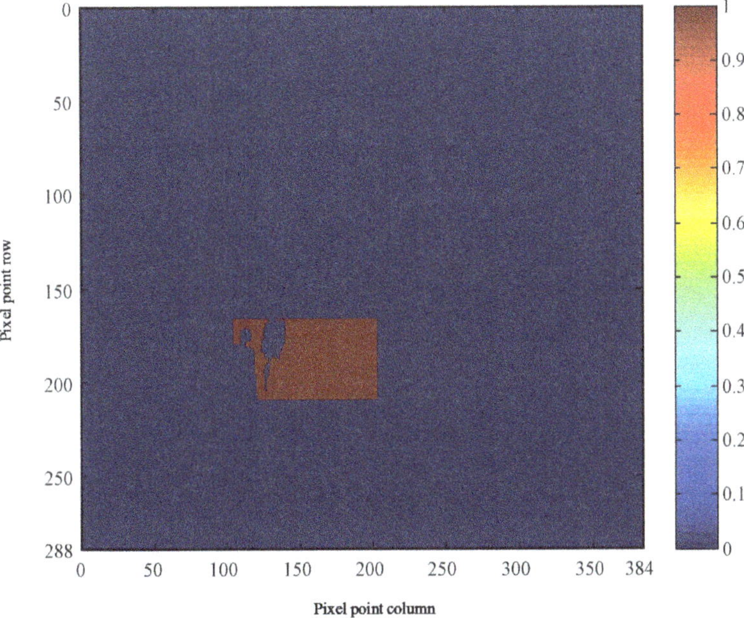

Fig. 3.45. Pump head body segmentation matrix imaging.

each area/component of the monitored equipment and the gray level of the corresponding pixel in the infrared thermal imaging map is not exactly the same. Each region/component forms a specific and independent functional relationship (or conversion rule) according to its own attributes such as the magnitude and rate of temperature change. The conversion rules are shown in Eq. (3.29):

$$n_g = f_H[g_H(T)], \quad T \in C_H, \quad C_H \subseteq C, \tag{3.29}$$

where n_g is the gray value level; $f_H(\cdot)$ denotes the function relationship between the gray value level and infrared radiation intensity in the pump head region after segmentation and extraction; $g_H(\cdot)$ denotes the function relationship between the infrared radiation intensity in the pump head region after segmentation and extraction and the surface temperature value; T is the surface temperature in the pump head region after segmentation and extraction; C_H is the surface temperature range in the pump head region after segmentation and extraction; C is the surface temperature range of the monitored device as a whole.

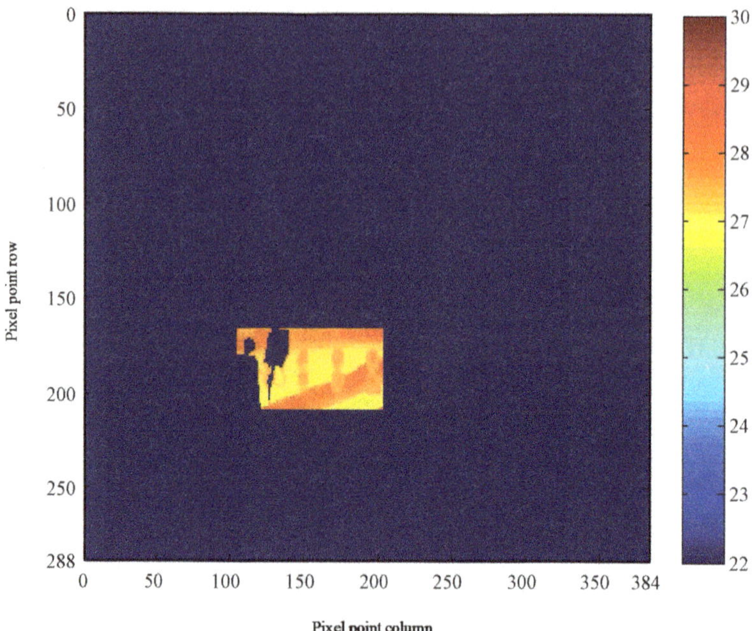

Fig. 3.46. Infrared thermogram of the pump head body at the initial moment.

After the transformation is completed, the gray level of the corresponding pixels in the pump head part is extracted, and the distribution interval is counted to generate a bar chart.

At the initial moment of equipment operation, there are a total of 3445 pixel points in the area where the pump head body part is located in the infrared thermal imaging map. The minimum gray value is 149.81, the maximum gray value is 210.38, and the arithmetic mean value is 174.72; the gray value spans the interval from 140 to 220. After running for a period of time, among the pixel points in the area of the pump head body, the minimum grayscale value is 89.25, the maximum grayscale value is 156.19, and the arithmetic mean value is 130.48; the grayscale value spans the interval from 80 to 160.

(v) Comparative analysis of main gray level displacement before and after enhancement treatment

Before segmentation and extraction, it can be seen from Fig. 3.48 that, at the initial moment of operation, there are 3445 pixels in the

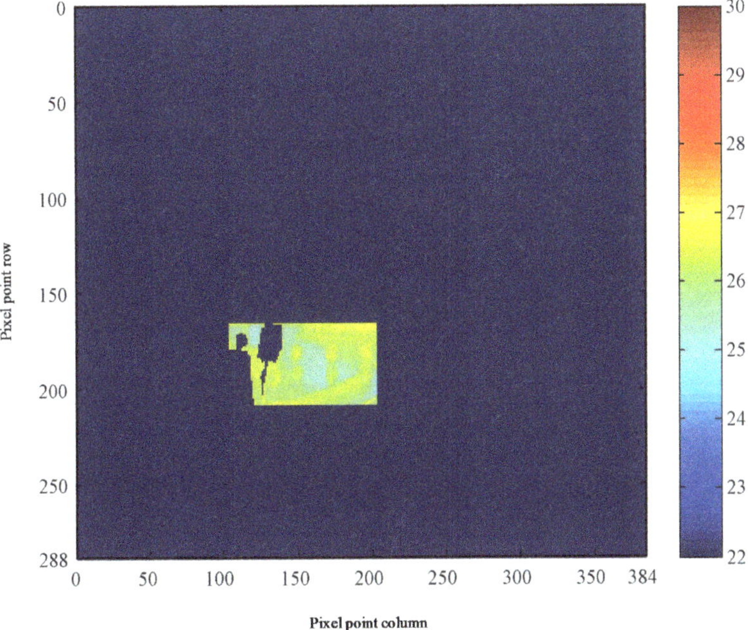

Fig. 3.47. Infrared thermography of the pump head after running for a period of time.

area where the pump head is located in the infrared thermal imaging map, of which 2926 pixels (84.93% of the total number of pixels) are in the region of 55–70 grayscale value. In other words, the main gray level of the pump head is concentrated in the 55–70 region at the initial time. After running for a period of time, the number of pixels with a gray level between 45 and 50 is the highest, with 2009 pixels (58.32% of the total). That is to say, after running for a period of time, the main gray level of the pump head part is concentrated in the range of 45 to 50.

After the segmentation and extraction process, it can be seen in Fig. 3.49 that at the initial moment of the equipment operation, there are 3445 pixels in the area where the pump head is located in the infrared thermal imaging map. Among them, the number of pixels with gray values between 160 and 200 is the largest, totaling 2649 pixels (76.89% of the total). In other words, at the initial moment, the main gray level of the pump head part is concentrated in the range of 160 to 200. After running for a period of time, among the pixels in

Table 3.1. Comparison of temperature ratios of average gray values of images before and after enhancement treatment.

Monitoring objects	Temperature range		k		Comparison of before and after treatment
	Pre-treatment	Post-treatment	Pre-treatment	Post-treatment	
Pump head	22.1~45.0°C	24.8~28.6°C	11.18	67.37	6.03 times
Power side	22.1~45.0°C	26.9~38.4°C	11.18	22.26	1.99 times
Electrical machinery	22.1~45.0°C	28.1~45.0°C	11.18	15.15	1.36 times
Transmission case	22.1~45.0°C	28.5~32.1°C	11.18	71.11	6.36 times
Input and output terminals	22.1~45.0°C	25.3~29.9°C	11.18	55.65	4.98 times

the pump head area, the number of pixels with gray levels between 120 and 140 is the largest, with a total of 1,817 pixels (52.74% of the total). That is to say, after running for a period of time, the main gray level of the pump head part is concentrated in the 120–140 area.

(2) Comparison of the effect of image enhancement before and after processing

(i) Mean gray value temperature ratio evaluation index

As can be seen from Table 3.1, before the enhancement treatment, the temperature of the infrared thermogram corresponding to the monitored equipment as whole ranges from 22.1 to 45.0°C, and the average gray value temperature ratio $k = 11.18$.

After the segmentation extraction and independent imaging enhancement processing, the temperature of the infrared thermal imaging map corresponding to the pump head body region ranges from 24.8 to 28.6°C, and the average gray value temperature ratio $k = 67.37$, which is about 6.03 times the k value of the image before processing. Similarly, the corresponding infrared thermographic map temperature of the power end region takes values in the range of 26.9 to 38.4°C, and the average grayscale value temperature ratio $k = 22.26$, which is about 1.99 times the k value of the image before processing. The value of the motor region is about 1.36 times the pre-processing value, the k value of the transmission box region is about 6.36 times the preprocessing value, and the k value of the input and output end region is about 4.98 times the preprocessing value.

In summary, the smaller the temperature change range of a component/region during the running time, the larger the difference with the overall temperature range of the monitoring object, the stronger the expansion effect of the said enhancement processing method on the k-value, and the better the enhancement effect on the color distinguishability of the temperature difference. The larger the range of temperature change of a part/region during the running time, the closer it is to the range of values of the overall temperature of the monitoring object, the weaker the expansion effect of said enhancement processing method on the k-value is, and the smaller the enhancement effect on the color distinguishability of the temperature difference is.

(ii) Gray value span evaluation metrics

As can be seen from Table 3.2, before the enhancement treatment, the initial moment of each component/region grayscale value is more concentrated in the low-value area (minimum value 48.78, maximum value 106.43), the overall visual effect of the grayscale image is darker, and the corresponding pseudo-color image main color is blue. After running for a period of time, the temperature of the pump head body and the input and output end area under the effect of water flushing appear to decrease slightly, and the span of the grayscale value moves to the low numerical value area. Power end, motor, and other components/regions involved in the conversion of electrical energy and internal energy, mechanical energy, and internal energy have a larger temperature change and a faster rate, resulting in a significant shift in the span of the gray value to the high-value area, the visual effect of the grayscale image is changed from dark to bright, and the corresponding pseudo-color image main color is changed from blue to yellow or red.

After the enhancement process, the grayscale value concentration area of the pump head body, transmission box, and input/output area at the initial moment obviously moves to the high value area (e.g., the minimum grayscale value of the pump head body rises from 52.11 to 149.81; the minimum grayscale value of the transmission box rises from 72.07 to 111.56; the minimum grayscale value of the input/output area rises from 48.78 to 163.20), and the overall visual effect of the grayscale image is changed from dark to bright, and the

Table 3.2. Comparison of the span of gray values of images before and after enhancement processing.

Monitoring objects		Gray value span		Comparison of before and after treatment
		Pre-treatment	Post-treatment	
Pump head	Starting moment	52.11~73.17	149.81~210.38	2.88
	Running for a while	31.04~54.33	89.25~156.19	2.87
	Scope of change	31.04~73.17	89.25~210.38	2.88
Power side	Starting moment	54.33~106.43	87.98~147.90	1.15
	Running for a while	68.74~181.83	104.55~234.60	1.15
	Scope of change	54.33~181.83	87.98~234.60	1.15
Electrical machinery	Starting moment	67.73~78.72	82.62~92.82	0.92
	Running for a while	166.30~255.00	173.40~255.00	0.92
	Scope of change	67.63~255.00	82.62~255.00	0.92
Transmission case	Starting moment	72.07~88.70	111.56~159.38	2.88
	Running for a while	83.15~111.98	143.44~226.31	2.87
	Scope of change	72.07~111.98	111.56~226.31	2.88
Input and output terminals	Starting moment	48.78~87.59	163.20~252.45	2.30
	Running for a while	36.59~84.26	135.15~244.80	2.30
	Scope of change	36.59~87.59	135.15~252.45	2.30

main tone of the corresponding pseudo-color image is changed from blue to green, yellow or even red.

The concentrated area of grayscale value at the initial moment of the power end and motor area also moves to the high-value area, which optimizes the visual brightness of the grayscale image. After running for some time, the pump head body, input and output end regions show a small decrease in temperature under the action of water flow, and the span of gray value moves to the low-value area. Compared with the initial moment, the main color of the pseudo-color image changes significantly (e.g., the pump head body

Table 3.3. Comparison of the main gray level displacement of the image before and after the enhancement process.

Monitoring objects		Gray value span		Comparison of before and after treatment
		Pre-treatment	Post-treatment	
Pump head	Starting moment	55~70	160~200	117.5
	Running for a while	45~50	120~140	82.5
	Scope of change	−15	−50	35
Power side	Starting moment	40~80	80~100	30
	Running for a while	120~140	160~180	40
	Scope of change	70	80	10
Electrical machinery	Starting moment	67.5~80	75~95	11.25
	Running for a while	230~255	235~255	2.5
	Scope of change	168.75	160	−8.75
Transmission case	Starting moment	75~80	120~140	52.5
	Running for a while	95~110	200~240	107.5
	Scope of change	25	80	55
Input and output terminals	Starting moment	70~80	215~235	150
	Running for a while	50~60	165~185	120
	Scope of change	−20	−50	30

part changes from yellow-orange to blue-green; the input and output ends change from red to yellow), i.e., the color (visual) legibility of the temperature difference is improved.

Power end, motor and other parts/areas with large amplitude and fast rate of temperature change have a large span of grayscale values moving to the high-value area, the grayscale image changes from dark to bright, and the corresponding pseudo-color image main color changes from light blue (or blue hue) to red, with obvious color differences and high visual legibility.

In terms of the scale attribute of the span of gray values, the enhancement process extends the available gray range for parts/areas with small temperature variations (pump head body by a factor of 2.88; transmission by a factor of 2.88; and inputs/outputs by a factor of 2.30), amplifying the visual differences in color used to characterize small temperature differences. For components/areas with large temperature variations, which already take up a large number of available gray levels, the enhancement does not significantly expand the available gray range but even compresses it (e.g., motors are compressed to 92% of their preprocessed value).

(iii) Main gray level displacement evaluation metrics

As shown in Table 3.3, before the enhancement treatment, the main gray level of the pump head body, transmission box, input and output, and other components/areas with small temperature changes moved less. After running for a period of time, the main gray level of the pump head moves 15 gray levels to the lower value area compared to the initial moment, the main gray level of the gearbox moves 25 gray levels to the higher value area, and the main gray level of the input/output area moves 20 gray levels to the lower value area. After the enhancement process, the main gray level of the pump head body after running for a period of time is shifted by 50 gray levels to the lower value area compared to the initial moment, and similarly, the main gray level of the gearbox is shifted by 80 gray levels to the higher value area, and the main gray level of the input/output side is shifted by 50 gray levels to the lower value area. The larger the shift in the main gray level, the larger the change in the main gray level caused by the temperature change at different moments, i.e., the larger the visual difference in the image, the easier it is for the monitor to recognize and understand. The enhancement process magnifies the difference in visual effect corresponding to the temperature difference of certain parts/areas and improves the operability of infrared thermography monitoring of parts/areas with small temperature changes.

For components/areas with large temperature variations, such as power ends and motors, the enhancement process will not significantly increase or even reduce the magnitude of the main gray level shift (e.g., the main gray level shift of the motor after the process is

8.75 gray levels less than that before the process), since the visual effect differences corresponding to the temperature variations are sufficiently recognizable by the monitors.

(3) Analysis and summary

Aiming at the problems of reduced local contrast and reduced visualization of detailed information when multiple subsystems, components or regions with large differences in temperature attributes are transformed into imaging based on a uniform functional relationship in the process of infrared thermal imaging monitoring of shale gas fracturing field equipment, an image enhancement method for infrared thermal imaging monitoring of shale gas fracturing equipment based on the optimization of the distribution of gray values is proposed, which can be used to improve the visualization of equipment detailed information and enhance the recognition of temperature differences or changes on the surface of the equipment based on the visual differences in image color.

(i) The case selected five components/areas of pump head body, power end, motor, transmission box, and input/output end as the research objects, defined three indexes, namely, average gray value temperature ratio based on gray value, gray value span, and main gray level displacement, as the evaluation criteria, and compared and analyzed the visual effects of infrared thermal imaging maps of the monitoring objects at different operation stages before and after enhancement processing and the visual legibility of the parameter changes. The results show that the described image enhancement method is more effective in optimizing the parts/regions with smaller temperature changes and larger differences between the local temperature range and the overall temperature range of the equipment.

(a) The average gray value temperature ratio can be increased to a maximum of 6.36 times (transmission box) and a minimum of 4.98 times (input and output) before processing, i.e., each temperature level within the range of temperature values of the monitoring object can theoretically be used to obtain more gray levels to characterize the differences or changes in temperature parameters, which improves the visualization of detailed information.

(b) The maximum gray value span can be extended to 2.88 times (pump head body, transmission box) before processing, and the minimum can be extended to 2.30 times (input and output) before processing, i.e., the temperature range of the monitoring object can be converted into a larger range of gray values of the image, so that the difference or change of brightness and color of the infrared thermal imaging map of different stages and objects is more significant, and it is easier for the monitoring personnel to identify and understand.

(c) The main gray level displacement can be expanded to a maximum of 3.33 times before processing (pump head body), and a minimum of 2.50 times before processing (input and output), that is, monitoring the object of different operating phases of temperature changes triggered by changes in the image gray value (the main gray level of the magnitude of the movement) is more significant, to enhance the level of the image of the parameters of the weak changes in the visualization of the level of the image.

(ii) For components/regions with large temperature changes and local temperature ranges close to the overall temperature range of the equipment, the differences in image visualization caused by the differences in their own parameters are sufficient for the monitor to recognize, and the enhancement method described above will not significantly amplify the evaluation indexes, and may even reduce them to a certain extent.

(a) The temperature ratio of the average grayscale value of the motor grows only 1.36 times that of the preprocessing one, and that of the power end grows 1.99 times that of the preprocessing one, which is significantly lower than the minimum growth multiplier of the other components (4.98 times that of the input and output ends).

(b) The span of gray values at the power end only expanded to 1.15 times the preprocessing level, and the span of gray values at the motor shrunk to 92% of the preprocessing level, which is significantly lower than the minimum expansion multiplier of the other components (2.30 times at the input and output ends).

(c) Power end of the main gray level displacement amplitude is only expanded to 1.14 times before treatment, the motor main gray level displacement amplitude is reduced to 95% before treatment, significantly lower than the other components of the minimum expansion multiplier (input and output end of 2.50 times). That is, the optimization effect of each index is weaker than that of the components/regions with smaller temperature change amplitude and larger difference between the local temperature range and the overall temperature range of the equipment.

3.4 Early Accident Monitoring and Identification of Water Leakage at the Output End of Fracturing Plunger Pump Based on Infrared Thermal Imaging

Shale gas fracking equipment mainly includes fracking pump trucks, sand mixing trucks, instrumentation trucks, and fluid distribution and sand supply equipment. Among them, the plunger pump system of the fracturing pump truck is the main functional module that converts the fracturing fluid of a certain viscosity under atmospheric pressure into high-pressure and high-flow fracturing fluid required for production, and its equipment performance and operation status will have a direct impact on the quality of the fracturing production operation, and therefore it is the key focus of the shale gas fracturing field equipment safety management and monitoring work. In view of the common early incidents of water leakage at the output end of fracturing plunger pumps, this section proposes a monitoring and identification method based on the statistics of RGB value distribution of infrared thermal imaging maps and proves the feasibility and validity of the mentioned method by designing relevant experiments and comparing and analyzing the results and experimental phenomena.

3.4.1 *Overview of the principles of early accident monitoring and identification methods based on RGB value distribution statistics*

RGB value is the identification code of any color under the RGB color standard, which consists of R value that characterizes the red luminance order, G value that characterizes the green luminance order, and B value that characterizes the blue luminance order, and R, G, and B can be regarded as independent numerical variables of each other. Therefore, any color in nature can be represented by a three-dimensional vector ($\boldsymbol{R, G, B}$).

In the case of a color image, the rendering is determined by the RGB values at each pixel and their distribution. The more RGB values an image "takes up", the richer the color, and the opposite, the more monotonous the color. The greater the difference in RGB values between neighboring pixels, the greater the visual contrast, and vice versa. It is known that the infrared thermal imaging monitoring technology characterizes the temperature of the corresponding temperature measurement point by the color of each pixel, the distribution of the surface temperature of the monitoring object by the color distribution of each pixel, and the temperature difference of the monitoring object by the color difference of the adjacent pixels. Therefore, the more RGB values the processed infrared thermogram "occupies", the more comprehensive the characterization of the temperature distribution of the monitoring object and the more obvious the response to the temperature difference.

Therefore, the statistical analysis of the position distribution of RGB value points in the color space of the infrared thermal imaging map can be used to evaluate the visual effect of the image, and then evaluate the feasibility and effectiveness of the early accident monitoring and identification method of shale gas fracturing equipment based on infrared thermal imaging. The more discrete the distribution of point positions is, the better the visual effect of the image is, the larger the color difference used to characterize the small temperature difference is, the stronger the distinguishability is, and the better the feasibility and effectiveness of the monitoring and identification method is; The more concentrated the distribution of point locations, the more homogeneous the color effect of the image, the smaller the color differences used to characterize small temperature differences or even the phenomenon of color fusion in which multiple temperature levels share the same RGB value, the worse the discriminability, and the worse the feasibility and effectiveness of the said monitoring and identification method.

Secondly, statistical analysis of the number of pixels contained in each RGB value can be used to evaluate the current temperature state of the monitoring object, and thus evaluate the operation state of the shale gas fracturing equipment, and identify early accidents related to abnormal temperature changes. The more pixels a certain RGB value contains, the wider the distribution of its corresponding temperature value T_{RGB} on the surface of the monitoring object, which means that the current operating status of the monitoring object may be related to the temperature value T_{RGB}. On the contrary, it means that the operating state of the monitoring object at the current moment is less related to the temperature value T_{RGB}. In view of the fact that the operation process of shale gas fracking equipment involves the conversion of a large amount of electric energy, mechanical energy, and internal energy and that there are many accident modes related to abnormal changes in temperature parameters, the temperature value T_{RGB} can be used as an important basis for the monitoring and identification of early accidents of shale gas fracking equipment.

3.4.2 *Implementation steps*

Step 1: experimental design of water leakage accident at the output end of piston pump.

Simulation of water leakage at the high-pressure output of shale gas fracking pump based on the liquid leakage caused by the loosening of pipeline connection bolts at the output end of the piston pump failure simulation experimental platform. Design related experiments and collect infrared thermography monitoring data of the water leakage experiment.

Step 2: construct the segmentation matrix of the output end of the piston pump.

Do image segmentation on the infrared thermal imaging map of the piston pump, extract the image edges of the output end flange port pipe section, and label the relative positions of each pixel point on the edge line in the image data matrix layer. In other words, the external contour of the pipe segment at the output flange port is outlined on the transformed data matrix of the infrared thermogram by transforming the coordinates of the image pixel points to the positions of the matrix elements. The contour line and its internal elements are defined with a value of 1, and the external elements of the contour line are defined with a value of 0. The output end flange port pipe segmentation matrix $\boldsymbol{F}$ is constructed.

Step 3: segmentation of piston pump output position image.

Do the Hadamard product of the segmentation matrix of the pipe segment at the output end of the piston pump and the transformed data matrix $\boldsymbol{I}$ of the infrared thermal imaging image, and construct the extracted data matrix $\boldsymbol{I}'_F$ of the pipe segment at the output end of the piston pump.

Step 4: independent imaging of the region at the output end of the piston pump.

The extraction matrix $\boldsymbol{I}'_F$ of the pipe segment at the output end of the piston pump is imaged. All elements of the extraction matrix $\boldsymbol{I}'_F$ except the output flange segment of interest has a value of 0 and can be left out of the imaging process.

Step 5: the RGB value distribution statistics of the infrared thermal imaging map at the output end of the piston pump.

Statistical analysis of the RGB value distribution of the infrared thermal imaging map of the flange section at the output end of the piston pump: Label the three-dimensional coordinates of the RGB values of all pixels of the infrared thermal imaging map in the three-dimensional space consisting of the variables R, G, and B, and construct a statistical map of the spatial distribution of the RGB values; The number of pixel points contained in each RGB value is counted, which is converted into a volume parameter and assigned to the corresponding RGB value points to construct a statistical map of spatial-scalar distribution of RGB values. The spatial distribution of RGB value points and their changes in the statistical map can be used to characterize the range of values and fluctuation range of the surface temperature of the equipment related to the tonal richness of the infrared thermal imaging map, and the size of each RGB value point and its changes in the statistical map can be used to characterize the current level and future trend of the surface temperature of the equipment related to the main color of the infrared thermal imaging map.

Step 6: early accident monitoring and identification of water leakage in the output end region of the plunger pump.

Based on the changes in the distribution of RGB values of the infrared thermographic map at different moments, analyze the changes in the temperature parameters of the flange port pipe section at the output end. If the magnitude and rate of temperature change is greater than the conventional value or set threshold, the monitoring system will determine the current area of failure or accident. By the infrared thermal imaging map color and visual level of the parameter description that is when the magnitude and rate of change in color values is greater than the conventional values or set thresholds, the monitoring system will determine the current area of failure or accident.

3.4.3 *Case study*

In this section, the leakage experiment of the pipe section of the flange port at the output end of the piston pump will be taken as an example to illustrate the practicality and effectiveness of the described method.

In order to better carry out the related scientific research work, a set of plunger pump fault simulation experimental platforms is designed and built, as shown in Fig. 3.48. It mainly includes a plunger pump system module, an operating parameter monitoring module, a central control module, and an infrared thermal imaging monitoring module. The components and main functions of each module are described as follows:

(1) Plunger pump system module: It contains a plunger pump, transmission box, variable frequency motor, water tank for circulation, water pipeline, valve, etc. It is used to simulate the working form of a shale gas fracturing plunger pump and the operating characteristics of the equipment in terms of mechanical, power, liquid, and electric circuits.

(2) Operation parameter monitoring module: It contains an input flow meter, output flow meter, output pressure gauge, lubricant oil temperature measuring device, lubricant oil pressure gauge, etc., which is used to monitor the operation parameters of the plunger pump system (flow rate, pressure, lubrication status, etc.).

(3) Central control module: It contains a frequency conversion motor controller, parameter display, and parameter output device,

Fig. 3.48. Piston pump failure simulation experiment platform.

which is used to regulate the operating frequency of the piston pump system and read, store, and expand the operating parameters.

(4) Infrared thermal imaging monitoring module: contains infrared thermal imager, computer, etc., used for the acquisition, processing, and analysis of infrared thermal imaging information of piston pump.

The functions and relationships of each module are shown in Fig. 3.49. Among them, the plunger pump system structure sketch is shown in Fig. 3.50, the variable frequency motor output power, through the transmission box to the plunger pump power end.

As can be seen in Fig. 3.52, the output power of the inverter motor is transferred to the power end of the piston pump through the gearbox.

An infrared thermal imaging monitoring module is a peripheral device, its setup position, shooting angle, etc., can be flexibly designed according to the experimental needs or site conditions. The

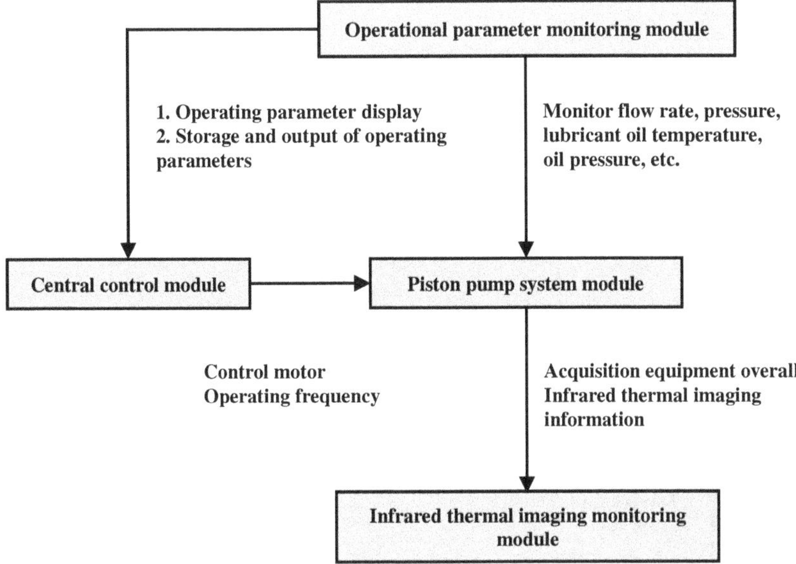

Fig. 3.49. Description of the functions and relationships of each module.

monitoring imaging results are transmitted to the supporting computer in real time, and the computer software realizes the storage, reading, analysis, and other related operations of the infrared thermography data.

The experimental scenarios and processes involved in the water leakage experiment of the flange port section of the output end of the piston pump described in the case study are described as follows:

(1) Start the plunger pump failure simulation experimental platform, adjust the operating frequency to 25Hz, select any time point to start timing after the output flow is stable, and set the timing moment as the starting moment of the experimental process ($t = 0$s), and all the subsequent descriptions of the time are the relative time with the timing moment as the zero point.

(2) Adjust the motor operating frequency from 25Hz to 30Hz, and record the adjustment operation moment as $t = 160$s.

(3) Adjust the motor operating frequency from 30Hz to 50Hz, and record the adjustment operation moment as $t = 334$s.

(4) Start loosening the pipe connection bolts at the output end at $t = 465$s. The operation continues until about $t = 490$s when

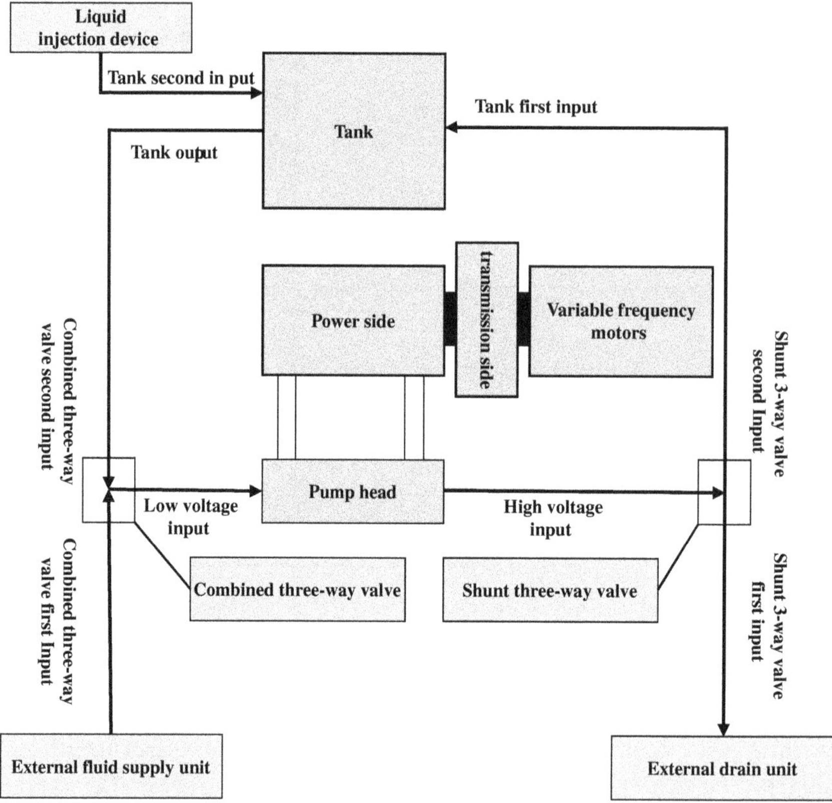

Fig. 3.50.   Plunger pump system structure sketch.

the output flow rate begins to drop significantly, as shown in Fig. 3.51.

(5) Set $t = 480$s as the start time for leakage experimental flow monitoring and $t = 660$s as the end time for leakage experimental flow monitoring.

(6) In order to avoid irreversible damage to the pipeline caused by the continuous leakage of high-speed fluid $t = 717$s when adjusting the motor operating frequency, from 50Hz to 30Hz.

(7) After continuously observing the leakage and the temperature change of the output end for a period of time, adjust the motor operation frequency from 30Hz to 50Hz, and record the moment of adjustment operation as $t=810$s.

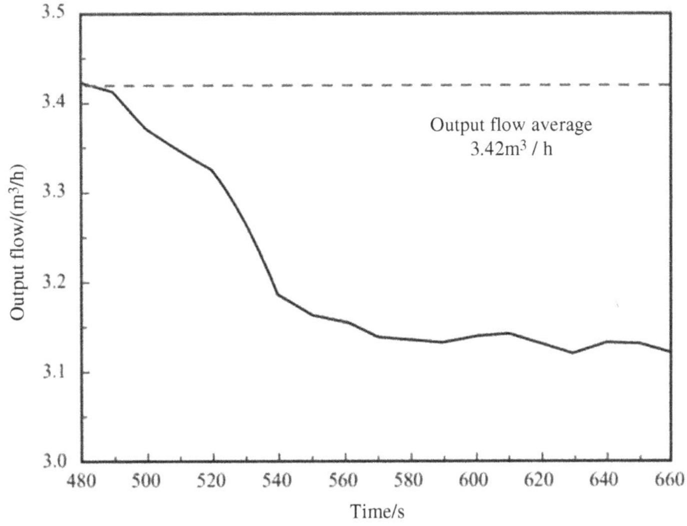

Fig. 3.51. Leakage accidental flow change at output.

(8) Adjust the motor operating frequency at the same time ($t \approx$ 810s) and begin to tighten the output end of the pipeline connection bolts, until the initial state is restored.

(9) The flow rate gradually returns to normal.

From the information in the manual of the equipment and the statistics of flow parameters under normal operation conditions, it can be seen that the average value (standard value) of the output flow of the piston pump at the operation frequency of 50 Hz is 3.42 m³/h. The changes of the output flow after the leakage are shown in Fig. 3.51 (the time period of the flow monitoring is chosen from the 480th to the 660th after the experiment starts).

Any point in the infrared thermographic map of the output end of the piston pump is selected as the reference temperature measurement point, which is used to observe the change of the surface temperature of the output end with time.

The time period from $t = 344$ s (the highest value of 29.52°C at the temperature measurement point) to $t = 1089$ s (the lowest value of 26.72°C at the temperature measurement point) was selected as the research object, and 9 infrared thermal imaging maps were

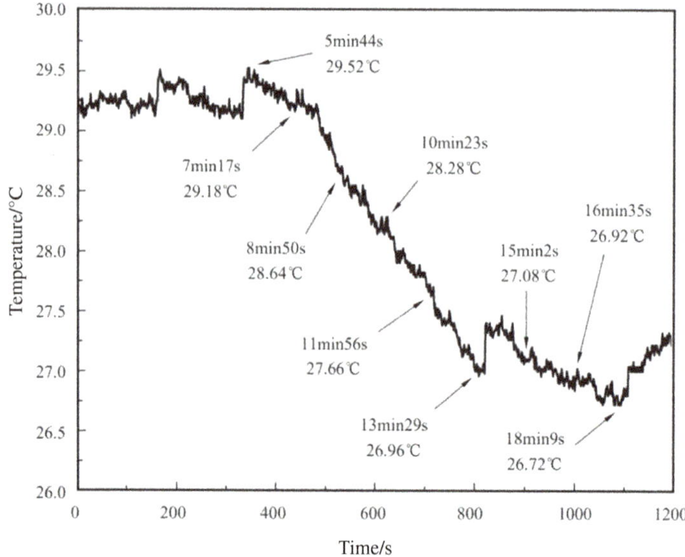

Fig. 3.52. Schematic diagram of the moment of video interception for infrared thermal imaging monitoring.

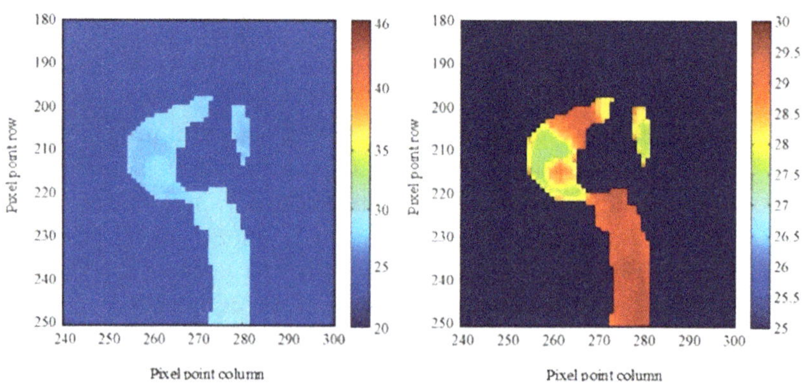

Fig. 3.53. 5 min 44 s infrared thermal imaging monitoring video screenshot enhanced processing before and after comparison.

intercepted in the infrared thermal imaging monitoring video at an interval of 93 s, $(745 \div 8 = 93 \ldots 1$, the remaining 1 s to make up to the last group, that is, the 8th and 9th interval of 94 s), as shown in Fig. 3.52, which are $t = 344$s (5 min 44 s), $t = 437$s (7 min 17 s),

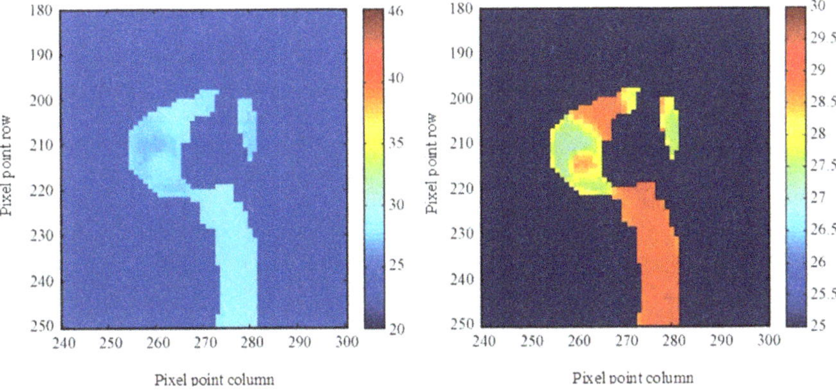

Fig. 3.54. 7 min 17 s infrared thermal imaging monitoring video screenshot enhanced processing before and after comparison.

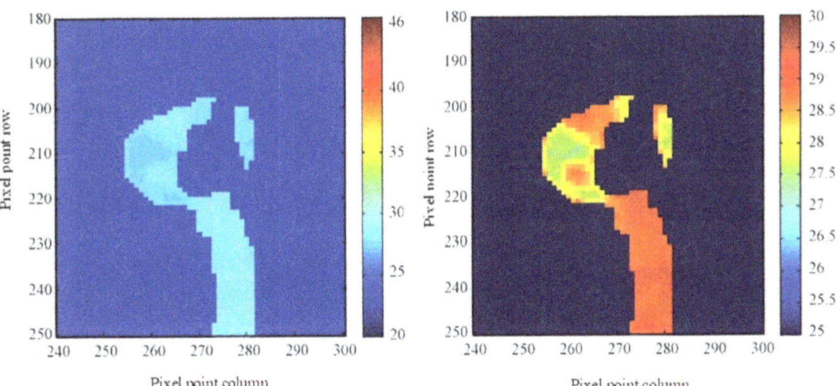

Fig. 3.55. 8 min 50 s infrared thermal imaging monitoring video screenshot enhanced processing before and after comparison.

$t = 530$ s (8 min 50 s), $t = 623$ s (10 min 23 s), $t = 716$ s (11 min 56 s), $t = 809$ s (13 min 29 s), $t = 902$ s (15 min 2 s), $t = 995$ s (16 min 35 s), $t = 1089$ s (18 min 9 s). The image enhancement method proposed in this section is used to segment and extract the nine intercepted infrared thermograms and process them independently. The visual effects of the localized infrared thermal imaging images at the output end before and after processing are shown in Figs. 3.53–3.61 (the left figure is before processing, and the right figure is after processing).

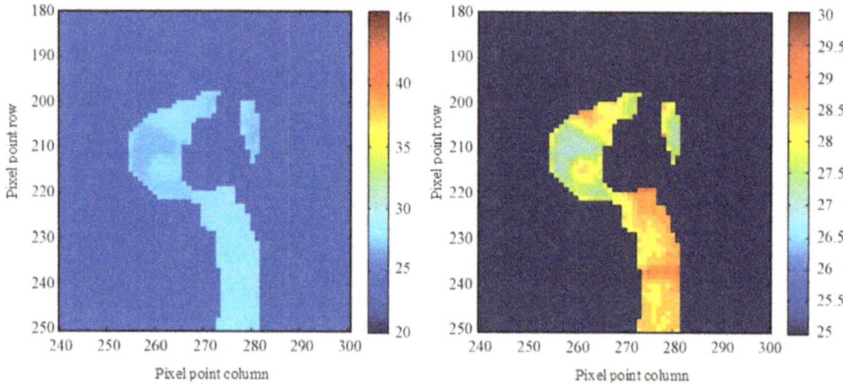

Fig. 3.56.   10 min 23 s infrared thermal imaging monitoring video screenshot enhanced processing before and after comparison.

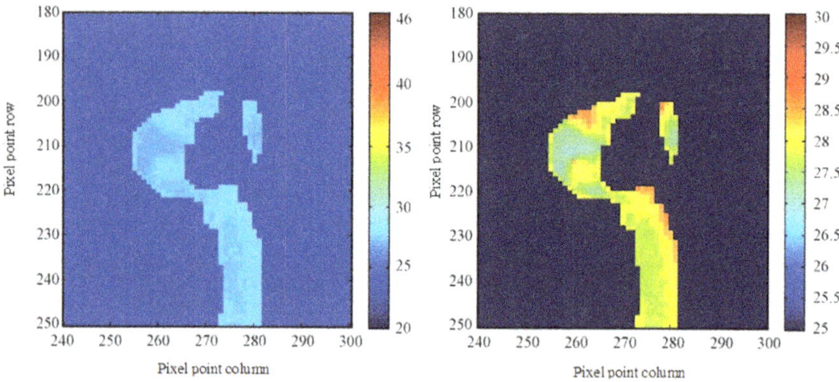

Fig. 3.57.   11 min 56 s infrared thermal imaging monitoring video screenshot enhanced processing before and after comparison.

The RGB value distribution statistics of 9 groups of infrared thermographic images before and after the enhancement treatment are shown in Figs. 3.62–3.71.

(1) 5 min 44 s infrared thermal imaging early accident monitoring and identification

As can be seen from Fig. 3.62, there are only 7 RGB value points in the first thermogram without enhancement processing (at 5min44s), which are used to represent 23 temperature value classes between

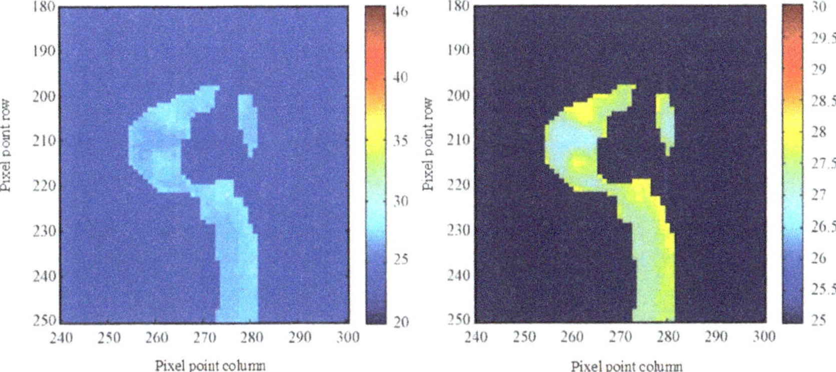

Fig. 3.58. 13 min 29 s infrared thermal imaging monitoring video screenshot enhanced processing before and after comparison.

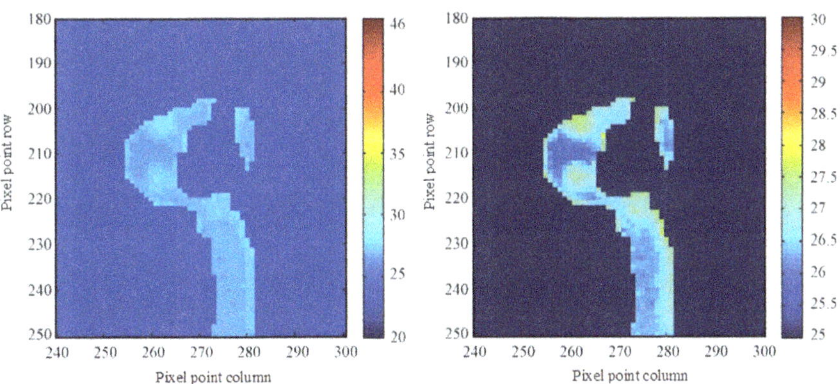

Fig. 3.59. 15 min 2 s infrared thermal imaging monitoring video screenshot enhanced processing before and after comparison.

27.3 and 29.6°C, i.e., each RGB value corresponds to 3.29 temperature value classes on average. And the R value is constant 0, the B value is constant 1, and the value change occurs only in the G value. It means that the main color of the image is blue and the minimum recognizable temperature difference based on the color difference should be 0.33°C.

As can be seen from Fig. 3.63, the first infrared thermography image (at 5min44s) after enhancement processing has a total of 23 RGB values for representing 23 temperature value classes, i.e., each temperature value class has a unique RGB value corresponding to it.

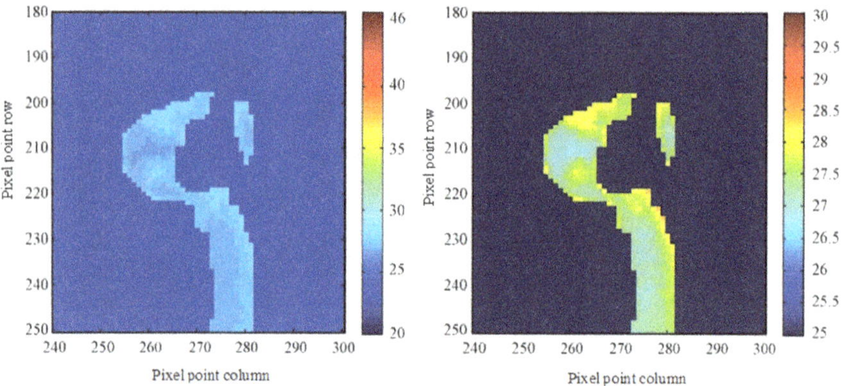

Fig. 3.60. 16 min 35 s infrared thermal imaging monitoring video screenshot enhanced processing before and after comparison.

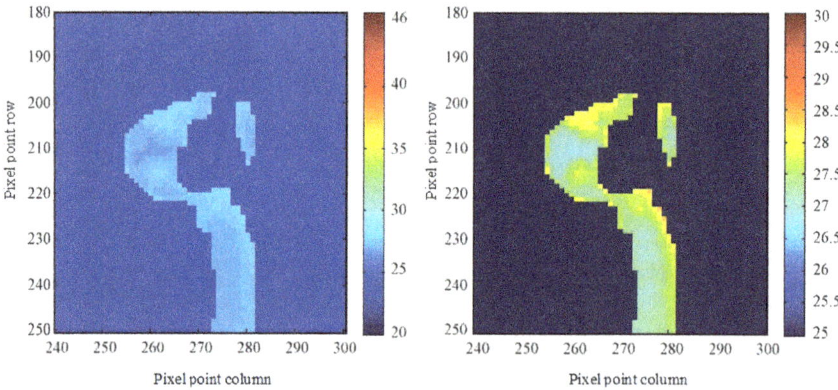

Fig. 3.61. 18 min 9 s infrared thermal imaging monitoring video screenshot enhanced processing before and after comparison.

The values change in all three values, R, G, and B, which indicates that the image hue is richer, and the minimum recognizable temperature difference based on the color difference is $0.10°$C.

From Fig. 3.63, it can be seen that there are 252 pixel points on the R-axis, i.e., there are 252 pixel points in the region of 29.2–29.6°C, which accounts for 45.32% of the total. There are 252 pixels in the R-axis, i.e., 29.2–29.6°C, accounting for 45.32% of the total number of pixels, and they are concentrated at two RGB values: (0.813, 0, 0) contains 80 pixels, (0.875, 0, 0) contains 70 pixels, which correspond to the temperature values of 29.5°C and 29.4°C, respectively.

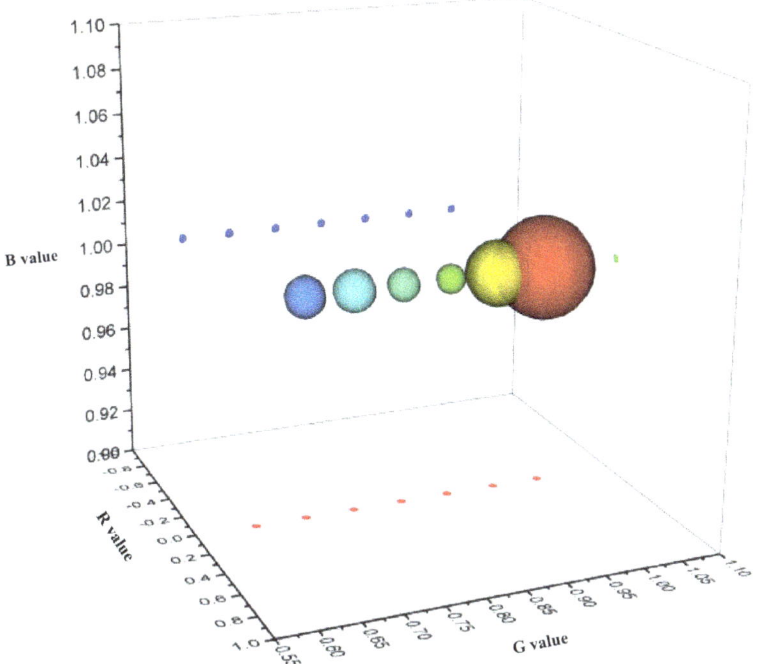

Fig. 3.62. 5 min 44 s RGB value distribution statistics of infrared thermal imaging map before enhancement processing.

There are 246 pixel points on the G-axis, i.e., there are 246 pixel points in the region of 27.7~29.1°C, accounting for 44.24% of the total number of pixel points, but they are evenly distributed, with an average of 17.57 pixel points for each temperature level, and thus there is no concentration of temperature points.

In summary, the temperature of the equipment surface in the first infrared thermal imaging map (at 5min44s) is concentrated at 29.4°C and 29.5°C, and the temperature of the measurement point is 29.52°C.

(2) 7 min 17 s infrared thermal imaging early accident monitoring and identification

As can be seen from Fig. 3.64, there are only 7 RGB value points in the second IR thermogram (at 7min17s) without enhancement processing, which are used to represent 24 temperature value classes between 27.1 and 29.5°C, i.e., each RGB value corresponds to 3.43

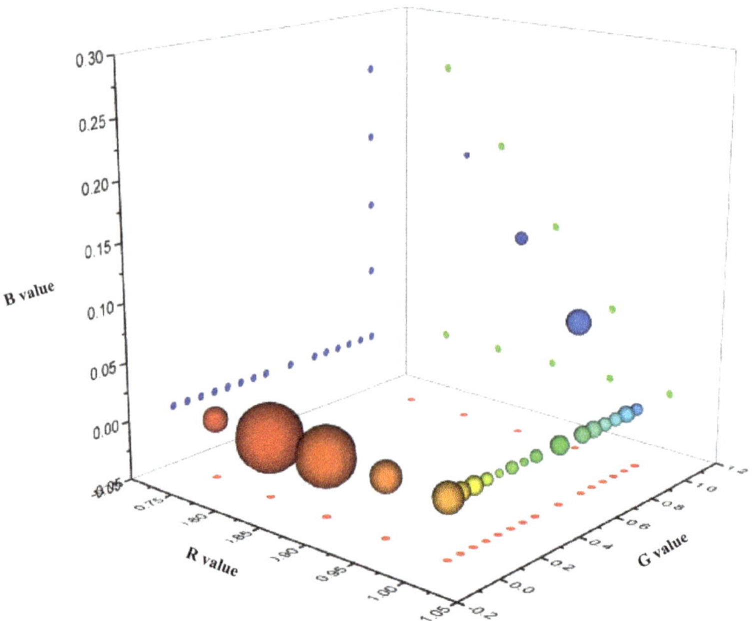

Fig. 3.63. 5 min 44 s RGB value distribution statistics of infrared thermal imaging map after enhancement processing.

temperature value classes on average. And the R value is constant 0, the B value is constant 1, and the value change occurs only in the G value. It indicates that the main color of the image is blue, and the minimum identifiable temperature difference based on the color difference should be 0.34°C.

As can be seen from Fig. 3.65, the second infrared thermography image (at 7min17s) after enhancement processing has a total of 24 RGB values for representing 24 temperature value classes, i.e., each temperature value class has a unique RGB value corresponding to it. And the values change in all three values, R, G, and B, which indicates that the image hue is richer, and the minimum recognizable temperature difference based on the color difference is 0.10°C.

From Fig. 3.65, it can be seen that there are 176 pixel points on the R-axis, i.e., there are 176 pixel points in the region of 29.2-29.5°C, which accounts for 31.65% of the total number of pixel points. There are 176 pixels on the R-axis, i.e., there are 176 pixels in the region of 29.2-29.5°C, accounting for 31.65% of the total number of pixels,

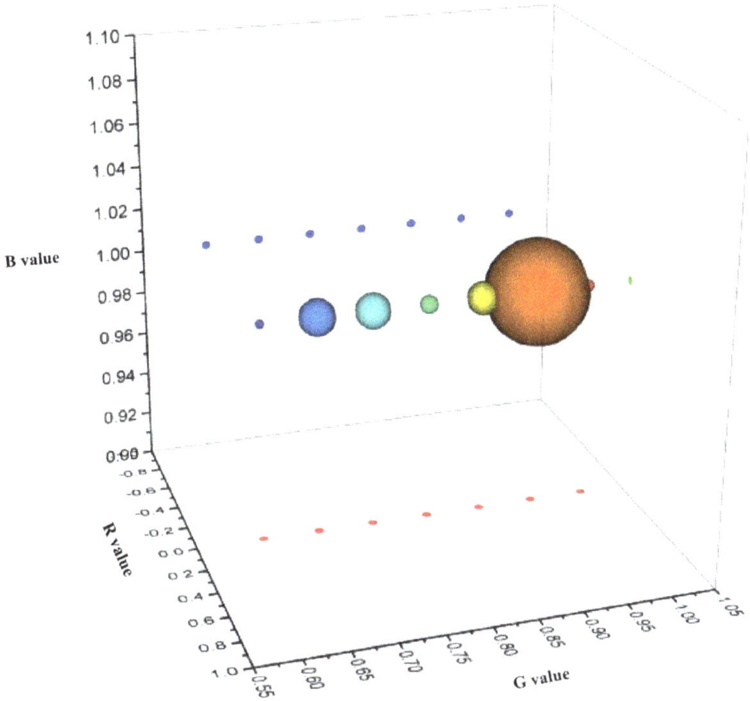

Fig. 3.64. 7 min 17 s RGB value distribution statistics of infrared thermal imaging map before enhancement processing.

and they are concentrated at two RGB values: (0.938, 0, 0) contains 70 pixels, and (1, 0, 0) contains 81 pixels, which correspond to the temperature values of 29.3°C and 29.2°C respectively. There are 299 pixels on the G-axis, i.e., there are 299 pixels in the region of 27.7–29.1°C, accounting for 53.78% of the total number, but they are evenly distributed, with an average of 21.36 pixels for each temperature level, and thus there is no concentration of temperature points.

To summarize, the infrared thermal imaging map (at 7 min 17 s) shows that the surface temperature of the equipment is concentrated at 29.2°C and 29.3°C, and the temperature of the measurement point is 29.18°C

(3) 10 min 23 s infrared thermal imaging early accident monitoring and identification

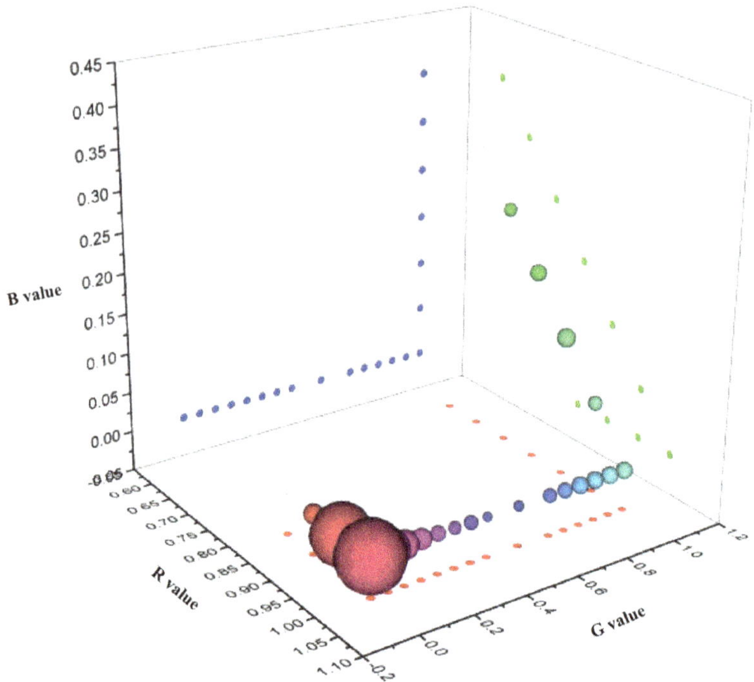

Fig. 3.65.   7 min 17 s RGB value distribution statistics of infrared thermal imaging map after enhancement processing.

As can be seen from Fig. 3.66, there are only 6 RGB value points in the fourth IR thermogram (at 10min23s) without enhancement processing, which are used to represent 22 temperature value classes between 27.0 and 29.3°C, i.e., each RGB value corresponds to 3.67 temperature value classes on average. And the R value is constant 0, the B value is constant 1, and the value change occurs only in the G value. It indicates that the main color of the image is blue, and the minimum identifiable temperature difference based on the color difference should be 0.37°C.

As can be seen from Fig. 3.67, the fourth infrared thermography image (at 10min23s) after enhancement processing has a total of 22 RGB values for representing 22 temperature value classes, i.e., each temperature value class has a unique RGB value corresponding to it. The values change in all three values, R, G, and B, which indicates that the image hue is richer, and the minimum recognizable temperature difference based on the color difference is 0.10°C.

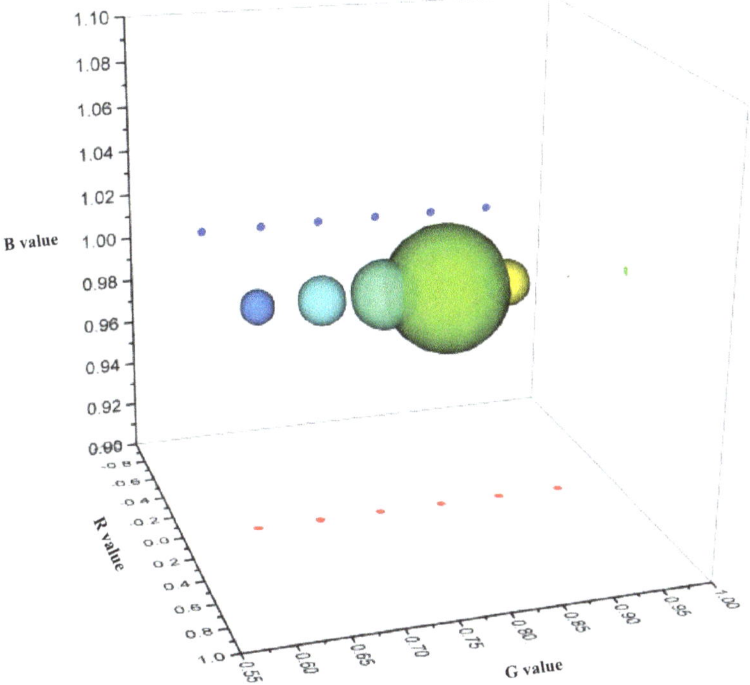

Fig. 3.66. 10 min 23 s RGB value distribution statistics of infrared thermal imaging map before enhancement processing.

From Fig. 3.67, there are 435 pixels on the G-axis, i.e., there are 435 pixels in the region of 27.7~29.1°C, accounting for 78.24% of the total number of pixels, and they are concentrated in the two RGB values: the point of (1,0.563,0) contains 90 pixels, and the point of (1,0.688,0) contains 81 pixels, which correspond to the temperature values of 28.4°C, 28.2°C, etc. The RB plane has 120 pixels, i.e., there are 120 pixels in the region of 27.0-27.6°C, accounting for 21.58% of the total number of pixels. In summary, the surface temperature of the equipment in the infrared thermal imaging map (at 10min23s) is concentrated at two places, 28.20°C and 28.40°C, and the temperature of the measurement point is 28.28°C.

(4) 11 min 56 s infrared thermal imaging early accident monitoring and identification

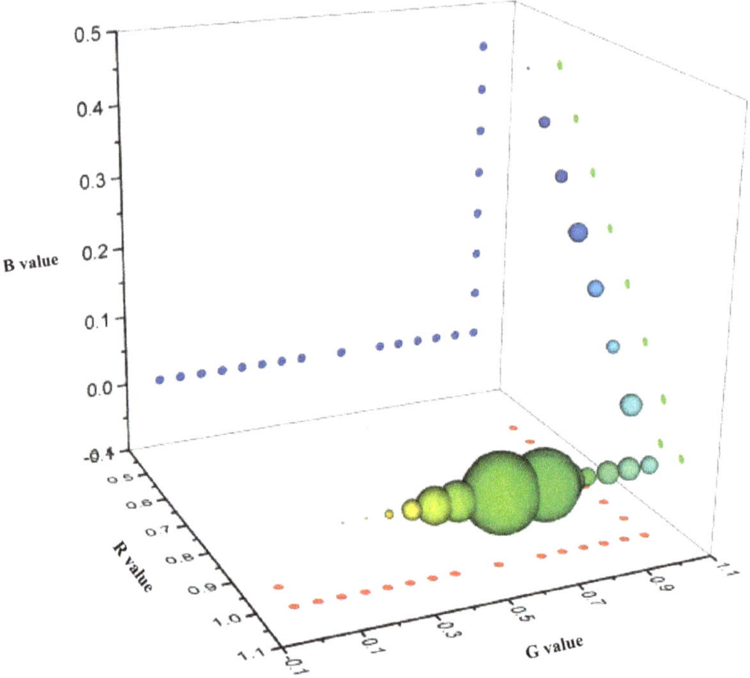

Fig. 3.67. 10 min 23 s RGB value distribution statistics of infrared thermal imaging map after enhancement processing.

As can be seen from Fig. 3.68, there are only 5 RGB value points in the fifth thermogram (at 11min56s) without enhancement processing, which are used to represent 17 temperature value classes between 27.0 and 28.7°C, i.e., each RGB value corresponds to 3.40 temperature value classes on average. The R value is constant 0, the B value is constant 1, and the value change occurs only in the G value. It indicates that the main color of the image is blue and the minimum identifiable temperature difference based on the color difference should be 0.34°C.

As can be seen from Fig. 3.69, the fifth infrared thermography image (at 11min56s) after enhancement processing has a total of 17 RGB values, which are used to represent 17 temperature value classes, i.e., each temperature value class has a unique RGB value corresponding to it. The values change in all three values, R, G, and B, which indicates that the image hue is richer, and the minimum

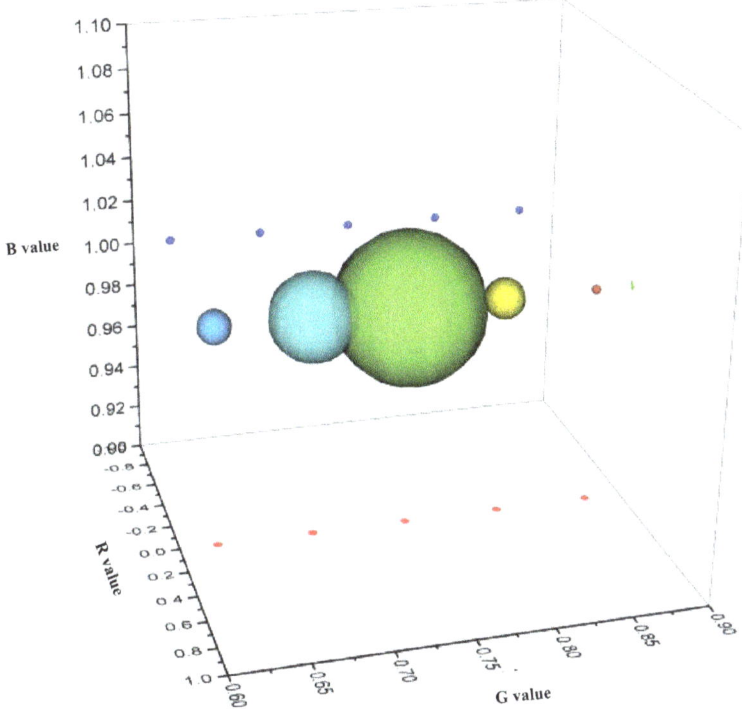

Fig. 3.68. 11 min 56 s RGB value distribution statistics of infrared thermal imaging map before enhancement processing.

recognizable temperature difference based on the color difference is 0.10°C.

From Fig. 3.69, there are 441 pixel points on the G-axis, i.e., there are 441 pixel points in the region of 27.7–28.7°C, accounting for 79.32% of the total number, and concentrated at two RGB values: the point of (1, 0.938, 0) contains 86 pixel points, and the point of (1, 1, 0) contains 94 pixel points, which correspond to the temperature values of 27.8°C and 27.7°C, respectively. RB plane contains 115 pixel points, i.e., there are 115 pixel points in the region of 27.0 to 27.6°C, which accounts for 20.68% of the total.

To summarize, the infrared thermal imaging map (at 11 min 56 s) shows that the surface temperature of the equipment is concentrated at 27.7°C and 27.8°C, and the temperature of the measurement point is 27.66°C.

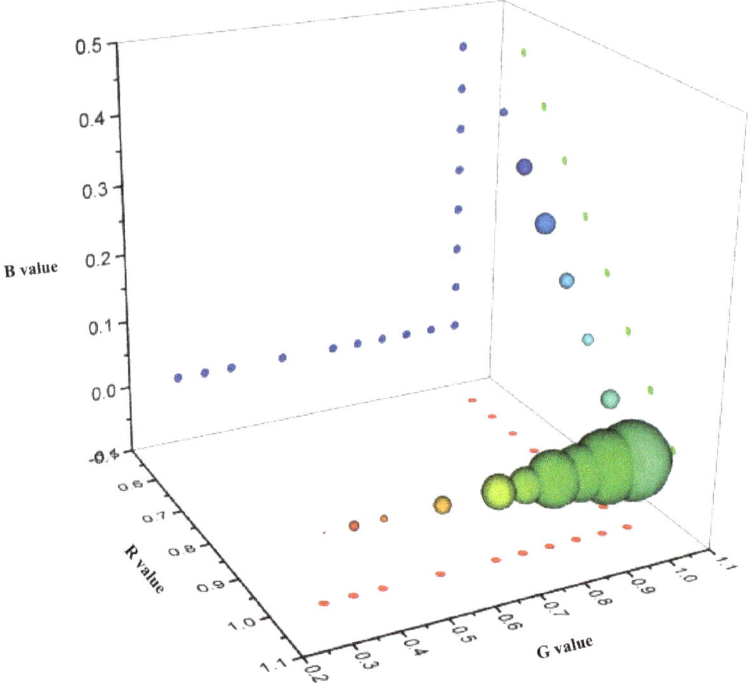

Fig. 3.69. 11 min 56 s RGB value distribution statistics of infrared thermal imaging map after enhancement processing.

(5) 13 min 29 s infrared thermal imaging early accident monitoring and identification

As can be seen from Fig. 3.70, there are only five RGB value points in the sixth IR thermogram (at 13min29s) without enhancement processing, which are used to represent 16 temperature value classes between 26.7 and 28.2°C, i.e., each RGB value corresponds to 3.20 temperature value classes on average. The R value is constant 0, the B value is constant 1, and the value change occurs only in the G value. It indicates that the main color of the image is blue, and the minimum identifiable temperature difference based on the color difference should be 0.32°C.

As can be seen from Fig. 3.71, the sixth infrared thermography image (at 13min29s) after enhancement processing has a total of 16 RGB values, which are used to represent 16 temperature value classes, i.e., each temperature value class has a unique RGB value

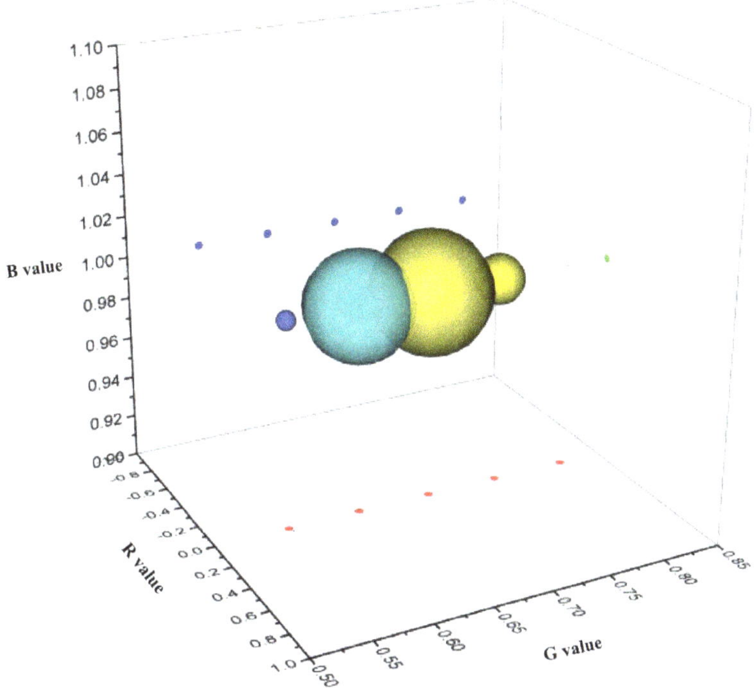

Fig. 3.70. 13 min 29 s RGB value distribution statistics of infrared thermal imaging map before enhancement processing.

corresponding to it. The values change in all three values, R, G, and B, indicating that the image hue is richer, and the minimum recognizable temperature difference based on the color difference is $0.10°$C.

From Fig. 3.71, there are 141 pixel points on the G-axis, i.e., there are 141 pixel points in the region of 27.7–$28.2°$C, accounting for 25.36% of the total number of pixel points. There are 415 pixel points on the RB plane, i.e., there are 415 pixel points in the region of 26.7–$27.6°$C, accounting for 74.64% of the total number of pixel points. And it is concentrated at one RGB value: the $(0.75, 1, 0.25)$ point contains 99 pixel points, which corresponds to the temperature value of $27.30°$C.

In summary, the infrared thermography (13 min 29 s) shows that the surface temperature of the equipment is concentrated at $27.30°$C, and the temperature at the measurement point is $26.96°$C.

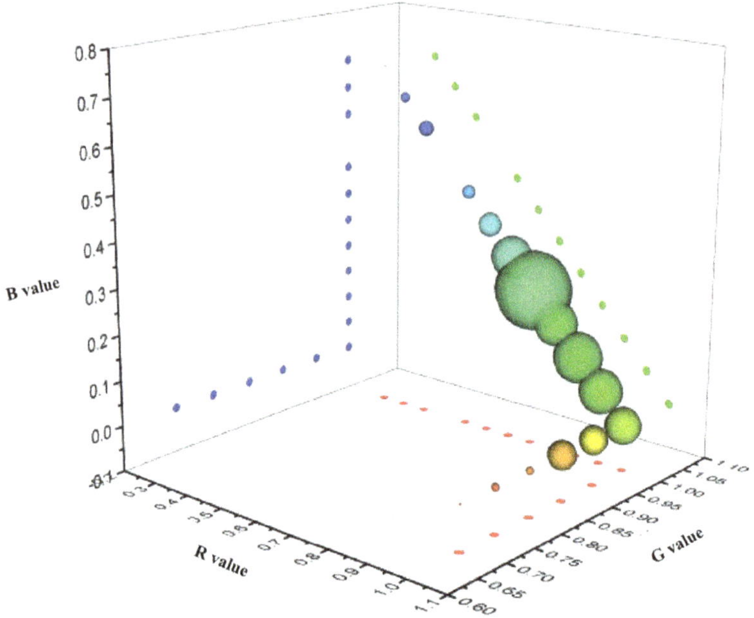

Fig. 3.71. 13 min 29 s RGB value distribution statistics of infrared thermal imaging map after enhancement processing.

(6) Output leakage early incident monitoring and identification

From the experimental results, it can be seen that the surface temperature of the output end of the piston pump changes less when it is running smoothly and tends to be close to the steady state. When adjusting the frequency of motor operation to change the output flow, the surface temperature of the output end will jump and return to a stable state after a period of operation. When the output end of the leakage accident, the pipeline components in the leakage of the liquid scouring effect of the surface temperature will appear to persistent reduction, until the high-pressure liquid and pipe wall friction of heat and the leakage of the liquid scouring cooling effect of the heat dissipation to re-establish the balance.

Comparison of the experimental phenomenon and the output temperature measurement point temperature trend can be seen, after the occurrence of leakage accidents, the output end of the surface temperature quickly dropped to a steady state (average temperature of 29.20°C) below the 29.20°C as a criterion for determining, when the output end of the surface temperature continues to be lower than the steady state and keep the downward trend, it can be deduced that at the current moment of the output end of the leakage accident may have occurred (after the data cleaning of the infrared thermal imaging monitoring video can exclude the phenomenon of abnormal temperature reduction caused by foreign objects blocking the output end area).

Converting $T = 29.20°C$ to the RGB value of the image in the set imaging rule, it can be obtained that $R=1$, $G=0$, $B=0$, i.e., $T = 29.20°C$ in the temperature space is converted to $(1, 0, 0)$ in the color space, as shown in Fig. 3.72.

When $T > 29.20°C$, the temperature decreases, the R value increases, and the G and B values remain unchanged, i.e., the RGB value point in the RG plane moves to the right along the R axis. When $27.70°C < T < 29.20°C$, the temperature decreases, the R value and B value are unchanged, the G value increases, that is, the RGB value point in the RG plane along the G axis moves upward.

When $T = 27.70°C$, the corresponding RGB value point is $(1, 1, 0)$. Then when $T < 27.70°C$, the G value is unchanged, the R value on the RB plane decreases, the B value increases, and the two follow the numerical relationship of $R + B = 1$. That is, the temperature decreases, the G value remains unchanged, and the point of the RGB value moves upward in the RB plane along the $R = 1 - B$ straight line.

From Fig. 3.72, the RGB value concentration point $(1, 0.156, 0)$ of the third infrared thermography map (at 8min50s) is higher than the critical point $(1, 0, 0)$ in the direction of the G-axis, which indicates that leakage may have occurred at 8min50s. Thereafter, the RGB value concentration point $(1, 0.622, 0)$ at the fourth infrared thermal imaging map (10min23s) continues to move upward along the G-axis direction, i.e., the corresponding surface temperature value of the equipment continues to decrease, indicating that a leakage accident can be confirmed at 10min23s.

After the leak was repaired by tightening the pipe connection bolts at the output end, the rate of decrease of temperature values at the temperature measurement points slowed down compared with that during the leak. The main temperature values of the monitored objects in the three intercepted infrared thermograms (at 15 min 2 s, 16 min 35 s, and 18 min 9 s) were $27.10°C$, $27.20°C$, and $27.00°C$,

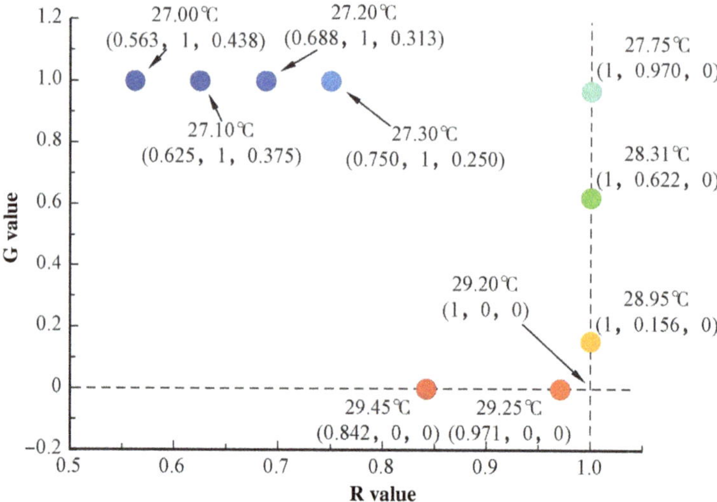

Fig. 3.72. Infrared thermal imaging map RGB value points RG plane projection map.

respectively. The concentration points of the corresponding RGB values are (0.625, 1, 0.375), (0.688, 1, 0.313), (0.563, 1, 0.438), as shown in Fig. 3.75. It shows that the leakage accident has been gradually repaired at this time, and the output surface gradually reestablishes the balance between heat dissipation and heat generation. As can be seen from the experimental phenomenon, about the 1080s when the output surface temperature dropped to the lowest value, that is, the current moment the output liquid and pipe wall friction of the heat generation capacity is about equal to the output end of the pipeline's own heat dissipation capacity, and thereafter, the output surface temperature will begin to gradually increase until it returns to the normal level.

3.5 Precise Identification and Early Warning of Fracturing Equipment Faults Based on Infrared Thermal Imaging and CNN

In recent years, deep learning methods have been developed rapidly, and the representative branch of convolutional neural network (CNN) has given rise to many types of network structures, which are widely used in the fields of image and video target detection, segmentation, and recognition. The faulty parts of the fracturing pump are mainly the output end, the pump head body, and the input end, which are not obvious on the infrared thermograms due to the cooling effect brought by the liquid flow and the thickness of the external casing. In order to overcome the limitations of the traditional infrared image analysis method when the temperature characterization is small, and to solve the difficulties in identifying the faults of shale gas fracking equipment represented by fracturing pumps, we carry out research on the typical fault characteristics of fracturing pumps, the preprocessing of infrared thermography images and the intelligent identification algorithm at the later stage. A holistic preprocessing method of infrared thermal image is established, and a convolutional neural network is introduced to extract and classify the fault features in the infrared thermal image of fracking equipment and optimize the parameters, so as to realize the identification and early warning of the operating faults of shale gas fracking pumps.

3.5.1 *Basic theory*

(1) Preprocessing of infrared thermograms

Infrared thermal imagers in the use of the process are susceptible to ambient temperature, light, emissivity, and wind speed, and supporting software palette settings for these factors makes the formation of infrared thermal images different. In order to make the infrared thermal image analysis results more accurate, the establishment of a holistic infrared thermal image preprocessing method. Firstly, all the images are grayscaled to eliminate the differences caused by different color palettes and further noise reduction, and edge sharpening, and finally, the image size is normalized according to the need, as shown in Fig. 3.73.

(i) Median filtering and noise reduction

The main methods currently used in image noise reduction are mean filter, wavelet transform, median filter, and related improvements. However, mean filtering only distributes the noise intensity of a certain point evenly on the surrounding data, although it reduces the amplitude and plays the role of noise reduction, it increases the

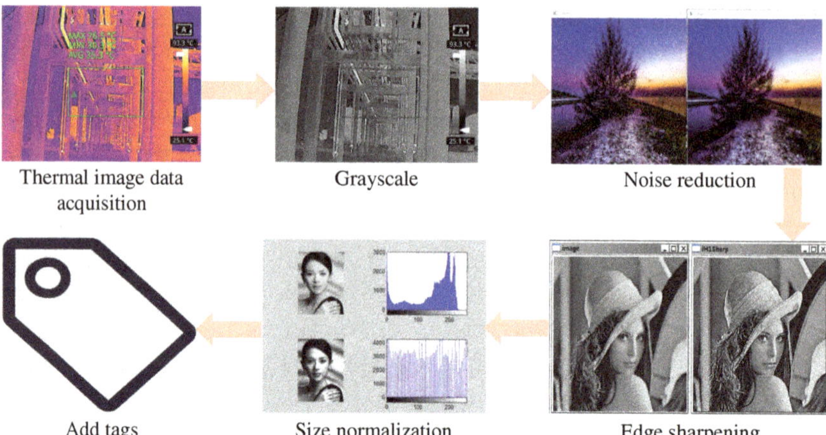

Fig. 3.73.   Infrared thermal image preprocessing process.

particle area of the noise point and makes the edge of the image become fuzzy, which is unfavorable to the next image processing. Wavelet transform for image noise reduction needs to be sacrificed at the expense of resolution, which is something that needs to be avoided at all costs in the analysis and study of thermal imaging images of fracturing pumps. The median filter is a kind of noise reduction method based on the statistical theory of ordering, which can better deal with the randomly distributed noise on the points of the "pretzel" class, and can maintain good clarity, so the median filter is used for the image noise reduction.

(ii) Laplace edge sharpening algorithm

After noise reduction, ensure the image contrasts with the edges take the Laplace algorithm for sharpening and enhancement. The Laplace operator is a class of isotropic differential operators with rotational invariance. The Laplace transform of a two-dimensional image function is an isotropic second-order derivative, which is defined as Eq. (3.30):

$$\nabla^2 f(x, y) = \frac{\partial^2 f}{\partial x^2} + \frac{\partial^2 f}{\partial y^2}. \tag{3.30}$$

To make the processing of the image easier, the equation is represented using a discrete form, denoted as Eq. (3.31):

$$\nabla^2 f = f(x+1, y) + f(x-1, y) + f(x, y+1)$$
$$+ f(x, y-1) - 4f(x, y). \tag{3.31}$$

Image sharpening is based on the principle of making a blurred image clearer by enhancing the grayscale contrast. Due to the application of the Laplace differential operator can attenuate the slow-changing gray areas in the image and on the other hand it can enhance the gray areas in the image with sudden gray changes. Therefore, the Laplace operator can be chosen to sharpen the original image to produce an image that describes the sudden grayscale changes, and then the image can be superimposed with the original image to produce a sharpened image. The basic method of Laplace sharpening can be expressed as Eq. (3.32):

$$g(x, y) = \begin{cases} f(x, y) - \nabla^2 f(x, y), \text{Negative mask centre factor} \\ f(x, y) + \nabla^2 f(x, y), \text{Positive mask centre factor} \end{cases},$$
$$\tag{3.32}$$

where $g(x,y)$ is the output; $f(x, y)$ is the original 2D image.

(2) Optimized CNN models

CNN is a special deep neural network model containing convolutional layers, which can effectively reduce the number of weights and the complexity of the network structure by virtue of its weight sharing, local awareness, down-sampling, etc., reduce the number of preprocessing image steps and have better generalization performance, and is therefore widely used in the field of speech and image recognition. The structure of a CNN model generally consists of five parts: input layer, convolutional layer, subsampling layer, fully connected layer, and output layer. Commonly used CNN models include LetNet, AlexNet, VGG, and GoogleNet.

Since the size of infrared thermal imaging images is larger compared to photos in various online image datasets, the increase in network depth will undoubtedly increase the training time, so the most classic convolutional neural network structure, LetNet-5, is selected, and its structure is shown in Fig. 3.74.

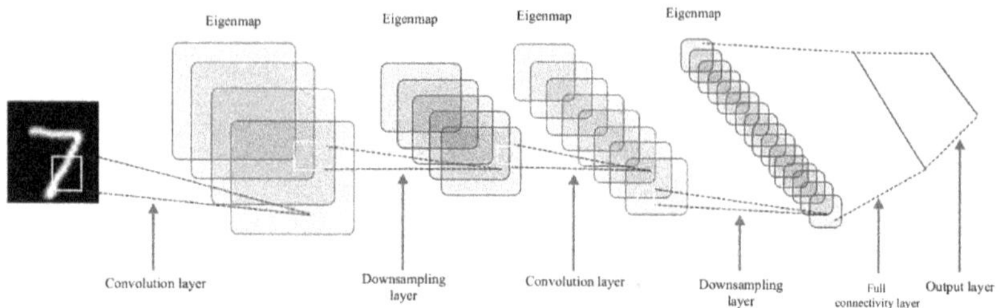

Fig. 3.74. LetNet-5 structure schematic.

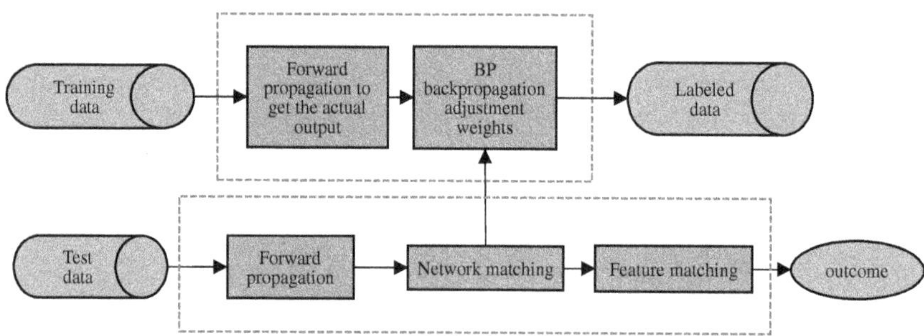

Fig. 3.75. CNN processing flow.

Among them, the convolutional layer and the subsampling layer are arranged alternately, different 2D feature maps are extracted by different convolutional kernels, the subsampling layer ensures the scaling invariance of the features and the sharing of the weights of the same feature map, and finally, the 2D feature map is converted to a 1D output by the fully-connected layer through the dot-product operation. The CNN processing flow is shown in Fig. 3.75.

In order to further improve the training and computing speed of the previous LetNet-5 network, as well as to obtain better accuracy and reduce the occurrence of overfitting in the case of small samples of fracturing pump operation faults, the Relu activation function and the Dropout layer are introduced for network optimization.

(i) Activation function Relu introduced

In a multilayer neural network structure, the inputs are weighted and summed and then also acted upon with a nonlinear activation function to approximate an arbitrary nonlinear function, otherwise

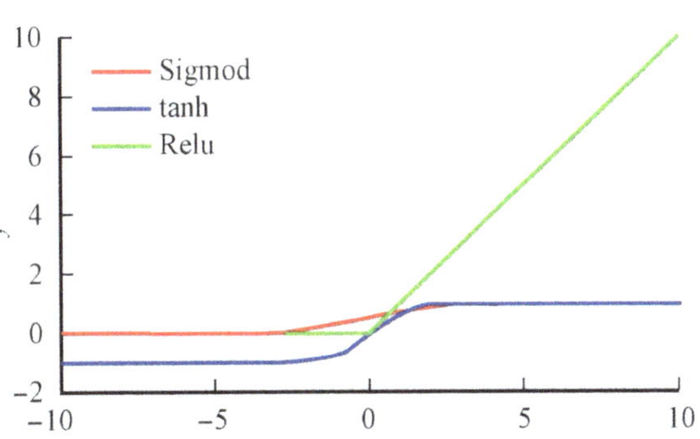

Fig. 3.76. Three types of activation function images.

the output is a linear combination of the inputs no matter how many layers the network has. Commonly used nonlinear activation functions are the Sigmoid function and tanh function. When using the Sigmoid function, the gradient of the function is almost zero once the input is far from the origin. The neural network back-propagation process is to chain law to calculate the differential of each weight W, when the back-propagation through more than one Sigmoid function will lead to the weight W on the loss function is almost zero, the phenomenon of gradient dispersion occurs, and the Sigmoid function needs to carry out the exponential operation, the speed of the image processing is slower. The tanh is a hyperbolic tangent function, the output is almost smooth when the input is very large or very small, and the gradient is very small, which is not favorable for weight updating. Therefore, the Relu function is introduced into the network as an activation function to solve the above problems. As shown in Fig. 3.76, Relu is a segmented function with segmented linear properties, which makes the output of a part of neurons to be 0, reduces the parameter interdependence, which increases the sparsity of the network, reduces the overfitting significantly, and improves the convergence speed during the training process.

(ii) Dropout layer setup

When training a convolutional neural network model, a large amount of data is needed as training samples, and too few training samples will cause overfitting of the model, resulting in a low

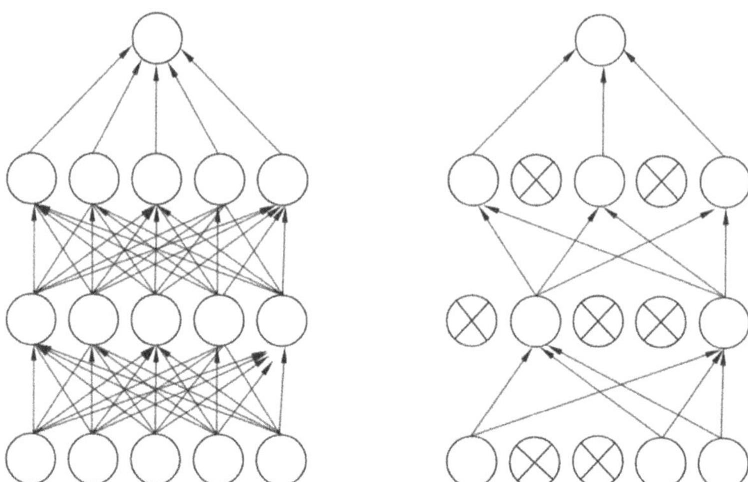

Fig. 3.77. Dropout layer schematic.

accuracy of the final classification results, and failure to ensure good robustness. To address this situation, a dropout layer (Fig. 3.77) is used, where the weights of certain hidden layer nodes of the network do not work in accordance with a certain probability during the training of the model. The non-working nodes can be temporarily considered as not being part of the network structure, but their weights are retained, which for stochastic gradient descent is a random selection, so that each mini-batch is training a different network, thus effectively preventing the occurrence of overfitting phenomenon.

3.5.2 *Steps for accurate identification and early warning of fracking pump operational faults*

(1) Typical failure analysis of fracturing pumps

The types of fracking pump failures and possible causes are as follows, based on field research data from fracking construction sites and fracking equipment manufacturers:

(i) The suction end is empty. Mostly caused by the failure of the sand mixer or the leakage, clogging, and settling of the pipeline at the suction end.

(ii) Pump head body leakage. As a result of long-term high-pressure and cyclic load caused by the action, the construction of the process parameters, and the nature of the material, no obvious signs of such failures need to replace the pump head body.

(iii) high-pressure output leakage. Mostly caused by uneven bolt fastening force, radial vibration amplitude being too large, seal failure, and other reasons.

(iv) Oil circuit misfire. The oil circuit leaks and the heat sink fails to dissipate the heat in time.

(v) Power end abnormality. Failure or abnormality of the power part or transmission device causes power abnormality or even stops the pump.

(2) Specific steps for fault identification and warning

(i) Training phase

Step 1: use infrared thermal imaging acquisition equipment for data collection if the intelligent focus mode will automatically adjust the focal length; if not, then manually adjust to monitor the regional equipment outline clearly.

Step 2: according to the actual need to select the infrared thermal imaging equipment, temperature width, automatic temperature width options for the picture of the highest and lowest temperature limit, and requires a difference of 8°C or more; intelligent temperature width can be removed from some of the inconspicuous temperature points to improve the display contrast; fixed temperature width can be customized temperature width upper and lower limits, and the temperature difference also requires a difference of 8°C or more.

Step 3: set the acquisition frequency f. Generally, the range of f is from 3 frames/frame to 0.5 frames/frame, which can be adjusted according to the length of time from the sign of fault to the occurrence of fault.

Step 4: the data collected by the infrared thermal imaging device is in video format, and it is necessary to transmit the video and convert it into a picture as an input for the next step at certain intervals, which is adjusted according to the hardware configuration, and the shorter the transmission interval, the better.

Step 5: preprocess the image, the infrared thermal imaging image obtained in step 4 will be grayed out, median noise reduction and edge sharpening in turn. The preprocessed infrared thermal imaging image is size normalized as needed, and is used as a training set and a test set data sample.

Step 6: repeat steps 4 and 5 to obtain at least 200 normal and faulty infrared thermal imaging image samples.

Step 7: network training, using the images in step 6 as input, is the next step of network training. CNN network in addition to the input and output there are many parameters, such as the learning rate, step size, the number and size of the convolution size, etc., in the introduction of the Relu function and the Dropout layer also need to be set to discard the rate, etc., the adjustment of these parameters has a direct impact on the degree of convergence of the network and the speed, generalization performance and accuracy. Therefore, the network needs to be adjusted within the common range to select the best combination of parameters suitable for fracturing equipment fault identification represented by fracturing pumps.

(2) Testing and application phase

Step 1: sample infrared thermal imaging images of different types of faults and trained models are obtained and then tested for accuracy using a randomly selected test set.

Step 2: after the accuracy rate meets the requirements, the initial moment of fault occurrence and after the single infrared thermal imaging image data for a single test, the initial warning time. If the single fault data from the time of fault occurrence to 10s after the occurrence of the test, if it can only recognize the image data from 4 to 10s after the occurrence of the fault, the warning time will be 4s, and the test will be conducted several times to ensure the accuracy of the results.

Step 3: repeat steps 1 to 7 in the training phase, install and debug the infrared thermal imaging acquisition equipment at the shale gas fracturing construction site and set up the parameters, obtain the infrared thermal imaging images of the fracturing equipment, and then use them as inputs to the trained CNN model after preprocessing to determine whether the fracturing equipment is in a normal

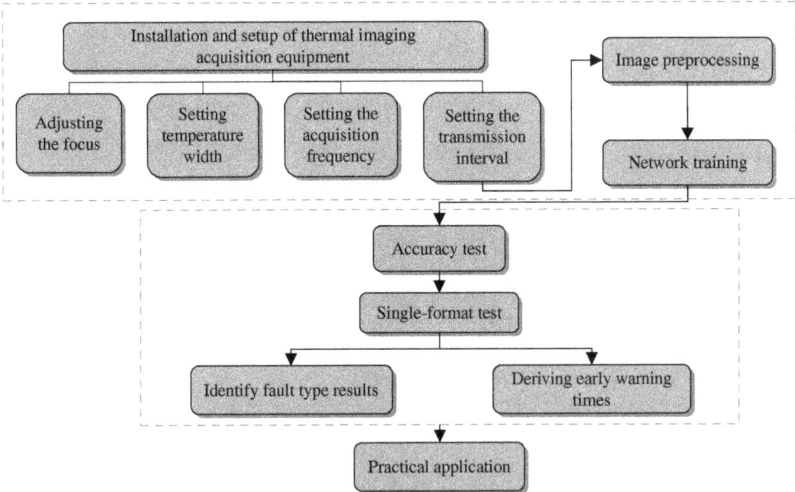

Fig. 3.78. Fault recognition and early-warning step-by-step process.

state or not, and output the fault category if it is recognized as a fault.

The overall flow is shown in Fig. 3.78.

3.5.3 *Cases and analysis*

(1) Data acquisition and preprocessing

Set the relevant parameters, in which the focus mode is set to automatic, the temperature width is set to intelligent mode, the acquisition frequency is 3 pictures/s, the video sending interval is 5s, and the size of the infrared thermal imaging map obtained is 384*280 pixels. The final image size used for training and validation is normalized to 180*180 pixels.

Taking the high-pressure output terminal leakage fault as an example, the fault was simulated by loosening the output terminal bolts under the operating frequency of 50Hz and high pump pressure of 7.7MPa. The infrared thermal image data from 510 to 640s were selected for analysis, and the corresponding temperature changes of the output end and the ground temperature reference point are shown in Fig. 3.79, in which the leakage started at the 604th time. The temperature at the ground reference point Sp1 is 26.5°C

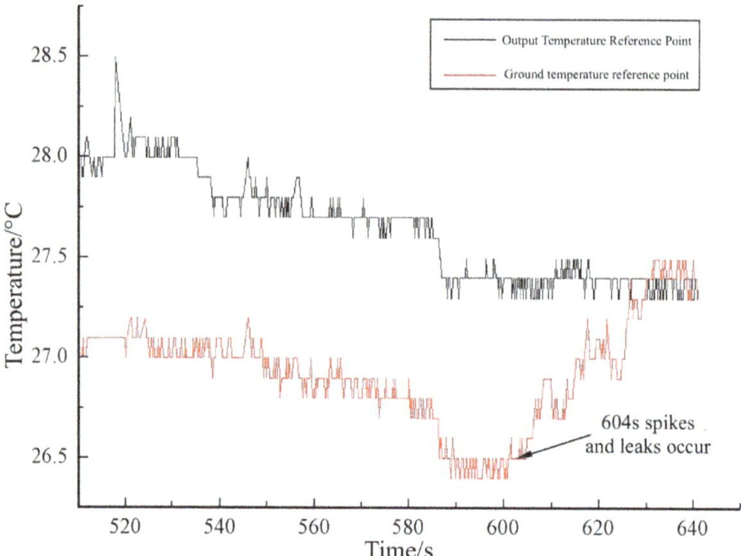

Fig. 3.79.   Temperature change at the reference point for temperature measurement.

and the temperature at the output reference point Sp2 is 27.3°C before the leakage puncture occurs. After the leakage, the temperature of Sp1 gradually rises to 27.7°C due to the higher temperature of the leaking liquid than that of the ground and the environment; at the same time, the surface temperature of the output end also rises to 27.7°C due to the scouring of the leaking liquid and the effect of heat conduction, and finally both of them tend to be stabilized.

The corresponding localized infrared thermal images before and after the puncture leakage fault are shown in Fig. 3.80.

It can be seen that the output and ground temperature changes are very small, reflected in the thermal image of the color difference is small, and due to changes in light, emissivity and other changes in the infrared thermal image of the color level has also changed, relying solely on the human eye to judge the easy fatigue and misjudgment. According to the grayscaling, denoising, and sharpening preprocessing process of the resulting image processing, the results are shown in Fig. 3.81. Then the size normalization, the size of 180 * 180 pixels of the dataset, of which 110 leakage fault samples, the corresponding normal class samples for 120, as shown in Fig. 3.82.

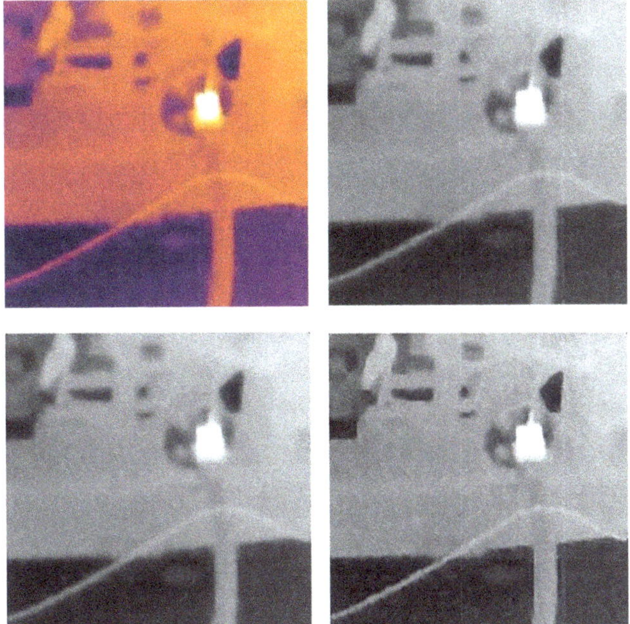

(a) Infrared thermal image of a puncture leak before it occurs (b) Infrared thermal image of a puncture leak

Fig. 3.80. Localized infrared thermograms before and after leak stabbing.

Fig. 3.81. Schematic diagram of the preprocessing results of the infrared thermal image.

2. CNN network training

The improved CNN network model adopts a five-layer network structure, selects Relu as the activation function, adds a dropout layer, takes max-pooling as the means, and uses cross-entropy to define the loss, and after many times of adjusting the parameters,

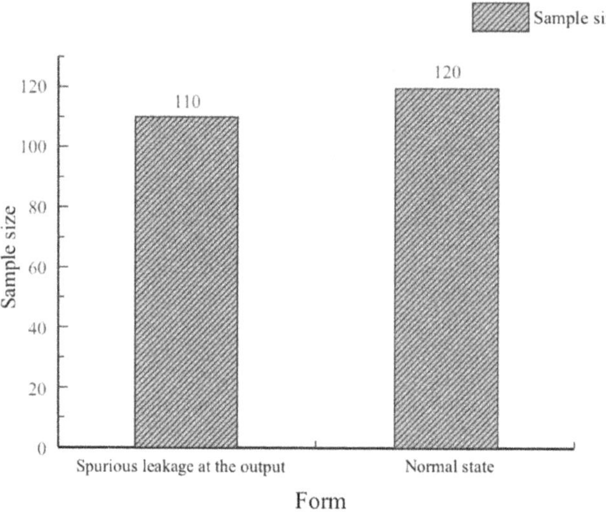

Fig. 3.82.   Fault dataset.

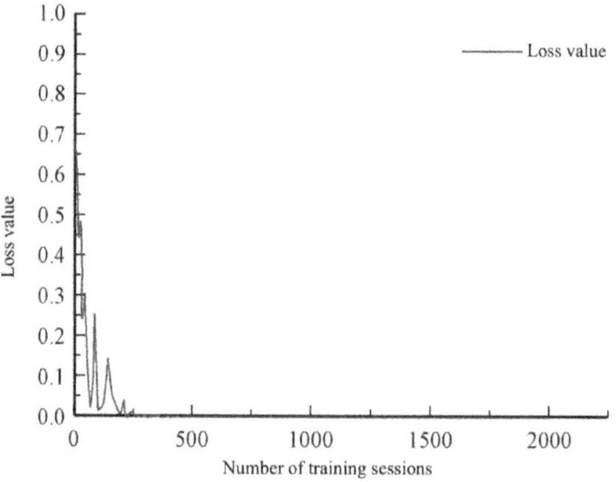

Fig. 3.83.   Loss function variation curve.

it is more effective when the learning rate is selected to be 0.0001, and the dropout rate is selected to be 0.25. The change of loss value when the number of training times is 2000 is shown in Fig. 3.83, and the loss value has been reduced to very small after about 300 times of training. The change in training accuracy is shown in Fig. 3.84, after about 300 times of training, the accuracy is also close to 100%.

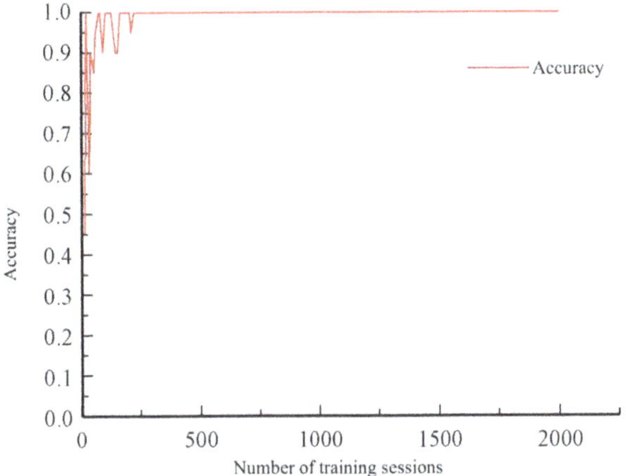

Fig. 3.84. Training accuracy change curve.

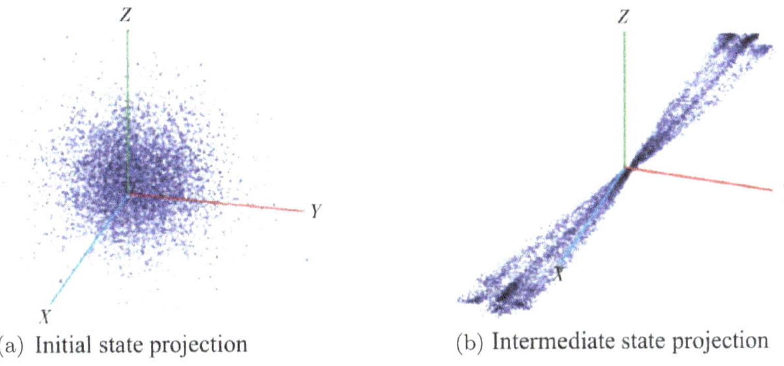

(a) Initial state projection   (b) Intermediate state projection

Fig. 3.85. Data 3D spatial projection.

Using the Tensorboard tool to visualize the training process can get the tensor changes of convolutional layer 1, convolutional layer 2, fully connected layer, and the classifier, while the high-dimensional vector inputs are projected into 3D space by the embedding projector in the tool, the initial state is shown in Fig. 3.85(a), and the state of the intermediate 200 iterations is shown in Fig. 3.85(b), a clear positional distribution can be seen.

3. Fault identification results

Select 110 spiked fault samples and 120 normal samples for recognition, and compare the method in this section with common image

Table 3.4. Recognition results of different methods on the spiked leakage fault dataset.

Methods	Recognized as faulty/ sheet	Number of actual faulty/ sheet	Recognized as normal/ sheet	Actual normal/ sheet	Accuracy /%
K-means	67	40	163	92	57.4
LBP+SVM	75	53	155	103	67.8
Hog+SVM	82	79	148	110	82.2
CNN	99	98	131	120	94.8

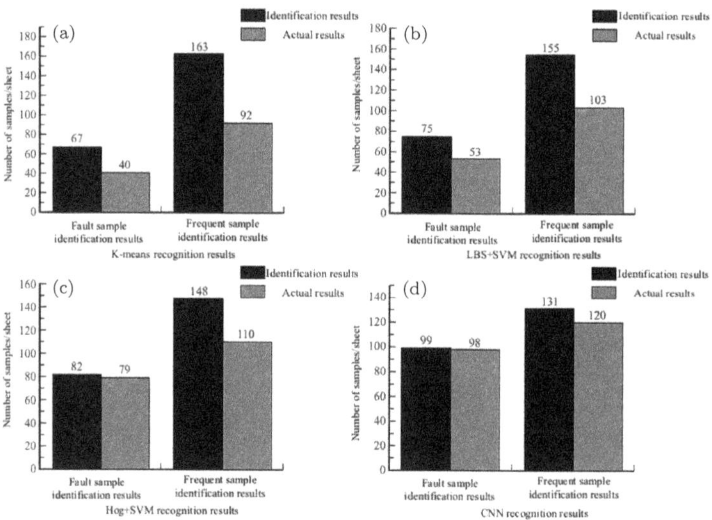

Fig. 3.86. Spur leakage fault identification results under different methods.

recognition classification methods such as LBP+SVM, HOG+SVM, and clustered K-means, and take the number of correctly recognized normal and spiked fault images (in sheets) as a percentage of the total number of images in the test set as the accuracy rate, and the results are shown in Table 3.4 and Fig. 3.86.

The accuracy of the proposed method reaches 94.8%, which is greatly improved compared with the traditional method. The random single infrared thermography test shows that the trained network model can accurately identify the infrared thermography image at the 604th, i.e., the initial moment of the spiked leakage fault. According to the field monitoring range, transmission time and the delay caused by the personnel to respond to the preliminary estimate can be 10s ahead of the early warning, and with the fatigue of the monitoring staff, this gap will still increase. The advance time allows operators to shut down the faulty equipment or adjust the operating conditions, thereby reducing the severity of the consequences of the accident.

References

[1] Li S. Research on early accident monitoring and recognition method of shale gas fracturing pump based on infrared thermal imaging. PhD Thesis, *China University of Petroleum (Beijing)*, 2019. (In Chinese)

[2] Liu H, Hu J. An adaptive defect detection method for LNG storage tank insulation layer based on visual saliency. *Process Safety and Environmental Protection*, 2021, 156: 465–481.

[3] Hu J, Li S, Zhang L. Parameter effects analysis of erosion wear of fracturing pipelines based on an orthogonal experimental design. *Petroleum Science Bulletin*, 2018, 4: 466–474. (In Chinese)

[4] Hu J, Liu H. Intelligent monitoring and remote early warning distribution system for downhole accidents in shale gas fracturing operations. In: *China Automation Conference and International Intelligent Manufacturing Innovation Conference.* Chinese Society of Automation, Jinan, 2017: 6–13. (In Chinese)

[5] Hu J, Wang Q, Zhang L, *et al.* Fatigue failure evolution of fracturing pump under changing pressure condition. *Petroleum Machinery*, 2017, 45(4): 67–73. (In Chinese)

[6] Kalpana G, Jyoti S. Texture-based self-adaptive moving object detection technique for complex scenes. *Computers & Electrical Engineering*, 2018, 70: 275–283.

[7] Hu WC, Chen CH, Chen TY, *et al.* Moving object detection and tracking from video captured by moving camera. *Journal of Visual Communication and Image Representation*, 2015, 30: 164–180.

[8] Zhang YG, Zheng J, Zhang C, *et al.* An effective motion object detection method using optical flow estimation under a moving camera. *Journal of Visual Communication and Image Representation*, 2018, 55: 215–228. (In Chinese)

[9] Zheng Y. Research and application of data cleansing in multi-radar data fusion algorithm. PhD Thesis, *Beijing University of Posts and Telecommunications*, 2018. (In Chinese)

[10] Zhang L, Zhang X, Hu J, *et al.* A comprehensive method for safety management of a complex pump injection system used for shale-gas

well fracturing. *Process Safety and Environmental Protection*, 2018, 20: 370–387.

[11] Wang Q, Zhang L, Hu J, *et al.* A dynamic and non-linear risk evaluation methodology for high-pressure manifold in shale gas fracturing. *Journal of Natural Gas Science and Engineering*, 2016, 29: 7–14.

[12] Zhang X, Zhang L, Hu J. Real-time diagnosis and alarm of down-hole incidents in the shale-gas well fracturing process. *Process Safety and Environmental Protection*, 2018, 116: 243–253.

[13] Fu W, Shen Q, Liu Y, *et al.* Engine fault diagnosis based on infrared thermal imaging technology. *Journal of Shanghai Maritime University*, 2016, 37(4): 65–69. (In Chinese)

[14] Liu Y, Pei S, Liu Y, *et al.* Deep learning based target detection method for abnormal hot spots infrared images of transmission and transformation equipment. *Southern Grid Technology*, 2019(2): 27–33. (In Chinese)

[15] Zhou K, Liao Z, Xiao Y, *et al.* Construction of infrared image classification model for power equipment based on improved CNN. *Infrared Technology*, 2019, 41(11): 1033–1038. (In Chinese)

Chapter 4

Early-Warning Techniques for Accident Events Based on Text Data Mining

4.1 Text Data Mining and Correlation Analysis for Safety Management in Oil and Gas Production Systems

In recent years, with the gradual establishment and improvement of the enterprise accident hidden danger investigation and management information system and the government's accident hidden danger investigation and management network management platform, enterprises in the management of the HSE audit process have accumulated a large number of records of hidden danger events, monitoring and testing data, and other information and have a large number of types of hidden dangers, heterogeneous data from multiple sources, and other characteristics. However, most of the production units used only one side of these data. The asymmetric development between the improvement of data collection and storage capacity and the insufficient level of data processing and utilization strongly restricts further improvement of enterprise safety management capacity.

Mining the potential value in safety management data, finding out the association rules between hidden dangers, and realizing the early warning of accident events will become the key to the safe development of enterprises. In order to improve the enterprise's hidden danger management work, experts and scholars at home and abroad have carried out relevant research work, such as using the

277

B/S mode to design a coal mine accident early-warning system, to achieve management, early warning, and monitoring of hidden dangers; data mining has been widely applied in early warning, such as using Apriori method to extract association rules from highway traffic accident data, mining the relationship between accident patterns and weather, and providing a basis for traffic accident warning; Using Bayesian networks to identify the relationships and probabilities of factors affecting hazardous chemical transportation accidents, and confirming that personnel errors and enterprise management are key to reducing accident rates; A food safety warning model was established based on an improved Apriori filtering algorithm.

Through the research, it has been found that the application of data mining technology in the investigation and management of potential accidents is still in its infancy, and there is less research on the early warning of accidental events; therefore, for the large amount of unstructured or semi-structured textual hidden danger data stored in the enterprise, in order to achieve potential knowledge discovery and effectively utilize the value of the hidden danger textual data, this section combines the data mining technology to establish the hidden danger in an early-warning and visualization system. Through text mining technology and data association mining technology, one can analyze the hidden accident data and mine the required knowledge information from it; at the same time, using visualization technology, the boring text or table data can be visualized and displayed in the form of Graphical User Interface (GUI), so that the management personnel can clearly see the useful information processed by the data mining technology, which facilitates the analysis and analysis of the data. It makes managers clearly see the useful information processed by data mining technology, facilitate the analysis and management, and realize human–computer interaction.

4.1.1 *Accident hidden danger text data feature extraction method*

Text mining belongs to the category of Natural Language Processing (NLP), and the document objects processed by mining usually have the characteristics of being massive, heterogeneous, and distributed. The objects handled by traditional data mining are structured, but most of the enterprise security management text exists in a semi-structured or unstructured form. Therefore, the first problem faced by text mining is how to appropriately and accurately characterize the document content in the computer, so that it should not only contain sufficient information to reflect the characteristics of the document but also not be too complex to make the learning algorithms difficult to handle.

The characterization of a document and the selection of its features are the issues to be considered in text mining and information retrieval, through the extraction of feature words in the document and mathematical quantification to characterize the content of the document. The original document from the unstructured mode into a structured computer can be directly understood and processed; that is, the document is scientific abstraction, and the construction of its mathematical model is a way to characterize and replace the content of the document, so that the computer can use this mathematical model to calculate and operate to achieve the purpose of document recognition.

On the other hand, if the structured text is processed directly, its large dimension will generate huge computational overheads for the subsequent work, which not only wastes memory and time, reduces the efficiency of the whole processing, but also affects the precision and accuracy of the processing results. Therefore, the text vectors need to be filtered at a deeper level to find out the features that best represent the text categories on the basis of ensuring the meaning of the original text. To address this problem, the most effective method is to select the feature volume to achieve the purpose of dimensionality reduction, i.e., keyword extraction. Usually, the concept of scoring ordering is adopted to score each feature, and the score of each feature is calculated according to a certain feature evaluation function, and then these features are processed in descending order based on the score value, from which a certain number of the highest-scoring features are selected as feature words.

There are four ways to perform feature selection: (1) using mapping or transformation to convert the initial features into fewer new features; (2) selecting certain most representative features from the initial features; (3) selecting the most influential ones according to the expert's knowledge of the features; and (4) using mathematical methods to find the features with the most classification information, which is more accurate and especially suitable for the application of an automatic text classification mining system due to its fewer human interference components.

The steps of text feature extraction include text preprocessing, obtaining a keyword candidate set, and keyword extraction.

4.1.1.1 *Text preprocessing*

(1) Data cleaning

Usually, the enterprise safety management information system stores a large number of accident events and hidden danger data reports, in order to follow up the data processing in a more convenient, effective and accurate manner. Most original data cleanup work is done to remove the "noise", redundancy, or contradictory data, including the removal of extraneous information. This includes removing irrelevant information, empty transactions, and standardizing data specifications. Irrelevant information refers to information that has nothing to do with the purpose of this mining, such as number, time, and location, and deleting irrelevant data can save storage space to a greater extent. A null transaction is a transaction that does not contain any itemset, and its existence is not only not helpful for the subsequent extraction of keywords and the mining of correlation rules but also occupies more memory space and processing time, so it is necessary to remove null transactions to reduce the processing time. Null transactions need to be removed to reduce processing time; unified data specification refers to the unification of the type, unit, and format of the data to be analyzed.

(2) Segmentation

There are three common methods for word separation: dictionary matching, semantic analysis, and probabilistic statistical models. At present, probabilistic statistical model-based algorithms are more effective, semantic analysis-based algorithms are too complicated, and dictionary matching-based algorithms are relatively easy to implement, but the results are not accurate enough. At present, there are a number of publicly available tools for word separation that can be applied, such as Jieba word separation, Tsinghua University's THULAC, and Harbin Institute of Technology's LTP. Some research institutes have tested 540 data points from news, Weibo,

Table 4.1. Common Chinese segmentation tools.

Tool name	Segmentation granularity	Error status	Lexical annotation	Authentication method	Interface
BosonNLP	Multiple choice	None	Exist	Token	Rest API
IkAnalyzer	Multiple choice	None	Exist	None	Jar
NLPIR	Multiple choice	Exist	Exist	None	Multi-language interface
SCWS	Multiple choice	None	Exist	None	PHP libraries/ command line tools
Jieba	Multiple choice	None	Exist	None	Python
Pangu subtext	Multiple choice	None	None	None	none
dismember an ox as skillfully as a butcher	Multiple choice	None	None	None	Jar
So gou subwording	Few	Exist	Exist	None	Supports uploading documents but with a high failure rate
Tencent Wen zhi	Few	Exist	Exist	Signature	REST API
Sina cloud	Big	None	Exist	Sina warehouse needed	REST API
Language cloud	Moderate	None	Exist	Token	REST API

and Volkswagen Dianping using common word segmentation tools, and the test results are shown in Table 4.1.

In this section, the Python environment is adopted to mine the textual data of accident events, and the Jieba participle tool is used to realize the participle work of textual data of accident hazards. Jieba participle belongs to the participle method based on probabilistic statistical model, and the principle of participle is as follows: Based on the statistical lexicon, a prefix lexicon is built to obtain all the participle possibilities, based on the location of the participle, a Directed Acyclic Graph (DAG) build on the DAG graph using dynamic programming is utilized to calculate the maximum probability path (the most likely result), and finally one based on the maximum probability path is also used to achieve segmentation.

4.1.1.2 *Obtaining a candidate set of keywords*

After the textual data of accident and hidden danger is divided into words, a very important step is to "deactivate" the data after the division of words. Deactivated words are those function words that cannot represent the main content of the document. For example, "had", "only", "and" are auxiliary words, and "because" and other words can only indicate the structure of the statement. Not only does this not reflect the main idea of the document but it also reduces the efficiency of keyword extraction, so is necessary to filter it out. The removal of deactivated words is usually carried out by traversing the deactivated word list; this section uses the "Baidu deactivated word list", with a total of about 1850 fields.

"De-duplication" can be regarded as a deeper level of data cleaning, removing the data that cannot be cleaned up directly, which can effectively improve the speed of data processing and the accuracy of the results. The "deactivated words" processed word set is used as the keyword candidate set for the next step of keyword extraction.

4.1.1.3 *Keyword extraction*

Keywords are usually extracted by constructing an evaluation function, evaluating each feature to get the respective evaluation values, also called weights, and then ranking all the features according to the size of the weights and extracting the specified number of optimal features as the keyword set of the processed text. Currently, popular keyword extraction methods include TF-IDF algorithm, Page-Rank algorithm, Text Rank algorithm, and Mutual Information.

(1) The TF-IDF algorithm

Term Frequency-Inverse Document Frequency (TF-IDF) is a feature evaluation algorithm that can be used in the field of information retrieval and text mining, which uses the idea of probabilistic statistics to evaluate word criticality in a document set. TF-IDF is based on the assumption that if a word appears frequently in one document but less frequently, or even not at all, in other documents, then it means that the word can represent the topic of this document well, has better differentiation performance (i.e., the word is more critical), and it makes it easy to differentiate this file from other documents. The core calculation idea is that the weight of the feature word in the document or the frequency of the feature word in the document is inversely proportional to the number of documents containing the feature word.

The meaning of *IDF* is that if the number of documents containing feature word t in a corpus is less, the greater the differentiation of feature word t, the higher the *IDF* value, and the greater the possibility of it being used as a keyword. The calculation method is shown in Eq. (4.1):

$$IDF = \lg\left(\frac{N}{n} + 1\right), \tag{4.1}$$

where N is the total number of documents in the corpus; $n\cdot$ is the number of documents in which feature word t occurs.

If the feature word t has a high frequency of occurrence in a document, even though it also has a high frequency of occurrence in other documents, it cannot be 100% denied that it has the probability of becoming a keyword. In order to reduce the influence of *IDF*, the concept of *TF* is added to it to correct this problem.

TF denotes the frequency of occurrence of feature word t in the specified document, which is calculated in Eq. (4.2):

$$TF = \frac{m}{M}, \tag{4.2}$$

where m is the number of times the feature word t appears in the specified document; $M\cdot$ is the total number of words in the specified document.

Finally, the weight formula of TF-IDF is shown in Eq. (4.3):

$$\omega_{\text{TF–IDF}} = TF \times IDF. \tag{4.3}$$

(2) Textrank algorithm

Influenced by the PageRank method in the field of information retrieval, Rada Mihalcea proposed a graph-based text keyword extraction algorithm, TextRank, which treats a document as a graph, and the candidate keywords in the document as nodes, and assumes that the size of a window is N. It is determined that words appearing in the same window in the same sentence are co-occurring, and undirected edges are constructed between these words. Assuming a window size of N, the algorithm determines that the words appearing in the same window are co-occurring in the same sentence, so it constructs two connected undirected edges between these words, and improves the original PageRank, and the final recursive formula is shown in Eq. (4.4):

$$WS(V_i) = (1-d) + d \times \sum_{V_i \in In(V_i)} \frac{\omega_{ji}}{\sum_{\nu_k \in Out(V_i)}} WS(V_j), \qquad (4.4)$$

where ω_{ji} denotes the weight of the edge pointing to node V_i from node V_j; $In\ (V_i)$ denotes the set of nodes pointing to node V_i; $Out\ (V_i)$ denotes the set of nodes pointing to node V_i; and d is the damping coefficient. The TextRank algorithm constructs an undirected unweighted graph, where each node is assigned an initial value of 1, and then substituted into Eq. (4.4) to iteratively compute the weights until the final result is smooth or reaches the set threshold, and then the word with the highest weight is selected as the keyword. Compared with general statistics-based keyword algorithms, the TextRank algorithm is more complex to implement and requires multiple iterations, usually numbering 20–30.

4.1.2 *Accident and hazard data association analysis technology*

Association analysis is a method used to discover meaningful associations hidden in large-scale datasets, and the discovered associations are usually characterized by association rules or frequent item sets. In this section, the Apriori algorithm will be used for association analysis of accident hazard data.

4.1.2.1 *Introduction to the problem*

In Association Analysis, a dataset is viewed as a collection of transactions, each of which is in turn viewed as a collection of distinct data. $I = \{i_1, i_2, i_3, \ldots, i_d\}$ is the set of all items in the transaction set, and $T = \{t_1, t_2, t_3, \ldots, t_N\}$ is the set of all transactions. Each transaction t_i contains a set of items that are subsets of I.

(1) Basic definition

    (i) *Item*: Each data in a dataset is called an item.

(ii) *k-itemset*: A set containing more than 0 items is called an itemset.

    If an itemset contains k items, it is called k-itemset.

(iii) *Support count of itemset*: It means the number of transactions that contain a particular itemset. Mathematically, the support count of an itemset $\sigma(A)$ can be expressed as Eq. (4.5):

$$\sigma(A) = |\{t_i | A \subseteq t_i, t_i \in T\}|, \tag{4.5}$$

where the symbol $|\cdot|$ indicates the number of elements in the set.

(iv) The support of the term set is calculated in Eq. (4.6):

$$S = \sigma(A)/N, \tag{4.6}$$

where N is the number of transactions.

(v) Frequent itemset: The set of all items that satisfies the minimum support (min-sup) is called frequent itemset.

(vi) Association rule: This is an implication expression of the form A → B, where A and B are disjoint sets of terms, i.e., A∩B= Ø.

(vii) The support calculation of the association rule is shown in Eq. (4.7):

$$S(A \rightarrow B) = \sigma(A \cup B)/N. \tag{4.7}$$

(viii) The confidence of the association rule is calculated in Eq. (4.8):

$$C(A \rightarrow B) = \sigma(A \cup B)/\sigma(A). \tag{4.8}$$

(2) Importance measure

Support S is an important metric for itemsets and association rules, which determines how frequent a particular itemset or rule is in a given dataset. Rules with low support may have some chance, and rules with low support are mostly meaningless in terms of value, so S is usually used to remove those meaningless itemsets or rules.

Confidence C Mining of strongly association rules is an important metric that measures the reliability of reasoning through rules. For example, for a rule $A \rightarrow B$, the higher the C, the higher the probability that a transaction containing A also contains B. The higher the C, the higher the confidence. Clearly, the confidence level C can be viewed as the conditional probability of B given A.

(3) Basic strategy of association rule mining

One traditional approach to mining association rules is to calculate the support and confidence of each rule, but this approach is costly and daunting because the number of rules mined from a dataset can be exponential. Therefore, most association rule mining algorithms are usually divided into the following two steps:

(i) Frequent item set generation: Its goal is to calculate the support size of each set, and find all frequent itemsets by setting the minimum support.

(ii) Generation of rules: The goal is to calculate the confidence level of each rule, by setting the minimum confidence level, and filter out all the high confidence rules from the frequent itemsets found in the previous step; these rules are called strong rules.

4.1.2.2 *Generation of frequent itemsets*

The Lattice Structure is often used to enumerate all possible sets of terms. Figure 4.1 shows the lattice of itemsets for $I = \{a, b, c, d, e\}$. In general, a dataset containing k items may yield $2^k - 1$ frequent itemsets (excluding the empty set). In many real-world cases, the values will be particularly large, so the itemset search space to be examined may be exponential.

Initially, the method used to discover frequent itemsets in transactions was to compute the support counts of the candidate itemsets in the lattice structure. In order to achieve this goal, each candidate itemset needs to be compared with each transaction, and obviously the computational overhead of this approach can be very high. Therefore, some later algorithms try to reduce the computational complexity of generating frequent itemsets. One of these algorithms is the a priori principle, which effectively removes certain candidate itemsets without calculating the support value. The basic idea of the a priori principle is that if a term set is frequent, then all its subsets must also be frequent. Consider the itemset lattice shown in Fig. 4.2. Suppose there is an itemset $\{c, d, e\}$. Clearly, any transaction containing $\{c, d, e\}$ must also contain its subsets. Thus, if $\{c, d, e\}$ is a frequent itemset, then all of its subsets (the itemsets in the dotted line of Fig. 4.2) must also be frequent. Conversely, if an item set is infrequent, all its supersets must be infrequent. As shown in Fig. 4.2,

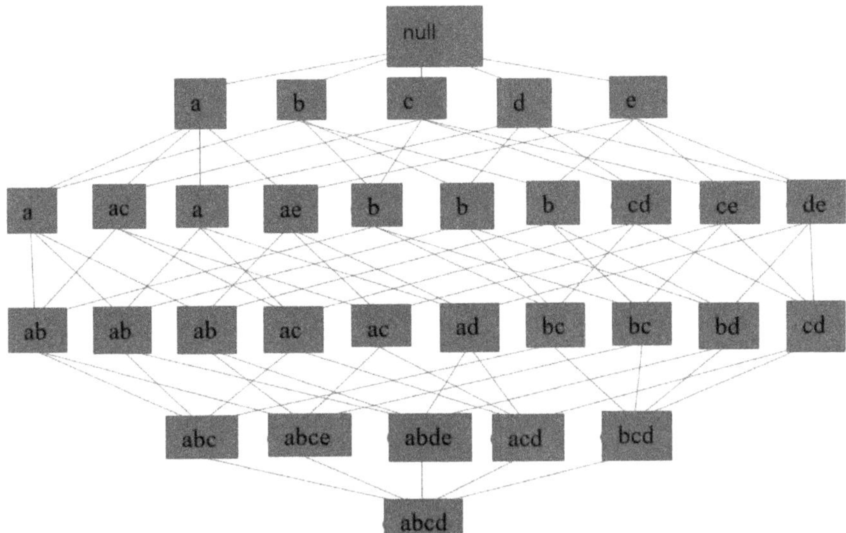

Fig. 4.1. Lattice of term sets.

once the itemset $\{a, b\}$ is non-frequent, the entire superset containing $\{a, b\}$ can be pruned immediately. This strategy of pruning the exponential search space based on support is called "support-based pruning".

4.1.2.3 *Generation of association rules*

Similar to the principle of frequent itemset generation, only the objects are different. Association rules are extracted from a given frequent itemset and up to $2^k - 2$ association rules can be generated for each frequent k-itemset. Association rules can be extracted in this way: The frequent itemset Y is partitioned into two non-empty subsets X and Y–X such that the rule $X \rightarrow Y$–X satisfies a confidence threshold.

In order to reduce the computational overhead, the rules can be pruned by the confidence level. If rule $X \rightarrow Y - X$ does not satisfy the minimum confidence, then a rule of the form $X \rightarrow Y - X'$, where X' is a subset of X, must also not satisfy the minimum confidence.

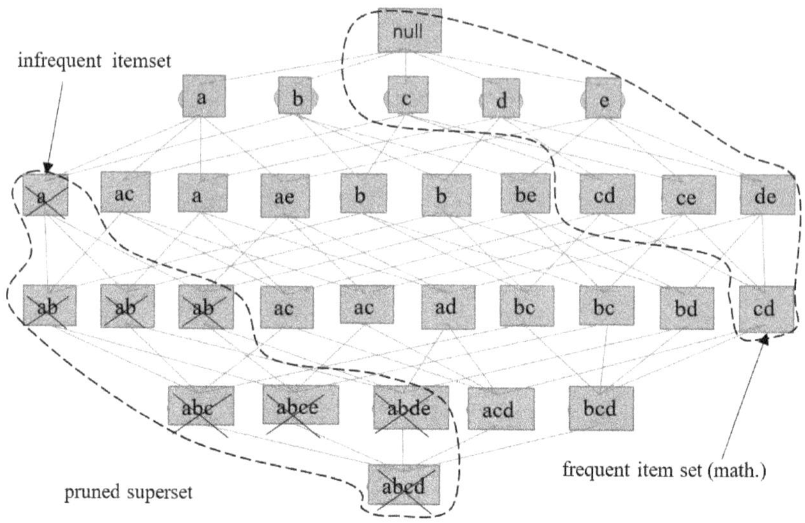

Fig. 4.2. A priori principles and support-based pruning.

4.1.2.4 *Apriori algorithm*

The Apriori algorithm pioneered the use of support-based pruning techniques to control the exponential growth of the candidate itemsets in all aspects. It has two important features in generating frequent itemsets: First, it is a level-wise algorithm, i.e., it traverses one level of the itemset lattice at a time from the frequent 1-item set to the longest frequent item set. Second, it adopts a "generate-test" strategy to discover frequent itemsets. After each iteration, the new candidate itemsets are composed of the frequent itemsets generated in the previous iteration, and then the support is calculated for the new candidate itemsets and compared with the minimum support. This algorithm requires a total of $k_{\max} + 1$ iterations, where $k_{\max}$ is the maximum length of the frequent itemset. The Apriori algorithm also takes a layer-by-layer approach to generate association rules, where each layer corresponds to the number of items in the rule posterior. First, all high-confidence rules with only one term in the post rule are extracted; then, these rules are used to generate new candidate rules; finally, according to the pruning method, the final desired strong association rules can be obtained efficiently.

4.1.3 *Accident hazard text preprocessing and keyword extraction*

Accurate and clear results are usually based on raw data with a good format, and when the raw data do not have good processing quality, even the best processing method cannot get accurate results. Therefore, it is necessary and important to preprocess the raw data of accident hazards, which will determine whether your results are reliable or not. In the face of thousands of data, people often cannot quickly lock the focus; compared with the inconvenience of manually analyzing a large amount of text data judgment and huge time consumption, the use of computer keyword extraction can be more effective and faster to dig out more valuable information.

In this section, we combine nearly 3,000 accident and hidden danger data records of an enterprise in each month from 1999–2017 and establish a customized dictionary to improve the accuracy of word separation. The TF-IDF algorithm is utilized for keyword extraction, and the extracted keywords can well help managers locate the key objects and improve the efficiency of enterprise safety management and the accuracy of management measures. The original information of hidden accidents is stored in the form of Excel files, and the Python3.5 language environment is used to develop a suitable program to achieve the purpose of keyword extraction, which effectively reduces the feature dimension of the textual data of hidden accidents and facilitates the further analysis of the subsequent warning.

4.1.3.1 *Text preprocessing of accident hazards based on custom dictionaries*

In the text preprocessing stage of word separation, it is often too limited to use only the statistical dictionary that comes with the Jieba word separation tool, because the statistical dictionary contains only the basic words, which are not targeted, resulting in the inability to identify most of the highly specialized words; therefore, it is more difficult to obtain the desired results. This problem can be solved by creating user-defined dictionaries to increase the directionality of the participle effect.

(1) Accident and hazard data cleaning and segmentation

Clean up the original data and delete the empty transactions and irrelevant date and personnel information in the Excel file in preparation for the split word. Jieba split word is very good to increase the user's authority, can be customize to build a professional dictionary of enterprise safety accidents and hazards, that is, the subject of the thesaurus or the professional thesaurus, so that the split word results are more closely related to the actual terminology. Jieba split word can be installed directly using the command "pip install Jieba" to install and "import Jieba" to reference. Take a hidden danger data "没有安全操作规程或不健全" as an example; the code is as follows:

```
import jieba
data = '没有安全操作规程或不健全'
Seg_list = jieba.cut ( data )
Print ( '/' .join ( Seg_list ) )
```

The output is 没/有/安全/操作/规程/或/不/健全.

It can be clearly seen from the output results that if only based on the built-in dictionary that comes with the Jieba participle for word separation, the results are somewhat unsatisfactory, especially for the more specialized data; it is more difficult to get the exact field. On this basis, if we introduce the implementation of the built-in professional dictionary, the result will be more accurate, and the code for loading the professional dictionary will be as follows:

```
import jieba
data = '没有安全操作规程或不健全'
jieba.load_userdict ( 'D: /Python/Lib/site-packages/jieba/userdict.txt' )
Seg_list = jieba.cut ( data )
Print ( '/' .join ( Seg_list ) )
```

The output is 没有/安全操作规程/或/不健全.

User-defined dictionaries need to be stored in "utf-8" encoding in order to be recognized by the Jieba Segmentation Tool. After loading the customized professional dictionary, the split result is significantly improved, effectively identifying "安全", "操作" and "规程" three times as one word output, which is in line with the real situation. Obviously, if you want more accurate results of word separation, a more detailed custom dictionary needs to be established.

(2) De-duplication processing

In the process of text processing, in order to simplify text analysis, a pre built stop word list is introduced, which contains some words that will not contribute useful information in the analysis. By comparing with this stop word list, the system can delete these stop words, thereby reducing unnecessary data interference and improving the efficiency of text processing. Taking the above example as an example; the code is as follows:

```
import jieba
data = '没有安全操作规程或不健全'
jieba.load_userdict ( 'D: /Python/Lib/site-packages/jieba/userdict.txt' )
Seg_list = jieba.cut ( data )
c = movestopwords ( Seg_list )
print ( c )
```

The output is 安全操作规程不健全.

Obviously, after traversing the list of deactivated words, words such as "没有", "或", which are not very directional and very common, are accurately removed. In this way, it can effectively reduce the amount of data storage, as well as the computational difficulty and time of the subsequent processing work. Among them, the movestopwords() function is a custom deactivation function, which completes the deactivation work by traversing the deactivation word list.

4.1.3.2 *TF-IDF algorithm-based keyword extraction for hidden accident hazards*

Jieba comes with the TF-IDF algorithm and Textrank algorithm. By comparing the keywords obtained after processing the hidden accident dataset with these two algorithms, it is found that the keywords extracted by the TF-IDF algorithm are more accurate; so, this method is chosen for the extraction of hidden accident keywords in this section.

The TF-IDF algorithm can be used to extract keywords through the command code "import jieba. analyse", which is as follows:

```
import jieba.analyse
fh3 = open("E:/Processed data/initial processing - keywords.txt", "w")
k = self.spinBox.value()
tag = jieba.analyse.extract_tags(cutdata,k)
kword = "
for word in tag:
    kword = kword+word+'\n'
    fh3.write(kword)
fh3.close()
```

Here, k is the number of keywords extracted from the user input, obtained from the screen.

The extracted keywords will be written to the text file "Initial Processing - Keywords. txt" under the path "E:\ Processed Data".

4.1.4 *Association analysis and early warning of accident hidden danger data*

On the basis of the keyword extraction, the data association analysis method is introduced to calculate the probability of occurrence of hidden dangers, which helps enterprises achieve early warning of accidental events, so as to avoid the sudden occurrence of accidents and improve the level of safety management.

In this section, based on a total of 211 data of a petroleum enterprise's accident and hidden danger management data, Apriori algorithm is used to perform association analysis to find out the strong association rules contained in the transaction set. The association rules can be divided into frequent antecedents and frequent consequents; i.e., in the rule "$A \to B$", A is a frequent antecedent and B is a frequent consequent. With the mined strong association rules, it is easy to predict the probability of occurrence of the corresponding frequent subsequence B when the frequent antecedent A occurs, so as to achieve the purpose of early warning of accident events. At the same time, the mined rules can be used to assess the risk of the frequent posterior terms, in order to extend the role of the association rules for early warning and enhance the usefulness of the association warning.

4.1.4.1 *Apriori-based association analysis of potential accident hazards*

Based on the Apriori algorithm, the algorithm is programmed in a Python language environment.

The logical steps of the algorithm can be summarized as follows:

(1) The dataset is first scanned once and the support for each item is calculated. Once this is done, the set of all frequent 1-item sets will be available.
(2) Next, the algorithm uses the frequent $(k-1)$-itemsets generated by the previous iteration to generate new candidate k-itemsets.
(3) The dataset is scanned again to calculate the support of the candidate items.
(4) After calculating the support of the candidate itemsets, the algorithm will filter out all the candidate itemsets with support less than min_sup.
(5) This part of the algorithm will end when no new frequent itemsets are generated.
(6) The process of generating association rules is similar to the process of generating frequent itemsets, iterating over each frequent k-item set, but instead of having to repeatedly scan the dataset to determine the confidence of the candidate rules when the rules are generated, it is sufficient to use the support computed during the generation of frequent itemsets to obtain the confidence of each rule.

In the end, by inputting 211 enterprise accident hidden danger transactions, the custom minimum support (min_sup) is 0.3, and the minimum confidence (min_conf) is 0.6. In order to improve the usefulness of hidden danger warning, only association rules whose frequent posterior items contain only one item are selected, and a total of 47 association rules are mined in the end. Some of the association rules are shown in Table 4.2, and the data in the table are sorted in descending order by the value of confidence level.

The mined multiple strong association rules are analyzed, from which several representative rules are filtered out as shown in the following:

Rule 1: "lack of safety operation knowledge $\rightarrow$ insufficient education and training", S $= 0.527363184$, C $= 1$.

Table 4.2. Accident hazard association rules (partial).

Num	Association rules		Degree of support	Confidence level
	Frequent antecedent	Frequent postulates		
1	Lack of knowledge of safe practices	Insufficient education and training	0.527363184	1
2	Insufficient education and training	Lack of knowledge of safe practices	0.527363184	1
3	Lack of knowledge of safe operation, violation of operating procedures or workforce discipline	Insufficient education and training	0.512437811	1
4	Insufficient education and training, violation of operating procedures or workforce discipline	Lack of knowledge of safe practices	0.512437811	1
5	Lack of knowledge of safe operation, defective equipment, facilities, tools, accessories	Insufficient education and training	0.373134328	1
6	Insufficient education and training, defective equipment, facilities, tools and accessories	Lack of knowledge of safe practices	0.373134328	1
7	Poor production site environment, violation of operating procedures or labor discipline	Defective equipment, facilities, tools, accessories	0.368159204	0.74
8	Insufficient education and training, violation of operating procedures or labor discipline	Defective equipment, facilities, tools, accessories	0.373134328	0.72815534

9	Lack of knowledge of safe operation, violation of operating procedures or labor discipline	Defective equipment, facilities, tools, accessories	0.373134328	0.72815534
10	Insufficient education and training, lack of knowledge of safe operation, violation of operating procedures or labor discipline	Defective equipment, facilities, tools, accessories	0.373134328	0.72815534
11	Defective equipment, facilities, tools, accessories	Poor environmental conditions at production sites	0.383084577	0.620967742
12	Poor production site environment, violation of operating procedures or workforce discipline	Insufficient education and training	0.303482587	0.61
13	Poor production site environment, violation of operating procedures or workforce discipline	Lack of knowledge of safe practices	0.303482587	0.61
14	Defective equipment, facilities, tools, accessories	Insufficient education and training	0.373134328	0.60483871
15	Defective equipment, facilities, tools, accessories	Lack of knowledge of safe practices	0.373134328	0.60483871

Rule 2: "insufficient education and training → lack of knowledge about safety operations", S = 0.527363184, C = 1.

Rule 3: "insufficient education and training, defective equipment, facilities, tools, accessories → violation of operating procedures or labor discipline", S = 0.373134328, C = 1.

Rule 4: "lack of knowledge of safe operation, defective equipment, facilities, tools, accessories → violation of operating procedures or labor discipline", S = 0.373134328, C = 1.

Rule 5: "poor production site environment, violation of operating procedures or labor discipline → defective equipment, facilities, tools, accessories", S = 0.368159204, C = 0.74.

Rule 6: "insufficient education and training, violation of operating procedures or labor discipline → defective equipment, facilities, tools, and accessories", S = 0.373134328, C = 0.72815534.

Rule 7: "lack of knowledge of safe operation, violation of operating procedures or labor discipline → equipment, facilities, tools, accessories are defective", S = 0.373134328, C = 0.72815534.

Rule 8: "defective equipment, facilities, tools, accessories → poor production site environment", S = 0.383084577, C = 0.620967742.

Rule 9: "defective equipment, facilities, tools, accessories → Inadequate education and training", S = 0.373134328, C = 0.60483871.

Rule 10: "Defective equipment, facilities, tools, accessories → lack of knowledge of safe operation", S = 0.373134328, C = 0.60483871.

Based on these rules, the following explanations can be obtained from this part of the accident hazard data:

(1) From rules 1 and 2, it can be seen that "lack of knowledge of safe operation" and "insufficient education and training" will definitely appear together. It is possible that the lack of safety operation knowledge is due to the fact that the employees do not learn it in the education and training stage, which leads to their unskilled mastery of the relevant operation knowledge; and these two items appear most often in the data of hidden accidents in these years, which reminds the enterprise management personnel in the education and training that the employees should be paid extra attention to.

(2) Rules 3 and 4 show that when there are "defects in equipment, facilities, tools and accessories", and when there is "insufficient

education and training" or "lack of knowledge of safe operation", there will be "violation of safety regulations" by employees. The situation of "violation of operating procedures or workforce discipline" will definitely occur. These two rules are different from the previous two rules in that there is no logical cause-and-effect relationship because correlation analysis only analyzes from probability and does not consider cause and effect.

(3) Same as (2), there is no logical causal relationship between the latter frequent items of rules 5, 6, and 7, "defective equipment, facilities, tools and accessories" and the former frequent items, but it may be due to the lack of training of the employees, which leads to their weak awareness of safety, coupled with some improper operation and the influence of the environment in the normal course of events, which leads to the failure to inspect the equipment and facilities on time, thus resulting in defects. The defects may be due to the fact that the equipment and facilities were not inspected on time.

(4) Rules 8, 9, and 10 reveal that when "defects in equipment, facilities, tools, and accessories" occur, there is a 60% probability that "poor environment at the production site," "insufficient education and training," and "lack of training" will occur. Inadequate education and training" and "Lack of knowledge of safe operation" are reasons.

Based on these explanations, the next step is to formulate targeted risk control measures.

4.1.4.2 *Early warning of potential accidents combined with risk assessment*

Through the above-mentioned analysis, we can clearly see the frequent antecedents and frequent consequents of each association rule. In the management process, once the problem of frequent antecedents of association rules occurs, it will naturally be associated with the frequent antecedents that may occur, so as to achieve the role of accident hidden danger warning. In order to increase the practicability of accident hazard warning and better utilize the extracted association rules in the enterprise, it can be combined with the risk evaluation method to carry out the risk assessment on the frequent posterior items, so as to expand the early-warning effect of the association rules.

For example, with reference to the definition of risk, the assessment of the risk of accident hazards is decomposed into three indicators: the probability of occurrence L, the severity S, and the adjustment factor δ of the hazard. There are many ways to measure the probability of occurrence L, but most of the traditional measures are highly subjective in the classification of probability levels and the setting of benchmark minutes, and lack an objective quantitative system. From the formula of confidence Conf in the Apriori algorithm, it can be seen that the confidence is actually the conditional probability of the frequent occurrence of the latter item when the frequent occurrence of the former item in a certain rule, so the confidence of the association rule can be used to calculate the probability of occurrence of the hidden danger, and the value range is in the range of $[0, 1]$.

Consequence severity S reflects the severity of the consequences of the hidden danger, and the existing evaluation methods usually establish a suitable comprehensive evaluation index system in terms of the degree of harm, the difficulty of management, the degree of influence, etc., and use hierarchical analysis, fuzzy evaluation, BP neural network, DHGF integration, and other methods to determine the weights of the indicators, so as to determine the final severity of the consequences.

The adjustment coefficient δ is designed to adjust the previous two items to a certain extent to enhance the reliability of the results. The subjective judgment is made by the personnel responsible for the management of hidden accidents and incidents based on their own experience and the actual situation of the enterprise, so as to make the assessment results more in line with the real situation. The final model can be expressed as follows:

$$R = Conf \times S \times \delta, \tag{4.9}$$

where "$\times$" only means that R is related to *Conf*, S, and δ is not a mathematical product.

Risks are usually classified in the form of a risk matrix combined with the ALARP principle. Generally, the risk level of hidden danger can be classified into four levels, low, medium, high, and serious, which correspond to low warning level, medium warning level, high warning level, and major warning level. For different warning levels, enterprises can better take measures against the emergence of hidden dangers and more effectively carry out safety supervision work.

According to the association rules, the frequent antecedents of the rules are found, and the frequent antecedents are evaluated for early warning; the risk value is calculated, and the warning level of hidden danger is finally determined, so that the safety management personnel can designate the corresponding early-warning measures and the corresponding warning level for different levels of hidden danger, thus realizing the pre-control of hidden danger.

4.2 Causal Extraction of Risk Factors Based on Text Augmentation

With the gradual development of AI technology, deep neural network models are widely used in NLP to improve the level of intelligence and computational efficiency of analysis and processing, while at the same time, the requirements on the data size have gradually increased, and the excellent effect of deep learning methods depends to a large extent on the amount of effective data. Although deep learning has been widely used in general-purpose NLP tasks, in the field of oil and gas production safety management text, there is a "small sample" problem, so the existing data volume cannot support the training requirements of deep learning models. The direct application of deep learning methods is prone to problems such as model overfitting and poor generalization ability, which is not conducive to the implementation of subsequent NLP tasks in specific scenarios.

Therefore, in order to solve the small-sample learning problem that exists in the accident texts of some specific oil and gas production sites (e.g., the LNG reserve site), and to reduce the unfavorable impact of the lack of effective texts on the deep learning method, this section proposes a text-enhanced risk factor causality extraction method. First, based on the corpus expansion of LNG reserve depot safety management texts using relevant texts in the petrochemical field, feature enhancement and word splicing methods are combined with bidirectional long- and short-term memory networks and conditional random field algorithms, so as to obtain more textual semantic expressions. Improve the feature extraction capability of the causality extraction method and get the causality of accident evolution by extracting the causality nodes in the safety management text. Finally, the proposed causality extraction method for risk factors of LNG reserves is practically applied in a case study, and compared and evaluated with the causality extraction method without text enhancement, and the causality extraction method with different text enhancements to prove the effectiveness and accuracy of the proposed method.

4.2.1 *Fundamentals of textual causality extraction*

4.2.1.1 *Text enhancement methods*

In the case of insufficient or unbalanced sample data, text augmentation is an effective way to expand the size of data samples and improve the robustness of the model. The larger the size and higher the quality of the data, the better the generalization ability of the deep learning model, the reduction of the occurrence of overfitting in the model training process, so that the model pays more attention to the semantic information of the text, and is no longer sensitive to the local noise of the text. Taking the text of safety management of LNG reserve as an example, its text enhancement can be realized by the following three methods:

(1) Corpus expansion

A corpus is a collection of various types of texts needed in a natural language processing task, which can usually be divided or labeled correspondingly depending on the target task. Corpus expansion in this section refers to the method of expanding the number of texts needed in the desired domain by utilizing the domain corpus that has a strong correlation with the desired domain in terms of content and constructing texts that have a similar distribution of textual features as those of the desired domain to realize the expansion of the number of texts needed in the desired domain. In the case that the training corpus of the model of the required domain is small, extracting templates from the text of the required domain through syntactic analysis and lexical annotation, and artificially constructing texts with similar features to those of the text of the required domain can realize the increase of the number of training texts and reduce the occurrence of overfitting. At the same time, the use of a domain corpus with strong relevance to the desired domain for text construction can ensure that the constructed text is relevant to the desired domain in terms of content, and the training results of the deep learning model are less affected, so it has reliability. Compared with the singularity of language and content brought about by directly utilizing templates for text generation on the existing corpus, more diverse sample contents can be introduced to achieve better learning results for the model, thus effectively increasing the size of the corpus and supporting the training of deep learning model parameters.

(2) Feature enhancement

On the basis of the original single word, more feature dimensions are added to the original text data through the methods of word boundary delineation, lexical analysis, radicals' extraction, pinyin extraction, syntax analysis, etc., so as to enrich the text features and semantic information and realize the purpose of text feature enhancement.

(3) Sentence segment splicing

On the basis of the initial sentence segments, N to N splicing of sentence segments is carried out to realize the increase in the number of sentence segments, enrich the linguistic expression of the original text, and produce a more diverse number of long sentence segments to achieve better training of the deep learning neural network's feature extraction ability for long text and improve the performance of the model.

4.2.1.2 *Bidirectional long- and short-term memory network algorithm*

Compared to the problem of gradient vanishing or gradient explosion that exists in recurrent neural networks, Long Short Term Memory (LSTM) is able to better capture the long-term dependency effect. The main idea is to store historical information in memory units, and the actions of updating, decaying, inputting, and outputting in the memory units are controlled by multiple gates to control the corresponding memory units, and the parameters of these gates are learned to decide whether the information in the memory units is preserved or forgotten. The internal structure of the LSTM unit is shown in Fig. 4.3, and it contains the memory unit, update gate, forget gate, and output gate.

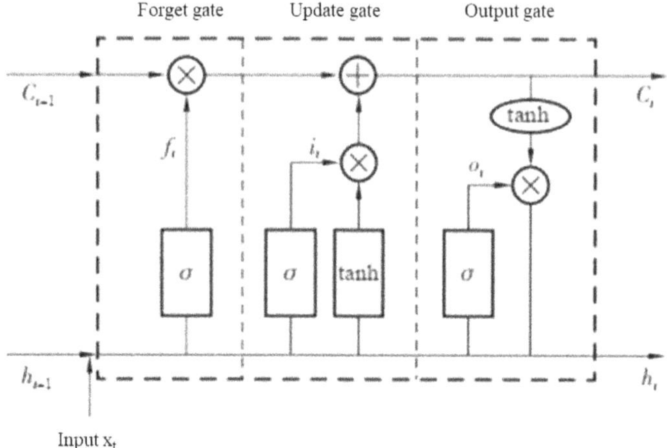

Fig. 4.3. LSTM cell.

Let the input at time t be u_t, and the hidden layer and memory unit at time $t-1$ be a_{t-1} and C_{t-1}, respectively, and the computation process is as follows:

(1) Calculate gate information for controlling information in the memory unit.

Update gate:

$$u_t = \sigma(W_{xu}x_t + W_{au}a_{t-1} + W_{cu}C_{t-1} + b_u). \tag{4.10}$$

Output gate:

$$O_t = \sigma(W_{xo}x_t + W_{ao}a_{t-1} + W_{co}C_{t-1} + b_o). \tag{4.11}$$

Forget gate:

$$f_t = \sigma(W_{xf}x_t + W_{af}a_{t-1} + W_{cf}C_{t-1} + b_f). \tag{4.12}$$

(2) Computational memory units:

$$C_t = f_t C_{t-1} + u_t \tanh(W_{xC}x_t + W_{aC}a_{t-1} + b_C). \tag{4.13}$$

(3) Calculate the value of the hidden layer at time t:

$$a_t = o_t \tanh(C_t), \tag{4.14}$$

where W and b both denote parameter matrices; σ is generally taken as a sigmoid function.

As can be seen from Fig. 4.3 and Eqs. (4.10)–(4.14), the update gate is multiplied with the value of the memory cell at the time of no gate to input the input information into the memory cell. The forgetting gate and the value at the moment $t-1$ are multiplied to get the decay of the memory cell. The output gate is multiplied with the memory cell at moment t to output the information from the memory cell to the hidden layer, which affects the output of each gate at moment $t+1$. After the control of the information by the memory cells, the long- and short-term memory network can save the most useful information for the task.

Since the unidirectional LSTM model cannot process the contextual information at the same time, the basic idea of Bidirectional Long-Short Term Memory (Bi-LSTM) network is to take forward and backward LSTM for each word sequence separately, and then merge the outputs of the same moment. Thus, for each moment, it corresponds to forward and backward information; the specific structure is shown in Fig. 4.4, where the output is shown in Eq. (4.15):

$$ht = [\overrightarrow{h}_t,\ \overleftarrow{h}_t].\tag{4.15}$$

4.2.1.3 *Conditional random field algorithm*

In sequence annotation, Bi-LSTM is able to extract semantic information over long distances, but it cannot consider the influence relationship between output labels; at this time, the Conditional Random

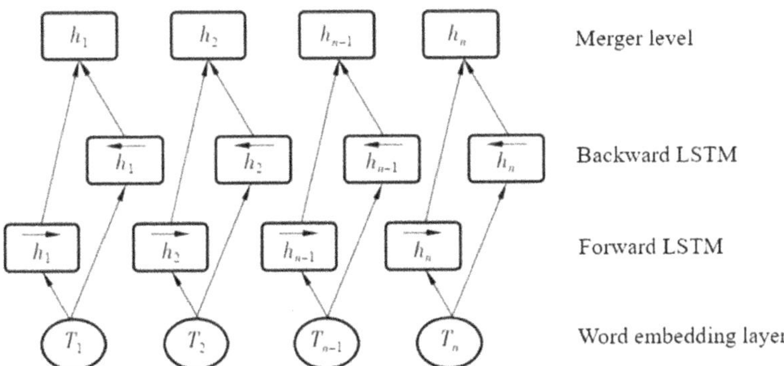

Fig. 4.4. Bi-LSTM model structure.

Field (CRF) algorithm can solve this problem well. This algorithm can obtain an optimal prediction sequence through the relationship of neighboring labels, which can make up for the shortcomings of Bi-LSTM. For any observation sequence $X = (x_1, x_2, \ldots, x_n)$ of length n, for the prediction sequence $Y = (y_1, y_2, \ldots, y_n)$, its score function formula Eq. (4.16) is obtained:

$$s(X, Y) = \sum_{i=0}^{n} A_{y_i, y_{i+1} + \sum_{i=1}^{n} P_{i, yi}},$$

(4.16)

where A denotes the transfer score matrix of size $k + 2$, $A_{y_i, y_{i+1}}$ represents the transfer probability of label y_i to label y_{i+1}, where k is the number of labels; P denotes the firing score matrix of size $n \times k$, where n is the number of words, and $P_{i, yi}$ denotes the firing probability of the y_i label of the ith word. The probability that the predicted sequence Y is generated is calculated by Eq. (4.17):

$$P(Y|X) = \frac{e^{s(X,Y)}}{\sum_{\tilde{Y} \in Y_X} e^{s(X, \tilde{Y})}}.$$

(4.17)

The likelihood function of the predicted sequence is obtained by taking logarithms at both ends to get Eq. (4.18):

$$\ln[P(Y|X)] = s(X, Y) - \ln\left[\sum_{\tilde{Y} \in Y_x} s(X, \tilde{Y})\right],$$

(4.18)

where Y_x denotes all possible labeling sequences. Decoding yields the output sequence Y^* that maximizes the conditional probability [Eq. (4.19)]:

$$Y^* = \arg\max_{\tilde{Y} \in Y_X} p(\tilde{Y}|X).$$

(4.19)

4.2.2 *A text enhancement-based causal extraction method for risk factors of LNG reserves*

In this section, the causality extraction task of the safety management text of the LNG reserve is converted into the sequence annotation task of extracting the causal nodes (including risk factor nodes

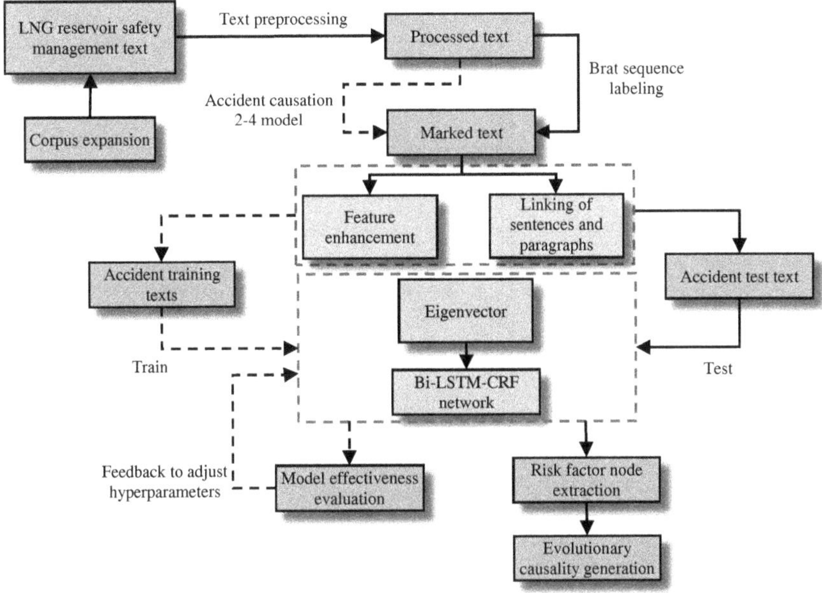

Fig. 4.5. Text enhancement-based causality extraction methodology process for risk factors.

and accident type nodes) in the safety management text, so as to obtain the causal relationships among the risk factors of the LNG reserve. The text enhancement-based causal relationship extraction method for LNG reserve risk factors adds a text enhancement layer on top of the Bi-LSTM feature extraction layer and takes the conditional random field as the loss calculation layer. Compared with the Bi-LSTM-CRF method, the proposed method in this section inputs more semantic feature information into the deep neural network through the method of "feature enhancement plus segment splicing", which is able to improve the effect of the causal relationship extraction task. The flow of the method is shown in Fig. 4.5.

4.2.2.1 *Security management text corpus expansion*

Since the LNG reserve is still in the early stage of development in China, there is not much safety management text information that can be collected, which is not conducive to the training of the deep learning model; therefore, the corpus is expanded by the safety

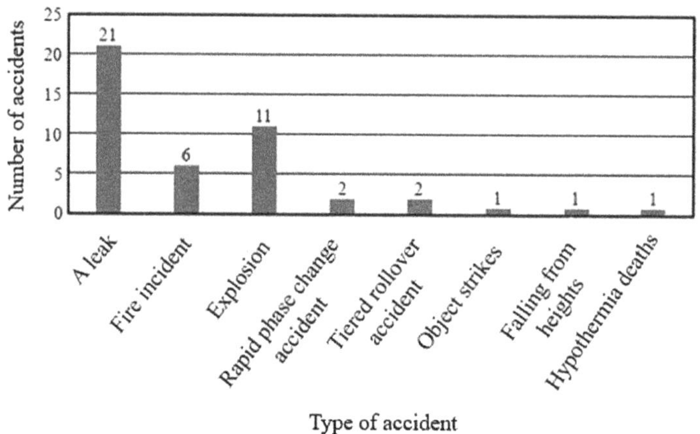

Fig. 4.6.   Types of LNG reserve accidents and number of occurrences.

management text of mature fields similar to the LNG reserve. By reviewing the literature related to LNG accidents, the possible types of accidents and the number of occurrences in the LNG reserve are obtained as shown in Fig. 4.6, and the possible causes and proportions of accidents are shown in Fig. 4.7. As can be seen from the figure, leakage, explosion, and fire are the main types of accidents in LNG reservoirs; the main causes of accidents are failure of appurtenances (35%), personnel operation error (31%), operation violation (7%), equipment leakage (7%), and equipment failure (7%). It can be seen that the accident characteristics of the LNG reserve have commonality with the types and causes of accidents that occur in the production and operation of petrochemical enterprises, and the structure and function of the equipment in the LNG reserve system are similar to those in the petrochemical system, and the intrinsic pattern of accidents in the two systems is the same. Therefore, in this section, we extract linguistic description templates from the LNG reserve texts by syntactic analysis and manual analysis of sentence structure, and construct texts with similar characteristics to the LNG reserve safety management texts based on the safety management texts accumulated in the production and operation process of petrochemical enterprises, so as to supplement and improve the training corpus and to ensure the normal progress of the subsequent analysis and processing work.

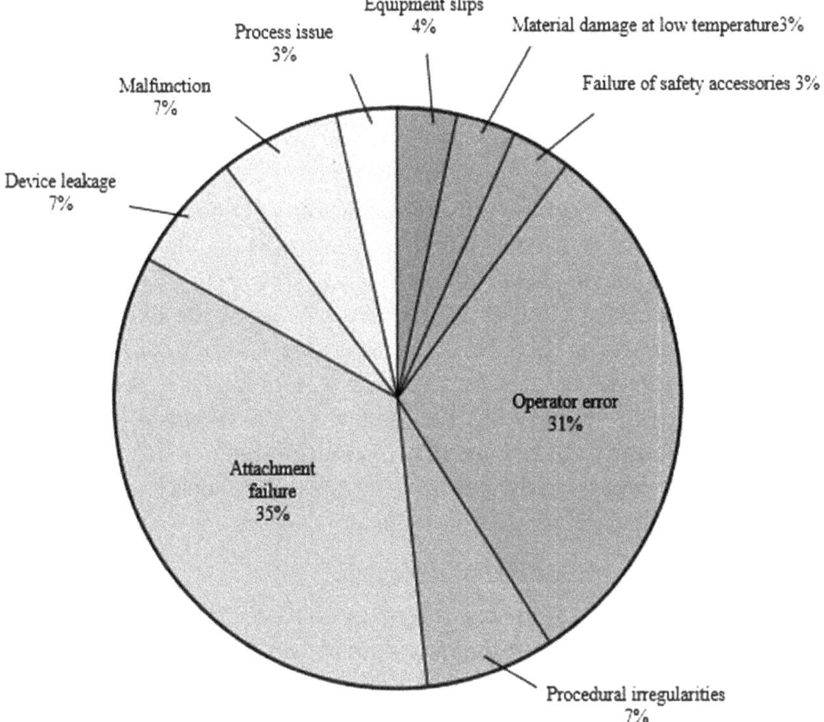

Fig. 4.7. LNG reservoir accident causes and proportions.

4.2.2.2 *Security management text preprocessing*

In the process of collecting safety management texts for actual LNG reserves and related petrochemical fields, it was found that most of the existing safety management texts have the following characteristics due to the lack of corresponding record specification standards:

(1) Invalid information

Since there is no unified standardization of safety management records, there is a lot of invalid information, such as names of personnel, conversations of personnel, and descriptions of irrelevant events.

(2) Complicated terminology

Due to the record of different personnel, there are no standardized terminology norms, so there will be more than one expression, abbreviation shorthand, and so on.

(3) Incomplete accident information

Since there is no unified standardization of safety management records, there are problems such as imperfect safety management information and lack of logical nodes.

To summarize, in view of the characteristics of the existing safety management texts in the LNG reserve and petrochemical fields, it is necessary to preprocess the text corpus to remove the noise information and logically supplement the text with the experts' opinions to ensure the accuracy of the causality extraction results before carrying out the subsequent causality extraction task. The safety management text preprocessing process is as follows:

Step 1: Deletion of invalid information.

Most of the existing safety management texts are recorded in a more colloquial way, with redundant words and too much invalid information, which seriously affects the quality of the text, so it is necessary to appropriately delete useless information in the text, such as the date, conversations between employees, descriptions of operating procedures, descriptions of the daily passages, and other texts that do not contain causal information, as shown in Table 4.3.

Step 2: Completion of key information.

Due to the lack of corresponding standards in some safety management texts, key information on the evolution of risk factors is missing in the existing texts, which seriously affects the subsequent extraction of causal information. Therefore, this part combines the experience of experts to supplement key information in the existing texts, so as to improve the causal chain of risk factor evolution, which will help the subsequent work, as shown in Table 4.4.

Step 3: Segmentation of safety management text.

Before conducting quantitative representation of text, by establishing a customized professional field dictionary and combining it with the Jieba vocabulary tool, the text can be accurately divided into different fields.

Table 4.3. Example of deleting invalid information from security management text.

Serial number	Original security management text	Safety management text after the first pre-treatment
1	On October 20, 1944, LNG liquid leaked out of Tank 4 in the Cleveland LNG storage area, but the staff did not notice anything unusual at that time. After some time, as there was no cofferdam installed in the tank, the LNG liquid and evaporated gas flowed into the drain or was blown by the wind and dispersed into the surrounding streets, resulting in an explosion and a fire.	Sudden rupture and leakage of LNG storage tanks in the LNG storage area, and leakage of LNG liquid. Since there was no cofferdam installed at the time, the LNG liquid and evaporated gas flowed into the drainage pipe or was blown by the wind and dispersed to the surrounding streets, resulting in an explosion and a fire.
2	In 2008, a fire and explosion occurred at a pipeline company due to inadequate employee safety training. In the early morning of the same day, the pressure of a pipeline section suddenly increased, and the local gas transmission office immediately organized the inspection and leakage detection of the whole line. At 9:00 a.m., Zhang, a staff member of the local pipeline maintenance station, inspected and checked the leakage of the pipeline section, and no abnormality was found. In the early morning of the next day, the pipeline staff carried out the last inspection leakage before the incident, and found no anomalies. At 4:00 a.m., a leakage explosion occurred in the pipeline, producing a loud noise, and the valve room process room of the pipeline section caught fire.	A pipeline company suffered a fire and explosion due to inadequate employee safety training. As the employees did not detect the abnormality in time, the pipeline leaked and exploded, creating a loud noise and a fire in the process room of the valve room of the pipeline section.

4.2.2.3 *Security management text labeling*

The steps for implementation are as follows:

Step 1: Classification of risk factors for security management texts. In order to obtain the training and testing text dataset for the subsequent causal relationship extraction model, the risk factors in the

Table 4.4. Examples of additional improvements to key safety management information.

Serial no.	Safety management text after the first pre-treatment	Second pre-treatment security management text
1	Sudden rupture and leakage of LNG storage tanks in the LNG storage area, and leakage of LNG liquid. Since there was no cofferdam installed at the time, the LNG liquid and evaporated gas flowed into the drainage pipe or was blown by the wind and dispersed to the surrounding streets, resulting in an explosion and a fire.	A fire and explosion occurred at the LNG storage area due to loopholes in routine inspections. LNG liquid leaked from LNG storage tanks due to the failure of tank materials at low temperatures, and because the tanks were not cofferdam, LNG liquid and evaporated gas flowed into drainage pipes or were blown by the wind and dispersed into the surrounding streets, where fires and explosions occurred in the presence of sparks.
2	A pipeline company suffered a fire and explosion due to inadequate employee safety training. As the employees did not detect the abnormality in time, the pipeline leaked and exploded, creating a loud noise and a fire in the process room of the valve room of the pipeline section.	A fire and explosion occurred at a pipeline company due to inadequate employee safety training. As a result of daily inspections, leakage checks were not in place, no anomalies were detected, making the pipeline leak, and in the event of an open fire, an explosion was generated.

causal nodes in the safety management text related to the LNG reserve are categorized, and with reference to the Accident Causation 2-4 model, the risk factors are classified into the following categories: fundamental factors, indirect factors, and direct factors, which are shown in Table 4.6 as an example.

By dividing the risk factor nodes in the safety management text and combining the accident type nodes, the accident causal evolution path in the development of risk factors can be obtained. As shown in Fig. 4.8, when there are leaks in the daily inspection, the tank material is prone to failure at low temperatures, which leads to the leakage of the LNG liquid, and due to the lack of cofferdams, the LNG flows into the drainage pipes, and the evaporated natural gas is dispersed

Table 4.5. Example of security management text segmentation.

Second pre-treatment security management text	Accidental text after disambiguation
A fire and explosion occurred at the LNG storage area due to loopholes in routine inspections. LNG liquid leaked from LNG storage tanks due to the failure of tank materials at low temperatures, and because the tanks were not cofferdam, LNG liquid and evaporated gas flowed into drainage pipes or were blown by the wind and dispersed into the surrounding streets, where fires and explosions occurred in the presence of sparks.	LNG storage area/due to/loopholes in daily inspection/fire and explosion/LNG storage tanks/due to/failure of tank materials at low temperatures/leakage of/LNG liquid/due to/lack of cofferdams for/storage tanks/causing/LNG liquid/and/evaporating gases/to flow/into/drains/or/blown by the wind/to/neighboring streets/sparks/fire/explosion

by the wind, which ultimately leads to a fire and explosion when it meets with the wind. Eventually, a fire and explosion occurred in the event of a spark.

Step 2: Securely manage text sequence annotation.

In the sequence labeling task, the deep learning model classifies each field of the input sequence and then outputs its corresponding predetermined label, which represents the field's category and boundary. Therefore, before training the deep neural network, the label categories need to be predefined to label the data. An example of labeling the causal nodes in the security management text using Brat natural language text annotation software is shown in Fig. 4.9. The green labeling represents "root factor", purple labeling represents "accident type", yellow labeling represents "indirect factor", and blue labeling represents "direct factor".

Samples of training sequences for subsequent deep learning neural network training can be obtained based on the Brat labeling tool. The training sample label set adopts the BIO standard, where B denotes the head of the causal node, I denotes inside the causal node, and O denotes of the non-causal node. The labeled label set is shown in Table 4.7.

Table 4.6.   Example of risk factor classification for security management text.

Pre-treated safety manage-ment texts	Risk factor nodes			
	Underlying factor nodes	Indirect factor nodes	Direct factor nodes	Incident type node
LNG storage area/due to/loopholes in daily inspection/fire and explosion/LNG storage tanks/due to/failure of tank materials at low temperatures/leakage of/LNG liquid/due to/lack of cofferdams for/storage tanks/causing/LNG liquid/and/evaporating gases/to flow/into /drains/or/blown by the wind/to/neighboring streets/sparks/fire/explosion	Loopholes in daily inspection	Tank material failure at low temperatures, LNG leaks, no cofferdams	Flows into drains, is dispersed by wind, meets sparks	Fire and explosion

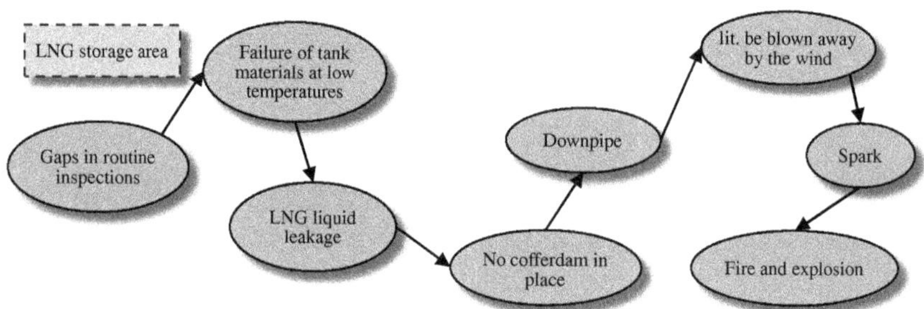

Fig. 4.8.   Example of causal evolutionary paths for risk factors in security management texts.

4.2.2.4 *Security management text causality extraction*

Step 1: Text feature enhancement.

In this section, on the basis of the single textual feature of the security management text, the feature information of four dimensions, namely, word boundary, lexicality, radicals, and pinyin, is added, and finally, the security management text containing five-dimensional features is formed, which enriches the corpus information, can help the causal relationship extraction model learn more diversified feature information, and improves the model's generalization ability.

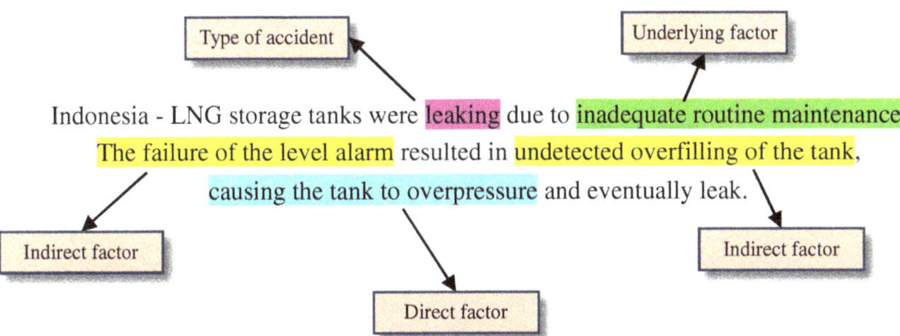

Fig. 4.9. Example of text labeling for security management.

Table 4.7. BIO tag set.

Causal node categories	Corresponding tag name	Start marker	Middle marker	End marker
Underlying factor	GEN	B-GEN	I-GEN	I-GEN
Indirect factors	JIAN	B-JIAN	I-JIAN	I-JIAN
Direct factors	ZHI	B-ZHI	I-ZHI	I-ZHI
Type of accident	ACI	B-ACI	I-ACI	I-ACI
Else	OTHER	O	O	O

Step 2: Segment splicing.

The N to N splicing operation on the segments of the safety management text containing risk factors increases the number of safety management texts and the proportion of long texts in the overall corpus, which can optimize the processing capability of the causal relationship extraction model on long text sequences and improve the robustness of the model.

Step 3: Parameterization of the causal extraction model.

In the initialization phase of the model, a total of 11 hyperparameters in the model need to be set, and the required setting parameters are shown in Table 4.8. By setting appropriate parameters, it can help the model to have good performance and effect.

Step 4: Construction of causal relationship extraction model for risk factors of LNG reserve.

Table 4.8. Parameters of the causal extraction model.

Serial number	Parameter name	Serial number	Parameter name
1	Dimension of a word vector	7	LSTM dimension
2	Word Boundary Vector Dimension	8	Lexical vector dimension
3	Dimensionality of a vector of radicals	9	Dimensionality of a pinyin vector
4	Batch size	10	Activation function
5	Learning rate	11	Pruning gradient
6	Number of training sessions		

The LNG reserve depot risk factor causality extraction model is mainly composed of four layers, namely, feature text input layer, feature text embedding layer, bidirectional LSTM feature extraction layer, and CRF causal node output layer, as shown in Fig. 4.10.

(1) Feature text input layer

Since the input of each batch of neural network is usually required to be of fixed length, and the length of each security management text is usually inconsistent, it is firstly necessary to extend each sentence segment input in each batch to the same length, and the sentence segments whose lengths are less than the fixed lengths are made up with "$\langle PAD \rangle$" labels; at the same time, for the non-dictionary base of the words is replaced by "$\langle UNK \rangle$" tags.

(2) Feature text embedding layer

Feature text embedding refers to the vectorized representation of the feature texts of the five dimensions of the input causal extraction model. The initial feature vector is obtained by mapping all feature texts in a randomly initialized feature space.

(3) Bi-directional LSTM feature extraction layer

The bi-directional LSTM layer is composed of two layers of LSTM network structure, one forward LSTM and one reverse LSTM. Through the special gating structure of LSTM, it can capture the contextual information of the security management text and extract high-dimensional features of the security management text.

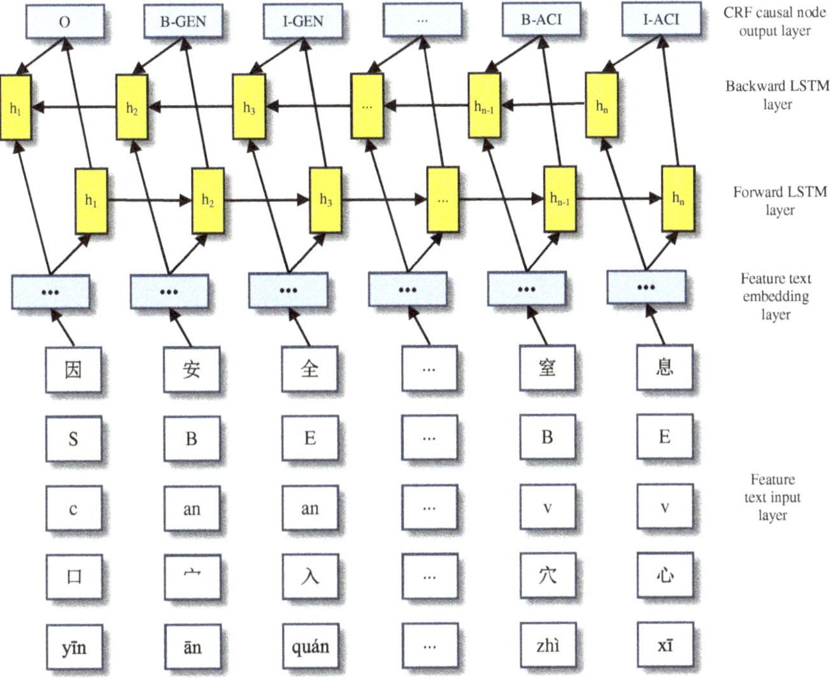

Fig. 4.10. Causality extraction model for LNG reservoir risk factors.

(4) CRF causal node output layer

This model adopts the CRF method as the loss function. By using the Bi-LSTM layer to extract high-dimensional features from the data, and combining the characteristics of CRF global optimization, the model learns more strong constraints at the security management sentence level and maximizes the probability of the output sequence, so as to improve the accuracy of the model's causality extraction results, and to make up for the shortcomings of the Bi-LSTM local optimization.

Step 5: LNG reservoir risk factor causality extraction model training.

After text enhancement and encoding of the labeled safety management training text dataset, all the sentence segments are converted into corresponding feature vectors, which are used as inputs to the model, and the parameters of the Bi-LSTM-CRF deep learning model are trained. The specific steps are as follows: (1) Input the text-enhanced five-dimensional feature text into the feature text input layer of the model, perform the length complementation operation, and transform it into the corresponding feature vectors through the feature text embedding layer; (2) realize the feature learning of the model through the two-layer Bi-LSTM layer; and (3) back-propagate using the output layer of the CRF causal node, and normalize at the sequence level to realize the gradient descent.

Step 6: LNG reservoir risk factor causality extraction model testing.

The trained risk factor causality extraction model is tested using the safety management test text dataset, and the advantages and disadvantages of the model are evaluated by comparing the gap between the predicted labels and the actual labels. The basic implementation process is the same as shown in step 5, but at this time, there is no need to calculate and update the gradient of the model parameters.

4.2.3 *Case study*

4.2.3.1 *Testing and validation*

With the task of annotating the sequence of causal nodes of safety management text containing causal information as the target task, the text enhancement-based causal extraction model of risk factors for LNG reserve depot is practically applied.

Step 1: Safety management text corpus expansion.

Based on the 31 LNG reserve safety management texts, 331 petrochemical-related safety management texts are added to realize corpus expansion. The final LNG reserve safety management text dataset is formed, which contains 362 unstructured safety management texts and nearly 2,000 risk factor causal relationships, and the dataset is divided into the training set and the test set according to the ratio of 8:2.

Step 2: Safety management text preprocessing.

Firstly, the invalid information is deleted manually, and the key information is supplemented and improved; then based on the Python language environment, combined with the Jieba lexical toolkit, the lexical operation is carried out on the safety management text, and the risk factor segments and accident type segments are cut out from the text.

Step 3: Safety management text labeling.

After the safety management text is divided into risk factor nodes and accident type nodes, based on the Ubuntu system environment, the Brat text annotation tool is utilized to annotate all kinds of causal nodes on the safety management text of the LNG reserve, and finally 1870 causal nodes are marked.

Step 4: Safety management text feature enhancement.

Based on the Python language environment, the four features of word boundary, lexical properties, radicals, and pinyin are extracted from the safety management text to form the safety management text corpus of the LNG reserve containing five-dimensional features.

Step 5: Segment splicing of safety management text.

Based on the Python language environment, based on the feature-enhanced corpus, 2-to-2 and 3-to-3 splicing operations are carried out on the segments of the safety management text containing risk factors to form the required corpus for the risk factor causality extraction model of the LNG reserve.

Step 6: Deep learning model parameter setting.

According to the results of several experiments, the parameter settings of the LNG reservoir risk factor causality extraction model are shown in Table 4.9. The dimension of word vector is set to 100 dimensions, the dimension of word boundary vector is 20 dimensions, the dimension of radical vector is 50 dimensions, the dimension of lexical vector is 50 dimensions, the dimension of pinyin vector is 50 dimensions, the dimension of LSTM unit is 128 dimensions, the activation function adopts Relu function, the batch size is 32, the learning rate is 0.0001, the pruning gradient is set to $[-5, 5]$, and the number of training times is set to 50. The same initialization model is used on each dataset during the experiment, the algorithm is run 3 times each and the results are averaged.

Step 7: Causal extraction modeling of risk factors for LNG reservoirs.

Based on the Python language environment, TensorFlow deep learning framework, combined with cnradical toolkit, we build a text

Table 4.9. Model parameter settings.

Parameter name	Parameter value	Parameter name	Parameter name
Dimension of a word vector	100	LSTM dimension	128
Word Boundary Vector Dimension	20	Lexical vectordimension	50
Dimensionality of a vector of radicals	50	Dimensionalityof a pinyin vector	50
Batch size	32	Activation function	Relu
Learning rate	0.0001	Pruning gradient	$[-5, 5]$
Number of training sessions	50		

enhancement-based causal extraction model of LNG reservoir risk factors, and the structure of the model is shown in Fig. 4.11.

In order to assess the effectiveness of the causal extraction method for risk factors of LNG reserve, the results of the method application are evaluated from three aspects, and the closer the evaluation indexes converge to 100%, the better the effectiveness of the method.

Firstly, from the coarse-grained point of view, whether the proposed method can return four types of causal nodes as expected (i.e., there is no situation in which a certain type of causal node is not returned) is evaluated, which is reflected by calculating the causal node recall rate, and the calculation is shown in Eq. (4.20):

$$R = \frac{c_1}{c_2} \times 100\%, \qquad (4.20)$$

where c_1 is the number of correct ones in the returned results; c_2 is the number of results that should be returned.

Secondly, from a fine-grained perspective, determine whether the extracted causal nodes meet the requirements, and evaluate the effectiveness of the proposed method by calculating the recognition accuracy A for the task of labeling the sequence of causal nodes of the

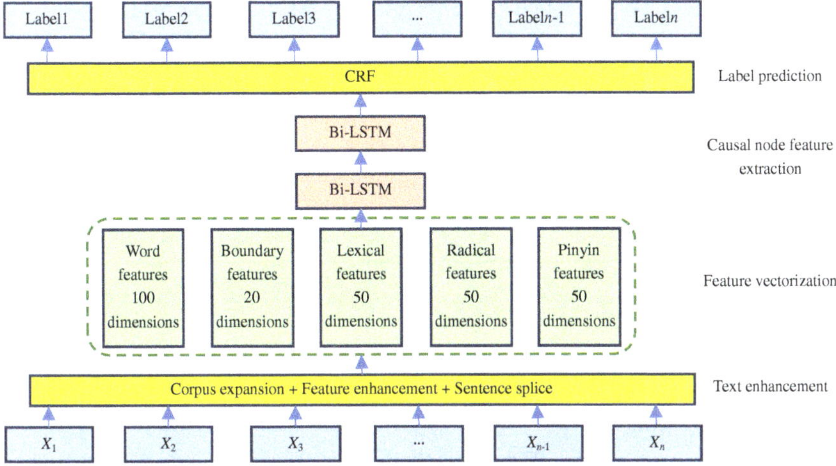

Fig. 4.11. LNG reservoir risk factor causality extraction model structure.

security management text, which is shown in Eq. (4.21):

$$A = \frac{c_1}{c_3} \times 100\%, \tag{4.21}$$

where c_1 is the number of correct ones in the returned results; c_3 is the number of all returned results.

Finally, it is evaluated from the balance point of view to examine whether the proposed method can better balance the effects of both coarse-grained and fine-grained to achieve the comprehensive optimization of the risk factor extraction results, which is calculated in Eq. (4.22):

$$F1 = \frac{2A \times R}{A + R} \times 100\%, \tag{4.22}$$

where A is the causal node recognition accuracy; R is the causal node recall rate.

On the four types of causal node texts, namely, underlying factors, indirect factors, direct factors, and accident types, the causal extraction method based on text enhancement for risk factors of LNG reserve storage proposed in this section is applied to analyze the advantages and disadvantages of its extraction effect on different types of causal texts. According to Eqs. (4.20)–(4.22), the changes of recall, accuracy, and F1 value of the model on the four types of

Table 4.10.   Evaluation metrics for the 50th training result of the method in this section on the four types of causal nodes.

Causal node type	Recall rate/%	Accuracy/%	F1 value/%
Type of accident	99.48	95.74	97.57
Underlying factor	94.38	93.07	93.72
Indirect factors	56.75	35.63	43.77
Direct factor	49.54	33.84	40.21

causal labels are shown in Figs. 4.12–4.14, respectively; among them, the situation of the model evaluation index after the 50th training is shown in Table 4.10.

As can be seen from Table 4.10, the method proposed in this section performs best in extracting accident-type nodes, with a recall of 99.48%, accuracy of 95.74%, and F1 value of 97.57%. This is followed by the root factor node (94.38% recall, 93.07% accuracy, and 93.72% F1 value), the indirect factor node (56.75% recall, 35.63% accuracy, and 43.77% F1 value), and the worst performance is the direct factor node (49.54% recall, 33.84% accuracy, and 40.21% F1 value). This result may be due to the fact that compared with the accident type text and the underlying factor text, the indirect factor text and the direct factor text have more diverse forms of representation, with fuzzy boundaries, variable text lengths, and other characteristics, which lead to the fact that the risk factor causality extraction method cannot extract and learn the semantic features in a better way when extracting the indirect factor nodes and the direct factor nodes.

From Fig. 4.12, it can be seen that there is a big difference in the trend of the recall rate of the proposed method in this section in the task of extracting four types of causal nodes. Accident-type nodes have the fastest pre-improvement, and the pre-improvement rate of fundamental factor nodes is slightly smaller than that of accident-type nodes. The indirect factor nodes and direct factor nodes have a smaller advancement speed than the accident-type nodes and the underlying factor nodes, and the growth is slower and accompanied by certain fluctuations, of which the fluctuations of the indirect factor nodes are the most obvious. The four curves gradually converge with the increase in the number of training times and tend to a stable level.

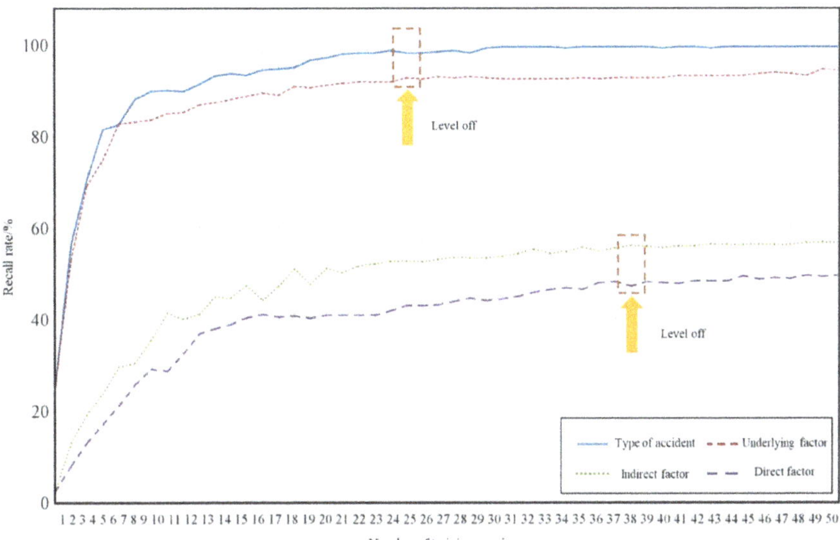

Fig. 4.12. Variation of recall of the method in this section with the number of trainings on four types of causal nodes.

Comparing and analyzing the four curves, the following conclusions can be obtained:

(1) The method proposed in this section in extracting four types of causal nodes: The recall rate of extracting accident-type nodes and root factor nodes is much higher than that of extracting indirect factor nodes and direct factor nodes.
(2) The initial recall of the method proposed in this section is about the same when extracting accident-type nodes and root factor nodes, which is about 25%, and when extracting indirect-type nodes and direct factor nodes, which is about 3%.
(3) The method proposed in this section converges faster in extracting accident-type nodes and root factor nodes, with the recall leveling off at about the 23rd training, and in extracting indirect cause nodes versus direct cause nodes, the recall leveling off at about the 37th training.

From Fig. 4.13, it can be seen that the trend of accuracy and the trend of recall are approximately the same in the task of extracting four types of causal nodes with the text enhancement-based risk

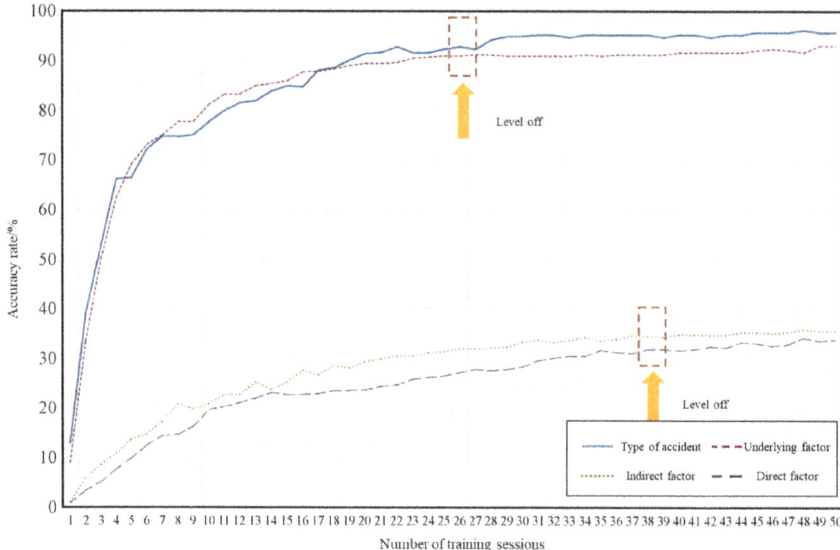

Fig. 4.13. Variation of the accuracy of the method in this section with the number of training sessions on the four types of causal nodes.

factor causality extraction method. The following conclusions can be obtained by comparing and analyzing the four curves:

(1) The accuracy of the text-enhanced risk factor causality extraction method is much higher than that of the indirect cause nodes and direct cause nodes when extracting accident-type nodes and root cause nodes.

(2) In terms of extracting accident-type nodes and root factor nodes, the accuracy of the method proposed in this section converges to the level at about the 27th training, which is faster compared to extracting indirect factor nodes and direct factor nodes, which converge to the level at the 38th training.

By comparing and analyzing the four curves in Fig. 4.14, the following conclusions can be drawn:

(1) The F1 value of this method is much higher when extracting accident-type nodes and root factor nodes than extracting indirect factor nodes with direct factor nodes.

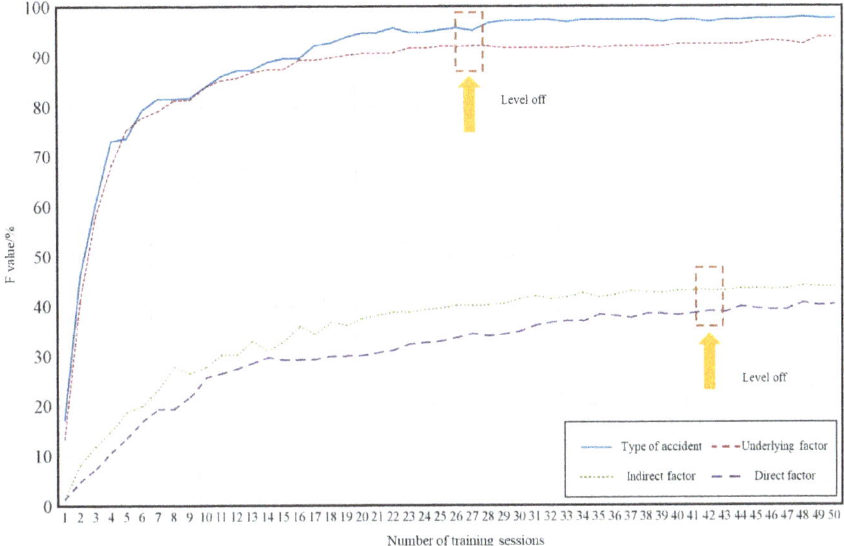

Fig. 4.14. Variation of the method's F1 value with the number of training sessions on the four types of causal nodes.

(2) When extracting accident-type nodes and root factor nodes, the F1 value converges faster and levels off at about the 27th training; while when extracting indirect factor nodes and direct factor nodes, the F1 value levels off at about the 42nd training.

4.2.3.2 *Comparative analysis*

In order to verify the extent of text enhancement for the causal extraction methods of risk factors of the LNG reserve, a comparative analysis is conducted with three groups of different causal extraction methods on the same safety management texts related to the LNG reserve, as shown in Table 4.11, and the effects are quantified using Eqs. (4.21) and (4.22).

The change of loss value with the number of training times for the four risk factor causality extraction methods on the LNG reserve-related safety management text is shown in Fig. 4.15.

From Fig. 4.15, it can be seen that the loss of the four methods on the LNG reserve-related safety management text shows the same trend of change: After a sudden drop, it gradually converges and

Table 4.11. Scheme design for comparison of causal extraction methods.

Serial number	Method name	Validation text
1	Initial Bi-LSTM + CRF method	LNG reservoir-related safety management texts
2	Feature enhancement + Bi-LSTM + CRF method	
3	Segment splicing + Bi-LSTM + CRF method	
4	Text enhancement + Bi-LSTM + CRF methods	

tends to a stable level. Comparative analysis of the four curves can lead to the following conclusions:

(1) In terms of the speed of loss decline: text enhancement + Bi-LSTM + CRF method > sentence segment splicing + Bi-LSTM + CRF method > feature enhancement + Bi-LSTM + CRF method > initial Bi-LSTM + CRF method, so it can be concluded that the text enhancement method is able to make the training loss value decline to convergence more quickly.

(2) In terms of the lowest loss value: text enhancement + Bi-LSTM+ CRF method > sentence segment splicing + Bi-LSTM + CRF method > feature enhancement + Bi-LSTM + CRF method > initial Bi-LSTM + CRF method; therefore, it can be concluded that the addition of text enhancement method can better reduce the loss value of the causal node extraction task so that the prediction results are closer to the real results.

(3) Among the text enhancement methods, the sentence segment splicing method is better than the feature enhancement method in terms of loss reduction.

After 50 times of training of the four risk factor causality extraction methods on the safety management text related to the LNG reserve, the recall, accuracy, and change of the F1 value of the methods are shown in Figs. 4.16–4.18, respectively, of which the model

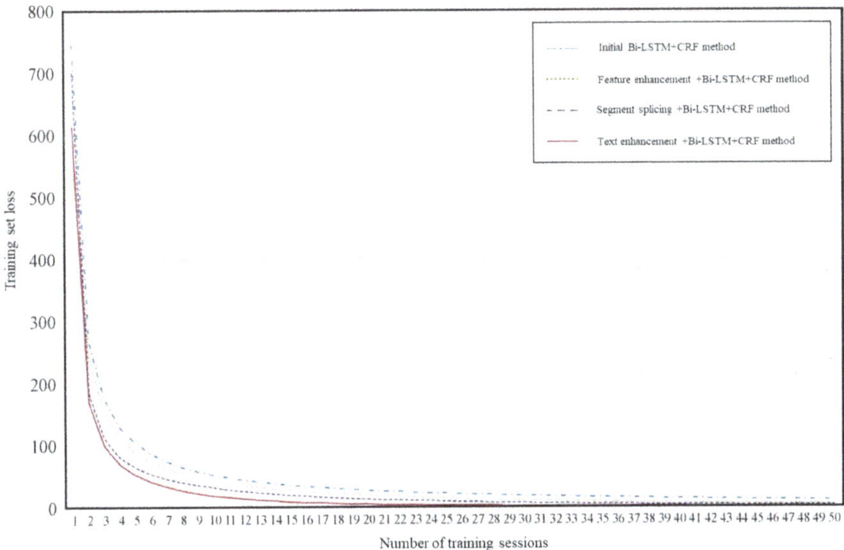

Fig. 4.15. Losses as a function of the number of training sessions for the four risk factor causality extraction methods.

evaluation indexes and the improvement of the indexes' effects at the end of the 50th training are shown in Table 4.12.

From Table 4.12, it can be seen that the text enhancement-based causality extraction method for LNG reservoir risk factors performs better in all aspects, and has been optimized compared to the initial Bi-LSTM + CRF method, with the model's recall, accuracy, and F1 value improved by 15.78%, 17.40%, and 17.41%, respectively.

As can be seen from Fig. 4.16, the four methods show roughly the same trend in the text of safety management related to LNG reserves: The recall rate rises faster in the early stage, and then gradually converges with the increase of the number of training times. Comparing and analyzing the four curves, the following conclusions can be drawn:

(1) As shown in the solid line box, the initial recall of the text enhancement + Bi-LSTM + CRF method is higher, which is nearly 10% higher compared to the initial Bi-LSTM + CRF method, in order of accuracy, followed by the sentence segment splicing + Bi-LSTM + CRF method, feature enhancement + Bi-LSTM+CRF method, and the initial Bi-LSTM + CRF method.

Table 4.12. Model evaluation values and improvements for the 50th training of the four risk factor causality extraction methods.

Causality extraction methods	Recall rate $R/\%$	Accuracy $A/\%$	F1 value	$+R/\%$	$+A/\%$	+F1
Initial Bi-LSTM + CRF method	54.80	35.42	43.01	0	0	0
Feature enhancement + Bi-LSTM + CRF method	63.91	41.27	50.15	+9.11	+5.85	+7.14
Segment splicing + Bi-LSTM + CRF method	69.45	52.01	59.47	+14.65	+16.59	+16.46
Text enhancement + Bi-LSTM + CRF method	70.58	52.82	60.42	+15.78	+17.40	+17.41

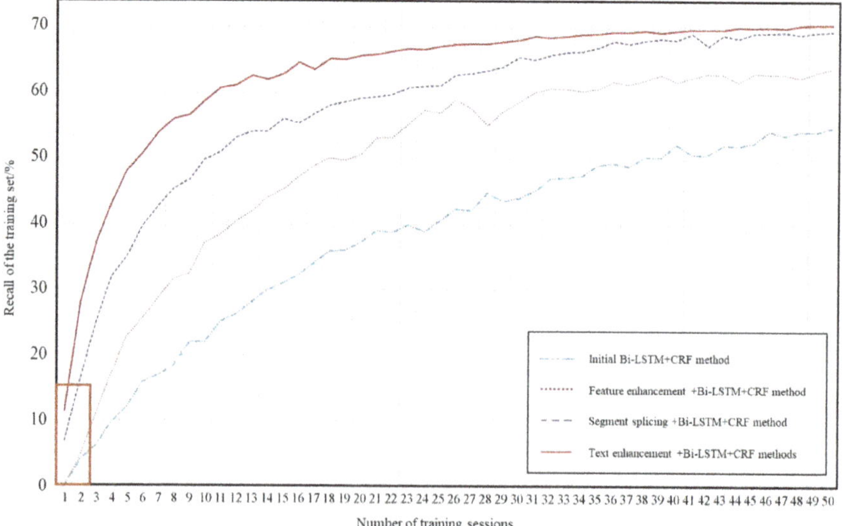

Fig. 4.16. Variation of recall with number of training sessions for four risk factor causality extraction methods.

(2) Regarding the increase in speed of recall, Text enhancement + Bi-LSTM + CRF method > Segment splicing+Bi-LSTM + CRF method > Feature enhancement + Bi-LSTM + CRF method > Initial Bi-LSTM + CRF method; therefore, it can be concluded

that after text enhancement, the risk factor causality model can rise to convergence more quickly with the increase in the number of training times.

(3) Regarding the highest recall rate, text enhancement+ Bi-LSTM + CRF method > sentence splicing + Bi-LSTM + CRF method > feature enhancement + Bi-LSTM + CRF method > initial Bi-LSTM + CRF method; therefore, the text-enhanced risk factor causality extraction model is able to better return the nodes of various types of risk factors and types of accidents according to the expectation.

(4) Among the text enhancement methods, the segment splicing method can improve the recall rate better than the feature enhancement method.

From Fig. 4.17, it can be seen that the accuracy of the model shows roughly the same trend for the four training sets: The accuracy rises faster in the early stage, and gradually converges with the increase of the number of training times. Comparing and analyzing the four curves, the following conclusions can be obtained:

(1) The initial accuracy of the text-enhanced risk factor causality extraction method has nearly 20% improvement compared with

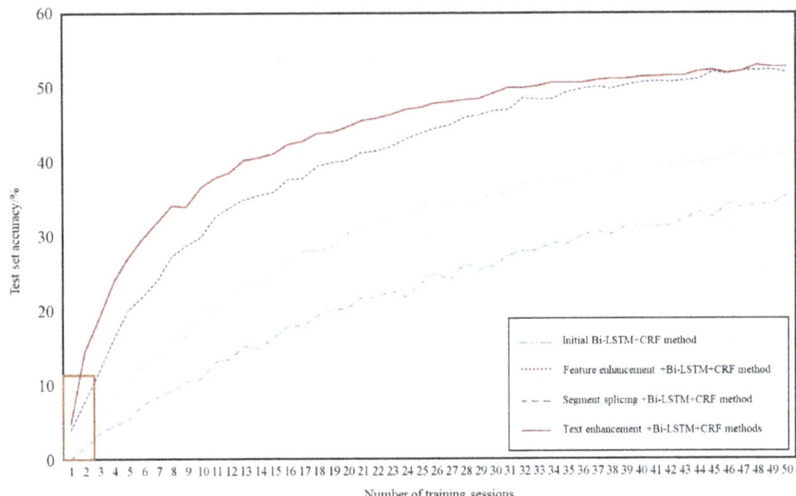

Fig. 4.17. Variation in the accuracy of four risk factor causality extraction methods with number of training sessions.

the initial Bi-LSTM+CRF method, as shown in the solid line box. In order of accuracy size, followed by sentence segment splicing + Bi-LSTM+CRF method, feature enhancement + Bi-LSTM+CRF method and initial Bi-LSTM+CRF method.

(2) In terms of the rising speed of accuracy, text enhancement + Bi-LSTM + CRF method $\geq$ sentence segment splicing + Bi-LSTM + CRF method > feature enhancement + Bi-LSTM + CRF method > initial Bi-LSTM + CRF method; therefore, it can be concluded that the accuracy of the methods mentioned in this section can rise to convergence faster, but the convergence speed of the models with sentence segment splicing and feature enhancement is not much different. The models do not differ much in convergence speed.

(3) In terms of the highest accuracy, text enhancement + Bi-LSTM + CRF method > sentence segment splicing + Bi-LSTM + CRF method > feature enhancement + Bi-LSTM + CRF method > initial Bi-LSTM + CRF method, so it can be concluded that the causality extraction model is able to return all kinds of causal nodes more accurately after adding the text enhancement method.

(4) Among the text enhancement methods, the accuracy of the sentence segment splicing method still works better than the feature enhancement method.

From Fig. 4.18, it can be seen that the F1 values of the four risk factor causality extraction methods show roughly the same trend with the number of trainings: The F1 values rise faster in the early stage and gradually converge with the increase of the number of trainings. Comparison and analysis of the four curves can be obtained from the following conclusions:

(1) As shown in the solid line box, compared with the initial Bi-LSTM+CRF method, the initial accuracy of the proposed method in this section has been improved by nearly 5%, and in the order of accuracy, the following are in the following order: sentence segment splicing + Bi-LSTM+CRF method, feature enhancement + Bi-LSTM+CRF method and the initial Bi-LSTM+CRF method.

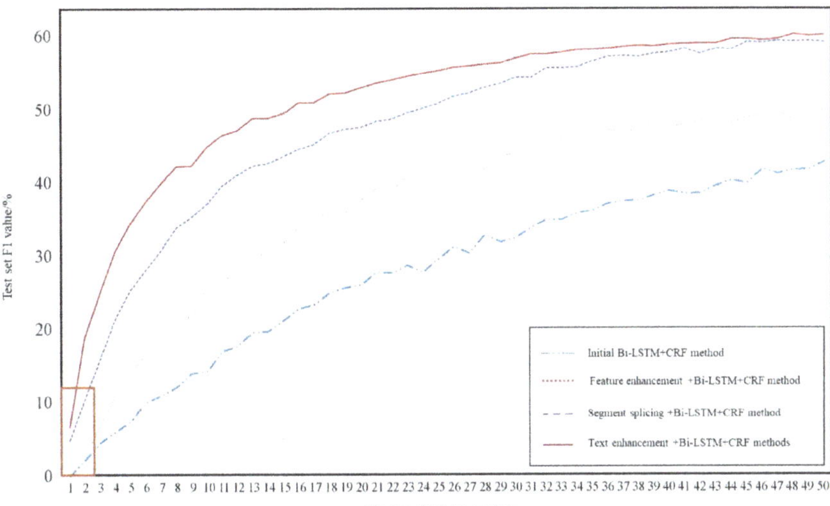

Fig. 4.18. Variation of F1 values with number of training sessions for four risk factor causality extraction methods.

(2) In terms of the increase in speed of model accuracy, text enhancement + Bi-LSTM + CRF method > sentence segment splicing+ Bi-LSTM + CRF method > feature enhancement + Bi-LSTM + CRF method > initial Bi-LSTM + CRF method. Therefore, it can be concluded that the addition of textual enhancement allows the F1 values of the causal extraction model to rise to convergence more quickly, but there is not much difference in the speed of convergence between the model with sentence segment splicing and the model with feature enhancement

(3) In terms of the highest F1 value of the model, text enhancement+ Bi-LSTM + CRF method > sentence segment splicing + Bi-LSTM + CRF method > feature enhancement + Bi-LSTM + CRF method > initial Bi-LSTM + CRF method, so it can be concluded that the methods proposed in this section are better balanced.

(4) Among the text enhancement methods, the sentence segment splicing method can enhance the F1 value better than the feature enhancement method.

4.3 Construction of a Knowledge Map for Risk Factor Evolution in Oil and Gas Production Systems

The establishment of a knowledge graph for the evolution of risk factors in LNG reservoirs not only integrates the causal relationships embedded in unstructured safety management texts but also intuitively reflects the developmental process of risk factors, providing a basis for the subsequent realization of intelligent early warning of accidents in LNG reservoirs. The generalization of specific risk factors is crucial for the establishment of the knowledge graph of risk factor evolution in LNG reserve, which can reduce the node size of the knowledge graph of risk factor evolution, improve the logical description ability and universality of the knowledge graph, and effectively characterize the development law of the risk factors in LNG reserve by combining the risk factors with similar meanings but with different textual expressions into the same abstract factor class. At the same time, the causal relationship of risk factors after generalization can help managers better understand the hidden causal patterns behind specific events.

Therefore, in order to improve the summarization and inductive nature of the knowledge graph of risk factor evolution of LNG reserve, to mine more semantic information, to solve the limitations of the conventional text generalization method on text expression, to reduce the impact of the Chinese word separation error on the accident generalization results, and to address the characteristics of the complex and variable linguistic expressions in the text of safety management, this section proposes a knowledge graph construction method for risk factor evolution of LNG reserve based on Char-Word feature based Agglomerative Nesting (CW-AGNES) method. In order to introduce richer semantic information and improve the effect of risk factor generalization, text features and binary phrase features are added to the feature vector representation, and the AGNES algorithm is used to generalize the figurative risk factors in the safety management text obtained by the risk factor causality extraction method, so as to obtain the general risk evolution causal law and transfer probability in the operation process of the LNG reserve, so as to realize the construction of the risk factor evolution knowledge graph of the LNG reserve. The effect of the generalization method is evaluated on the nodes of the figurative risk factors in the case study to verify the accuracy and reliability of the proposed risk factor evolution knowledge graph construction method.

4.3.1 *Fundamentals of risk factor generalization*

4.3.1.1 *Word2Vec model*

Word2Vec is a classical distributed word embedding representation, the main idea is to map each feature word into the corresponding feature space and to vectorize the semantic information of the text to some extent. Word2Vec provides two models, one is the CBOW model and the other is the Skip-Gram model, both of which are three-layer networks including an input layer, projection layer, and output layer, which can be efficiently trained on a large amount of data to obtain a distributed representation. The CBOW model predicts the occurrence probability of the center word through the contextually distributed representation of the center word and outputs the distributed representation of the center word. The Skip-Gram model, contrary to the idea of the CBOW model, predicts the distributed representation of the context by means of the center word, using the Huffman tree Softmax as the probability calculation method for the center word, and predicts the words of the context based on the path from the root node to the specified node. Compared with CBOW, Skip-Gram can better solve the problem that the relationship between words outside the window and the current word cannot be reflected in the model due to the small window, and the structure of the Skip-Gram model is shown in Fig. 4.19.

If given the center word w_t and its context word $c_t = w_{t-k}, \ldots, w_{t-1}, w_{t+1}, \ldots, w_{t+k}$, where $2k$ denotes the window size, the objective function of the Skip-Gram algorithm is computed in Eq. (4.23):

$$P(c_t|w_t) = \sum_{j=1}^{2k} \lg P(w_j|w_t), \tag{4.23}$$

where $P(w_j|w_t)$ is the conditional probability of occurrence of word w_j under a given center word w_t.

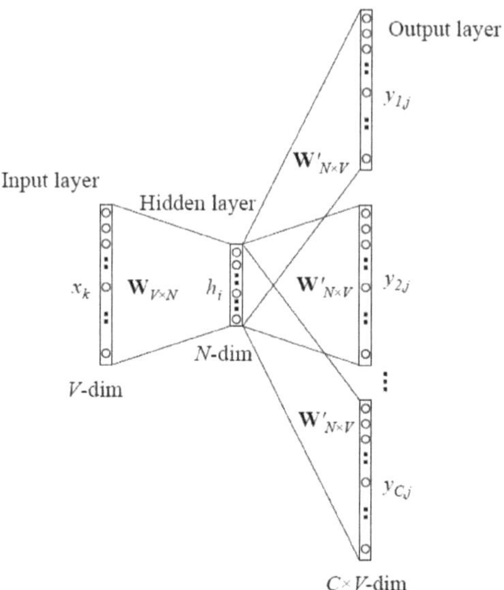

Fig. 4.19. Skip-Gram model structure.

The Skip-Gram model can avoid the problems of semantic ambiguity and inability to accurately represent text information compared with traditional text representation methods.

4.3.1.2 *Agglomerative Nesting algorithm*

The Agglomerative Nesting (AGNES) algorithm is a bottom-up cohesive hierarchical algorithm, which takes each initial object as a cluster and merges the cluster with the smallest distance between two objects belonging to different clusters by calculating the distance between each object, and merges the cluster with the closest distance by continuously calculating the distance until the set termination condition is reached or the merging of all objects is completed. The steps of AGNES algorithm are described as follows:

Input: a dataset S containing n sample data points.

Output: clustering results.

Step 1: Consider each data point as a separate cluster.

Step 2: Calculate the spatial distance $dist(i, j)$ between all data points, and establish the initial similarity matrix between data points.

Step 3: Merge the two clusters with minimum distance and maximum similarity.

Step 4: Recalculate the distances between the newly merged clusters and other clusters to update the similarity matrix.

Step 5: Repeat steps 3 and 4 until you have merged into one class or reached the specified number of clusters.

4.3.1.3 *T-distributed Stochastic Neighborhood Embedding algorithm*

The T-distributed Stochastic Neighborhood Embedding (TSNE) algorithm is a typical stream learning method, which has received much attention from scholars and experts in various fields in recent years. The basic idea of the TSNE algorithm is to map the points in the high-dimensional space to the low-dimensional space while keeping the probability of the distributions of each other unchanged, using the Gaussian distribution to transform the distance into a probability distribution in the high-dimensional space, and using the T-distribution to transform the distance into a probability distribution in the low-dimensional space, using the joint probability to represent the similarity corresponding to points, and obtaining the sample distributions in the low-dimensional space by optimizing the KL dispersion of the distance between the two distributions.

Let the high-dimensional data points be $\boldsymbol{X} = (x_1, x_2, \ldots, x_n)$ and the low-dimensional mapping points $\boldsymbol{Y} = (y_1, y_2, \ldots, y_n)$, and the KL dispersion is calculated in Eq. (4.24):

$$C = KL(P \parallel Q) = \sum_i \sum_j p_{ij} \lg \frac{p_{ij}}{q_{ij}}, \tag{4.24}$$

where p_{ij} is the joint probability of the sample distribution in the high-dimensional space; q_{ij} is the joint probability of the sample distribution in the low-dimensional space, then the calculation of the joint probability function of the high-dimensional data points x_i, x_j is shown in Eq. (4.25):

$$p_{ij} = \frac{\exp[-\parallel x_i - x_j \parallel^2 /(2\sigma)^2]}{\sum_{k \neq l} \exp[-\parallel x_k - x_j \parallel^2 /(2\sigma)^2]}. \tag{4.25}$$

The joint probability can be found in Eqs. (4.26) and (4.27):

$$p_{ij} = \frac{p_{j|i} + p_{i|j}}{2n},\qquad (4.26)$$

$$p_{j|i} = \frac{\exp[-\parallel x_i - x_j \parallel^2 /(2\sigma_i)^2]}{\sum_{k \neq i} \exp[-\parallel x_i - x_k \parallel^2 /(2\sigma_i)^2]},\qquad (4.27)$$

where σ is the Gaussian variance centered at x_i, for which the best σ is obtained by performing a dichotomous search with a pre-set complexity factor. The joint probability function of the data points y_i, y_j is computed in the low-dimensional space as shown in Eq. (4.28):

$$q_{ij} = \frac{(1+ \parallel y_i - y_j \parallel^2)^{-1}}{\sum_{k \neq l} (1+ \parallel y_k - y_l \parallel^2)^{-1}}.\qquad (4.28)$$

The TSNE algorithm minimizes the loss function by the gradient descent method, and the logical steps of the algorithm are as follows:

(1) Calculate the joint probability for a given complexity according to Eqs. (4.27) and (4.28).
(2) Randomly initialize Y with a normal distribution N to obtain the initial solution $Y^{(0)}$.
(3) An iterative operation is performed using the gradient descent algorithm to finally obtain a low-dimensional representation of the high-dimensional input vector.

TSNE uses a more long-tailed T-distribution in the low-dimensional space, so that clusters with larger distances in the high-dimensional space have larger distances in the low-dimensional space, thus solving the data crowding problem.

4.3.2　*Construction of LNG reservoir risk factor evolution knowledge graph based on char-word features*

Aiming at the richness of semantic information and utterance representations in the text of LNG reserve risk factors, this section adds a char-word feature vector layer on the basis of the AGNES algorithm and proposes a knowledge graph construction method based on CW-AGNES to help better realize the establishment of the knowledge graph for the evolution of the risk factors in the LNG reserve.

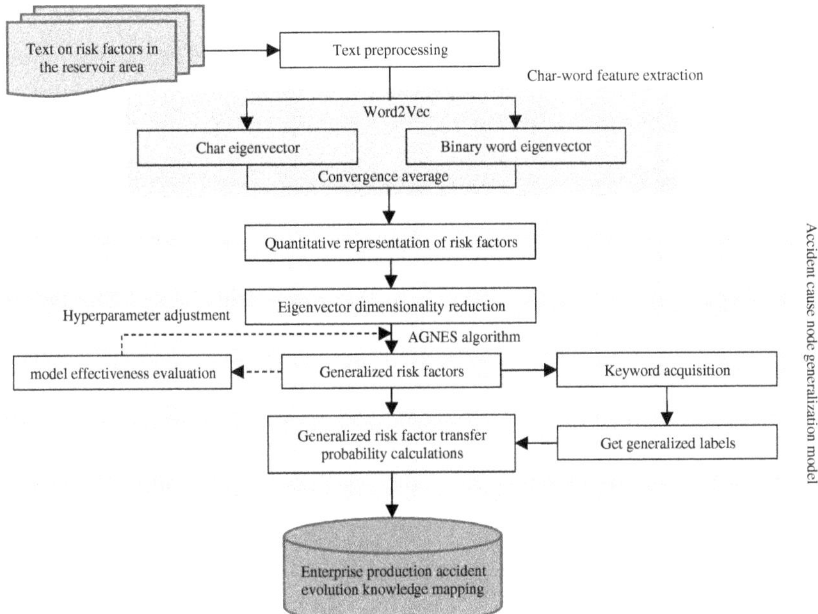

Fig. 4.20. LNG reservoir risk factor evolution knowledge graph construction methodology flow.

Compared with the traditional AGNES algorithm, the CW-AGNES method proposed in this section can better introduce rich effective information by extracting the char feature vectors and binary word feature vectors of the risk factors in the security management text, making the feature mapping space of the risk factors more accurate and preserving more semantic information, and guaranteeing the accuracy of the generalization of the risk factors. The specific flow of the LNG reservoir accident evolution knowledge map construction method based on the CW-AGNES algorithm is shown in Fig. 4.20.

4.3.2.1 *Word2Vec-based text feature vector acquisition for security management*

Step 1: Text preprocessing.

The text related to the safety management of LNG reserves is divided into features, and the text is cut into different feature datasets in the form of chars and binary phrases, respectively.

Step 2: Text vectorization based on char features.

Using the Word2Vec tool, the char feature dataset is used to learn semantic information to characterize the word feature information in the security management text in the form of high-dimensional vectors.

Step 3: Text vectorization based on binary word features.

Using the Word2Vec tool, the binary phrase feature dataset is used to learn semantic information to characterize the binary phrase feature information in the security management text in the form of high-dimensional vectors.

4.3.2.2 *A generalization method for risk factors based on char-word features*

Risk factor generalization refers to the process of generalizing the specific risk factor nodes in the causal relationship into abstract meaning risk factors by clustering. The risk factors and their causal relationships obtained by the risk factor causality extraction method are between specific events, and due to the richness of Chinese grammatical expressions and numerous vocabularies, it is not possible to find out the general law of risk factor evolution. For example, it is only possible to obtain a specific risk factor causal evolution relationship like "Inadequate equipment inspection → internal leakage of the recondenser → increase in compressor temperature → LNG leakage"", but it is not possible to summarize and find a more generalized and generalized causal evolution pattern, such as "imperfect management of equipment operation → equipment leakage → equipment temperature exceeds the standard → leakage accident". Therefore, generalization of specific risk factors can make the risk factor causality chain more representative and regular. The schematic diagram of risk factor generalization for LNG reservoirs is shown in Fig. 4.21, in which the lower layer is the causal network of specific risk factors composed of concrete risk factors, and the upper layer is the causal network of abstract risk factors composed of abstract risk factors generalized from the bottom layer.

As shown in Fig. 4.21, the directed edges between the nodes of the solid line and the solid nodes constitute the concrete causal network, the directed edges between the nodes of the dashed line and the

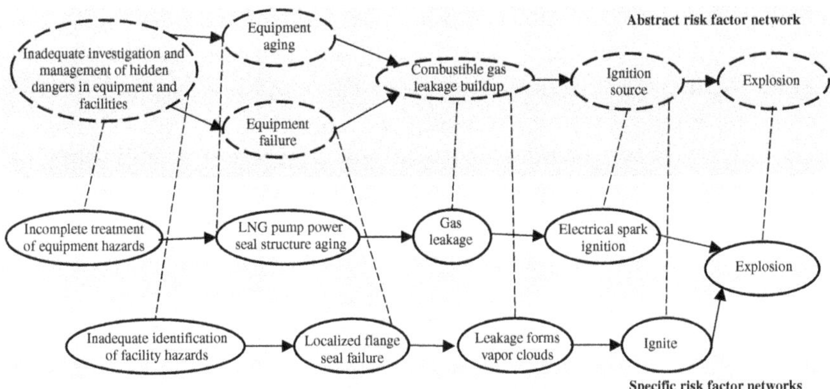

Fig. 4.21. Schematic diagram of risk factor node generalization.

Table 4.13. Required setup parameters for the generalization method.

Serial no.	Parameter name	Serial no.	Parameter name
1	Char vector dimension	6	Batch size
2	Word vector dimension	7	Learning rate
3	Training models	8	Trainings number
4	Min. word frequency	9	Window sizes
5	Clustered linking approach	10	Distance calculation method

dashed nodes constitute the abstract causal network, and the dashed undirected edges between the nodes of the solid line and the dashed nodes represent the transformation process between the concrete risk factors and the abstract risk factors. The generalization process of risk factors for LNG reserve is shown in the following:

Step 1: Parameterization of the generalization method.

In the initialization phase of the method, it is necessary to set the parameters involved in the method, and the required setting parameters are shown in Table 4.13. By setting the appropriate parameters, the method can help to have good performance and results.

Step 2: Node representation of risk factors based on char-word features.

Texts can consist of a combination of chars or binary phrases. In order to more effectively utilize the valuable information embedded in safety management texts, risk factors are divided into char groups based on individual chars, and into binary phrases, except for indivisible terms and individual adverbs. For example, the indirect factor node "Safety marking of energized equipment is not obvious" is split into "Safety marking of energized/equipment/is not/obvious" and "Safety marking of energized/equipment/is not/obvious" and "Safety marking of energized/equipment/is not/obvious".

On this basis, the vector representation of a specific risk factor node is obtained using the method of feature vector averaging, i.e., the average of the vectors of all the chars in the risk factor node, and the vectors of the binary phrase are taken as the numerical representation of the risk factor node in the feature space.

Step 3: Risk factor node feature downscaling.

Under the circumstance of ensuring less loss of semantic information of risk factor nodes, in order to improve the efficiency of the model, the TSNE algorithm is chosen to downsize the feature vectors of risk factor nodes in order to reduce the computational complexity of the subsequent steps.

Step 4: The number of generalized cluster classes is determined.

The cluster categories for generalization are determined based on expert experience based on the actual content of the risk factor text.

Step 5: Risk factor node generalization model construction.

First, based on the proposed CW-AGNES method, the generalization operation is performed on the risk factor nodes to merge the nodes with similar distances in the semantic space into one class; then, the keywords in such risk factor nodes are obtained by statistically counting the word frequency information of the risk factor nodes clustered into one class; and finally, the generalized labels of the risk factors in this class are obtained based on the keyword information.

Based on the spatial distribution of the risk factor node vectors, the Euclidean distance is chosen as a measure of similarity between the nodes, and the calculation is shown in Eq. (4.29):

$$dist_{ij} = \sqrt{(i_1 - j_1)^2 + (i_2 - j_2)^2 + \cdots + (i_n - j_n)^2}, \qquad (4.29)$$

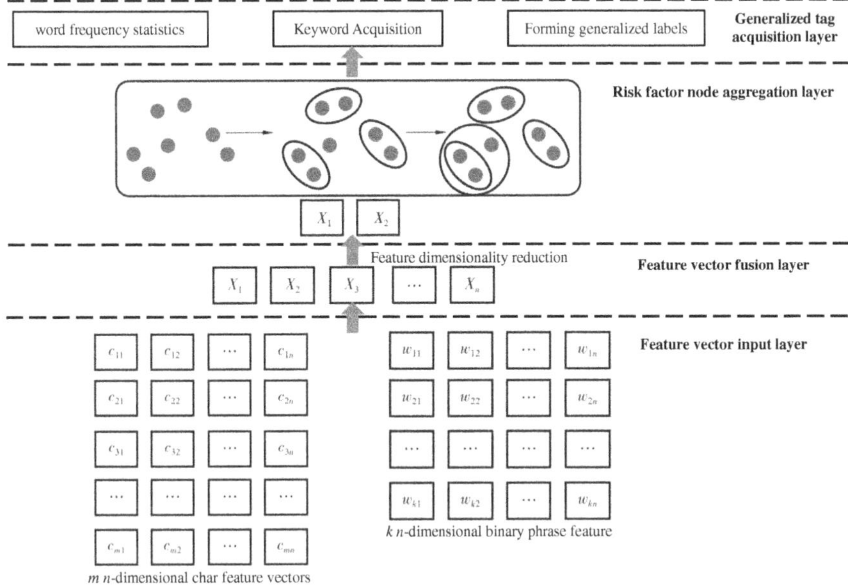

Fig. 4.22. Nodal generalization model architecture for LNG reservoir risk factors.

where $dist_{ij}$ is the distance between nodes; i_n, j_n are the coordinates in the semantic space of nodes.

Eventually, the formed risk factor node generalization model of LNG reserve consists of four layers, which are feature vector input layer, feature vector fusion layer, risk factor node aggregation layer, and generalization label acquisition layer, and the specific generalization model architecture is shown in Fig. 4.22.

4.3.2.3 *Construction of causal evolutionary knowledge graph of risk factors for LNG reserve*

The LNG reserve risk factor evolution knowledge graph is in the form of $G = VE$, where V is the set of points, in which each node represents a risk factor, and E is the set of edges, which represents the evolutionary causal relationship between risk factors. In order to measure the likelihood of the causal development of the risk, the causal transfer probability should be indicated on the edges of the knowledge graph of the evolution of risk factors in the LNG reserve. The transfer probability of risk factors is calculated in Eq. (4.30):

$$p(E_j|E_i) = \frac{count(E_i, E_j)}{\sum_k count(E_i, E_k)}, \tag{4.30}$$

where $count\,(E_i, E_j)$ denotes the frequency of occurrence of E_j when E_i occurs; $\sum_k count(E_i, E_k)$ denotes the total number of all possible events when E_i occurs.

After the generalization of risk factors, specific risk factors with different expressions or similar meanings will be normalized, so that most of the extracted risk factors have the same causal nodes in their causal relationships, thus allowing multiple causal chains to be crosslinked by one or some common abstract risk factors to form an abstract knowledge graph of the evolution of risk factors in LNG reservoirs.

4.3.3 *Case study*

4.3.3.1 *Testing and validation*

The text data in the case are divided into two parts: (1) Dataset A, LNG reserve and petrochemical field-related texts, with a total of 2972 entries, is used for the training of char-word feature vectors; (2) Dataset B, risk factors in the safety management texts related to LNG reserve, with a total of 773 entries, is used for the testing of the risk factor evolution knowledge graph construction method.

Step 1: Parameterization of the LNG Reservoir Risk Factor Node Generalization Method.

As shown in Table 4.14, the initialization parameters of the method are set, where the word vector dimension is set to 100 dimensions, the word vector dimension is set to 100 dimensions, the training model is set as Skip-Gram model, the minimum word frequency is set to 1, the size of each batch is set to 1000, the learning rate is set to 0.0001, the number of training times is set to 50, the window size is set to 10, the linkage method for clustering is the sum of squares of deviations method, and the distance calculation method is the Euclidean distance calculation.

Table 4.14. Required setup parameters for the generalization method.

Parameter name	Parameter value	Parameter name	Parameter value
Char vector dimension	100	Batch size	1000
Word vector dimension	100	Learning rate	0.0001
Training models	Skip-Gram	Trainings number	50
Min. word frequency	1	Window sizes	10
Clustered linking approach	Sum of Squares of Deviations	Distance calculation method	European distance

Step 2: Security management text feature vector acquisition.

Based on the Python language environment, combined with the gensim toolkit, the vector model of char features and binary word features is built for dataset A. Char2Vec feature vector model for char features (containing 2979 100-dimensional char feature vectors), and Word2Vec feature vector model for binary word features (containing 29576 100-dimensional word feature vectors) were built.

Step 3: Nodal representation of risk factors for LNG reservoirs.

The risk factors in dataset B are divided into chars and binary words, and the corresponding pre-trained feature vector model is used to obtain all the char feature vectors and binary word feature vectors of each risk factor; the feature vectors of all the chars and binary words of the risk factor nodes are averaged, and they are used as a characterization of the cause of the accident node to obtain the corresponding 100-dimensional feature vectors.

Step 4: Determination of the number of generalized cluster categories of risk factors for the LNG reserve.

Based on the textual content of the underlying, indirect and direct factors, the number of generalization clusters for the underlying factors was set at 9, for the indirect factors at 42, and for the direct factors at 31.

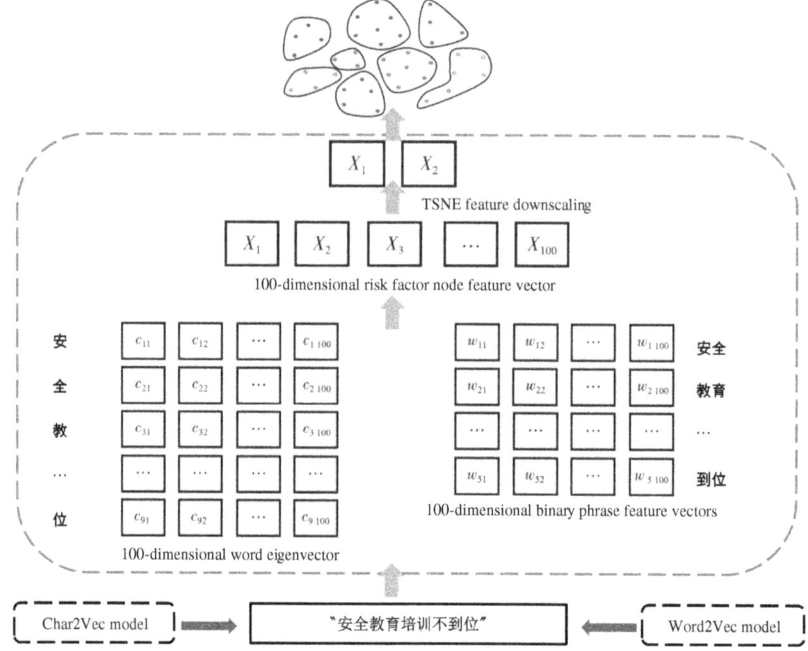

Fig. 4.23. LNG reservoir risk factor nodal generalization model flow.

Step 5: Nodal generalization modeling of LNG reserve risk factors.
Based on Python language environment, combined with sklearn and
scipy toolkit, we build a node generalization model of risk factors of
LNG reserve based on the CW-AGNES method and take the input of
the root cause node "safety education and training are not in place"
as an example, the flow of the model is shown in Fig. 4.23.

The generalization model of risk factor nodes of LNG reserve
based on CW-AGNES is generalized on fundamental factor nodes,
indirect factor nodes, and direct factor nodes, respectively, and the
generalization results are as follows:

(1) Results of the generalization of the underlying factors

On dataset B, the 41 root factor nodes are generalized, and the scatter plot of the generalization results in the 2D plane when the number
of cluster categories is 9 is shown in Fig. 4.24. From the 9 colors of
cluster categories in Fig. 4.24, it can be seen that the points with
similar distances in the semantic space are roughly clustered into
one category, and the keywords of each cluster category are obtained

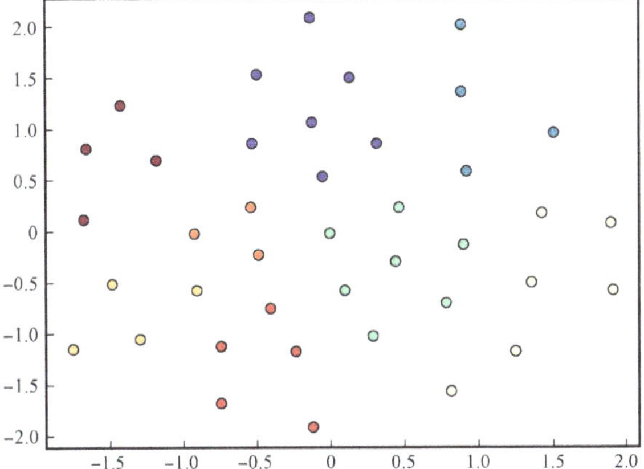

Fig. 4.24. Plot of the generalization results of the underlying factors on the 2D plane.

Table 4.15. Keywords for each cluster of the underlying factors.

Categories	Cluster keywords
Category 1	Operating procedures, imperfections, stoppages
Category 2	Inadequate, management, operations
Category 3	Inadequate, routine, inspection
Category 4	Inadequate, exhaustion, hidden dangers
Category 5	Training, solid, insufficient
Category 6	Security, lack, training
Category 7	Inadequate, training, education
Category 8	Security management, down to earth, vulnerabilities
Category 9	Not in place, identification, risks

by counting the word frequency information of the root factor nodes clustered into one category, and the specific results are shown in Table 4.15.

Summarizing the keyword information of each cluster category in Table 4.15, the main meanings of the root factor nodes in the nine cluster categories are obtained: Category 1 is imperfect operating procedures, category 2 is not in place for operation management, category 3 is not in place for daily inspection, category 4 is not in place for hidden danger investigation, category 5 is not solid enough for training, category 6 is deficient in safety training, category 7 is

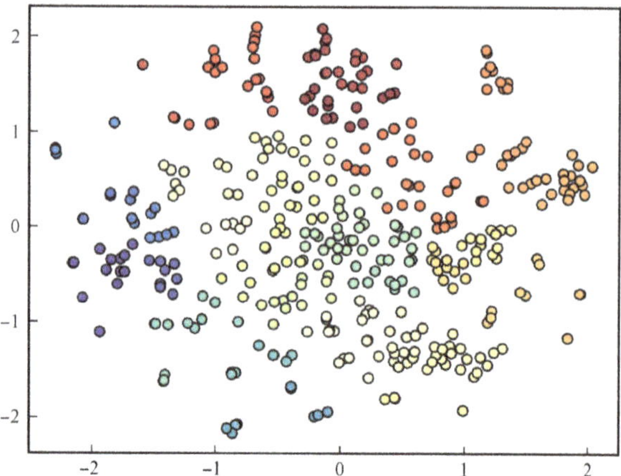

Fig. 4.25. Plot of generalization results for indirect factors in the 2D plane.

not in place for education and training, category 8 is not in place for safety management, and category 9 is not in place for risk identification. It can be seen that categories 5, 6, and 7 overlap in content, and the risk factor node generalization model fails to successfully cluster these three categories into one, resulting in errors in the generalization results.

(2) Indirect factor generalization results

On dataset B, 400 indirect cause nodes are clustered, and the scatter plot of the generalization results in the 2D plane when the number of cluster categories is 42 is shown in Fig. 4.25.

From the cluster classes of the 42 colors in Fig. 4.25, it can be seen that the points that are close to each other in the semantic space are roughly clustered into one class, but due to the fact that some of the points are very close to each other, they cannot be divided into the cluster boundaries well, which makes the generalization results have some errors. The keyword information of each cluster class is shown in Table 4.16.

Summarizing the keyword information for each cluster category in Table 4.16, the main meanings of the indirect factor nodes in the 42 cluster categories are obtained: Category 1 is corrosion of storage tanks, pipelines, and other equipment; category 2 is rupture of elbows and other attachments; category 3 is leakage of flanges and other attachments; category 4 is failure of flanges, valves, and other

Table 4.16. Keywords for each cluster of indirect factors.

Categories	Cluster keywords
Category 1	Corrosion, storage tanks, pipeline valves, heat exchangers, bolts
Category 2	Elbow, pipe, rupture, sticking, over time
Category 3	Flanges, leaks, coolers, joints, check valves
Category 4	Aging, flanges, failures, valves, heads
Category 5	Overflow, wellbore, sudden
Category 6	Control valve, polymerization reactor, drum, abnormal, new line
Category 7	Pipeline, rush, high pressure, hose, shutdown
Category 8	Pressure, excessive, propylene tower, air cooler, exceedance
Category 9	Flash blast, buffer tanks, heat exchangers, shutters, explosion proofing
Category 10	Non-compliance, errors, inadequate, inspections, personnel
Category 11	Unreasonable, placement, placing, drilling births, drill pipe
Category 12	Ventilation, indoor, poor, bottom of cesspool, charcoal burning
Category 13	Combustible gas, undetected, toxic gas, oxygen content, concentration
Category 14	Aged, stressed, uneven, pumping unit, four wood tower
Category 15	Oversize, exhaust port, overflow, pipe spacing, poison control equipment
Category 16	Irregularity, operation, not in place, personnel, emergency drills
Category 17	Improper operation, operating errors, inappropriate, personnel, employees
Category 18	Safety awareness, dress code, inappropriate, driller, standardized
Category 19	Failure to wear, safety harness, safety protective equipment, no, safety equipment
Category 20	Safety guards, failure to, safety guards, pulleys, safety warnings
Category 21	Safety range, operation, safety warning, safety management, safety distance
Category 22	Fixed, lift trucks, gates, tanks, blowout preventers
Category 23	Pipe hoist, unbalance, brake, pipe sleeve, short pipe
Category 24	Breakage, loss of control, ceiling, subplate valves, netting
Category 25	Failure, pipe fitting, axial pump, floating head, storage tanks
Category 26	Tank, unauthorized, gasifier, LNG, overfilling
Category 27	Tanks, internal counterparts, oil, inside tanks, frost
Category 28	Operating environment, slippery, wet, ice, transformer base

(Continued)

Table 4.16. (*Continued*)

Categories	Cluster keywords
Category 29	Shot blasting machine, gas valve, abnormalities detected, ignition, cigarettes
Category 30	Failure to confirm safety, operating environment, timely, inspection, power test
Category 31	Personnel, violation, operation, exchange of information, valve manifolds
Category 32	Non-professional, person, equipment, operation, work
Category 33	High-voltage power line, pipeline, crossing, pump truck, approaching
Category 34	Pumping rods, drilling tools, disconnecting, dislodging, dumping
Category 35	Too close, incline, high voltage line, ground, lifting
Category 36	High winds, heavy rain, sudden
Category 37	Soil, rain, pipe trench, pipe trench wall, collapse
Category 38	Ticket, failure, quality, welds, bursting box
Category 39	Dislodgement, traveling slide, large hook, lower hook frame, deformation
Category 40	Tubing, lighter workpieces, lifting tubing, upward movement, large hooks
Category 41	Beam, drop, rig, hose, rover
Category 42	Wire rope, breakage, aging, hook, spreader

attachments; category 5 is sudden overflowing of the bottom of wells; category 6 is abnormalities of control valves and other equipment; category 7 is rapid increase in the pressure of pipelines and other equipment; category 8 is exceeding of the pressure of the equipment; category 9 is flash explosion of buffer tanks, heat exchangers, and other equipment; category 10 is irregular operation and inspection; category 11 is unreasonable placement of facilities; category 12 is poor indoor ventilation; category 13 is failure to detect the concentration of combustible and toxic gases; category 14 is old and dilapidated, uneven stress on equipment; category 15 is overflowing exhaust ports and excessive spacing of pipes; category 16 is non-standardized operation; category 17 is improper and faulty operation; category 18 is non-standardized safety attire; category 19 is failure to wear safety protective equipment; category 20 is not having safety protection equipment; category 21 is working outside the safety range; category 22 is fixing of lifting trucks and other equipment;

category 23 is imbalance of pipeline cranes; category 24 is broken and out-of-control parts; category 25 is equipment failure; category 26 is too much LNG filling; category 27 is tanks with oil and frost; category 28 is slippery working environment; category 29 is abnormal equipment; category 30 is failure to confirm the working environment; category 31 is illegal operation by personnel; category 32 is operation of equipment by non-professionals; category 33 is pipeline crossing high-voltage wires; category 34 is disconnection and detachment of pumping rods, drilling tools; category 35 is too close to high-voltage wires; category 36 is sudden wind and rainstorm; category 37 is collapse of pipeline trench; category 38 is failure of operation ticket and quality of welds; category 39 is dislodging and deformation of travelling slide; category 40 is upward movement of oil pipe; category 41 is dropping of beams and other heavy objects; and category 42 is aging and breakage of wire ropes and other tools. It can be seen that the meanings of categories 6 and 29, 7 and 8, 27 and 28, 33 and 35 are similar; the meanings of categories 39 and 41 overlap; and there are two different meanings for categories 14, 15, and 38. Apart from these categories, the textual approach can better aggregate indirect factors with similar meanings into one category.

(3) Direct factor generalization results

On dataset B, 332 root cause nodes are clustered and when the number of cluster categories is 31, the scatter plot of the generalization results in the 2D plane is shown in Fig. 4.26.

As can be seen from the cluster classes of the 31 colors in Fig. 4.26, the clustering results are able to satisfy the requirement that the points in the same cluster class are close to each other in the semantic space to a certain extent, but the existence of different categories of points that are too close to each other will make the generalization results inaccurate. The keyword information of each cluster is shown in Table 4.17.

Summarizing the keyword information for each cluster in Table 4.17, the main meanings of the root factor nodes in the 31 clusters are as follows: Category 1 is natural gas leakage, category 2 is insufficient oxygen content, category 3 is LNG leakage forming a vapor cloud, category 4 is flying out of attachments, category 5 is failure to wear anti-virus equipment, category 6 is compressor

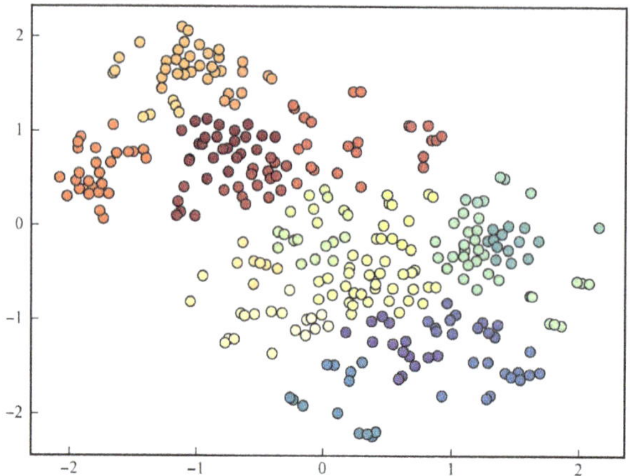

Fig. 4.26. Plot of generalization results for direct causes in the 2D plane.

and other equipment with excessive temperatures and bursting, category 7 is sparks from operations, category 8 is exposure to ignition sources, category 9 is hydrogen sulfide concentration, category 10 is combustible toxic gases, category 11 for ignition and smoking, category 12 for oil and gas accumulation, category 13 for electrocution and electrification, category 14 for gaskets popping off, category 15 for high-speed rotation, category 16 for entanglement and slipping, category 17 for heavy equipment dropping and tilting, category 18 for being crushed by the equipment, category 19 for equipment breaking or losing balance, category 20 for failure to wear safety equipment, category 21 for equipment leakage, category 22 for tripping, equipment slipping, category 23 for falling through the air, category 24 for inadvertent slipping and falling, category 25 for rubber core and donkey head popping out, category 26 for losing footing, category 27 for pipe trench collapsing, category 28 for sliding of large-scale equipment, category 29 for sliding of oil pipes and other instruments, category 30 for category 29 is the sliding of oil pipes and other equipment, category 30 is the collapsing and slipping of steel pipes, and category 31 is the tipping of supports and other devices. It can be seen that the meanings of categories 1 and 3, 7 and 8, 23, 24, and 26 are similar, while the meanings of categories 10 and 12, 7, 8, and 11 overlap. It can be seen that the method proposed in this section tends to divide semantically similar direct factors into different clusters when generalizing direct factors, but in general it can better cluster semantically similar factors into one category.

Finally, by combining the keyword information of each cluster category of the generalization results with the original risk factor text, the generalized labels of all risk factors are obtained as shown in Table 4.18.

Step 5: LNG Reservoir Risk Factor Evolution Knowledge Graph Construction.

Based on the generalization model of risk factor nodes of LNG reserve based on the CW-AGNES method, the risk factor nodes are generalized; the transfer probability between the cause nodes of enterprise production accidents is calculated by using Eq. (4.30), and 476 pairs of causality relationships with effective transfer probability are obtained. Finally, the knowledge graph of risk factor evolution of LNG reserve is obtained, and the causal evolution relationships of some risk factors are shown in Table 4.19.

An analysis of the constructed knowledge map of the evolution of risk factors in LNG reservoirs shows that "safety education and training are not in place," "safety supervision mechanism is not perfect," "equipment operation and management are not perfect," "routine inspection and maintenance are not in place," "hidden danger investigation and treatment are not in place" are the root cause of accidents in LNG reservoirs. Among the indirect causes of LNG reserve accidents are "unauthorized operation by personnel," "irregular operation by personnel," "equipment in an unsafe condition," "personnel not wearing safety protection devices" "equipment malfunction, failure" "equipment breakage". Among the direct causes of LNG reserve accidents, it is necessary to pay attention to the situation of "flammable gas leakage and accumulation" in the workplace, to pay attention to the situation of "equipment movement" for movable equipment to prevent crushing, and prevent "object strikes" for high-speed equipment and equipment with a lot of attachments.

4.3.3.2 *Comparative analysis*

In order to evaluate the generalization effect of the method proposed in this section on the risk factors of LNG reserves, the actual risk factor texts are classified according to the generalization labels in

Table 4.17. Keywords for each cluster of direct factors.

Categories	Cluster keywords
Category 1	Leakage, natural gas, air cooler, LNG, pipeline, oil and gas
Category 2	air, oxygen content, oxygen content, oily, inhalation, liquid
Category 3	Leakage, vapor cloud, LNG, overpressure, pressure station, pressure gauge
Category 4	Flying, injury, head, pin, tank top, release buckle
Category 5	Failure to wear, storage tank, residue, cable tank, gas mask, air respirator
Category 6	Fresh hydrogen compressor, burst, temperature, excessive, piping, furnace
Category 7	Sparks, generated, during operation, kinetic sparks, oxy-coal sparks, airbursts
Category 8	Electrical sparks, open flame, sparks, static electricity, ignition, exposure to fire
Category 9	Hydrogen sulfide, concentration, limit, explosion, nitrogen, reaches
Category 10	Combustible gases, toxic gases, accumulation, aggregation, detection, deposition
Category 11	Ignition, smoking, lights, dormitory
Category 12	Oil, gas, buildup, leakage, pump room, spray-in, sand and gravel
Category 13	Contact, electric shock, electrified, welding rod, high voltage discharge, copper core
Category 14	Gaskets, catch basins, burglar-proof doors, popping open, dropping, bending
Category 15	High-speed rotation, clothing, pulley, scraping, strangulation, clothesline
Category 16	Coil, slip, pump, pulley, fall, safety rope
Category 17	Falling, tilting, overhead crane, large hook, roller square filler shell, rope suspender
Category 18	Crush, pipe hoist, personnel, vehicle, abdomen, gantry crane
Category 19	Breakage, loss of balance, wire rope, swing, improper, personnel evacuation
Category 20	Failure to wear, insulating devices, inside-to-outside, safety equipment, outside eaves, safety ropes
Category 21	Leakage, equipment, LNG, pipeline, loss of control, washer
Category 22	Tripping, blowout preventer, electric pole, slip, continuous fuel line, blowout preventer kava
Category 23	Fall, step-off, platform, second floor, inadvertent, work platform

(*Continued*)

Table 4.17. (*Continued*)

Categories	Cluster keywords
Category 24	Slip, tank, fall, inadvertent, fall, drop
Category 25	Ejection, rubber core, donkey's head
Category 26	Fall, slip, elbow, man with, connecting pipe, top drive
Category 27	Collapse, pipe trench, pool wall, mound of dirt, lifting frame, crushing collapse
Category 28	Slip, foundation block, tanker, collapse, pipe, loose
Category 29	Slip, pipe, pipe hoist, drill chain, scrape down, strut
Category 30	Steel pipe, collapse, slip, crush, short pipe, manifolds
Category 31	Dumping, stanchion, flatbed, casing, crane, rotation

Table 4.18, and then the adjusted RAND coefficients and adjusted Mutual Information, the FMI values, and the V-measure values are selected as the assessment indexes of the proposed method, as follows:

(1) Adjusted Rand Index

Adjusted Rand Index (ARI) is used to measure the similarity between the predicted cluster vectors of the model and the real vectors of the samples, with a value in the range of $[-1, 1]$, and the larger the value, the more the clustering results match with the real situation, which is calculated in Eq. (4.31):

$$ARI = \frac{\sum_{ij} \binom{n_{ij}}{2} - \left[\sum_i \binom{a_i}{2} \sum_j \binom{b_j}{2}\right] / \binom{n}{2}}{\frac{1}{2}\left[\sum_i \binom{a_i}{2} + \sum_j \binom{b_j}{2}\right] - \left[\sum_i \binom{a_i}{2} \sum_j \binom{b_j}{2}\right] / \binom{n}{2}}, \tag{4.31}$$

where i, j denote the real and predicted cluster classes respectively; n_{ij} denotes the number of real cluster class i and predicted cluster class j; a_i denotes the same meaning as n_i; b_j denotes the same meaning as n_j.

(2) Adjusted Mutual Information

Adjusted Mutual Information (AMI) utilizes a mutual information-based approach to measure the degree of correlation between two cluster vectors, with values ranging from $[-1, 1]$, where a larger value means that the clustering results are more consistent

Table 4.18. LNG reservoir risk factors generalized label categories.

Risk factor type	Generalized tag category name
Root factor	Inadequate safety education and training; Imperfect safety monitoring mechanism; Imperfect operating procedures; Risk and hazard identification and control are not in place; Imperfect equipment operation management; Operation management is not in place; Daily inspection and maintenance are not in place; Hidden danger investigation and management is not in place; Safety management system is not perfect.
Indirect factors	Personnel operation is not standardized; Equipment in an unsafe condition; Unsafe behavior of personnel; Hazardous substances spread; No warning signs at hazardous locations; Collapse; Equipment without safety guards; Failure of personnel to check oxygen concentration; Equipment used beyond design limits; Explosion of equipment; Deformation of equipment; Generation of ignition sources; Improper operation; Cold damage to equipment; Unauthorized operation by personnel; Insufficient safety knowledge of personnel; Lack of on-site guidance and supervision; Poor operating environment; Equipment aging; Incorrect position of equipment and parts; Failure of personnel to wear safety equipment; Failure of personnel to test the concentration of flammable and hazardous gases; Equipment leakage; Malfunction or failure of equipment; Improper placement of equipment and facilities; Failure to conduct safety inspections prior to operation; Natural disaster; Equipment not turned off/on as required; Inadequate safety awareness among personnel; Errors in personnel communication; Loopholes in safety management; Poor ventilation; Corroded equipment; Failure of testing instruments to give alarm; Personnel slip and fall; Abnormal equipment parameters; Personnel working within safety limits; Equipment parts are broken or fractured; Equipment facilities are not fixed; Incorrect operation; Equipment does not meet requirements; LNG stratification/tumbling.

(*Continued*)

Table 4.18. (Continued)

Risk factor type	Generalized tag category name
Direct factor	Combustible gas leakage buildup; Sparks from operations; Dumping of equipment; Damaged or failed equipment components; High equipment temperatures; Building collapse; LNG in contact with water; Personnel crushed; Improper positioning of personnel; Personnel misjudgment; Hazardous gas leakage buildup; Failure to wear safety equipment; Equipment energized; Excessive pressure on equipment; Equipment running at high speed; Land collapse; Personnel electrocuted; Personnel scratched; Personnel operating errors; Lack of safety protection on site; Low ambient oxygen concentration; Equipment movement; Equipment falling; Loss of control of equipment; Flying parts; Personnel losing their footing; Personnel impacted; Person sucked into equipment; Contact with hot object; Layering of liquid in tank; Person slips and falls.

with the real situation, and the calculations are shown in Eqs. (4.32)–(4.36):

$$AMI = \frac{MI(U,V) - E[MI(U,V)]}{\max\{H(U), H(V)\} - E[MI(U,V)]}, \tag{4.32}$$

$$MI(U,V) = \sum_{i=1}^{|U|} \sum_{i=1}^{|V|} P(i,j)\,\mathrm{lb}\,P(i,j)/P(i)P(j), \tag{4.33}$$

$$H(U) = \sum_{i=1}^{|U|} P(i)\,\mathrm{lb}\,P(i), \tag{4.34}$$

$$H(V) = \sum_{j=1}^{|V|} P(j)\,\mathrm{lb}\,P(j), \tag{4.35}$$

$$P(i,j) = \frac{|U_i \cap V_j|}{N}, P(i) = \frac{|U_i|}{N}, \quad P(j) = \frac{|V_j|}{N}, \tag{4.36}$$

where U and V are two ways of dividing a set containing N samples.

Table 4.19. Knowledge mapping of the evolution of risk factors in LNG reservoirs (partial).

Serial no.	Causal node	Transfer probability 10^{-3}
1	Inadequate safety education and training	6.95
	Unauthorized operation by personnel	0.70
	LNG delamination/tumbling	0.70
	Abnormal equipment parameters	4.17
	Accumulation of flammable gas leaks	4.17
	Leakage accidents	—
2	Inadequate routine inspection and maintenance	10.43
	Failure of equipment	4.17
	Combustible gas leakage buildup	0.70
	Frostbite accidents	—
3	Inadequate identification and management of hidden dangers	0.70
	Use of equipment exceeds design limits	0.70
	Combustible gas leakage buildup	0.70
	LNG in contact with water	0.70
	Cold explosions	—
4	Inadequate operating procedures	0.70
	Equipment does not meet the requirements	0.70
	Personnel operation is not standardized	2.09
	Damaged and failed equipment parts	0.70
	Combustible gas leakage buildup	49.37
	Sparks from operations	25.73
	Explosion	—
5	Inadequate management of operations	0.70
	Equipment not turned off/on as required	0.70
	Combustible gas leakage buildup	49.37
	Sparks from work	4.17
	Fire and explosion	—
6	Inadequate safety education and training	6.95
	Unauthorized operation by personnel	2.09
	Failure of personnel to detect the concentration of flammable and hazardous gases	10.43
	Combustible gas leakage accumulation	49.37
	Sparks from operations	27.82
	Explosive accidents	—

(Continued)

Table 4.19. (*Continued*)

Serial no.	Causal node	Transfer probability 10^{-3}
7	Inadequate safety education and training	6.95
	Unauthorized operation by personnel	0.70
	Personnel operation is not standardized	0.70
	Equipment fall	0.70
	Parts flying out	0.70
	Failure to wear safety equipment	0.70
	Mechanical accidents	—
8	Inadequate routine inspection and maintenance	2.78
	Corroded equipment	4.87
	Accumulation of flammable gas leaks	49.37
	Sparks from work	19.47
	Fire accident	1.39
	Burning accident	—
9	Inadequate safety monitoring mechanism	1.39
	Cold damage to equipment	0.70
	Equipment without safety guards	2.09
	Leakage of flammable gas accumulation	49.37
	Sparks from operations	4.17
	Fire and explosion accident	—

(3) FMI value

FMI is the geometric mean of accuracy and recall, with a value in the range of $[0, 1]$, and a value close to 1 means that the predicted cluster class is more consistent with the real cluster class. The calculation is shown in Eq. (4.37):

$$FMI = \frac{TP}{\sqrt{(TP + FP)(TP + FN)}}.$$ (4.37)

(4) V-measure value

V-measure is the weighted average of homogeneity and completeness to evaluate the similarity between two cluster class vectors, with the value range of $[0, 1]$, and a V-measure value of 1 indicates the optimal similarity. Among them, homogeneity means that each cluster class contains only a single category of samples, the larger the value,

Table 4.20. Generalization effects of different generalization methods on underlying factors.

Different generalization methods	ARI/10^{-1}	AMI/10^{-1}	V-measure/10^{-1}	FMI/10^{-1}
Word2Vec +AGNES method	4.45	5.40	7.12	5.21
Binary word features + Word2Vec + AGNES method	3.70	4.99	6.88	4.54
Char-word features + Word2Vec + AGNES method	4.59	5.71	7.32	5.34
Methods in this section enhance the effect	+3.15%	+5.74%	+2.81%	+2.50%

the smaller the homogeneity; completeness means that the same category of samples are categorized into the same cluster, the larger the value, the smaller the completeness. The calculation is shown in Eq. (4.38):

$$V = \frac{2hc}{h + c},\tag{4.38}$$

where h is homogeneity; c is completeness.

The different generalization methods and the method proposed in this section were used to calculate the assessment indicators ARI, AMI, V-measure, and FMI on the textual generalization task for each type of risk factor, and the generalization effects on the underlying factors are shown in Table 4.20 and Fig. 4.27, the generalization effects on the indirect factors are shown in Table 4.21 and Fig. 4.28, and the generalization effects on the direct factors are shown in Table 4.22 and Fig. 4.29.

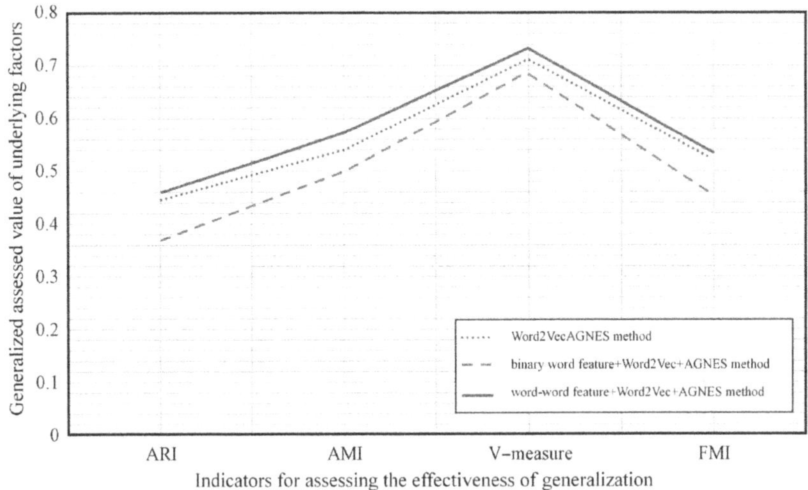

Fig. 4.27. Line graph of values of indicators for the generalized assessment of underlying factors.

As can be seen from Table 4.20, compared to the Word2Vec + AGNES method, the generalization method with the addition of char-word features shows a higher improvement in all four metrics. Among them, the most improvement is in the indicator AMI, which reaches 5.74%, indicating that the addition of char-word features can better help the generalization model to retain more real information in the underlying factors and improve the reliability and accuracy of the generalization results.

The three methods are ranked according to the generalization effect, and it can be seen from Fig. 4.27: the char-word feature+Word2Vec+AGNES method proposed in this section Word2Vec + AGNES method > binary word feature + Word2Vec + AGNES method. For the LNG reservoir root factor text, adding binary word features cannot improve the generalization effect of the generalization method, which may be due to the fact that the number of root factor text is small, and at this time, the segmentation of the root factor text into binary phrases cannot retain the original semantic information, so the generalization effect of the method is poorer, but if we add the char feature information, the effect of the method can have a better improvement.

Table 4.21.   Generalization effects of different generalization methods on indirect factors.

Different general-ization methods	ARI/10^{-1}	AMI/10^{-1}	V-measure /10^{-1}	FMI/10^{-1}
Word2Vec +AGNES method	2.42	4.27	6.47	2.67
Binary word features +Word2Vec +AGNES method	2.96	4.38	6.53	3.19
Word-Word features +Word2Vec +AGNES method	3.03	4.63	6.68	3.27
Methods in this section improve results	+25.21%	+8.43%	+3.25%	+22.47%

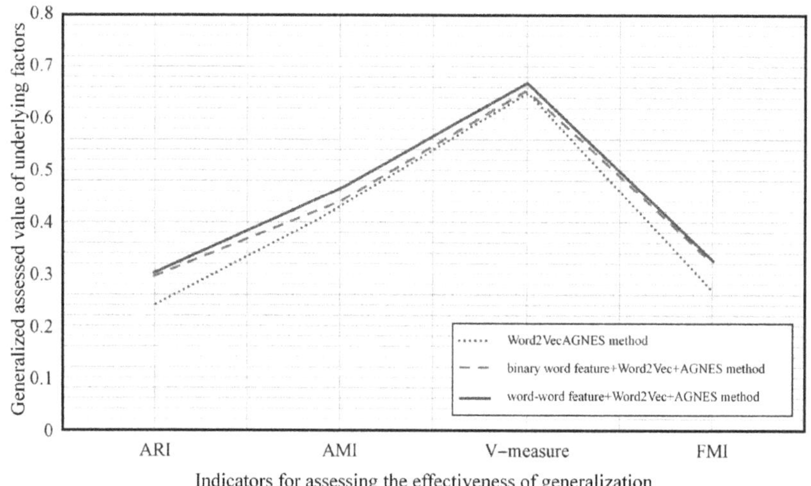

Fig. 4.28.   Line graph of indicator values for the assessment of generalization of indirect factors.

As can be seen from Table 4.21, compared to the Word2Vec + AGNES method, the generalization method with the addition of char-word features shows a significant improvement in all four metrics. Among them, the most improvement is in the metric ARI, which is 25.21%, and FMI, which is 22.47%. It shows that the proposed method in this section improves the degree of overlap between the prediction results of the generalization method and the real labels, retains more effective information in the text of indirect factors, and makes the generalization results more accurate and closer to the actual categories.

The three methods are ranked according to the generalization effect of the indirect factors, as shown in Fig. 4.28, which shows that the char-word feature +Word2Vec+AGNES method proposed in this section > binary word feature +Word2Vec+AGNES method Word2Vec+AGNES method. For the indirect factors of LNG reserve, adding binary word features and char-word features can improve the generalization effect of the method, and the original semantic information can be better preserved; the generalization method also has a certain enhancement effect after adding word features, but it is not obvious in the indexes AMI and V-measure.

From the results of the method indicators in Table 4.22, it can be seen that the generalization effect of the method proposed in this section has improved in all four indicators. Among them, the most

Table 4.22. Evaluated values of different generalization methods on direct factors.

Different generalization methods	ARI/10^{-1}	AMI/10^{-1}	V-measure/10^{-1}	FMI/10^{-1}
Word2Vec +AGNES method	2.62	4.50	6.19	3.08
Binary word features +Word2Vec +AGNES method	2.65	4.34	6.06	3.11
Char-Word features +Word2Vec +AGNES method	2.87	4.62	6.24	3.32
Methods in this section improve results	+9.54%	+2.67%	+0.81%	+7.79%

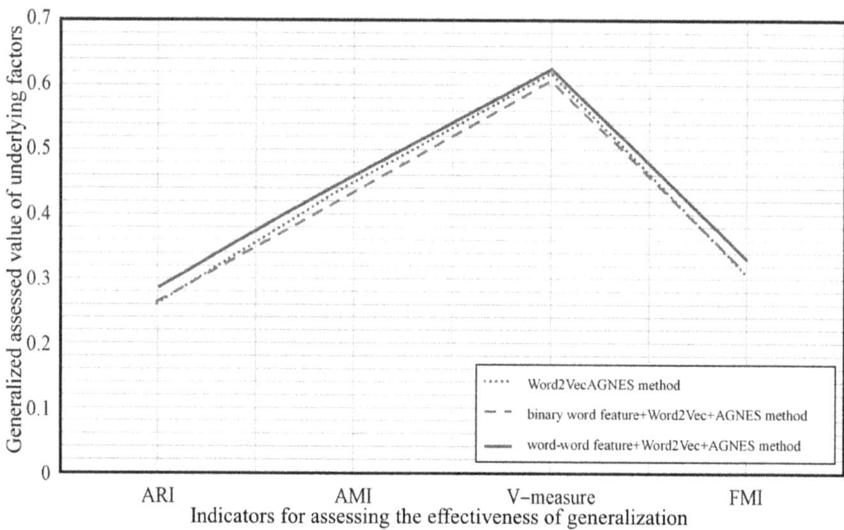

Fig. 4.29. Line graph of indicator values for the generalized assessment of direct factors.

improvement is in the ARI, which reaches 9.54%. The results show that the addition of char-word features improves the generalization ability of the method, increases the degree of overlap between the prediction results and the real labels, and better preserves the semantic information in the text of the direct factors, making the results of the text generalization method more accurate and effective.

For the three folds in Fig. 4.29, the three methods are ranked according to the generalization effect of the direct factors: the char-word feature +Word2Vec+AGNES method proposed in this section > Word2Vec+AGNES method > Binary word feature +Word2Vec+AGNES method, and the model's ability of extracting textual semantic information is from text and can collect more valuable information. For the direct factors of LNG reserve, adding binary word features cannot improve the generalization effect of the generalization method, the reason may be that the semantic information of the text of the direct factors cannot be reflected by binary phrases, but will introduce errors, but if we add the char feature information, the effect of the method can be improved to a certain extent.

4.4 Intelligent Early Warning of Accidents in Oil and Gas Production Systems Based on Knowledge Graphs

In order to more fully and effectively utilize the unstructured text accumulated in the process of safety management, and realize the multi-node risk factor evolution path prediction, this section is based on the knowledge graph of risk factor evolution in the LNG reserve, and finds out the evolution path with the maximum probability of the transfer of the existing risk factors in the depot area through the text-matching method and the local optimization search method, so as to realize the intelligent warning of accidents, and thus better help the depot area to complete the safety management, hidden danger management, and accident prevention, so as to ensure the safety of the LNG reservoir during operation.

4.4.1 *Accident intelligent early-warning fundamentals*

4.4.1.1 *Short text similarity calculation method*

Commonly used text similarity calculation methods can be categorized into: similarity methods based on segmented fields and similarity methods based on edit distance, where similarity methods based on segmented fields usually have cosine similarity and Jaccard similarity.

(1) Cosine similarity method

The cosine value of the angle between two vectors in the vector space is calculated by the cosine formula to represent the degree of similarity between two short texts, the closer the cosine value is to 1, the more similar the two short texts are. Cosine similarity is a commonly used similarity measure, the calculation results are accurate and suitable for short text processing, the calculation is shown in Eq. (4.39):

$$\cos(A, B) = \frac{A \cdot B}{\|A\|\|B\|} = \sum_{i=1}^{n} A_i B_i \Bigg/ \left(\sqrt{\sum_{i=1}^{n} A_i^2} \sqrt{\sum_{i=1}^{n} B_i^2} \right), \quad (4.39)$$

where A and B are the vector representations corresponding to the two short texts, respectively.

(2) Jaccard similarity method

Jaccard coefficient is equal to the ratio of the intersection and concatenation of two short texts' field sets, i.e., the fields shared by two short texts divided by all the fields contained in two short texts, which can be used to measure the correlation between two paragraph texts. The advantage of the similarity function based on Jaccard coefficient is that the set intersection operation has nothing to do with the order of the fields in the set, so the order of the fields has basically no effect on the result of the similarity measure, but at the same time, it also means that the method cannot reflect the semantic information contained in the order of the text. The similarity calculation is shown in Eq. (4.40):

$$\mathrm{Jaccard}(S_i, S_j) = \frac{S_i \cap S_j}{S_i \cup S_j},\qquad(4.40)$$

where S_i, S_j are the combinations of all the words of the short text used to calculate the similarity, respectively.

(3) Similarity methods based on edit distance

The similarity function based on edit distance is a measure of the similarity of two strings by considering the text strings to be matched as a whole, and by using the minimum cost of the editing operation required to convert the string into another string as a measure of the similarity of the two strings. The editing operations therein include insertion, deletion, substitution, position swapping, and so on. Its similarity calculation is shown in Eq. (4.41):

$$similarity = 1 - \frac{ED_{AB}}{\max(L_A, L_B)},\qquad(4.41)$$

where ED_{AB} is the minimum edit distance; L_A is the length of the string of short text A, L_B is the length of the string of short text B.

4.4.1.2   *Localized optimal search method*

Localized optimal search method is a heuristic search method, which is an improvement of the depth-first search method. The basic idea of the method: after a node is expanded, the estimated value is calculated for each child node according to the pre-specified evaluation function $f(x)$, and the smallest one is selected as the next node to be examined, as the next node to be examined is only selected within the range of child nodes each time, the range is relatively narrow, which is called the localized optimal search. The logical steps of the method can be expressed as follows:

(1) Place the initial node S_0 into the Open table and calculate its value $f(S_0)$.
(2) If the Open table is empty, the problem is unsolved and fails to exit.
(3) Take out the first node of the Open table and put it into the Closed table and note the node as n.
(4) Examine whether node n is expandable. If it cannot be extended, the task is complete and the search is exited.
(5) Expand node n, use the evaluation function $f(x)$ to calculate the estimated value of each child node, and put it into the first part of the Open table in ascending order of the estimated value, set a pointer to the parent node for each child node, and turn to step (2).

4.4.2 Knowledge graph-based intelligent early-warning method for accidents

Through the text matching method, the nodes that are most similar to the existing risk factors in the reservoir area are located in the risk factor evolution knowledge graph, and they are used as the starting position of the evolution path for evolutionary reasoning of the subsequent development, so as to realize the intelligent warning of accidents.

4.4.2.1 Risk factor text matching based on pre-trained feature vectors

In order to ensure the accuracy of the text matching results of the risk factor nodes of the LNG reserve, the effects of three text similarity calculation methods (pre-trained feature vector + cosine similarity method, Jaccard similarity method, and similarity method based on editing distance) on the text of the enterprise production accidents are compared and analyzed, and the method with the best effect is finally selected as the text matching method of the risk factors in this section.

Three sets of short text combinations with the same actual semantics and actual similarity of 1 are used: "safety education and training are not in place" and "training is not solid enough" "safety education and training are not in place" and "safety education is not put into practice", "safety education and training is not in place" and "daily training and education is not effective" as an example, through the three text similarity calculation methods to obtain the similarity results. The ratio between the similarity obtained by the methods and the actual similarity is taken as the accuracy rate of the similarity calculation methods, and the specific results are shown in Table 4.23.

From Table 4.23, it can be clearly seen that the average matching accuracy of the method based on pre-trained feature vector + cosine similarity in three sets of short text combinations reaches 83.33%, which is significantly better than the Jaccard similarity method, and the Similarity method based on the edit distance, therefore, in this section, we choose the method of pre-trained word vectors + cosine similarity as a text similarity for the risk factors of the LNG reserve bank calculation method, the pre-trained feature vector model used is Word2Vec model + Char2Vec model obtained from training in Chapter 3.

4.4.2.2 *Knowledge graph-based inference of accident evolutionary paths*

The knowledge graph-based evolutionary reasoning model for risk factors of LNG reserve is shown in Fig. 4.30.

Step 1: Similar node matching based on pre-trained feature vectors.

Text matching is performed by the pre-trained feature vector + cosine similarity method, which calculates the similarity between the existing risk factors of the LNG reserve pool and all the embodied risk factor nodes before generalization of the knowledge graph, and finds the most similar risk factor nodes.

Table 4.23. Comparison of three sample text similarity calculation methods.

Serial number	Different combinations of short texts	Pre-trained feature vector + cosine similarity	Jaccard similarity	Similarity based on edit distance	Actual similarity
1	"Inadequate safety education and training" "Training is not solid enough"	0.78	0.13	0	1
2	"Safety education and training are not in place" "Safety education is not in place"	0.88	0.29	0.44	1
3	"Inadequate safety education and training" "Ineffective daily training and education"	0.92	0.29	0.11	1
Average matching accuracy /%		83.33	23.67	18.33	—

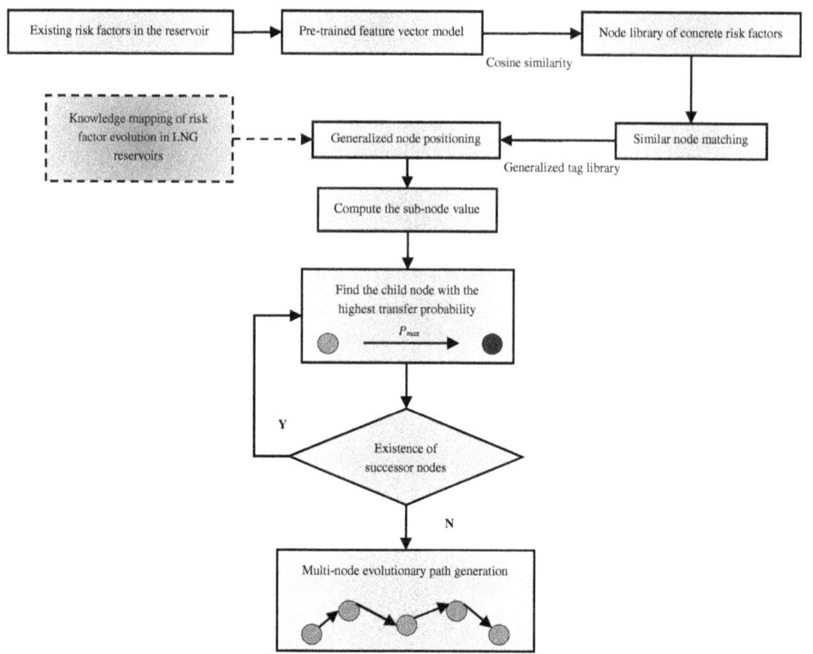

Fig. 4.30. LNG reservoir risk factor evolutionary inference modeling.

Step 2: Generalization node localization.

According to the pre-defined generalization label library, obtain the generalization labels of the risk factor nodes that are most similar to the existing risk factors in the reservoir area, and locate the nodes in the knowledge graph of risk factors in the LNG reserve reservoir that are the same as the generalization labels.

Step 3: Evolutionary path inference based on local optimization.

Based on the local optimal search method, the generalized nodes in the knowledge graph are used as the initial nodes for path inference, and the child nodes are determined through the causal edges of the knowledge graph, and the evaluation function $f(x)$ of the child nodes is calculated in Eq. (4.42):

$$f(x) = \frac{1}{p}, \tag{4.42}$$

where $f(x)$ is the child node evaluation value; p is the transfer probability between parent and child nodes.

Based on the evaluation values of all child nodes, select the child node with the highest transition probability as the subsequent node, and continuously repeat the process of searching for subsequent nodes until they no longer exist, generating a multi node evolutionary path of risk factors.

4.4.3 *Case study*

According to the intelligent early-warning method for accidents based on the knowledge graph of risk factor evolution of LNG reserve, proposed in this section, case applications and analysis are carried out for the risk factor scenarios existing in the three aspects of safety management, equipment and facilities, and process parameters of LNG reserve.

4.4.3.1 *Case 1: Intelligent accident alerting for inadequate education and training scenarios in depot areas*

When the staff of the LNG reserve depot audits the safety management of the depot, they find that the depot has deficiencies in the education and training of the employees, and find out the evolution path of this risk factor by utilizing the intelligent early-warning method for accidents proposed in this section, which is shown in the following process.

Step 1: Similar node matching.

According to the pre-trained eigenvector model and Eq. (4.39), the most similar risk factors to the existing risk factor "insufficient education and training in the depot" are matched in the node library of figurative risk factors, and the matching results are shown in Table 4.24.

Table 4.24. Similar node matching results for case 1.

Serial number	Similar node	Similarity/10^{-2}
1	Inadequate safety education and training	75.70
2	Inadequate education and training	75.39
3	Inadequate training and education	75.39
4	Ineffective daily training and education	74.90
5	Lack of safety culture	74.44

The highest similarity was 75.70×10^{-2}, followed by "inadequate education and training" 75.39×10^{-2}, "inadequate training and education" 75.39×10^{-2}, "ineffective daily training and education" 74.90×10^{-2}, and "lack of security culture" 74.44×10^{-2}. It can be seen that the most similar figurative risk factor node matches the semantics of the existing risk factors, so "inadequate safety education and training" is taken as the most similar figurative risk factor node of the existing risk factors.

Step 2: generalized node localization.

Obtain the generalized label of the figurative risk factor node "safety education and training are not in place" as "safety education and training is not in place", and locate the nodes with the same generalized label in the knowledge graph of risk factors of LNG reserve.

Step 3: evolutionary path inference based on local optimization.

Based on the local optimization search method, the node "safety education and training are not in place" is used as the initial node in the knowledge graph for path inference, and the multi-node evolution path of the risk factor "insufficient education and training in the storage area" is obtained, along with the transfer probabilities between the nodes, as shown in Table 4.25.

According to Table 4.25, the multi-node evolution path of the existing risk factor "insufficient education and training in the storage

Table 4.25. Evolutionary path of risk factors in case 1.

Causal node	Transfer probability/10^{-2}
Inadequate safety education and training	4.03
Failure of personnel to detect flammable and hazardous gas concentrations	1.04
Sparks from the operation	2.78
Splinter	1.46
Personnel operating irregularities	0.77
Combustible gas leakage buildup	4.94
Explosion	0.21
Object striking accident	—

area" of the LNG storage depot is finally obtained as follows: "Inadequate safety education and training → personnel operating irregularities → personnel failing to detect the concentration of flammable and hazardous gases → flammable gas leakage and accumulation → sparks from the operation → explosion → splinter → object striking accident". It can be seen that the accident early-warning method proposed in this section can effectively utilize the knowledge graph to synthesize the historical information of the risk factors of the LNG reserve, from which the most probable evolution path will be deduced in the event of the risk factor of insufficient education and training in the safety management category of the LNG reserve, and the results of the deduction are in line with the actual experience.

4.4.3.2 *Case 2: intelligent accident warning for aging scenarios of LNG storage tank valve elbows*

During the daily inspection, the staff of the LNG reserve depot found that the elbow of the LNG tank valve had aging problems, and used the intelligent warning method for accidents proposed in this section to find out the evolution path of this risk factor, which is described in the following process.

Step 1: Similar node matching.

According to the pre-trained feature vector model and Eq. (4.39), match the risk factors that are most similar to the existing risk factor "aging of the elbow of the LNG tank valve" in the figurative risk factor node library, and the matching results are shown in Table 4.26.

As can be seen from Table 4.26, the existing risk factors are most similar to the figurative risk factor node "valve elbow aging", with a similarity of 81.65×10^{-2}, followed by "pipeline valve aging"

Table **4.26.** Similar node matching results for case 2.

Serial number	Similar node	Similarity/10^{-2}
1	Valve elbow aging	81.65
2	Pipeline valve deterioration	77.28
3	Tank valve corrosion	73.62
4	Pipeline valve corrosion	73.41
5	Styrene flange aging	69.46

7.28×10^{-2}, "Tank valve corrosion" 73.62×10^{-2}, "pipeline valve corrosion" 73.41×10^{-2}, "styrene flange aging" 69.46×10^{-2}. It can be seen that the most similar "valve elbow aging" figurative risk factor node is semantically consistent with the existing risk factors, so it is taken as the most similar figurative risk factor node of the existing risk factors.

Step 2: generalized node localization.

Obtain the generalized label of the figurative risk factor node "valve elbow aging" as "equipment aging", and locate the nodes with the same generalized label in the knowledge graph of risk factors in the LNG reserve.

Step 3: evolutionary path inference based on local optimization.

Based on the local optimization search method, the node "equipment aging" is used as the initial node in the knowledge graph for path inference, and the multi-node evolution path of the risk factor "aging of LNG storage tank valve elbow" and the transfer probability between nodes are obtained as shown in Table 4.27.

According to the information shown in Table 4.27, the final multi-node evolution path of the existing risk factor "aging of LNG tank valve elbow" of the LNG reserve is: "equipment aging $\rightarrow$ combustible gas leakage buildup $\rightarrow$ sparks from the operation $\rightarrow$ explosion $\rightarrow$ splinter $\rightarrow$ object striking accidents". It can be seen that the accident early-warning method proposed in this section makes effective use of the historical information of the safety management text of the LNG reserve, and can effectively integrate the causal relationship between risk factors, so that when the relevant equipment in the LNG reserve appears similar to the problem of "aging of the LNG tank

Table 4.27. Evolutionary path of risk factors in case 2.

Causal node	Transfer probability/10^{-2}
Ageing of equipment	0.56
Sparks from the operation	2.78
Splinter	1.46
Combustible gas leakage buildup	4.94
Explosion	0.21
Object striking accident	—

valve elbows", it can be reasoned to get the most probable evolution path, and the results obtained by the method are consistent with the actual results. The results of the method are in line with practical experience.

4.4.3.3 *Case 3: intelligent accident warning for gas compressor over-temperature scenarios*

When the staff of the LNG reserve depot carries out their work normally in the depot area, an abnormal temperature rise of a gas compressor in the depot area is found from the monitoring instrument, and the evolution path of this risk factor is found out by using the intelligent warning method for accidents proposed in this section, the process is as follows.

Step 1: similar node matching.

According to the pre-trained feature vector model and Eq. (4.39), the most similar risk factors to the existing risk factor "gas compressor temperature is too high" are matched in the figurative risk factor node library, and the matching results are shown in Table 4.28.

Table 4.28 shows that the existing risk factor "high gas compressor temperature" has the highest similarity to the figurative risk factor node "new hydrogen compressor temperature exceeded" at

Table 4.28. Similar node matching results for case 3.

Serial number	Similar node	Similarity/10^{-2}
1	New hydrogen compressor temperature exceedance	75.58
2	High temperature of new hydrogen compressor	72.76
3	High compressor temperature	72.68
4	Pressure exceedance	66.47
5	Leakage from high pressure valves	66.07

75.58×10^{-2}, followed by "high temperature of new hydrogen compressor" 72.76×10^{-2}, "high compressor temperature" 72.68×10^{-2}, and "pressure exceeded" 66.47×10^{-2}, "high pressure valve leakage" 66.07×10^{-2}, it can be seen that the most similar figurative risk factor node "high temperature of the new hydrogen compressor" is roughly consistent with the semantics of the existing risk factors and has the same core meaning, so it is taken as the existing risk factor. Therefore, it is taken as the most similar figurative risk factor node.

Step 2: generalized node localization.

Obtain the generalized label "equipment temperature exceedance" for the specific risk factor node "new hydrogen compressor temperature exceedance", and locate the nodes with the same generalized label in the knowledge graph of risk factors of the LNG reserve.

Step 3: evolutionary path inference based on local optimization.

Based on the local optimization search method, the node "equipment temperature exceedance" is used as the initial node in the knowledge graph, and the multi-node evolution path of the risk factor "gas compressor temperature is too high" and the transfer probability between nodes are shown in Table 4.29.

According to the information in Table 4.29, it can be obtained that when the LNG reserve occurs "gas compressor temperature is too high", its multi-node evolution path is: "equipment temperature exceeds the standard $\rightarrow$ explosion accident $\rightarrow$ parts flying out $\rightarrow$ object striking accident". It can be proved that the accident early-warning method proposed in this section can effectively integrate the causal relationship between risk factors and fully utilize the value information hidden in the safety management text of the LNG reserve. When there is an abnormal process parameter like "gas compressor temperature is too high" in the LNG reserve, based on the

Table 4.29. Evolutionary path of risk factors in case 3.

Causal node	Transfer probability/10^{-2}
Equipment temperature exceedance	0.28
Splinter	1.46
Explosion accident	0.21
Object striking accident	—

method proposed in this section, the evolution path with the highest probability of occurrence of the risk factors in the LNG reserve can be deduced, and the results obtained by the method are based on the real text and in line with the actual experience.

4.4.3.4 *Analysis and summary*

(1) Aiming at the problems that conventional accident prediction methods need a large amount of a priori data support, cannot directly utilize unstructured text, and deep learning methods cannot realize multi-node risk factor evolution prediction. In order to realize the multi-node intelligent warning of LNG reservoir accidents, this section is based on the knowledge graph of risk factor evolution of LNG reservoirs, and finds out the multi-node evolution path with the largest probability of transferring the existing risk factors in the reservoir area through the text similarity matching method and the local optimal search method, so as to realize the intelligent warning of accidents.

(2) In this section, the knowledge graph is used as the fundamental support for intelligent early warning of accidents in LNG reservoirs, which can effectively integrate the existing textual knowledge of safety management and reflect the causal semantic associations among risk factors in the case of insufficient historical data and lack of a priori knowledge. From the case study, it can be seen that the proposed method can utilize the knowledge map information to obtain the evolution path of risk factors, and the reasoning results are in line with the actual experience. When "safety education and training are not in place" in LNG depots, it usually leads to "personnel operation irregularities", and safety education should be emphasized in daily safety management. When an explosion occurs, it is very easy to be accompanied by an object impact accident, so attention should be paid to the occurrence of such secondary accidents.

(3) Compared with the Jaccard similarity method and similarity method based on edit distance, the text matching method based on pre-trained feature vector and cosine similarity proposed in this section can well overcome the difficulties of rich grammatical vocabulary of risk factors of LNG reserve, and multiple representations of a single meaning, and the accuracy of text matching

results reaches 83.33%. Based on the knowledge graph to realize the intelligent warning of LNG reservoir accidents, it reduces the dependence on the amount of data and a priori knowledge and is able to obtain the multi-node accident development path and probability in the limited safety management text data.

References

[1] Zhang X, Hu J, Zhang L, *et al.* Textual generalization method of accident risk factors in oil & gas storage and transportation enterprises based on CW-AGNES. *Oil & Gas Storage and Transportation*, 2021, 40(11): 1242–1249. (In Chinese)

[2] Hu J, Zhang X, Wu Z. Research on associated early-warning and visualization of hidden danger in enterprise production based on TF-IDF. *China Safety Science Journal*, 2019, 29(7): 170–176. (In Chinese)

[3] Mikolov T, Sutskever I, Chen K, *et al.* Distributed representations of words and phrases and their compositionality. In: *Proceedings of the 27th Annual Conference on Neural Information Processing Systems (NIPS)*. NIPS, Lake Tahoe, 2013: 3111–3119.

[4] Mikolov T, Yih WT, Zweig G. Linguistic regularities in continuous space word representations. In: *Proceedings of the 2013 Conference of the North American Chapter of the Association for Computational Linguistics: Human Language Technologies (NAACL HLT)*. Association for Computational Linguistics, Stroudsburg, 2013: 746–751.

[5] Socher R, Bauer J, Manning CD. Parsing with compositional vector grammars. In: *Proceedings of the 51st Annual Meeting of the Association for Computational Linguistics*. Association for Computational Linguistics, Stroudsburg, 2013: 455–465.

[6] Socher R, Perelygin A, Wu J, *et al.* Recursive deep models for semantic compositionality over a sentiment treebank. In: *Proceedings of the 2013 Conference on Empirical Methods in Natural Language Processing*. Association for Computational Linguistics, Stroudsburg, 2013: 1631–1642.

[7] Naili M, Chaibi AH, Ghezala HHB. Comparative study of word embedding methods in topic segmentation. *Procedia Computer Science*, 2017, 112: 340–349.

[8] Mikolov T, Chen K, Corrado G, *et al.* Efficient estimation of word representations in vector space. arXiv:1301.3781, 2013.

[9] Sayer N. Google Code Archive-Long-term storage for Google Code Project Hosting. 2014. Available at: https://code.google.com/archive/p/open-evse/wikis/Hydra.wiki

[10] Mnih A, Hinton GE. A scalable hierarchical distributed language model. In: *Proceedings of the 23rd Annual Conference on Neural Information Processing Systems (NIPS)*. NIPS, Vancouver, 2009: 1081–1088.

[11] De Mulder W, Bethard S, Moens MF. A survey on the application of recurrent neural networks to statistical language modeling. *Computer Speech and Language*, 2015, 30(1): 61–98.

[12] Pham T, Tran T, Phung D, *et al.* Predicting healthcare trajectories from medical records: A deep learning approach. *Journal of Biomedical Informatics*, 2017, 69: 218–229.

[13] Evermann J, Rehse JR, Fettke P. Predicting process behavior using deep learning. *Decision Support Systems*, 2017, 100: 129–140.

[14] Greff K, Srivastava RK, Koutník J, *et al.* LSTM: A search space odyssey. *IEEE Transactions on Neural Networks and Learning Systems*, 2015, 28(10): 2222–2232.

Chapter 5

Early Intelligent Early-Warning Technology for Behavioral Safety of Oil and Gas Production Operators Based on Line-of-Sight Tracking

Herbert William Heinrich proposed in the theory of accident causation that unsafe human behavior or unsafe state of objects is the direct cause of accidents. He also pointed out that accidents are more often caused by unsafe human behavior, including personnel errors and operator negligence. According to the statistics published by the State Administration of Work Safety (SAWS) of China on its website, there were 50,673 work safety accidents in various industries in 2007, with a total of 101,480 deaths caused by these accidents. According to the statistics, 86 of them were very big accidents, with 1,525 deaths, and the investigation of the causes of the accidents revealed that the proportion of accidents due to the behavioral errors of the personnel was more than 70%. In the oil refining and chemical industries, personnel's behavioral errors are also the main reason for the occurrence of major production safety accidents. For example, some scholars of the U.S. chemical plants claimed that of the 190 accidents investigated, 34% were due to a lack of professional knowledge of personnel, 32% due to the design defects of the product or system, 24% due to procedural errors, and 16% due to personnel error. A survey of refinery accidents found that equipment and facility failures and personnel errors were the main causes, each accounting for 41%, while the remaining causes were procedural errors or lack of regular inspections of equipment and facility environments. The results of a large number of accidents show that 90% of hardware system failures in the chemical production process are caused by human error, so personnel behavioral errors can greatly reduce the safety and reliability of the redundant design of hardware systems. In the complex human–machine–environment system of the oil and gas production process, human activities play a dominant role, but due to the uncontrollability of personnel, this link is also the weakest and difficult to control in the oil and gas production process. Therefore, the research and development of identification and early-warning technology of operational behavioral errors in the oil and gas production process is very necessary.

In the domestic oil and gas production industry, the management mainly utilizes the methods of selecting operators or production-related personnel on the basis of merit, pre-service training, mutual supervision, and adopting appropriate incentives and penalties to reduce possible human errors, and emphasizes the importance of rules and regulations in daily production work. However, none of these control methods can intervene and control the abnormal operation of operators in the production process in real time to prevent or minimize the occurrence of human errors. For example, in 2005, the benzene unit of PetroChina Jilin Petrochemical Company exploded. Due to operational errors by the on-duty operator during the feeding process, the system temperature was too high, ultimately leading to a series of explosions. In 2011, the Anqing Xinfu Chemical plant experienced an explosion. Together, these two accidents created a bad social impact. Therefore, this section introduces line-of-sight tracking technology to realize real-time monitoring and tracking of operator eye movement data and further establishes an advanced sensing method and intelligent warning model for operator unsafe behavior and abnormal cognitive behavior, so as to avoid or reduce the occurrence of personnel operating errors.

Vision tracking technology collects information such as the real-time gaze position of the operator's eyes, the length of time spent in each position, and the trajectory of the eyeballs during the operation process, so as to provide researchers with judgment and decision-making on the operator's attention and possible future actions. This technique is able to monitor human eye movement characteristics in real time; during the operation process, the operator's eye movement characteristics can reflect his/her cognitive situation. Therefore, it is of great significance to use eye-tracking technology to recognize the abnormal cognitive behavior of the operator by taking the eye movement characteristics as the target of perception and nip the accidents that may be caused by human errors in the bud, so as to effectively prevent accidents caused by human errors and safeguard the process of oil and gas production.

5.1 A Technical Framework for Early Intelligent Warning of Personnel Behavioral Security Based on Line-of-Sight Tracking Technology

5.1.1 *Eye movement data acquisition for process operators based on line-of-sight tracking technology*

Aiming at the existing human error analysis methods, which mostly rely on historical data or subjective judgment and cannot realize real-time perception of cognitive behaviors in the production process, we establish a method for real-time collection of eye movement data from operators of oil and gas production processes based on line-of-sight tracking technology.

The eye movement data acquisition equipment includes four systems, i.e., the optical system, the pupil center coordinate extraction system, the view and pupil coordinate iteration system, and the image and data recording and analysis system. When a person's eyes look in different directions, there will be subtle changes in the eye, and these changes will produce features that can be extracted, and the eye movement data acquisition system can extract these features through image capture or scanning, signalize these features, and transmit them to a computer. At the front end of the eye movement data acquisition system, there is a camera device to capture the surrounding environment to form a real-time video. Through the connection of the equipment, the computer is used to prepare the corresponding software for the analysis and processing of eye movement data, so as to track the changes in the personnel's line of sight in real time, and to achieve the identification of the personnel's consciousness and psychological state, and the prediction of the possible behaviors in the next stage.

In the process of personnel eye movement monitoring, there are generally three basic ways of eye movement: gaze, eye hopping, and following movement. Eye movements can reflect the selection mode of visual information, which is important for revealing the psychological mechanism of personnel operating unsafe behaviors. The basic parameters used in the identification and early warning of unsafe behaviors using eye movement data include, but are not limited to, gaze parameters (e.g., the total number of gaze points, etc.), sweep parameters (e.g., the average sweep length, etc.), blink parameters (e.g., the blink frequency, etc.), and gaze point trajectory and hotspot diagrams, etc., with specific descriptions as shown in Table 5.1.

The front-end eye-tracker of the eye movement data acquisition system is categorized into two types, desktop type and head-mounted type, as shown in Fig. 5.1. The desktop type is suitable for operators in the central control room of oil and gas production sites, and is used to monitor and diagnose safety hazards such as errors, irregularities and fatigue in the operation of the process by internal operators. The head-mounted type is suitable for outdoor operation,

Table 5.1. Types of operator eye movement data.

Parameter classification	Parameter name	Parameter unit	Parameter description
Focusing parameters	F_{pn}	n	Total number of viewpoints
	F_{ps}	n	Number of gaze points per second
	F_t	s	Total gaze duration
	F_{td}	s	Average gaze duration
Sweep parameters	S_n	n	Total sweeps
	S_{lt}	px	Total sweep length
	S_{la}	px	Average sweep length
	S_t	s	Sweep total time
	S_{ta}	s	Average sweep time
	S_{va}	px/s	Average sweep speed
Blinking parameters	B_f	n/s	Blink frequency
	B_d	s	Blink duration

(a) Desktop eye movement data collector   (b) Head-mounted eye movement data collector

Fig. 5.1.   Eye movement data acquisition device.

inspection, maintenance, and safety supervision personnel, and is used for errors, mistakes, violations, inattention, and other safety hazards during outdoor operation and observation.

5.1.2   *Identification of safety-hazardous behaviors of offshore drilling operators based on TPE-LightGBM model and eye movement data*

The physiological and psychological aspects of operators in offshore drilling operations are complex and diverse, and human factors have become the main cause of frequent offshore drilling accidents. The methods of identifying the safety-hazardous behaviors of offshore drilling operators can be divided into two main methods. The first relies on manual inspection or post-accident investigation and analysis, and the second involves the use of video surveillance to collect data on the operating behavior of offshore drilling operators, followed by the use of artificial intelligence algorithms to identify hidden safety behaviors. However, both methods are unable to effectively identify safety hazards caused by the operator's current mental and physical state or fatigue level and do not provide real-time identification of safety-hazardous behaviors around the clock.

Introducing TPE optimization theory and LightGBM algorithm, real-time data collection was carried out through an eye movement sensor using wearable and non-invasive technologies. The application of eye-tracking technology to identify safety hazards among offshore drilling operators overcomes the lack of risk perception and monitoring methods for mitigating the safety hazards experienced by operators. A method utilizing composite hotspot mapping for eye movement to determine sensitive area divisions during offshore drilling operations was proposed, and in conjunction with qualitative selection and analysis of eye movement laws, a feature selection method for eye movement in the context of safety hazard behavior in offshore drilling was proposed. The TPE algorithm was introduced to enhance the LightGBM model, addressing the issue of inaccurate identification of safety hazard behavior in offshore drilling operators.

5.1.3 *Identification of safety-hazardous behaviors of process operators based on eye movement composite hotspot map*

Aiming at the unique characteristics presented by eye-tracking trajectories, including small trajectory scale, short duration, large degree of data jumping, weak regularity of turning (affected by subjective operation and objective physiological and psychological influences), and the difficulties in applying existing trajectory early-warning methods to early warning based on eye-tracking trajectories, we put forward the early-warning method for accidents of unsafe behaviors of offshore drilling operators based on eye-tracking trajectories.

According to the range of normal drilling operation eye-tracking trajectory node aggregation, the size and shape of the corresponding equipment unit, and the complexity and importance of the operation area, the scope of the region of interest of the drilling operation trajectory node is determined. In order to solve the adverse effect of eye-tracking trajectory node aggregation on trajectory serialization, an eye-tracking trajectory serialization method based on the IETTSM algorithm is proposed. The method processes the real-time collected eye-tracking trajectory data of offshore drilling operators by setting the region of interest and region of interest code, and then forms the eye-tracking trajectory sequence. Aiming at the current inability to realize real-time warning of accidents caused by unsafe behaviors of offshore drilling operators, an abnormal state prompting method based on the correspondence of trajectory nodes and a multi-node coupling-based accident warning method for unsafe behaviors of offshore drilling operators are proposed by using the

IETTSM-DLD model. The abnormal state prompt can prompt the abnormal behavior of offshore drilling operators in real time, so as to avoid the continuation of the abnormal state. The unsafe behavior warning method can realize real-time warning of accidents caused by unsafe behaviors of offshore drilling operators.

5.2 Line-of-Sight Tracking Technology-Based Experimental Platform for Operator Error Testing of Process Operators

By establishing a production process simulation and control platform required for the experiment, a number of possible interference scenarios in the chemical process are simulated. On the basis of this platform, relevant experimental programs are designed, and eye movement data are collected during the operator's interference suppression task in conjunction with line-of-sight tracking technology, which can be analyzed and counted to provide a data basis for the establishment of cognitive and error models.

5.2.1 *Process simulation platform design*

5.2.1.1 *Operating platform fundamentals*

In this eye movement analysis experiment, the operator's perception was studied through the control room operator's control of the ethanol production process. The entire production unit consists of a continuous stirred reactor (CSTR) where the reaction takes place and a distillation column where the product is processed. The main chemical reaction that occurs is the reaction between ethylene and water in the CSTR to produce the product ethanol, the reaction mechanism is shown in Eq. (5.1):

$$C_2H_4 + H_2O \rightleftharpoons CH_3CH_2OH \quad \nabla H = -45\,kJ \cdot mol^{-1}. \qquad (5.1)$$

The flow diagram of the production process is shown in Fig. 5.2. Since this reaction is exothermic, it is necessary to introduce circulating generated cooling water into the jacket of the continuously stirred reactor to maintain the continuity of the reaction. The distillation column for product distillation contains nine plates, and the mixture of reacted ethanol and unreacted ethylene generated in the CSTR

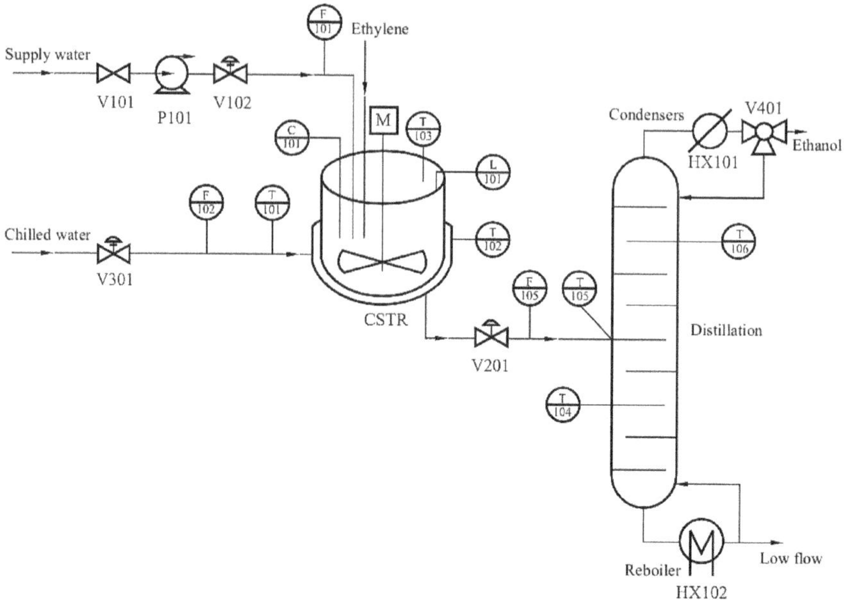

Fig. 5.2. Schematic diagram of ethanol production process.

enters the distillation column from the fifth plate, and the purer product ethanol is obtained in the distillate stream.

The reasons for choosing this process as the simulation process include (1) the overall process is relatively simple, which is convenient for the experimenter to master the process control method after a simple training; (2) this process also includes the continuous stirring reactor and distillation tower, which are more prone to failure and misuse in the actual production of chemical equipment. There is no automatic controller in this process, so all the monitoring and control processes need to be performed manually by the operator. During the design of the platform, we measured in real time the 11 measurement variables listed in Table 5.2, which are correlated with each other during the process, and configured high and low alarms for each of these variables, so that any deviation above the specified thresholds would be flagged by the alarms and then communicated to the operator.

The six disturbance scenarios listed in Table 5.3 were introduced in the design of the manipulation platform to analyze the cognitive behaviors of the operators, and all six scenarios (D1 to D6) were

Table 5.2. Process variables in the ethanol process.

Labels	Descriptive	Steady state value	Alarm limit	Lower limit of alarm
C101	Ethylene concentration in CSTR/(μmol/lt)	1378.5	1555.6	955.6
F101	CSTR feed flow rate/(lt/h)	700.7	950	550
F102	CSTR coolant flow rate/(lt/h)	130	200	70
F105	Distillation column feed flow rate/(lt/h)	733.4	993.7	575.3
L101	CSTR Level/m	1.25	1.8	0
T101	Cooling water inlet temperature/$^{\circ}$C	20	40	0
T102	Cooling water outlet temperature/$^{\circ}$C	29.4	32.5	15.2
T103	Temperature in CSTR/$^{\circ}$C	30.5	33	29.5
T104	Temperature of distillation plate 3/$^{\circ}$C	100	100.5	98.5
T105	Temperature of distillation plate 5/$^{\circ}$C	87.4	89.5	86.5
T106	Temperature of distillation plate 8/$^{\circ}$C	79.5	80.4	78.5

Table 5.3. Anomaly control scenarios in experiments.

Scene number	Scene description
D1	CSTR feed flow increased
D2	CSTR reduced jacket cooling water flow
D3	Reduced flow from CSTR to distillation columns
D4	Reflux ratio imbalance
D5	CSTR reduced feed flow
D6	CSTR jacket cooling water flow increase

related to the perturbations. During the experiment, the task to be performed by each tester is a randomized set consisting of these six scenarios. When a disturbance occurs, one or more variables will be significantly perturbed from their steady state, and an alarm will

Table 5.4.   Corrective actions to be taken in perturbation scenarios.

Scene number	Measures to be taken
D1	Adjust V102 to the slide to decrease the feed flow to the CSTR
D2	Adjust V301 to increase the cooling water flow to the slider
D3	Adjust V201 to increase the flow rate to the distillation column
D4	Adjust V401 to change the reflux rate
D5	Adjust V102 to increase the feed flow to the CSTR
D6	Adjust V301 to decrease the coolant flow to the slider

be generated. The operator will need to manipulate one or more of the four control valves to return the device to its normal state. Parameters associated with these conditions are the feed flow rate to the CSTR, F101, the cooling water flow rate, F102, the feed flow rate from the fractionator, F105, and the temperature at plate 5 of the distillation column, T105. For example, in Scenario D5, a sudden decrease in the feed flow rate to the CSTR results in a high buildup of ethylene in the reactor and generates a low alarm for F101 and a high alarm for C101. The operator needs to diagnose the situation and determine and implement the necessary corrective actions. In scenario D5, the disturbance can be eliminated by increasing the feed flow to the CSTR through valve V102. Details of the corrective action that should be taken for the remaining scenarios for disturbances are listed in Table 5.4.

5.2.1.2   *Human–machine interface*

The simulation platform used in this study was built by Aspen HYSYS software in conjunction with MATLAB. Firstly, the dynamic simulation of the ethanol production process was carried out using Aspen HYSYS software, and the process flow is shown in Fig. 5.3. The process flow is shown in Fig. 5.3. The established dynamic process is called in MATLAB, and the GUI interface is designed to control the process. The six disturbance scenarios mentioned above were set up in the program design, and the corresponding upper and lower alarm limits were set for the process variables in the production process.

The finalized experimental platform is shown in Fig. 5.4. Real-time values of the measured variables, a list of alarms, and trend

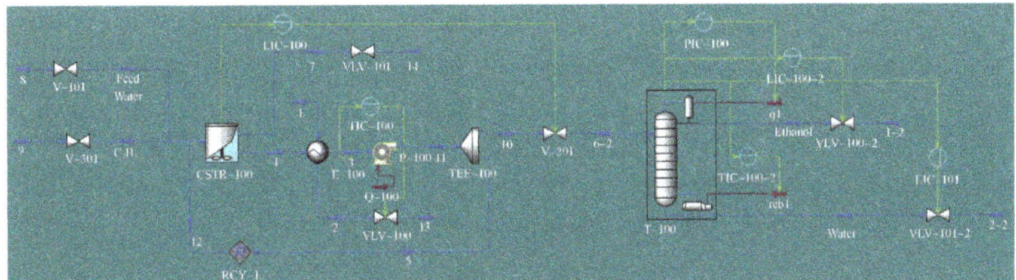

Fig. 5.3. Ethanol production simulation process.

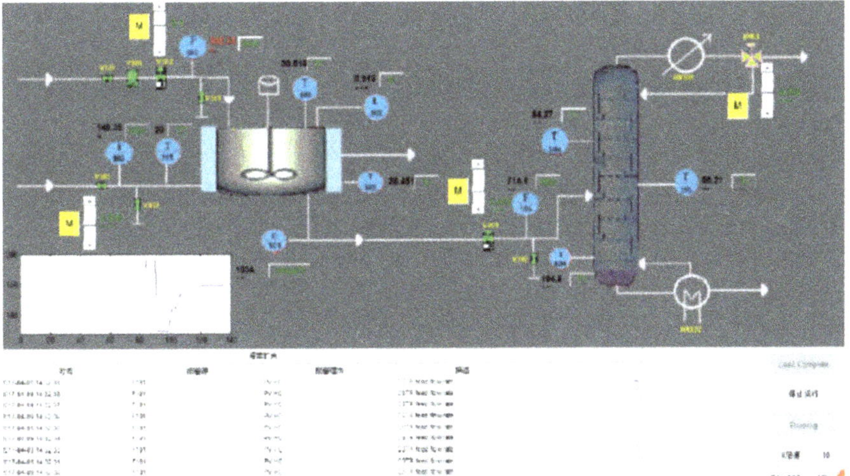

Fig. 5.4. Simulation operation experiment platform interface.

information for any of the variables (selected by the tester) can be seen in it. The details are as follows:

(1) The 11 blue circles in the center window represent the various process variables. The value near the circle is the real-time monitoring data value corresponding to that variable, and the color of the value indicates the current state of the variable: black indicates that the corresponding process variable is within its normal range, while red indicates that the variable is in alarm.

(2) The four sliders correspond to the four control valves, and the operator can manipulate any of the four valves in real time by moving the sliders of the control valves during the task.

(3) The lower Alarm Summary window contains information about all currently and historically flagged alarms, including their time of occurrence, the nature of the alarm (high PV or low PV), and a description of the alarm variable.

(4) The Trend window in the lower right corner contains trend information for a process variable. The operator can call up the detailed historical trend of any variable by clicking on the corresponding variable in the layout.

(5) The Stop Run button in the lower left corner can be used to end a running scenario at any time.

5.2.2 *Experimental program design*

5.2.2.1 *Overview of the experiment*

The experimental study was conducted in a controlled environment where the tester observed and controlled the process operation in real time by interacting with a human–computer interface. The operators were not informed of the true purpose of this study prior to the experiment to prevent influencing their cognitive behavior and to prevent them from developing any strategic responses that would aid or undermine the experimental hypothesis.

Each operator was provided with information prior to conducting the experiment to understand the technical details of the operation and the interaction with the HMI. The information included the following: (1) an overview section explaining the role of the tester as an operator and his/her responsibilities; (2) a technical section providing details of the ethanol production process; and (3) an introduction to the human–machine interface section that explains all display units and how to interact with the unit (including trend windows and operating valves). The tester will then simulate one of the illustrative scenarios to ensure that the tester has a detailed understanding of the task components prior to actual operation. The entire training phase lasts approximately 5–10 min.

Prior to the start of each task, the tester is informed of the operating instructions corresponding to that task, e.g., Interference D1 is to adjust the V102 valve position to reduce the feed flow rate to the CSTR (F101) and return the unit to its normal state in the event of an abnormality. The instructions for each disturbance direct the tester to react to the alarm caused by the disturbance and to use the slider to manipulate the control valves to return the unit to its normal state, and the particular control valves to be manipulated (Table 5.4) are explicitly mentioned in the instructions. After reading the instructions, the tester can start the task by pressing the "Start Run" button in the interface. Typically, around 10 s after the task is started, a perturbation is introduced without the tester's knowledge, and some process variables will deviate from their normal domain values and generate an alarm. When an alarm is flagged, the corresponding variable value changes from black to red in the Layout window, and the details of this alarm are listed in the Alarm Summary window.

At the start of the task, the tester will take such actions as he/she deems feasible to evaluate the abnormal condition and attempt to correct it, such as searching for information, observing the trend of the relevant parameter, opening/closing the control valve, etc. The tester will also take such actions as he/she deems appropriate to assess the abnormal condition and attempt to correct it. If the tester is successful in removing the interference, i.e., the abnormal condition is correctly handled, then all process variables will return to their normal ranges, i.e., the task is successful. If the tester is unable to return the entire process to a normal state within 240 s, the task is automatically stopped. The tester can also stop the operation at any time by pressing the "Stop Run" button in the schematic window.

The task flow in Fig. 5.5 represents the complete set of actions that the tester operates during the experiment. It represents all events from the beginning of the experiment to the end of the first task. With the exception of the very first training, all the other activities (starting with the task instructions) needed to be repeated by each tester as they performed the six tasks.

The experimental protocol has the following characteristics:

(1) Ensures that the tester has had some basic training prior to the start of the experiment.
(2) Increase the tester's familiarity with the process and interface through repetitive tasks.
(3) Prevent tester fatigue by limiting the total duration of the experiment to less than 30 min.
(4) Comprehensively test the cognitive processes of the testers by assigning scenarios with different difficulty levels and different parts of the process.

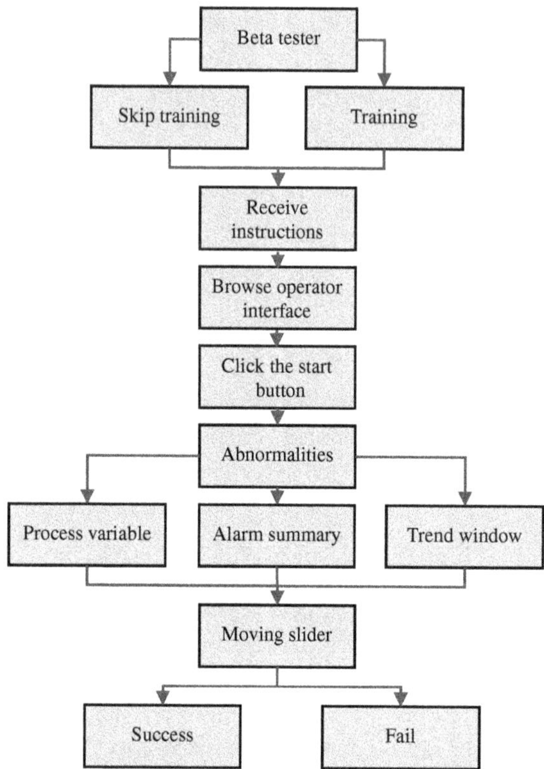

Fig. 5.5. Flow of tasks during the experiment.

(5) Randomize the assignment of scenarios to avoid any systematic bias.

5.2.2.2 *Specific experimental program*

(1) Experimental purpose

To obtain the eye movement feature parameters of operators when they perform operation behaviors in the simulated ethanol production operation platform, to explore the cognitive state of process operators when they perform operations, and to realize the eye movement feature extraction of process operators' behavioral errors.

(2) Experimental design

The experiment is conducted without informing the tester of the real purpose of the experiment, and the tester is only told to control the chemical process as an operator, so as to achieve the simulation effect of the actual process operation at the oil and gas production site.

(3) Experimental hypothesis

The generation of human errors in chemical operations is mainly related to the abnormal cognitive situation of the operator.

(4) Experimental tasks

The experimental task was focused on the tester's troubleshooting of disturbances (abnormal working conditions) generated in the ethanol production process. During the task, the introduced disturbances (randomly selected from D1 to D6) cause one or more process variables to deviate from the steady state marked by an alarm, and the tester needs to adjust the relevant sliders in order to return the production process to the normal state.

(5) Experimental apparatus and experimental platforms

Experimental apparatus: The Eyeso Ec80 telemetric eye-tracking device was used for the experiment (Fig. 5.6).

(6) Experimental procedure

Pre-experiment preparation stage: relevant information on the experimental process is distributed to the testers, including the relevant operation procedures that the testers should carry out and the introduction of the human–computer operation environment, and the experimenter gives a detailed explanation for the testers. Then, all the testers were trained, and one interference was selected for the testers to simulate the operation and act as an operator in it, to make

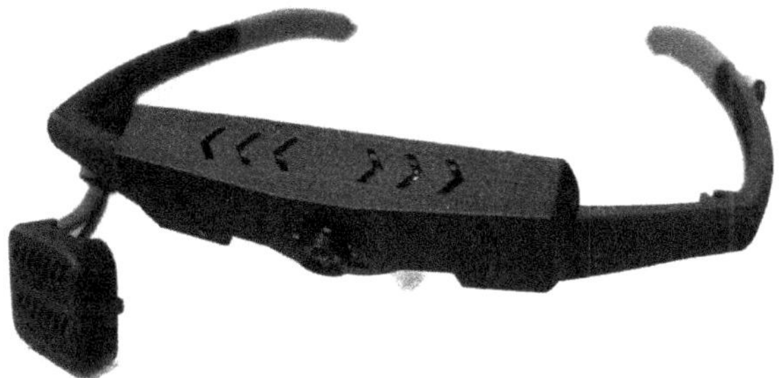

Fig. 5.6.   Eyeso Ec80 telemetric eye movement meter.

sure that the testers had carried out some basic training before the experiment started.

Eye movement data acquisition stage:

(1) The tester enters the laboratory, adjusts the experimental chair to the appropriate height and angle, and the experimenter tells the tester to relax, so that he/she is facing the display screen, keeping his/her body stable, and is 60 cm ± 10 cm away from the screen, as shown in Fig. 5.7.
(2) The experimenter adjusted the position and angle of the eye-tracking device so that it could clearly capture the eye movements and the trajectory of the tester's line of sight. This is shown in Fig. 5.8.
(3) Tester information collection. The tester fills in his/her personal information in the eye movement data analysis software, including age, specialty, and gender, and makes a note of his/her glasses-wearing situation and related characteristics.
(4) Nine-point calibration. Testers first entered the nine-point calibration interface of eye movement data acquisition, and were asked to "look at and follow the randomly moving red dots on

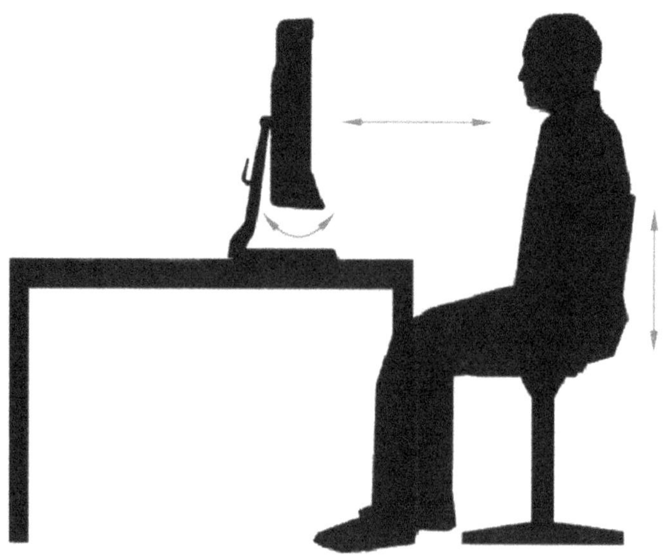

Fig. 5.7. Tester's experimental posture.

Fig. 5.8. Eye tracker angle.

the screen. During this process, the body and head are required to remain motionless". Formal experiments were conducted only after the calibrations had reached an acceptable level.

(5) Start of experiment. After the calibration was completed, the tester clicked the record button, and then the eye movement data acquisition and recording of the tester's operation process began, and the experiment officially started.

During the experiment, all the actions carried out by the tester during the operation of the interface are recorded by the eye movement data acquisition system, including the movement of the line of sight, the mouse click movement, and its operation of the various sliders and the changes in the process variables. Figure 5.9 shows the real scene of eye movement data acquisition during the operation of the tester.

The front-end of the data acquisition system uses the Eyeso telemetric eye movement sensor to capture the tester's line of sight, and at the same time, the acquisition system is also equipped with a screen recording function [Fig. 5.10(a)] that can record all the interactions between the tester and the human–machine interface in video. The video editing function [Fig. 5.10(b)] allows the video to be divided into several segments according to different features and cognitive conditions during the data analysis stage. Finally, the eye movement data [Fig. 5.10(c)] is exported for subsequent

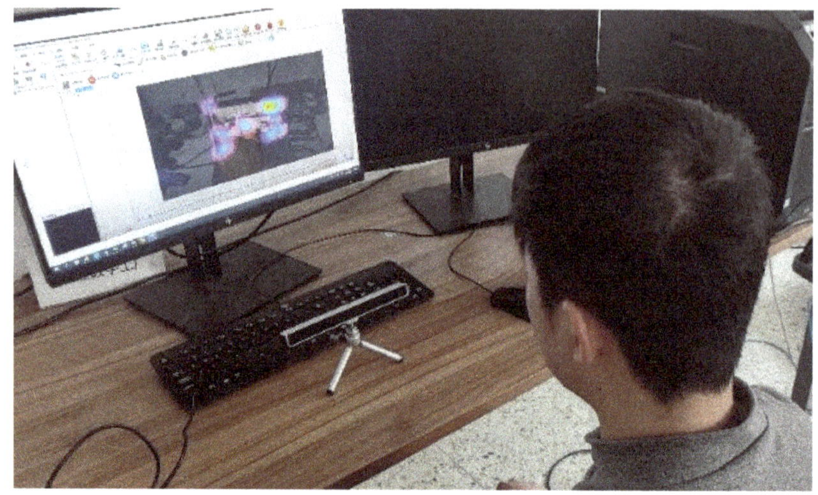

Fig. 5.9.   Tester experimental scenarios.

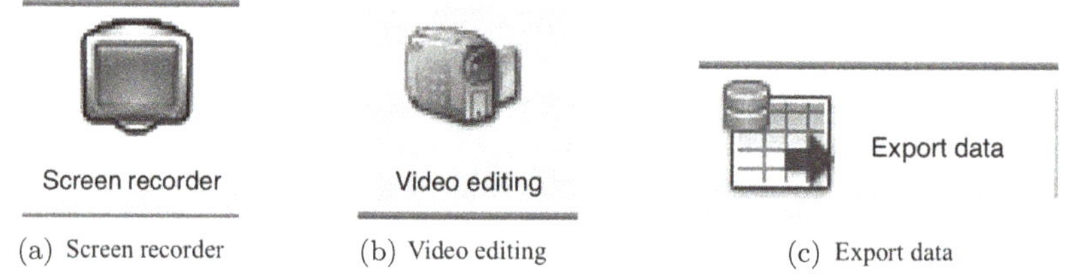

(a) Screen recorder   (b) Video editing   (c) Export data

Fig. 5.10.   Eye movement software part of the operation module.

computation and analysis of error pattern recognition, diagnosis, and early-warning models.

5.2.3   *Case study*

In order to fully reveal the connection between the cognitive state of the operator and his/her visual gaze, the visual gaze of the tester who successfully operated during the operation and his/her operations (mainly valve button control) were analyzed in relation to the changes in the process variables in the chemical process.

Figure 5.11 shows the trend of the main disturbance variable F105 flow rate in scenario D3, and the arrow below indicates the control measures taken by the operator at that corresponding moment. First,

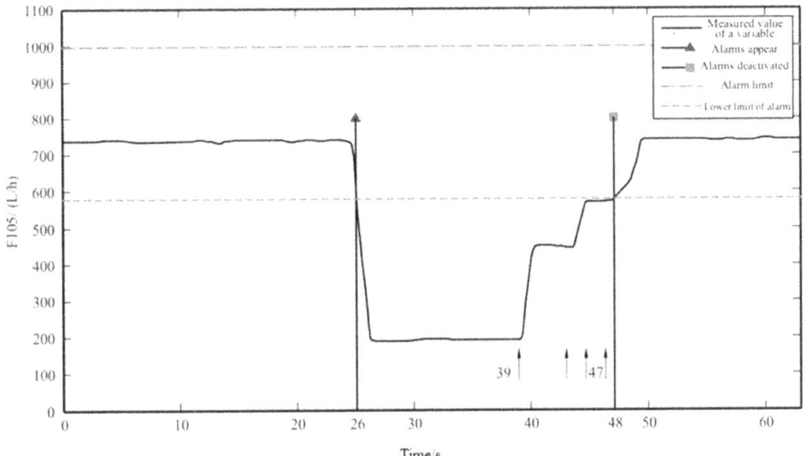

Fig. 5.11.   Trends in F105 flow rates in scenario D3.

from Tables 5.3 and 5.4, it can be seen that in Scenario D3, the abnormal disturbance is an unexpected drop in the flow rate from the CSTR into the distillation column, and the disturbance suppression action that the operator needs to take is to increase the flow rate into the distillation column using valve V201. As can be seen in Fig. 5.11, the above disturbance occurs at 25 s, when the flow rate of F105 suddenly drops and triggers an ultra-low alarm at 26 s. Therefore, the operator takes the first control measure to increase the flow rate through the V201 valve at 39 s, which increases the flow rate of F105 and stabilizes it at 440 L/h. Since the flow rate of F105 is still below the lower limit of the alarm at this time, the alarm is not lifted, so the operator increases the flow rate again between 44 s and 46 s, which brings the flow rate of F105 back to the normal range around 48 s and stabilizes the flow rate at around 50 s. This operation brings the F105 flow back to the normal range in about 48 s and to a stabilized level in about 50 s. At this point, the process is back to normal, all variable parameters are within the normal range, the operator's interference suppression task is successful, and the experiment is completed.

Next, the operator's line-of-sight stay during the experiment was analyzed to correspond with the change in flow rate in order to understand his perception during the operation. The variation of the curves in Fig. 5.12 shows the operator's line-of-sight dwellings during the task, i.e., how long the operator stayed in the various key labeled

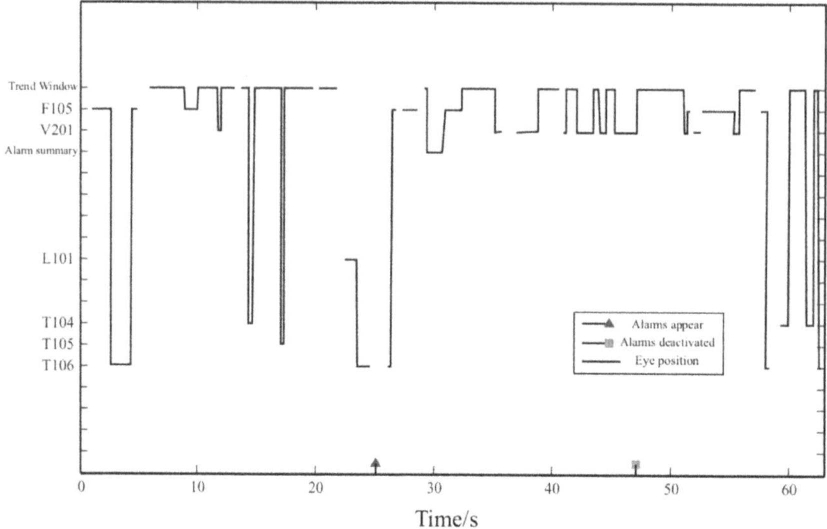

Fig. 5.12. Shift of the operator's line of sight.

areas. From Fig. 5.12, it can be seen that at the beginning of the task, the operator focuses on the F105 label and immediately shifts his sight to the T106 label, after which the operator observes F105, V201, T105, T106, and repeatedly jumps between the trend windows, but his dwell times are all relatively short, which leads to the conclusion that at this time, the operator is at the early stage of the task, and he is in the process of comprehending the operating environment. At 26 s, the first alarm occurs, and when the operator realizes that the alarm is generated, he starts to watch the T105 label at 28 s, and then stays longer between the F105 label and the trend window. From Fig. 5.11, we can see that the operator took the first control measure in 39 s, at this time, his line of sight was focused on the operation slider V201, and after taking the control measure, his line of sight returned to the trend window, and then he repeated the above operation, his line of sight moved back and forth between V201 and the trend window until the alarm was lifted in about 48 s, and the trend window flow rate reached the stabilized value. The above process reflects the situation of the operator's line of sight during the process control. 50 s or so, the process returned to a stable state, and the operator's line of sight began to jump between T104, T106, and F195, which is similar to the situation before the occurrence of the anomaly.

Therefore, through the analysis, it can be concluded that the operator's line of sight in the experimental process and its operation behavior corresponds to the change of process variables in the chemical process, which can effectively realize the understanding of the practical significance of the operation behavior and realize effective real-time perception of the operator's cognitive situation.

5.3 Process Operator Error Pattern Recognition Based on Learning Vector Quantized Neural Network

5.3.1 *Process operator safety hazard behavior and eye movement data collection*

In general, operators' work behavior is typically categorized as either normal or abnormal during identification. Several studies have delved into the classification of unsafe behaviors. For instance, categorized operators' unsafe behavior into human negligence and human error, while categorizing it into internal and external factors. Currently, there is no universally accepted classification method for operators' unsafe behavior.

When employing a classification method for workers' unsafe behavior, only errors that have already transpired are interpreted and categorized. Behaviors that have not yet occurred, but may occur at a later time and cause serious consequences, have not been defined. To rectify the limitations of the classification method for operators' unsafe behavior, a novel classification method for unsafe behavior among offshore drilling operators was proposed in this study, as illustrated in Fig. 5.13.

As depicted in Fig. 5.13, this study categorizes offshore drilling operation behavior into four types: normal operation, safety hazard behavior, error behavior, and violation behavior. Within the safety hazard behavior category, two subtypes are identified: fatigue operation and inexperienced operation. Error behavior is further classified into three subtypes: action error operation, omitted action operation, and wrong order operation. Error behavior refers to instances where operators make mistakes unintentionally, leading to certain consequences. Violation behavior, on the other hand, refers to instances where operators intentionally violate regulations, resulting in certain consequences.

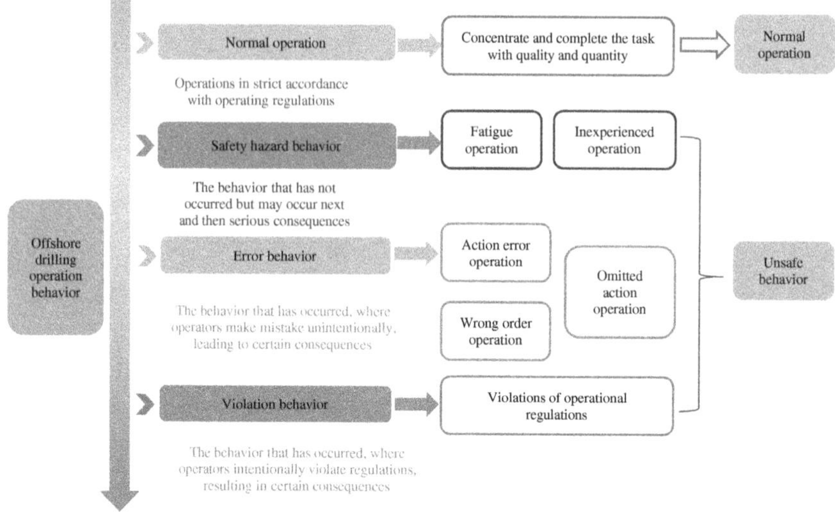

Fig. 5.13. Classification method for unsafe behavior among offshore drilling operators.

Notably, errors that have occurred can serve as reminders for operators to make corresponding corrections by inspecting personnel or equipment alarms. However, detecting safety hazard behaviors such as fatigue and inexperienced operation is often challenging. Therefore, enhancing the monitoring and identification of safety hazards among offshore drilling operators is of significant importance in ensuring the safety of offshore drilling operations. By incorporating human eye movement behavior, this study accurately reflects characteristics of unsafe behavior that may lead to accidents, including fatigue, mental activity, inexperienced operation, and tension operation. Consequently, eye-tracking data were employed in this study to identify safety hazards in offshore drilling operations and mitigate the occurrence of accidents.

5.3.1.1 *Experimental subjects*

This experiment was conducted in strict adherence to the ICMJE guidelines for the protection of research participants, as well as the Belmont Report and the Declaration of Helsinki. The subjects of the present experiment included a total of 42 trainees (24 males and

18 females) with normal or corrected vision, which met the experimental needs. The participants ranged from 21 to 30 years old, with an average age of 25 and a standard deviation of 2.15. Before the start of the experiment, all trainees underwent training and assessment by experienced professional technicians to ensure that each trainee had mastered enough basic offshore drilling processes and emergency response methods.

5.3.1.2 *Experimental equipment*

The experimental recording apparatus utilized in this study was the EYESO EG100 head-mounted eye movement data sensor, along with its associated eye movement data acquisition device. Operating at a sampling frequency of 100 Hz, this sensor boasts an average viewing angle error of 0.5°. Utilizing an infrared LED with a wavelength of 850 nm, it tracks human eye movements. It is crucial to highlight that excessive brightness of infrared radiation may pose a risk to human eyes; therefore, the safety of the sensor was evaluated prior to eye movement data acquisition. According to the manufacturer's damage assessment report on the infrared LED in the sensor, the radiation brightness is measured at $0.679\,\mathrm{W/cm^2 \cdot sr}$, well below the ISO15004-2-2007 specified limit of $6\mathrm{W/cm^2 \cdot sr}$, indicating that the sensor poses no risk to human eyes. A depiction of the head-mounted eye movement data sensor is presented in Fig. 5.14.

Fig. 5.14. The EYESO EG100 head-mounted eye movement data sensor.

Fig. 5.15.   The SD-JKFZ-7 offshore drilling simulation device.

The offshore drilling simulation test operating platform employed in this experiment was the SD-JKFZ-7 offshore drilling simulation device. The offshore drilling simulation device is shown in Fig. 5.15. Before the experiment commenced, proper labels were affixed to prominent positions on the buttons, operation handles, instruments, and other essential components of the experimental platform.

The equipment in Fig. 5.15 is as follows: 1-Wellhead operation scene; 2-Winch gear selection; 3-Maindrum speed selection; 4-spinner assembly (make up/break out); 5-Pneumatic spider (open/close); 6-Hook height, drilling parameters, etc.; 7-wt indicator, hook weight, etc.; 8-Winch direction selection; 9-Winch speed regulation; 10-Throttle speed regulation; 11-Make up drill stem; 12-Working brake; 13-Main clutch hanging; 14-Power and current meter; 15-Mud volume increment; 16-Stand pipe pressure; and 17-Torque meter.

5.3.1.3   *Experiment content*

The present experiment primarily aimed to simulate driller operation scenarios from offshore drilling sites in a laboratory setting, with a specific focus on trip operation procedures. Each participant underwent three rounds of data collection. In the initial round, participants executed the operation without prior training. In instances of operational errors, participants followed the emergency disposal method outlined in the operation manual until the trip experiment concluded. Subsequently, participants received training and underwent an examination to ensure proficiency in normal operation for the second round, followed by data collection. The third round

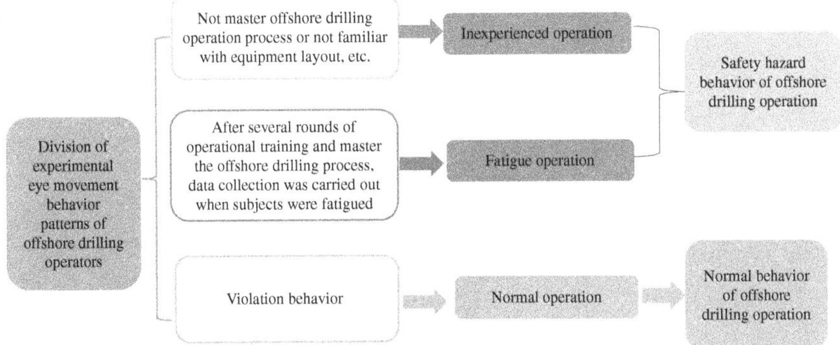

Fig. 5.16. Eye movement experiment process and the division of behavior patterns for offshore drilling operators.

involved gathering eye movement data during offshore drilling operations under conditions of fatigue. Fatigue levels were evaluated using the Fatigue Assessment Instrument (FAI), developed by Josopn E. Schwartz of the American Psychobehavioral Science Laboratory and Lina Jandorf of the Neurology Laboratory in 1993. Eye movement data collection during offshore drilling operations occurred when the fatigue assessment indicated moderate fatigue or higher. If the fatigue assessment did not meet the specified conditions, eye movement data collection for the subject was temporarily suspended.

The eye movement experiment process and the division of behavior patterns for offshore drilling operators are shown in Fig. 5.16. Simulated drilling operation experimental scenarios in offshore drilling sites are shown in Fig. 5.17.

5.3.1.4 *Eye movement data acquisition*

During the experiment, participants were instructed to wear a head-mounted eye movement data sensor throughout the entire process. All eye movement data of the subjects were recorded, including fixation data, saccade data, and pupil data. The first round of collected data consisted of eye movement data from inexperienced offshore drilling operators, while the second round included eye movement data from normal operations, and the third round comprised eye movement data from fatigue operations. Over 40 types of eye movement parameters were collected in this experiment, covering fixation

Fig. 5.17. Simulate drilling operation experimental scenarios in offshore drilling sites.

parameters, saccade parameters, and pupil parameters. Considering the repetitive nature of offshore drilling operations, parameters such as the total number of fixations and saccades, and the average pupil diameter data during a single operation process were selected. Based on these principles, five types of eye movement data parameters were chosen, including two types of fixation parameters, two types of saccade parameters, and one type of pupil parameter. The types of eye movement data collected for the three rounds of offshore drilling operators are shown in Table 5.5.

5.3.2 *Process operator safety-hazardous behavior identification method*

5.3.2.1 *Eye-tracking technology*

The amount of information that human beings acquire from the external world is largely dependent on visual input received by their eyes. Eye-tracking technology is based on the principle of detecting variations in the darkness or brightness of pupils. It utilizes sensors to capture and record real-time data of pupil movements, employing different capture methods based on the varying color depths of pupils among different races. When individuals are exposed to

Table 5.5.   Eye movement data types of offshore drilling operators.

Parameter grouping	Parameter name	Parameter units	Parameter description
Fixation parameters Tailing	F_{pn}	n	Total number of fixation points
	F_t	s	Total fixation time
Saccade parameters	S_{lt}	px	Total saccade length
	S_t	s	Total saccade time
Pupil parameters	$\mathrm{P_{ad}}$	px	Pupil average diameter

external stimuli, changes in pupil size and fixation positions accurately reflect their psychological, attentional, and recognition states.

Drilling operations are highly intricate and demanding tasks that necessitate close collaboration among multiple individuals. Monitoring and identifying operational behavior should align with the requirements of a relatively stable operating environment and a consistent operational mode. There are several justifications for the application of eye-tracking technology to enable real-time monitoring and identification of safety hazards associated with drilling operators. These justifications include the following reasons:

(1) Although drilling operations require close cooperation among various types of operators, each operator is assigned specific tasks on the drilling operation platform. For instance, the driller is responsible for operating the drilling equipment and overseeing its functionality, while the derrickman focuses on managing vertical movements and supporting the driller. Additionally, the assistant driller handles blowout preventer operations and assists the driller.

(2) Despite the overall complexity of offshore drilling operations, each type of operation has relatively singular attributes.

(3) The working environment for different types of offshore drilling operators remains relatively stable, with minimal position movement.

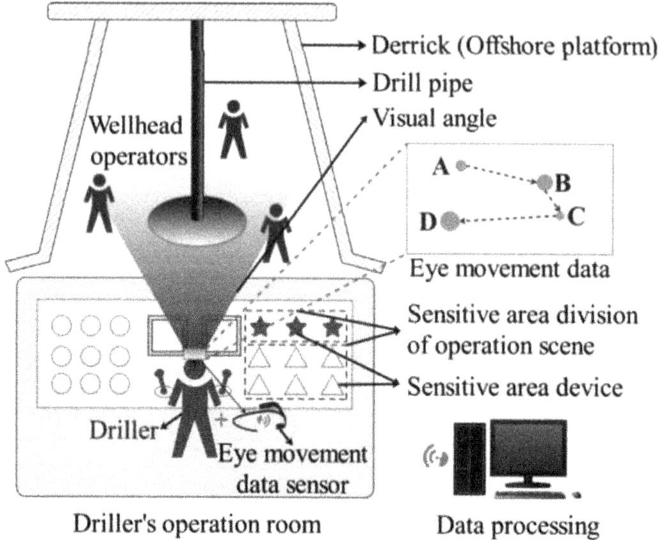

Fig. 5.18. Application of eye-tracking technology for identifying safety hazard behavior among offshore drilling operators.

The application of eye-tracking technology for identifying safety hazard behavior among offshore drilling operators is illustrated in Fig. 5.18. During offshore drilling operations, the driller is responsible for observing or operating the wellhead scene, the operating platform, and instruments. As depicted in Fig. 5.18, the buttons, handles, knobs, and the instruments of the driller's operating platform are distributed across different areas according to their respective functions. The driller must operate the device within the appropriate functional area based on field conditions. To mitigate the influence of personal factors (such as visual habits, sitting posture, height, and range of motion), equipment factors (including the size, shape, position, and operational mode of the device unit on the drilling equipment, as well as potential errors in eye movement data sensors), and environmental factors (such as noise and vibration) on the acquisition results of eye movement data, it is crucial to delineate sensitive areas within the operational scene. This approach entails creating a composite eye movement hotspot map during normal drilling operation behavior, achieved by gathering data from various drilling operators engaged in typical operations. Eye movement data collection from drilling operators involves designing the working scene to replicate real-site conditions, encompassing both environmental and equipment factors. Moreover, the composite eye movement hotspot map accounts for individual variances among different drilling operators comprehensively. Therefore, this composite eye movement hotspot map can effectively segment the sensitive areas of offshore drilling operation scenarios, making them suitable for the majority of operators. According to the results of the divided sensitive areas, the fixation and saccades data of operators within these areas during the operation are recorded separately. These data are then used for real-time identification of operator behavior in offshore drilling operations.

5.3.2.2 *Theory and construction procedure of Tree-structured Parzen Estimator-Light Gradient Boosting Machine model*

(1) Light Gradient Boosting Machine
The LightGBM algorithm is an improved algorithm for GBDT, primarily incorporating the idea of combining the histogram algorithm and the depth-constrained leaf-wise strategy. Its computation principle entails utilizing the negative gradient of the loss function as the residual approximation of the current decision tree and subsequently fitting the new decision tree. Each iteration of the model maintains the original model unchanged while adding a new function to the model, progressively aligning the predicted value with the actual value.

Before the emergence of the LightGBM model, XGBoost was the most commonly used enhanced GBDT algorithm, relying on the presorting method for decision trees. While XGBoost excels in determining segmentation points, it suffers from substantial space and time consumption during computation, particularly influenced by data size. The introduction of the LightGBM model addresses the shortcomings of GBDT algorithms like XGBoost when handling extensive data, providing solutions for large space consumption and extended calculation times.

During the process of gradient boosting, after the previous round of data is passed to the learner $F_{t-1}(x)$, the loss function of the learner can be obtained as $L(y, F_{t-1}(x))$. It is essential to identify a corresponding weak learner $h_t(x)$ that minimizes this loss function.

The corresponding loss function for this round is shown in Eq. (5.2):

$$f_0(x) = \arg\min \sum_{i=1}^{N} (L(y, F_{t-1}(x)) + h(x)), \quad h \in H. \tag{5.2}$$

The negative gradient of the loss function is shown in Eq. (5.3):

$$r_{ti} = -\left(\frac{\partial L(y, F_{t-1}(x_i))}{\partial F_{t-1}(x_i)} \right). \tag{5.3}$$

The square difference is used to fit Eq. (5.4):

$$h_t(x) = \arg\min \sum (r_{ti} - h(x))^2, \quad h \in H. \tag{5.4}$$

Finally, the strong learner of this round is generated: As shown in Eq. (5.5),

$$F_t(x) = h_t(x) + F_{t-1}(x). \tag{5.5}$$

The LightGBM algorithm introduces several enhancements compared to the traditional GBDT model, as outlined below:

(i) The LightGBM algorithm enhances leaf growth by utilizing the Gain formula to calculate splitting gain and guiding leaf nodes to split in the direction of maximum information gain, in contrast to the traditional GBDT model that splits leaf nodes deeply. This approach effectively reduces the loss function value and improves prediction accuracy.

(ii) The LightGBM algorithm employs the histogram algorithm and binning technique to optimize storage utilization and enhance training speed. When splitting leaf nodes, the original data is initially pre-sorted and compared. In this process, the histogram algorithm discretizes features into n-dimensional features for continuous types, forming a histogram of size n containing information on n groups. Simultaneously, the data is partitioned into a series of discrete ranges (bins), facilitating the traversal of discrete data to identify optimal division points. This simplification of data reduces memory usage and enhances the efficiency of the model.

(iii) The LightGBM algorithm employs the exclusive feature bundling algorithm to enhance its processing capabilities for big data and high-dimensional samples. This algorithm adopts the idea of building a graph, where features are treated as nodes and edges are connected between non-mutually exclusive features. Subsequently, the algorithm identifies all the bundled feature sets from the graph. By reducing the number of features used, this algorithm effectively reduces data scale, making it suitable for applications involving big data and high-dimensional datasets.

(2) Tree-structured Parzen Estimator-Light Gradient Boosting Machine

The LightGBM model involves several crucial hyperparameters that require adjustment, including the maximum tree depth, the number of leaves on each tree, the number of iterations, etc. Selecting appropriate values for these hyperparameters determines the speed and accuracy of prediction results as well as the successful completion of classification or regression tasks. Traditional parameter adjustment methods mainly include random search and grid search, among others. Grid search is an exhaustive method that explores the parameter space with a fixed step size, theoretically obtaining globally optimal results but at the cost of extensive computations and low computational efficiency. On the other hand, while random search exhibits high computational efficiency, its reliance on sampling introduces uncertainty and may lead to local optimum solutions. Therefore, there is a need to establish a hyperparameter optimization algorithm that ensures both global optimality and efficient execution in addressing multi-parameter optimization problems encountered in LightGBM.

The Bayesian optimization algorithm is a global optimization algorithm. It employs a probabilistic surrogate model to fit the objective function during the optimization process. The next evaluation point is then selected based on previous sampling results, facilitating the rapid attainment of the optimal solution. As shown in Eqs. (5.6) and (5.7),

$$p(f|H_i) = \frac{p(H_i|f)p(f)}{p(H_i)}, \tag{5.6}$$

$$H_i = \{\{(x_1, f(x_1)), \ldots, (x_i, f(x_i))\}, \tag{5.7}$$

where $p(f)$ and $p(H_i|f)$ are the prior probability distribution and likelihood distribution of f, respectively. $p(f|H_i)$ represents the conditional probability distribution of parameter f when the set H_i of observations has been given, namely the posterior probability distribution.

In this study, the Tree-structured Parzen Estimator (TPE) is chosen as the probability proxy model, and p in Eq. (5.6) is defined as given in Eq. (5.8):

$$p(x|y) = \begin{cases} l(x), & y < y^* \\ g(x), & y \geq y^* \end{cases},\qquad (5.8)$$

where $y^* = \min\{(x_1, f(x_1)), \ldots, (x_i, f(x_i))\}$ represents the optimal value at the observation threshold; $l(x)$ is the density estimate for which the loss function of the observed value x is less than y^*; and $g(x)$ represents the density estimate for which the loss function of the observed value $x \geq y^*$.

The TPE employs the expected improvement (EI) as the sampling function to select the next evaluation point that has the optimization effect on the value of the objective function. As shown in Eq. (5.9),

$$EI_{y^*} = \int_{-\infty}^{y^*} (y^* - y)p(y|x)dy = \int_{-\infty}^{y^*} (y^* - y)\frac{p(x|y)p(y)}{p(x)}dy, \quad (5.9)$$

when $p(y|x)$ is positive at $y < y^*$, setting the hyperparameter x for algorithm modeling will result in better outcomes compared to the optimal value above the observation threshold.

Letting $\gamma = p(y < y^*)$, the following can be established in Eqs. (5.10) and (5.11):

$$p(x) = \int_R p(x|y)p(y)dy = \gamma l(x) - (1 - \gamma)g(x), \qquad (5.10)$$

$$\int_{-\infty}^{y^*} (y^* - y)p(x|y)p(y)dy = \gamma y^* l(x) - l(x)\int_{-\infty}^{y^*} p(y)dy. \quad (5.11)$$

By substituting Eq. (5.9), Eq. (5.12) can be obtained:

$$EI_{y^*}(x) = \frac{\gamma y^* l(x) - l(x)\int_{-\infty}^{y^*} p(y)dy}{\gamma l(x) - (1 - \gamma)g(x)} \propto \left(\gamma + \frac{g(x)}{l(x)}(1 - \gamma)\right)^{-1}.$$

$$(5.12)$$

Equation (5.12) illustrates that when the hyperparameter x has the maximum probability $l(x)$ and the minimum probability $g(x)$, the maximum EI value is obtained. The TPE constructs a sample hyperparameter set through $l(x)$ and $g(x)$, and then evaluates x in the form of $g(x)/l(x)$. During each iteration, the algorithm returns the point x^* with the maximum EI value.

5.3.2.3 *Procedures for establishing the Tree-structured Parzen Estimator-Light Gradient Boosting Machine model*

Step 1: Collection of eye movement data for safety hazard behavior among offshore drilling operators.

Eye movement data for normal drilling operations, fatigue drilling operations, and inexperienced drilling operations are collected from selected offshore drilling operators.

Step 2: Division of the sensitive area of the operation scene for offshore drilling operators.

The objective of dividing sensitive areas is to capture effective eye movement data from operators during drilling operations. The process of dividing sensitive areas in the operational scenario can be divided into three parts:

(i) Collecting eye movement data during normal drilling operations from various drilling operators to create a composite eye movement hotspot map of normal drilling operations.

(ii) Determining the device units covered by each sensitive area based on the drilling operation scenario and the distribution of functional areas on drilling equipment.

(iii) Establishing the boundaries of each sensitive area according to the distribution of eye movement hotspots in the composite eye movement hotspot map of normal operations. When dividing sensitive areas, it is essential to ensure that the boundaries encompass both the devices and a significant majority of effective eye movement hotspots while excluding irrelevant or false eye movement hotspots.

Step 3: Screening eye movement parameters to characterize the operational behavior of offshore drilling operators.

The eye-tracking sensor facilitates real-time recording of eye movement data for offshore drilling operators. However, the collected data comprises various eye movement parameters, necessitating the identification of specific parameters that effectively characterize operator behavior. In previous studies, our team employed a screening method based on correlation analysis and sensitivity analysis of eye movement parameters to identify those that effectively characterize drilling operation behavior. The specific steps involved in implementing this method are as follows:

(i) Participants were divided into two groups, and the drilling well control simulation device was used to simulate drilling operations while simultaneously capturing and recording the eye movement data of each participant.

(ii) The laws and correlations of unsafe behavior eye movement data were investigated through analysis of experimental data and Pearson correlation analysis.

(iii) Eye movement parameters were ranked using the Morris, Sobol, and EFAST global sensitivity analysis methods to identify the most suitable parameters for characterizing unsafe behaviors among drilling operators.

The results revealed that four eye movement parameters could be screened from twelve parameters using the above-mentioned method. With an increase in experimental data, we expanded the eye movement parameters that could reflect operators' behavior to five. Table 5.5 presents the selected five eye movement parameters. Therefore, based on the research conclusions, the optimal eye movement features identifiable by the TPE-LightGBM model were determined according to the results of sensitive area division and the distribution of eye movement data.

Step 4: The hyperparameters of the LightGBM model are optimized based on the Tree-structured Parzen Estimator (TPE) algorithm.

The Hyperopt optimizer is widely recognized as one of the most commonly employed Bayesian optimizers. This optimizer integrates several optimization algorithms, such as Random Search, Simulated Annealing, and TPE. In this study, the TPE algorithm was employed to optimize the hyperparameters of the LightGBM model. The optimization process of hyperparameters can be divided into four parts.

(i) Preprocessing of the eye movement feature data.

The LightGBM model is an enhanced variant of the decision tree model, exhibiting the advantageous characteristic of insensitivity to feature dimensions. As demonstrated in Section 5.3.2, the LightGBM model employs the histogram method to identify the optimal feature for determining the splitting point of a leaf node, which is only related to the order of values. Therefore, it is unnecessary to normalize the eye movement features before recognizing the safety hazard behavior of offshore drilling operators.

(ii) Setting the TPE hyperparameter optimization range and generating an initial set of hyperparameter combinations through randomization.

Before generating the initial hyperparameter combination, it is necessary to set the optimization range for TPE hyperparameters. By consulting official guidance documents from the LightGBM model and relevant literature on setting methods, the hyperparameter range used in this model is determined. The range of hyperparameter values is shown in Table 5.6.

- Constructing a safety hazard behavior identification model for offshore drilling operators based on TPE-LightGBM.
- The np.random.default_rng module is used to initialize the hyperparameter space in accordance with the specified range of hyperparameters. Subsequently, a set of hyperparameter combinations

Table 5.6. The range of hyperparameter values and TPE optimization results.

Hyperparameters	Range of hyperparameter values	Optimized hyperparameters
max_depth	$[2, 15]$	14
n_estimators	$[1, 300]$	43
num_leaves	$[3, 90]$	39
learning_rate	$[0.001, 1)$	0.177
reg_alpha	$(0, 1]$	0.220
reg_lambda	$(0, 1]$	0.518
subsample	$(0, 1]$	0.757

$X = [x_1, x_2, x_3, \ldots, x_n]$ can be generated through this initialization process.

(iii) Determining the optimal combination of hyperparameters for the model using the TPE optimization algorithm.

- The EI value of the initial hyperparameters is calculated by determining the expected improvement using the sampling function applied to the initial hyperparameters.
- TPE functions as the probabilistic surrogate model, while EI serves as the sampling function. This facilitates the automatic selection of multiple evaluation points optimizing the objective function, followed by the computation of their corresponding EI values.
- The optimal value for each hyperparameter is determined by comparing the EI values of each evaluation point with the initial hyperparameter. This iterative process continues until all optimal values for hyperparameters are obtained. Consequently, these optimal values collectively form the combination of hyperparameters that yield the best performance.
- The optimal combination of hyperparameters, along with the eye movement dataset of safety hazard behavior of drilling operators, is utilized for model training, enabling the computation of the cross-validation evaluation metric accuracy (ACC) for the current model. As shown in Eq. (5.13),

$$ACC = \frac{TP + TN}{TP + TN + FP + FN}. \tag{5.13}$$

The true positive (TP) represents instances accurately classified as positive by the model. False negative (FN) refers to instances that are actually positive but mistakenly classified as negative by the model. False positive (FP) indicates instances that are actually negative but incorrectly classified as positive by the model. True negative (TN) signifies instances correctly classified as negative by the model.

- If the model achieves the desired recognition accuracy, the optimal combination of hyperparameters obtained represents the optimization outcome of the TPE algorithm. However, if the model fails to meet the design requirements, it is necessary to modify the range of hyperparameter optimization until it meets those requirements.

Step 5: Real-time identification of safety hazard behavior among drilling operators is accomplished based on the TPE-LightGBM model.

Utilizing LightGBM and the optimal hyperparameter combination, real-time eye movement data collected from drilling operators is evaluated to obtain behavior recognition results. Feature importance plays a crucial role in evaluating features in decision tree models such as LightGBM, XGBoost, Random Forest, and GBDT. All ensemble learning algorithms based on decision trees include a module for calculating feature importance. The decision tree serves as the foundation for establishing LightGBM, XGBoost, Random Forest, and GBDT models. When constructing a single decision tree model, the core idea is to select the most appropriate features as segmentation points for the leaf nodes. This information serves as an indicator to evaluate the importance of features. The higher the frequency a feature is utilized, the greater its importance, indicating its significant impact on model construction and optimization. Feature importance is calculated based on the total number of times a feature is utilized in constructing a decision tree.

The process of establishing the TPE-LightGBM model is shown in Fig. 5.19. The pseudocode of the TPE-LightGBM model is shown in Table 5.7.

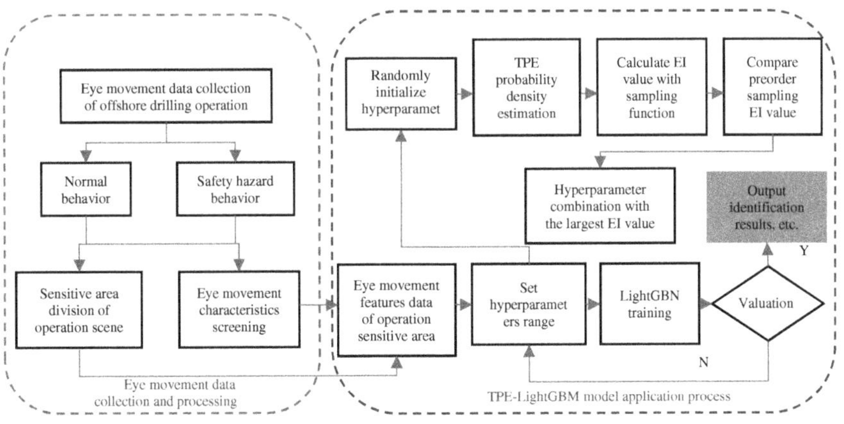

Fig. 5.19. The process of establishing the TPE-LightGBM model.

Table 5.7. The pseudocode of the TPE-LightGBM model.

Algorithm: TPE-LightGBM Algorithm
Input: Training data: X_train, y_train
Input: Testing data: X_text, y_text
Data: Eye movement data and hyperparameter space
/*Hyperparameter optimization */
1 params_best ← param_hyperopt()
/* Model training and prediction */
2 model ← LGBMClassifier(* * params_best, n_jobs =-1,random_state = 123)
3 model.fit(X_train, y_train)
4 $y\_pred \leftarrow model.predict(X\_test)$
/* Model evaluation */
$5score \leftarrow accuracy\_score(y\_pred, y\_test)$
6 $y\_pred\_proba \leftarrow model.predict\_proba(X\_test)$
7 $features \leftarrow X\_train.columns$
8 $importances \leftarrow model.feature\_importances\_$
/* Output the results */
Output 1: Optimized parameters: $params\_best$
Output 2: Predicted labels: y_pred
Output 3: Identification accuracy: $score$
Output 4: Feature names: $features$
Output 5: Feature importances: $importances$
Output 6: Predicted labels: $metrics.confusion\_matrix$

5.3.3 *Case study and analysis*

5.3.3.1 *Basic overview of offshore drilling operation scene and division of sensitive areas*

(1) Offshore drilling operation simulation experiment process

According to the experimental scheme for acquiring eye movement data on the safety hazard behavior of offshore drilling operators established in Section 5.3.1, the operation behavior of a driller in the offshore drilling operation site is simulated to obtain eye movement data. A total of 126 sets of experimental data were collected for this experiment, comprising 42 sets each of normal eye movement data, fatigue eye movement data, and inexperienced eye movement data. Each set of experimental data included fixation parameters (total number of fixation points in the sensitive area, total fixation time in the sensitive area), saccade parameters (total saccade length in the sensitive area, total saccade time in the sensitive area), and pupil average diameter.

Before commencing the simulation experiment, each subject is required to adhere to the regulations by wearing an eye-tracking sensor and calibrating their gaze and surroundings. Additionally, every subject operator had to possess a comprehensive understanding of the fundamental principles and operational procedures associated with offshore drilling simulation equipment. In case of an offshore drilling accident, emergency measures could be promptly implemented in accordance with the experimental safety operation procedures. The detailed descriptions of the normal tripping operation and the expected operation of the participants are as follows:

(i) *Offshore drilling simulation equipment start-up*: This includes the main clutch hanging and the opening of the BOP and other corresponding operations.

(ii) *Lifting drilling tools*: This involves winch gear selection, main drum speed selection, winch direction selection, winch speed regulation, throttle speed regulation, and working braking, among other corresponding operations.

(iii) *Working the brake, stop lifting drilling tools*: This includes working the brake, winch (gear, direction, and speed) closing, main drum closing, throttle closing, and other corresponding operations.

(iv) *Pneumatic spider close*: This includes the operations corresponding to pneumatic spider close.

(v) *Breaking out spinner assembly*: This involves the operations corresponding to breaking out the spinner assembly.

(vi) *Working the brake*: This includes the operations corresponding to working the brake.

(vii) *Making up the drill stem*: This involves the operations corresponding to making up the drill stem.

(viii) *Making up the spinner assembly*: This includes the operations corresponding to making up the spinner assembly.

(ix) *Pneumatic spider open*: This includes the operations corresponding to opening the pneumatic spider.

(x) *Lowering drilling tools*: This involves the operations corresponding to working the brake, meaning the "working brake" handle is kept in a natural state.

(xi) *Pneumatic spider close*: This includes the operations corresponding to pneumatic spider close.

The steps (i) to (xi) correspond to the operation process of the normal tripping operation. Failure to promptly identify safety hazard behaviors in offshore drilling operations can significantly contribute to operational errors, thereby resulting in accidents and injuries among operators.

(2) Division of sensitive areas

The mapping of eye movement hotspots is characterized by objectivity, as it provides both qualitative and quantitative insights into the focal areas and visual range of participants. An eye movement hotspot map visually represents the distribution of fixation intensity, depicting the specific functional relationship between visual fixation patterns and corresponding image parameters during operations. The rendering of the map is determined by factors such as the duration of fixation and the relative pixel positions within the image. Each color on the map corresponds to the length of fixation time in a particular region. Typically, red indicates the longest fixation time, followed by colors in the rainbow spectrum representing decreasing fixation times.

The composite hotspot map of normal tripping operations, illustrated in Fig. 5.20, highlights eye movement hotspots primarily clustering around the operation area (buttons, knobs, and handles), display area (parameter screen), and scene screen area. Throughout the normal tripping operation process (Steps i–xi), operators focus on the operation area while also monitoring changes in well control parameters displayed on the lower right small screen during tool lifting. This ensures the safety of wellhead operators and enables timely response to emergencies. In contrast, as shown in Fig. 5.21, fixation points in inexperienced operations are more scattered. Inexperienced operators focus on each operation area while also paying attention to non-operation areas, indicating unfamiliarity with the experimental bench layout and operation process. Conversely, the composite hotspot map of normal operations, as shown in Fig. 5.20, exhibits clearer fixation points, facilitating a better understanding of the operation's focal points. Regarding fatigue operations, depicted in Fig. 5.22, the difference in hotspot maps compared to normal operations is relatively minor. This suggests that operators remain familiar with the operation process and experimental bench layout even under fatigue conditions. Consequently, relying solely on eye

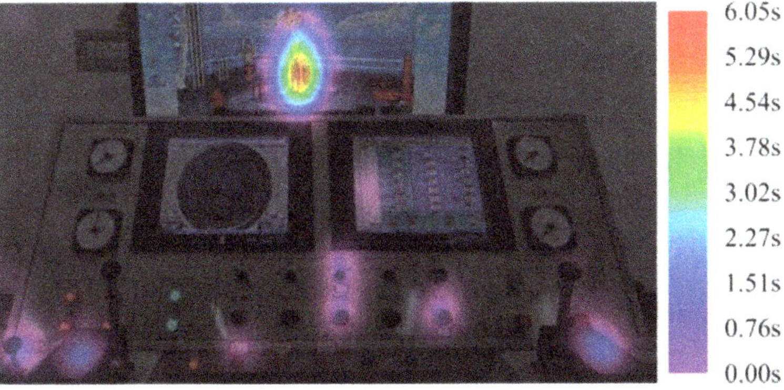

Fig. 5.20.   Composite hotspot map for normal offshore drilling operation.

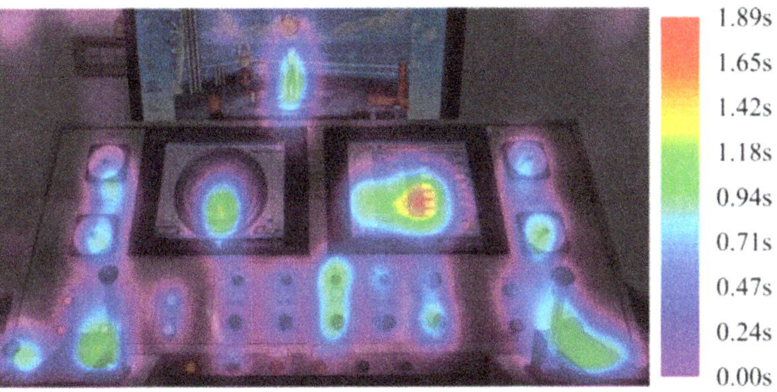

Fig. 5.21.   Composite hotspot map for inexperienced offshore drilling operation.

movement analysis for fatigue operations presents challenges, necessitating further analysis of alternative data types.

From the composite hotspot map of normal drilling operations, although the distribution of fixation points is relatively clear, there are still some issues in the statistical analysis and identification of fixation conditions:

(i) Fixation points are dispersed around buttons, handles, or screen parameters, rather than strictly centralized. This suggests potential variations in eye habits, sitting posture, height, and movement range among different operators.

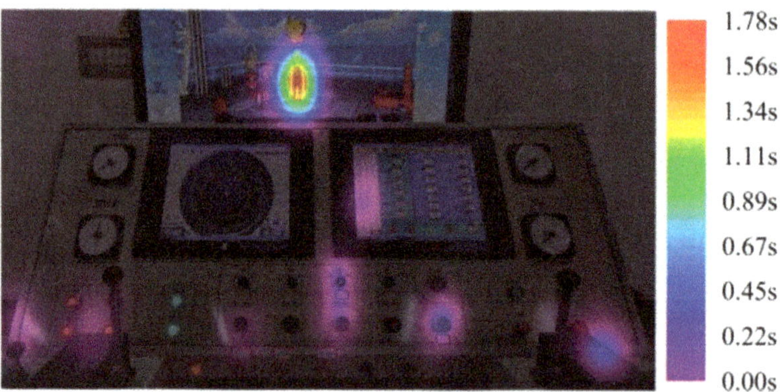

Fig. 5.22.   Composite hotspot map for fatigue offshore drilling operation.

(ii) Both normal and fatigue drilling operations exhibit irrelevant fixation points in the upper left corner. This implies that, despite operator familiarity with the operating process, there are still unrelated fixation points caused by distractions during drilling operations. If such irrelevant fixation information is included in the analysis, it may affect the identification results.

To address the issues mentioned earlier, the sensitive area of the offshore drilling operation scene needs to be divided. However, due to the large number of important equipment units, irregular shapes, and significant differences in placement positions in offshore drilling operation scenes, using a grid-based division may not provide good adaptability. This can result in inaccurate data recording or slow calculation speeds. Therefore, this study adopts an unshaped sensitive area division method to ensure precise recording of eye movement data and facilitate optimal positioning of important equipment units at the center of the sensitive area. The selected sensitive area division method is illustrated in Fig. 5.23.

According to the distribution of hotspot maps of fixation points in normal, fatigued, and inexperienced drilling operations, the offshore drilling simulation equipment was divided into three types of sensitive areas: the scene area, instrument area (1–4), and operation area (1–4). The scene area primarily encompassed wellhead equipment, wellhead operators, spinner assembly, joint, winch, main drum, pneumatic spider, and other related equipment. The instrument area primarily encompassed power and current meters, mud

Fig. 5.23. Sensitive area division of offshore drilling platform.

volume increment, weight indicator, hook height, drilling parameters, standpipe pressure, torque meter, etc. The operation area primarily encompassed status indicator lights, operation handles, operation buttons (switches and knobs), etc.

5.3.3.2 *Analysis of eye movement data laws in safety hazard behavior of offshore drilling operators*

Various parameter indices are employed to characterize the eye movement behavior of offshore drilling operators during their operations. As discussed in Section 5.3.1, these parameters effectively describe safety hazards associated with drilling operations. They include fixation parameters (total number of fixation points and total fixation time), saccade parameters (total saccade length and total saccade time), and pupil parameters (average pupil diameter). By analyzing the data laws associated with these eye movement parameters, their suitability in identifying safety hazards in drilling operations can be evaluated.

Analysis of Fig. 5.24 reveals notable differences in the average total number of fixation points across various operational conditions. Specifically, during normal operation, the average total number of fixation points is recorded at 35.52 times, slightly lower at

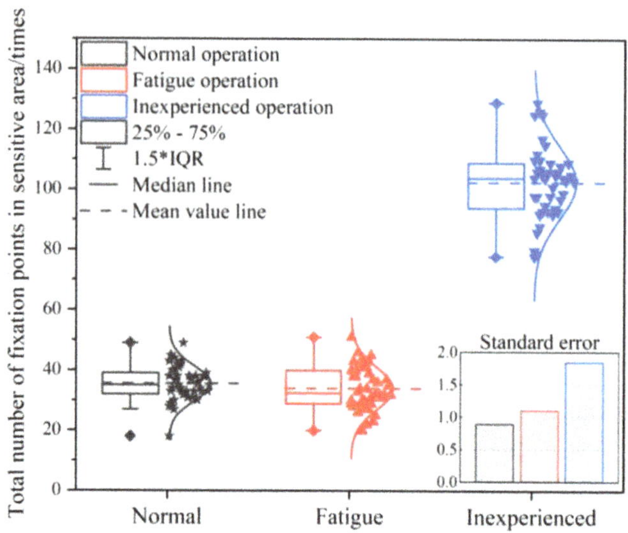

Fig. 5.24. Total number of fixation points in the sensitive area.

34.05 times during fatigue operation, and substantially higher at 102.55 times during inexperienced operation. Notably, the average total number of fixation points in the sensitive area is comparatively smaller between normal and fatigue operations, whereas it is significantly larger during inexperienced operations. These findings can be attributed to the familiarity of operators with the operation process and offshore drilling simulator during both normal and fatigue operations. In contrast, during inexperienced operation, operators may struggle to locate corresponding positions in the operation process or lack familiarity, resulting in a higher total number of fixation points.

Analysis of Fig. 5.25 reveals significant disparities in the average total fixation time among different operational conditions. Specifically, normal operation shows an average total fixation time of 129.37 seconds, whereas fatigue operation demonstrates a notably lower average of 52.94 seconds. In contrast, inexperienced operation exhibits the highest average total fixation time at 174.47 seconds. These distinct differences highlight the impact of operator experience and fatigue on fixation behavior. Inexperienced operators, grappling with the complexities of the task, tend to allocate more time to fixation, followed by normal operators. Conversely, fatigue

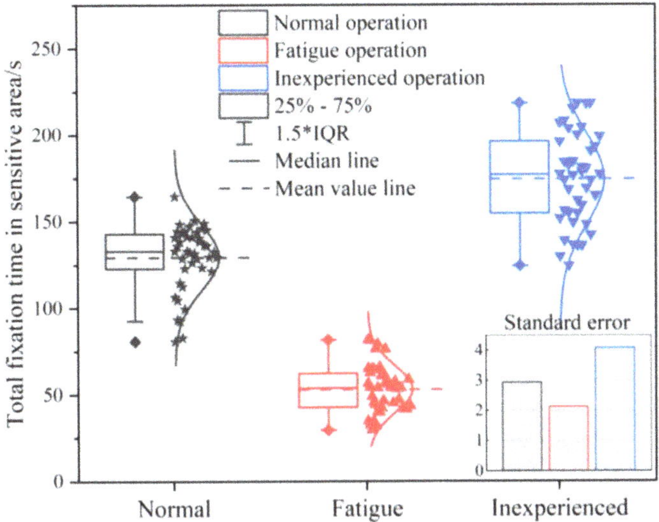

Fig. 5.25. Total fixation time in the sensitive area.

operators, likely experiencing cognitive and physical strain, exhibit shorter fixation times. Such variations may be attributed to factors such as increased blinking frequency due to fatigue, influencing the overall fixation duration.

Analysis of Fig. 5.26 reveals notable differences in the average total saccade length among various operational conditions. Specifically, normal operation shows an average total saccade length of 16,437.43px, while fatigue operation demonstrates a slightly lower average of 12,785.43px. In contrast, inexperienced operation exhibits a substantially higher average total saccade length at 36,376.83px. The minimal disparity between normal and fatigue operations in terms of saccade length within the sensitive area suggests proficient performance in both scenarios. However, during fatigue operation, increased blink frequency may contribute to minor shifts in fixation points upon re-engagement, resulting in longer overall saccadic distances compared to normal operation. Conversely, due to their limited familiarity with offshore drilling operations, inexperienced operators tend to spend more time exploring equipment displays, thereby covering a greater total distance with their saccades.

Analysis of Fig. 5.27 reveals distinct differences in the average total saccade time across various operational conditions. Specifically,

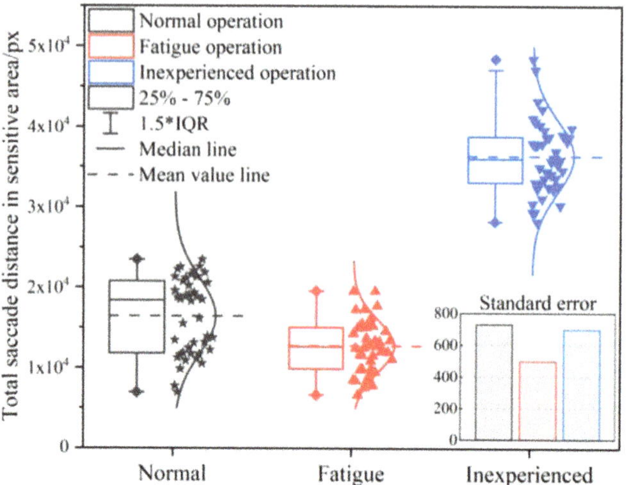

Fig. 5.26. Total saccade length in the sensitive area.

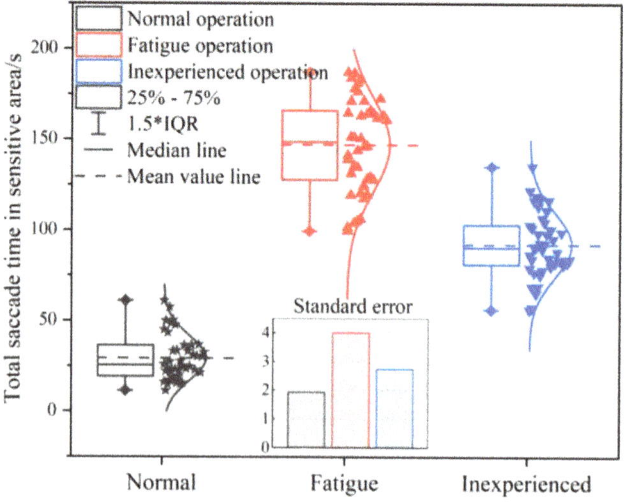

Fig. 5.27. Total saccade time in the sensitive area.

normal operation shows the shortest average total saccade time at 29.24 seconds, whereas fatigue operation exhibits the longest average time of 147.18 seconds. These findings can be attributed to factors such as excessive blinking or inadequate eye opening due to fatigue, which may have hindered precise tracking by the eye movement data

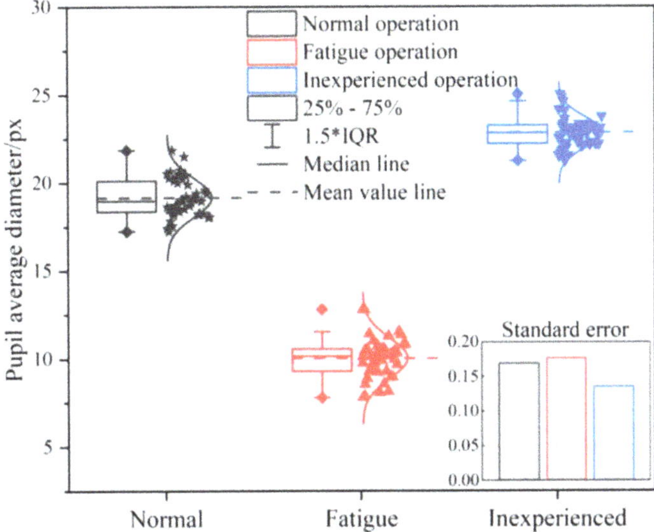

Fig. 5.28. Pupil average diameter.

sensor during rapid saccades. Consequently, this led to relatively higher values for total saccade time in both fatigued and inexperienced operators, highlighting challenges in accurate eye movement recording under such conditions.

Analysis of Fig. 5.28 reveals distinct differences in pupil average diameter across various operational conditions. For normal operation, the pupil diameter measures 19.16px, indicating a typical range consistent with expected fluctuations. In contrast, fatigue operation exhibits a significantly smaller diameter of 10.04px, likely due to operator fatigue leading to pupil constriction. Conversely, inexperienced operation displays the largest pupil diameter at 22.85px, reflecting heightened concentration levels as operators focus on specific operational aspects. These findings suggest that pupil diameter variations can serve as indicators of operator fatigue and engagement levels during offshore drilling operations.

Analysis of Fig. 5.29 reveals notable disparities in average experimental data acquisition time across different operational scenarios. For normal operation, the average acquisition time is relatively short at 174.38 seconds, indicating efficient task execution. In contrast, fatigue operation exhibits a longer acquisition time of 248.41

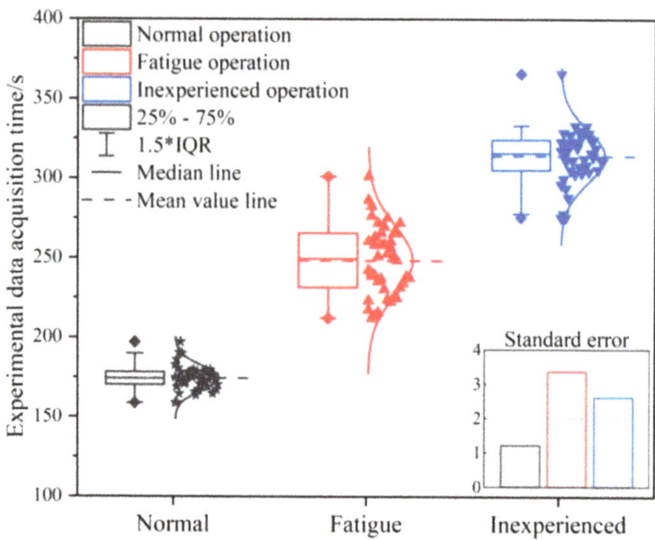

Fig. 5.29. Experimental data acquisition time.

seconds, attributed to the decelerated actions of fatigued operators, impairing their operational efficiency. The longest acquisition time is observed in inexperienced operations, totaling 313.98 seconds, indicative of the prolonged duration required by inexperienced operators to navigate through the operational process or familiarize themselves with simulation equipment. Fatigue-induced sluggishness in operators during fatigue operations can compromise their ability to make timely judgments and implement rescue measures in response to potential hazards. Similarly, inexperienced drillers' focus on locating buttons and handles rather than promptly observing wellhead operators' behaviors heightens the risk of misoperation and equipment damage, underscoring the importance of comprehensive training and familiarity with operational procedures.

During offshore drilling operations, drillers are required to manipulate various buttons, knobs, and handles according to the operation process. Simultaneously, they must ensure the safety of wellhead operators by closely monitoring the operation of wellhead equipment and personnel. The duration of fixation time can therefore indicate the level of attention given by drillers to specific areas during operations. To validate the effectiveness of the divided sensitive areas, a statistical analysis of drillers' fixation times in different areas is

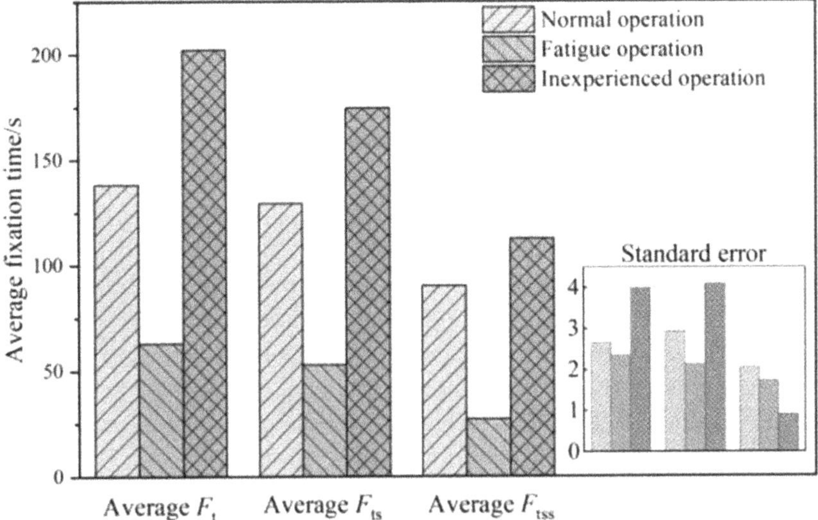

• Average F_{t} : Average total fixation time in the drilling operation scene

• Average F_{ts} : Average total fixation time in sensitive areas

• Average F_{tss} : Average total fixation time in the scene area

Fig. 5.30.   Comparison of average fixation time.

necessary. The comparison of average fixation times is presented in Fig. 5.30.

Based on the results of the sensitive area division, it can be inferred that the drilling operation scene comprises both sensitive and non-sensitive areas. These include the scene areas (the wellhead), the instrument area (1–4), and the operation area (1–4). Therefore, the total fixation time in the drilling operation scene (F_t) encompasses the total fixation time in the sensitive areas (F_{ts}), which in turn includes the total fixation time in the scene area (F_{tss}). As depicted in Fig. 5.30, the average total fixation time in the drilling operation scene (Average F_t), the average total fixation time in sensitive areas (Average F_{ts}), and the average total fixation time in the scene area (Average F_{tss}) are statistically analyzed across normal operation, fatigue operation, and inexperienced operation, respectively. Furthermore, examining the proportion of the average F_{tss} relative to the average F_t unveils crucial differences: normal operations exhibit a proportion of 65.2%, significantly higher than 55.7% for inexperienced operations and 43.1% for fatigue operations. This discrepancy suggests that fatigue and inexperienced operations are associated with a higher likelihood of accidents. The analysis results demonstrate significant variations in fixation times across different sensitive areas, indicating the effectiveness of the sensitive area division.

Based on the conducted data analysis, a distinct contrast emerges between the eye movement data linked to safety hazard behaviors during offshore drilling operations and those observed in normal drilling operations. This observation indicates the robust applicability of the selected eye movement parameters. Consequently, these eye movement data can serve as invaluable characteristic features for the TPELightGBM model intended for identifying potential safety hazard behaviors among offshore drilling operators.

5.3.3.3 *Safety hazard behavior identification of offshore drilling operators based on the TPE-LightGBM model*

The experiment involved collecting a total of 126 sets of data, comprising 42 sets each of normal, fatigue, and inexperienced eye movement data. These datasets were divided into 72% for training and 28% for testing purposes. Before optimizing the LightGBM model's hyperparameters, their range was established based on a comprehensive review of the literature and official documentation. Subsequently, the TPE algorithm, previously employed in a similar study, was utilized to optimize the hyperparameters of the LightGBM model. TPE, a Bayesian optimization algorithm rooted in a tree structure, is specifically designed to tackle global optimization challenges posed by black-box functions. Central to Bayesian optimization is the Gaussian process, which models the probability distribution of the objective function. This flexible approach is capable of effectively modeling functions of any dimension and is updated iteratively by estimating the probability distribution using known inputs and outputs. Consequently, the optimization process adheres to the principles of Gaussian distribution, also known as normal distribution. The optimization process for hyperparameters, based on the TPE optimization algorithm, is elaborated in detail in Section 5.3.2. The evaluation of the safety hazard behavior identification model for offshore drilling operators employed 10-fold cross-validation. The range of hyperparameter values and the optimization results achieved by TPE are summarized in Table 5.6.

A total of 36 groups of test samples were selected for this study, comprising 12 groups each of normal operation, fatigue operation, and inexperienced operation. Utilizing the optimal hyperparameter combination obtained from Table 5.6, the model underwent 45 iterations to identify safety hazards in offshore drilling operations. The behavior identification results based on the TPE-LightGBM model are depicted in Fig. 5.31. The x-axis represents the predicted category, the y-axis represents the actual category of the data, and the diagonal value represents the number of accurately predicted samples. A higher diagonal value indicates better model performance. Figure 5.31 illustrates that the model achieves optimal recognition effectiveness, with no occurrence of false positives or false negatives.

The evaluation of feature importance is a crucial metric for assessing the characteristics of the LightGBM model.

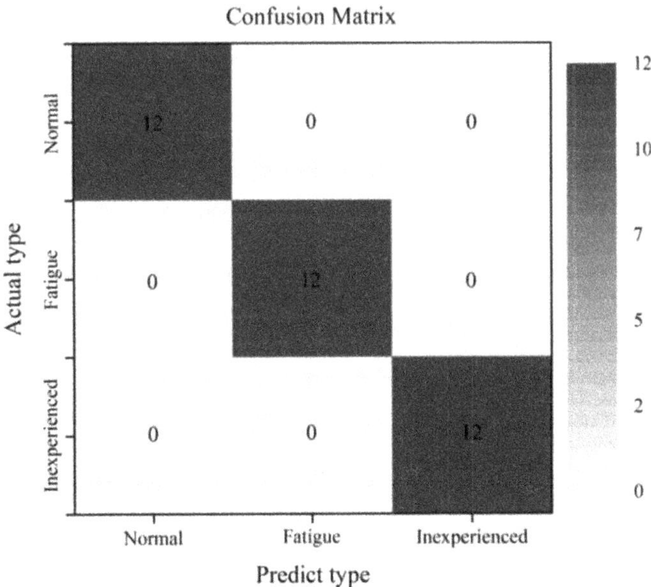

Fig. 5.31. Safety hazard behavior identification results of offshore drilling operation based on TPE-LightGBM model.

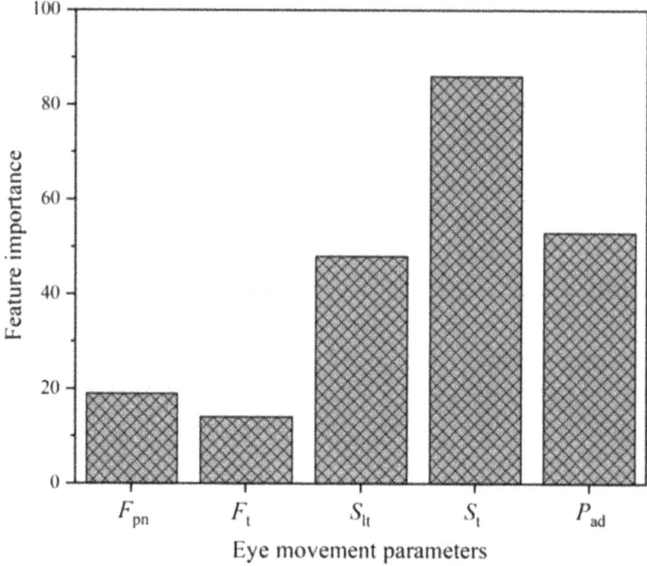

Fig. 5.32. Feature importance.

The model.feature_importance module in the LightGBM model is employed to calculate the importance of eye movement features of drilling operators. The calculation method of feature importance is based on the total number of times a feature is used in constructing a decision tree. A higher importance value indicates that the feature has a greater impact on model construction and optimization. The calculation results of feature importance are illustrated in Fig. 5.32.

The analysis based on feature importance enables the quantification of the impact of each eye movement feature on the identification of safety hazard behaviors of offshore drilling operators. As depicted in Figs. 5.24–5.28, substantial variations in the values of eye movement features are observed between normal operations and safety hazard behavior, indicating their critical role in accurately identifying safety hazard behaviors displayed by offshore drilling operators. Figure 5.32 further demonstrates that each eye movement feature holds relatively high significance. Among these features, the total saccade time in the sensitive area (S_t) holds the highest importance, followed by the pupil average diameter (P_{ad}), the total saccade length in the sensitive area (S_{lt}), and the total number of fixation points in the sensitive area (F_{pn}). The importance of total fixation time in the sensitive area (F_t) is comparatively lower, with a feature importance score of 14. The maximum ratio between the importance of each eye movement feature does not exceed 6.14, and there is no significant disparity in magnitude. This finding suggests that every eye movement feature plays a crucial role in identifying safety hazard behaviors among offshore drilling operators. The conclusion aligns with the findings presented in Figs. 5.24–5.28. Therefore, by quantifying the features' importance, the influence of the numerical changes of each eye movement feature on the identification of safety hazard behavior can be obtained.

5.3.3.4 *Model comparison*

The study conducted a comparison of three integrated learning algorithms — XGBoost, GBDT, and Random Forest — to further assess the efficacy of the proposed identification model for offshore drilling operators' safety hazard behavior. The results of this comparison are presented in Table 5.8.

As observed in Table 5.8, the current algorithm yielded the most optimal recognition results, evidenced by the lowest number of false alarms during the testing of samples. This algorithm not only showcases remarkably high identification performance but also displays exceptional suitability for real-time online data acquisition, facilitating efficient processing of big data. These characteristics position the algorithm for potential applications in large-scale offshore drilling operations.

Table 5.8. Comparative analysis of algorithms for identifying safety hazard behavior.

Model	Number of test samples	Number of false alarms
XGBoost	36	2
GBDT	36	3
Random forest	36	6
Proposed algorithm (TPE-LightGBM)	36	0

5.4 Behavioral Error Recognition of Process Operators Based on Eye Movement Hotspot Mapping

5.4.1 *Fundamental principle*

5.4.1.1 *Incidents of unsafe acts by operators of offshore production operations*

Offshore drilling operations are highly complex. In the offshore drilling operation, operators face a complex marine environment, extreme weather conditions, limited operating platform space, unstable seabed, and geological conditions. Operators must focus on drilling operations and emergency response under harsh environmental conditions. In the offshore operation, the frequency of offshore drilling platform accidents is relatively high, and the consequences are relatively serious. Offshore drilling accidents are usually caused by unsafe human operations, improper emergency disposal, or organizational failures. Therefore, strengthening the early warning of the unsafe behavior of drilling operators is the key to reducing offshore drilling accidents.

In the process of industrial production, the unsafe behavior of workers mainly includes three types: operator error, unskilled operation, and fatigue operation. Among them, human operation error refers to the behavior that has occurred and caused certain consequences, which belongs to the completed behavior. Unskilled operation and fatigue operation refer to behavior that has not yet occurred, but continuing the current state may lead to human operation errors and accidents, which is an attempted behavior. In the process of offshore drilling operations, there are many kinds of human operation errors. In order to realize detailed early warning of different types of human operation error, this chapter summarized human operation error into three types: action error, omitted action, and wrong order according to the characteristics of offshore drilling operations that need to be carried out in strict accordance with the operating process. Among them, the action error operation refers to the behavior of clicking or operating errors on buttons, knobs, or handles that need to be operated during offshore drilling operations. Omitted action operation refers to the behavior of missing operation of buttons, knobs, or handles that need to be operated in the process of offshore drilling. Wrong order operation refers to the behavior of

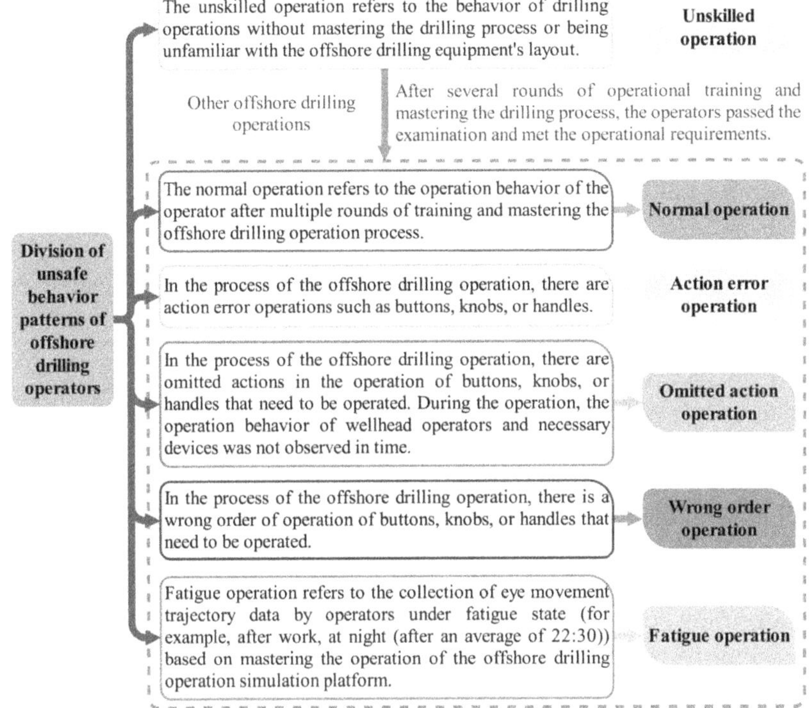

Fig. 5.33. The division method of unsafe behavior patterns of offshore drilling operators.

the wrong operation sequence of buttons, knobs, or handles that need to be operated during offshore drilling operations. In summary, this paper chooses to study six typical drilling operations (adding normal drilling operations) to explore real-time early-warning methods for unsafe behaviors of drilling operators.

There are six typical unsafe behaviors of drilling operators studied in this chapter: normal operation, unskilled operation, fatigue operation, action error operation, omitted action operation, and wrong order operation. The division method of unsafe behavior patterns of offshore drilling operators is shown in Fig. 5.33.

5.4.1.2 *Eye-tracking trajectory sequence method*

The eye-tracking trajectory is the physical scanning path for operators during their operation process. The trajectory can record the

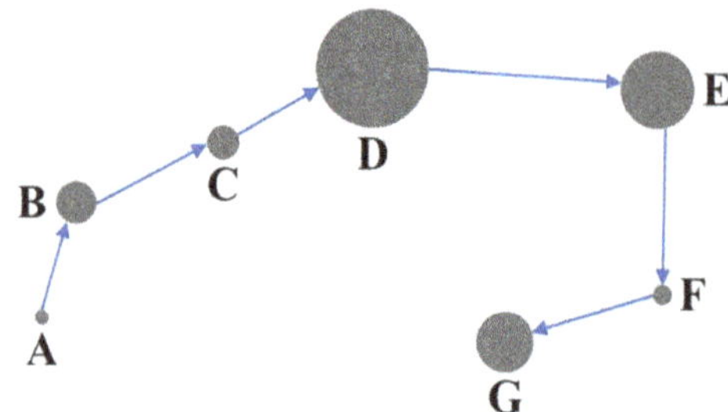

Fig. 5.34. Serialization of eye-tracking trajectory under ideal conditions.

order of the operator's concerns in carrying out a specific operation. In the operation process, sight has a significant guiding effect on the movement of operators. Therefore, the eye-tracking trajectory can accurately reflect the operational behavior of the current operator.

The serialization of the eye-tracking trajectory is to serialize and mark the sequence of the fixation point (eye-tracking trajectory node) position during the operation. The eye-tracking trajectory serialization adopted in this paper was to serialize trajectory nodes according to the distribution of the eye-tracking trajectory nodes in the working process of operators, form a string sequence without semantics, and study the early-warning method.

As shown in Fig. 5.34, the size of the trajectory node represents the length of the operator's fixation duration (the residence time of each trajectory node) at this node. The length of the blue line represents the length of the operator's eye-tracking transfer path. By encoding each eye-tracking trajectory node, the eye-tracking trajectory of the operator can be expressed as $A \rightarrow B \rightarrow C \rightarrow D \rightarrow E \rightarrow F \rightarrow G$. Therefore, the eye-tracking trajectory of the operator can be transformed into a sequence of strings composed of letters. Then the string can be processed to achieve the transformation of the eye-tracking trajectory problem.

The trajectory nodes of the eye-tracking trajectory under ideal conditions are independently distributed in the scene. However, the eye-tracking trajectory in the actual operation process is full of uncertainties (such as trajectory node aggregation). Among them, the aggregation of eye-tracking trajectory nodes means that the distribution of eye-tracking trajectory nodes is relatively concentrated in

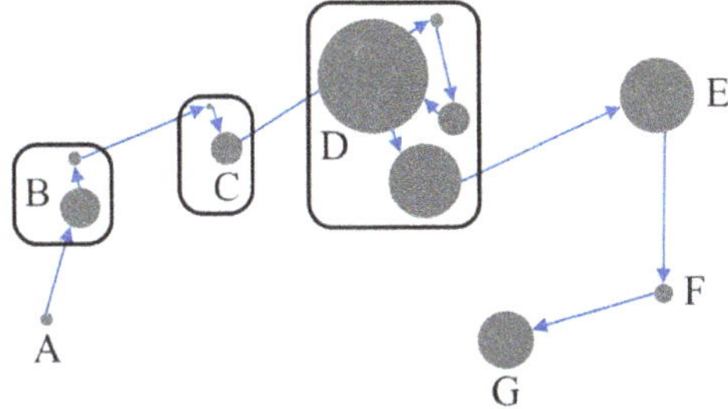

Fig. 5.35. Serialization of eye-tracking trajectory under actual conditions.

a certain area. For example, when subjects observe a button, due to personal factors (eye habits, sitting posture, operating status, etc.) and the influence of the external environment, their eye-tracking trajectory nodes are usually concentrated around the button rather than on a point. Therefore, it is necessary to solve the influence of trajectory node aggregation on operators' actual operation intention before serializing the operators' eye-tracking trajectory. The actual eye-tracking trajectory of the operators is shown in Fig. 5.35.

It can be seen from Fig. 5.35 that the actual eye-tracking trajectory is usually different from the ideal eye-tracking trajectory. As shown in Fig. 5.35, there are trajectory nodes in the B, C, and D regions, and the trajectory nodes of the actual eye-tracking trajectory are aggregation. The aggregation of trajectory nodes in the same period usually represents the existence of important operating equipment or units around the point. In the process of operation, due to external interference, equipment conditions, and personal physiological and psychological factors, the eye of the operator will jitter or drift. These factors lead to its trajectory nodes failing to focus on a single point but gathering around the equipment or unit that needs to be operated. The aggregation of trajectory nodes tends to cause the trajectory sequence to be too long and inconsistent with the actual operational intention.

In order to solve the adverse effects of trajectory aggregation on trajectory serialization, this paper proposes an improved eye-tracking trajectory serialization method (IETTSM). Compared with the collected eye-tracking trajectory node data sequence, the IETTSM proposed in this paper is mainly improved in the following two aspects.

(1) The trajectory sequence composed of (x, y) coordinate points is converted into a string sequence composed of A–Z letters.

(2) Set the trajectory node aggregation area TA:

$$\begin{cases} x_L \leq TA_{ix} \leq x_R \\ y_U \leq TA_{iy} \leq y_D \end{cases}, \tag{5.14}$$

where TA_i represents the ith trajectory node aggregation area, TA_{ix} represents the x-axis value range of the ith trajectory node aggregation area, TA_{iy} represents the y-axis value range of the ith trajectory node aggregation area, x_L and x_R represent the left and right limits of the x-axis value range of the ith trajectory node aggregation area, respectively. y_U and y_D represent the upper and lower limits of the y-axis value range of the ith trajectory node aggregation area, respectively. The trajectory sequence formed by serializing Fig. 5.35 with the above two improved methods is as follows: A → B (B trajectory node group) → C (C trajectory node group) → D (D trajectory node group) → E → F. The IETTSM trajectory serialization method can effectively reduce the impact of 41.67% of trajectory nodes on trajectory data applications.

5.4.1.3　*Damerau–Levenshtein distance*

The Levenshtein distance (LD), also known as edit distance, was proposed by scientist Vladimir Levenshtein. The LD algorithm uses the idea of dynamic programming to transform the solution of complex main objectives into the solution of multiple simple sub-objectives, thereby simplifying complex problems. The core of the algorithm is to construct the Levenshtein distance matrix (edit distance matrix), and calculate the value of each unit in the edit distance matrix by row until the value of the unit at the end of the last row is obtained, which is the result of the edit distance.

Each string consists of one or more characters, $S = s_1, s_2, \ldots, s_m$, $T = t_1, t_2, \ldots t_n$, where m and n are the lengths of the strings S and T, respectively. After determining the strings S and T, it is necessary to establish the edit distance matrix LDM, which is of order $(m + 1)(n + 1)$. Then the edit distance matrix can be expressed as

$$LDM_{(m+1)(n+1)} = d_{i,j}, \quad 0 \leq i \leq m, \quad 0 \leq j \leq n. \qquad (5.15)$$

The $d_{i,j}$ of each unit in the matrix represents the edit distance between the first i characters of the standard string S and the first j characters of the test string T.

In practical applications, the algorithm is usually used to calculate the degree of difference between two strings. The calculation method is how many edit operations need to be done on the string T to be converted into a string S. The string transformation methods used in this algorithm mainly include substitution, deletion, and insertion. It should be noted that each substitute, deletion, or insertion operation here can only perform a corresponding transformation on a character in the string. The value of the minimum number of string operations required to convert a string into another string is the value of the edit distance.

Damerau–Levenshtein distance (DLD) is an improvement of the LD algorithm. In addition to allowing, deletion, and insertion operations on strings, the algorithm also adds a transposition operation that allows two adjacent characters in a string. The algorithm is more in line with the practical application of gaze tracking. For example, when an operator is operating an adjacent device, its eye-tracking trajectory changes from A $\rightarrow$ B to B $\rightarrow$ A due to its own factors (sitting posture, physiological or psychological factors) and external factors (interference, environmental changes, or equipment factors). Using the transposition method of the DLD algorithm, the string B $\rightarrow$ A can be converted into A $\rightarrow$ B, which can effectively reduce the editing distance of the DL algorithm, so as to effectively avoid the calculation error caused by the eye backtracking.

In the DLD model, the substitution operation refers to replacing one character in a string with another. For example, the string S is MALL, and T is MAIL. The operation required to convert the string T to S is substituting I in the string T with L. Therefore, the value of the above string edit distance is 1.

The deletion operation refers to deleting a character in the string. For example, the string S is CAT, and T is CHAT. The operation required to convert the string T to S is the deletion of H in the string T. Therefore, the value of the above string edit distance is 1.

The insertion operation refers to inserting a character into a string to make it into another string. For example, the string S is READY, and T is READ. The operation required to convert the string T to S is to insert the Y character in the string T. Therefore, the value of the above string edit distance is 1.

The transposition operation refers to transposing adjacent characters in a string to become another string. For example, the string S is END, and T is EDN. The operation required to convert the string T to S is to replace the adjacent characters D and N in the string T. Therefore, the value of the above string edit distance is 1.

In the $DLDM_{(m+1)(n+1)}$ matrix, the first row represents the standard string S and the first column represents the test string T. $d_{i,j}$ represents the minimum number of edits required from string T to string S. The calculation method of $d_{i,j}$ is as follows:

$$d_{i,j} = \begin{cases} i, j = 0 \\ j, i = 0 \\ \min(d_{i-1,j-1}, d_{i-1,j}, d_{i,j-1}) + a_{i,j}, i, j > 0 \end{cases}, \qquad (5.16)$$

$$a_{i,j} = \begin{cases} 0 & s_i = t_j \\ 1 & s_i \neq t_j \end{cases} \quad (i = 1, 2, \ldots m; \; j = 1, 2, \ldots n). \qquad (5.17)$$

Therefore, the similarity between the standard string S and the test string T is

$$S_{\mathrm{DLD}} = 1 - \frac{d_{i,j}}{\max(m, n)}. \qquad (5.18)$$

Among them, m and n represent the length of the standard and the test strings, respectively. The larger the value of S_{DLD}, the higher the similarity of the two strings. Taking the two strings $S = $ "HCEF" and $T = $ "HCEG" as examples, Table 5.9 can be obtained by calculating according to Eqs. (5.16) and (5.17).

As can be seen from Table 5.9, $d_{4,4} = 1$. The conversion of string T to string S requires one modification operation. According to Eq. (5.5), $S_{\mathrm{DLD}} = 1 - 1/4 = 0.75$. Therefore, the similarity between strings T and S is 75%.

Table 5.9. Matrix table of strings S and T (example).

T \ S	Null	H	C	E	F
Null	0	1	2	3	4
H	1	0	1	2	3
C	2	1	0	1	2
E	3	2	1	0	1
F	4	3	2	1	1

5.4.2 *Early warning of production operators' unsafe behavioral accidents based on eye-tracking trajectories*

Experimental subjects, experimental equipment, offshore drilling simulation device, et al, are set up the same as in Section 5.3. Normal operation refers to the collection of eye-tracking trajectory data based on ensuring that the operator can master the operation of the SD-JKFZ-7 offshore drilling simulation device after several rounds of experimental operation.

Fatigue operation refers to the collection of eye movement trajectory data by operators under fatigue state (for example, after work, at night (after an average of 22:30)) on the basis of mastering the operation of the offshore drilling operation simulation platform. Fatigue Assessment Instrument (FAI) was used to determine the fatigue state of the operators.

Action error operation refers to the scenario where the operator executes step (7) resulting in a modification of the make-up process for the drill stem operation to that of the drill pipe operation while keeping the remaining drilling operation unaffected.

Omitted action operation refers to a situation where the operator, during step (3), closes the winch gear and the main drum before the working brake operation. This results in a drilling tool drop accident.

Wrong order operation refers to a situation where the operator carries out other operations in step (3) before the work braking operation. The working brake operation is subsequently performed after

Table 5.10. Eye-tracking trajectory data type for drilling operators.

Parametric classification	Parameter name	Parameter units	Parameters interpretation
Trajectory coordinate parameters	Trx	px	x-axis coordinates of each trajectory node
	Try	px	y-axis coordinates of each trajectory node
Trajectory time parameters	T_{trpb}	s	The start time of each trajectory node
	T_{trpd}	s	The residence time of each trajectory node
	T	s	The total time from the drilling experiment to completion (or accident)
Trajectory node number parameters	Trp	n	Total number of trajectory nodes

the remaining operations of step (3) are completed, resulting in a partial drop of the drilling tool.

Notably, each drilling operation was completed until the completion of the operation or the end of the experimental data collection when the drilling operation accident occurred. The types of eye-tracking trajectory data collected by drilling operators are shown in Table 5.10.

As shown in Table 5.10, there were 6 kinds of eye-tracking trajectory data collected in each experiment, including trajectory coordinate data (2 kinds), trajectory time data (3 kinds), and trajectory node number data (1 kind). In order to facilitate the understanding of eye-tracking trajectory data, this study replaced the nouns in eye-tracking technology. Among them, the meaning of the eye-tracking trajectory node and the fixation point is the same; the meaning of the eye-tracking trajectory and saccade path is the same; the meaning of the eye-tracking trajectory residence time and fixation time is the same. In total, 42 drilling operation trainers collected the experimental data used in the present study. Each drilling operation trainer collected a total of 252 groups of eye-tracking trajectory data of six drilling operations during the training process, including normal operation, unskilled operation, fatigue operation, action error operation, omitted action operation, and wrong order operation (42 groups of line of eye-tracking trajectory data were collected for each drilling operation behavior).

5.4.2.1   *Early warning of accidents involving unsafe behavior by operators of offshore production operations*

Step 1: Area of interest division of the offshore drilling operation scene based on the normal drilling eye-tracking trajectory and the standard drilling operation process.

Firstly, according to the eye-tracking trajectory of the normal drilling operation, the shape and size of the device unit, the complexity and importance of the operation area and the standard drilling operation process, and the areas of interest of the eye-tracking trajectory in the offshore drilling operation scene are divided.

Step 2: The serialization of the eye-tracking trajectory of drilling operators based on the IETTSM algorithm.

Using the proposed IETTSM algorithm, the x and y coordinate values range for each area of interest is set by combining the normal drilling operation trajectory node gathering area and the eye-tracking trajectory areas of interest, which are divided in step (1). Finally, the real-time acquisition of drilling operation eye-tracking trajectory data is serialized to become a non-semantic string sequence composed of A–Z letters.

Step 3: An evaluation model of the unsafe behavior of drilling operators is constructed based on eye-tracking trajectory.

Premised on the eye-tracking trajectory areas of interest in step (1) and the standard drilling operation process, a non-semantic standard drilling operation eye-tracking trajectory sequence composed of A–Z letters is established. Subsequently, based on the DLD algorithm, an evaluation model of the unsafe behavior of drilling operators is constructed based on the real-time acquisition of the eye-tracking trajectory sequence and the standard trajectory sequence.

Step 4: Abnormal behavior prompting of offshore drilling operators based on the trajectory node correspondence method.

Firstly, the eye-tracking trajectory sequence of drilling operators obtained in steps (1)–(2) is imported into the unsafe behavior evaluation model in real time. Then, the abnormal behavior of offshore drilling operators is prompted by means of the trajectory node correspondence method. When the node correspondence between the real-time eye-tracking track sequence and the standard track sequence does not match, the operator will be prompted about abnormal behavior and be urged to correct such behavior.

Step 5: Early warning of offshore drilling accidents caused by unsafe behavior based on the multi-trajectory node coupling method.

According to the eye-tracking trajectory of the normal drilling operation, normal drilling operators will also produce a small number of abnormal trajectory nodes. Hence, an infrequent occurrence of such deviations can merely suggest an abnormal status of the ongoing drilling operation and does not necessarily imply an unacceptable level of risk associated with a potential offshore drilling accident. To overcome such an issue, an early-warning method for offshore drilling accidents based on multi-trajectory node coupling was proposed in the present study. The trajectory node coupling similarity calculation method of real-time drilling trajectory sequence and standard trajectory sequence was adopted to predict accidents caused by drilling operators' unsafe behavior. When the coupling similarity of the input trajectory sequence is lower than a certain threshold, the accident warning is conducted for the drilling operators to reduce the occurrence of offshore drilling accidents.

The early-warning process of unsafe behavior accidents of offshore drilling operators based on the IETTSM-DLD model is shown in Fig. 5.36.

5.4.3 *Case study*

5.4.3.1 *Area of interest division of eye-tracking trajectory*

The eye-tracking trajectory exhibits the capability of continuously recording the position and duration of focus in an operator's gaze in real-time, which can reflect the whole process working state of the operator during the operation. In the eye-tracking trajectory diagram, each yellow line segment represents the gaze transfer process of the operator at two eye-tracking nodes. The nodes at both ends

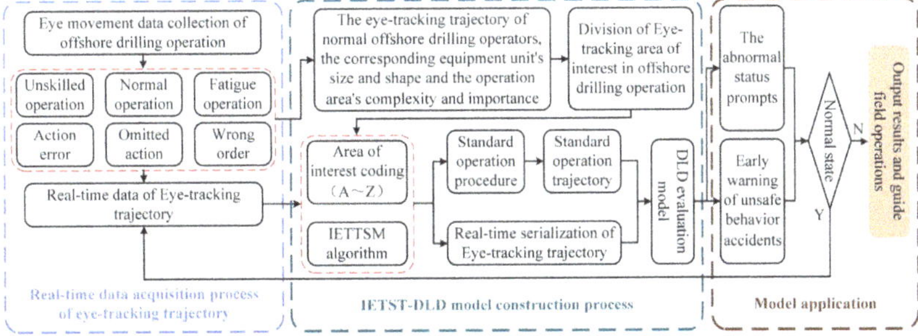

Fig. 5.36. The early-warning process of unsafe behavior accidents of offshore drilling operators based on IETTSM-DLD model.

Fig. 5.37. The superposition diagram of the eye-tracking trajectory of normal offshore drilling operators.

of the line segment are the areas that the operator focuses on. The superposition diagram of the eye-tracking trajectory of normal offshore drilling operators is shown in Fig. 5.37.

As shown in Fig. 5.37, the eye-tracking trajectory captured the operator's gaze points on various components such as buttons, knobs, handles, instruments, and wellhead operation scenes in the drilling operation scene. As presented in the standard offshore drilling workflow described in Section 5.4.2, operators are required to operate different devices during drilling operations. In order to ensure the safety of the wellhead operators, drillers need to pay attention to the dynamics of wellhead operators at all times. However, there are considerable differences in the eye habits, sitting posture, and physiological and psychological status of different drillers, which leads to significant differences in their eye-tracking trajectories. An observation can be made from Fig. 5.37 that although the distribution of trajectory nodes in the figure was relatively clear, there were still the following problems in the statistical analysis of the residence position of trajectory nodes:

(1) Despite being relatively scattered, the distribution of eye-tracking trajectory nodes was concentrated around the corresponding device to be operated. Such findings indicate that there were significant differences in the habits of eye use, sitting posture, and physical and psychological conditions among different operators.

(2) There were also trajectory nodes unrelated to the operation in the eye-tracking trajectory of normal drilling operations. The influence of the irrelevant trajectory nodes needed to be ignored when the areas of interest were divided. Since the irrelevant trajectory nodes were also part of the eye-tracking trajectory of drilling operations, such irrelevant trajectory nodes and their stay time needed to be recorded. Because the number of irrelevant trajectory nodes represents the concentration degree of the operator, if an operator does not concentrate, drilling accidents will easily occur.

In order to solve the difficulty of recording and analyzing the trajectory data of eye-tracking in drilling operations, the areas of interest of the offshore drilling operation scene needed to be divided. Due to a large number of important equipment units, irregular shapes, and large differences in placement position in drilling operation scenes, the grid adaptability is not good, and either the data recording is not accurate or the calculation speed is slow. In the present study, a rectangle was selected as the basic shape to divide the areas of interest of the important device units in the drilling operation scene. Combined with the range of normal drilling operation eye-tracking trajectory nodes aggregation, the corresponding equipment unit's size and shape, and the operation area's complexity and importance, the range of the areas of interest of the drilling

Fig. 5.38.   Area of interest division of eye-tracking trajectory nodes in offshore drilling operation.

operation trajectory nodes was determined. Figure 5.38 shows the area of interest division method of the offshore drilling operation eye-tracking trajectory node adopted in the present study.

The SD-JKFZ-7 offshore drilling simulator device is divided into 9 areas from A to I according to the distribution of eye-tracking track nodes in normal drilling operations, as shown in Fig. 5.38: A is the wellhead operation scene of the offshore platform; B is the display instrument of the hook height and drilling parameters; C is the winch gear selection handle, the main-drum speed selection handle, the spinner assembly handle, and the pneumatic spider handle; D is the make-up drill stem button; E is the winch direction selection knob; F is the winch speed regulation knob; G is the throttle speed regulation knob; H is the work brake handle and the main clutch hanging handle; and I is the drilling operation scene except for the other areas of A-H (irrelevant area). According to the statistics of the experimental data, 93.96% of the eye-tracking trajectory nodes are distributed in the divided areas of interest A–H.

5.4.3.2   *Serialization of eye-tracking trajectory of offshore drilling operators based on IETTSM algorithm*

According to the division method of areas of interest of eye-tracking trajectory nodes proposed in Section 5.1, combined with the proposed

IETTSM algorithm, the data cleaning and trajectory serialization of eye-tracking trajectory data of offshore drilling operators collected in real time were conducted.

As shown in Table 5.10, the eye-tracking trajectory parameters included trajectory coordinate parameters, trajectory time parameters, and trajectory node number parameters. According to the area of interest of the eye-tracking trajectory of the drilling operator, the coordinate values of the trajectory nodes collected in real time were read to determine the subordinate areas of interest of each trajectory node. Subsequently, with the completion of the set drilling process (or drilling accidents occurred), an eye-tracking trajectory sequence composed of the letters A–I was formed. Superposition diagrams of the eye-tracking trajectories of offshore different drilling operation behaviors are shown in Fig. 5.39. The serialization results of eye-tracking trajectories of different offshore drilling operations are shown in Fig. 5.40.

An observation can be made from Figs. 5.39 and 5.40 that the serialization of the eye-tracking trajectories of different offshore drilling operators can not only unify the trajectories but also preserve the details of the track node transfer during drilling operations. As presented in Fig. 5.40(a), there were irrelevant trajectory nodes and several error nodes in normal drilling operations, which is consistent with the actual drilling eye-racking trajectory (Fig. 5.39(a)). In Fig. 5.40, there were more irrelevant trajectory nodes and error

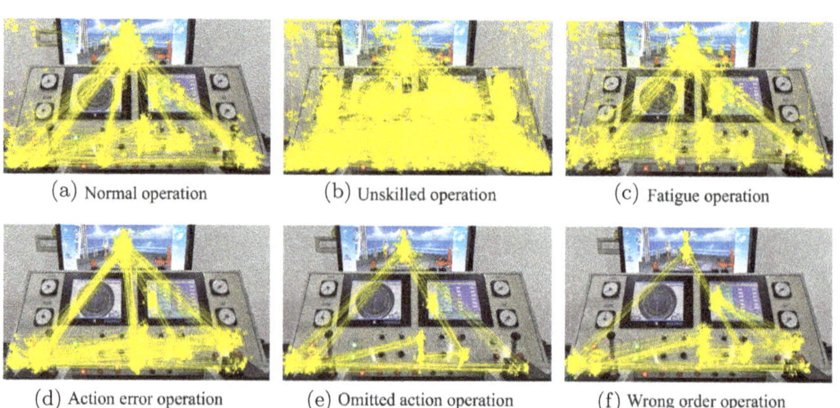

Fig. 5.39. Superposition diagrams of the eye-tracking trajectories of different offshore drilling operations.

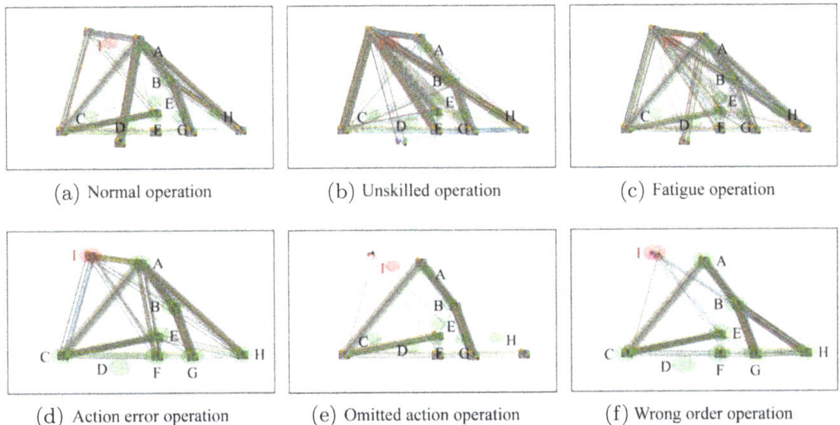

(a) Normal operation

(b) Unskilled operation

(c) Fatigue operation

(d) Action error operation

(e) Omitted action operation

(f) Wrong order operation

Fig. 5.40. The serialization results of the eye-tracking trajectory of different offshore drilling operations.

nodes in unskilled and fatigued operations than in normal operations. Such findings could be attributed to the unskilled operators being unfamiliar with the drilling process or the layout of the offshore drilling simulation device, resulting in a more scattered distribution of trajectory nodes during the operation. Because the fatigued operators were in a state of physical fatigue, the residence time of their trajectory node could not be prolonged. As a result, sight deviation occurred when operators operated the relevant operating device again, resulting in a large number of irrelevant trajectory nodes and error nodes. The number of irrelevant trajectory nodes with action errors was not much different from that of normal operations, while the number of irrelevant trajectory nodes with omitted action and wrong order operation was relatively small. Such findings indicate that the number of irrelevant nodes was positively correlated with the length of the operation under the same drilling status and proficiency. Additionally, there were no trajectory nodes in the area of interest D for action error, omitted action, and wrong order operation, which could be attributed to the setting of unsafe behaviors. Therefore, the IETTSM algorithm can realize the serialization of the eye-tracking trajectory of drilling operators, and effectively preserve the unsafe behavior state of offshore drilling operators in the process of operation. The changes in the number of trajectory nodes after eye-tracking trajectory serialization are shown in Fig. 5.41.

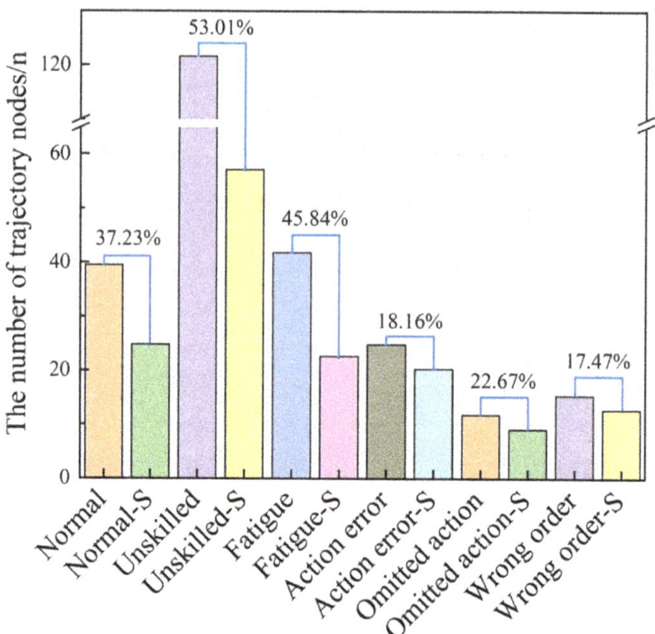

Fig. 5.41. The changes in the number of trajectory nodes after eye-tracking trajectory serialization.

Figure 5.41 shows that the number of trajectory nodes decreased after the serialization of eye-tracking trajectory for different types of offshore drilling operators. The number of eye-tracking trajectory nodes for unskilled and fatigue operations decreased relatively significantly, with decreases of 53.01% and 45.84%, respectively. The trajectory nodes with action error, omitted action, and wrong order operation have a relatively small decline. Such findings could firstly be related to the relatively short time of eye movement data collection (setting of wrong behavior in the experiment). Secondly, such behavior was based on the skilled operation of trajectory data collection, resulting in relatively few node errors.

From the number of nodes after serialization of eye-tracking trajectory nodes, the average number of nodes for unskilled jobs reached 57.14, which could be ascribed to the fact that unskilled operations needed to browse the drilling simulation device many times to determine the position of the relevant equipment. The average number of trajectory nodes in normal operation and fatigue operation were

24.81 and 22.64, respectively. Such results are consistent with the requirement of experimental data acquisition and operation on the basis of skilled operation. The average values of trajectory nodes of action errors, omitted actions, and wrong orders were 20.29, 9.1, and 12.71, respectively, which could be related to the wrong behavior steps designed in the experiment. By referring to the experimental steps in Section 1, an observation can be made that the average value of the trajectory nodes was consistent with the designed experimental process.

5.4.3.3 *Abnormal status prompting for offshore drilling operators based on the trajectory node correspondence method*

From the analysis results of Section 5.2, an observation can be made that there were irrelevant or abnormal nodes in the eye-tracking trajectories of different offshore drilling operators. Although a small number of irrelevant trajectory nodes would not have a significant impact on the normal completion of drilling operations, a large number of irrelevant nodes, especially abnormal nodes, would have a significant impact on drilling operations and even lead to offshore drilling accidents.

Based on the IETTSM-DLD model, the corresponding method of trajectory nodes was used to prompt drilling operators about abnormal operation status. According to the standard drilling operation process in Section 5.1 and the eye-tracking trajectory area of interest division method in Section 5.1, the standard trajectory node sequence of normal drilling tripping operation could be obtained as follows: $H \rightarrow C \rightarrow E \rightarrow F \rightarrow G \rightarrow B \rightarrow A \rightarrow H \rightarrow C \rightarrow E \rightarrow F \rightarrow G \rightarrow C \rightarrow A \rightarrow C \rightarrow A \rightarrow D \rightarrow A \rightarrow C \rightarrow A \rightarrow C \rightarrow A \rightarrow H \rightarrow A \rightarrow B \rightarrow H$ (26 steps in total). The eye-tracking trajectory node serialization model of offshore drilling operators shown in Section 5.2 was adopted to serialize the eye-tracking trajectory of drilling operators collected in real time. The DLD similarity between the serialized results and the standard trajectory node sequence was subsequently calculated, and the drilling operation status corresponding to the trajectory node could be obtained. Since the abnormal state prompting based on the trajectory node corresponding method was adopted, only the first 26 characters after serialization were calculated for DLD similarity (when the trajectory sequence was greater than 26, only the similarity of the first 26 characters was calculated; when the trajectory sequence was less than 26, only the string of corresponding characters was calculated for similarity, and the subsequent calculation would not be conducted). The abnormal status prompting of different offshore drilling operations based on the trajectory node correspondence method is shown in Fig. 5.42. The total number of trajectory nodes and the average number of abnormal prompts of different offshore drilling operations are shown in Fig. 5.43.

As can be seen from Figs. 5.42 and 5.43, the numbers of abnormal prompts of unskilled operation, fatigue operation, and action error operation were relatively large, and the numbers of abnormal prompts of normal operation, omitted action, and wrong order were relatively small. From the ratio of the total number of abnormal prompts/total number of eye-tracking trajectory nodes, the following pattern was observed: omitted action < normal operation < action error < wrong order < fatigue operation < unskilled operation. From the calculation results, an observation can be made that the proportion of abnormal prompts of normal operations was relatively low, and the main reason for the lowest ratio of omitted actions was that the drilling operation time was shorter, and the attention of the operators in the early stage of the operation was the highest. The average number of omitted action trajectory nodes was 9.1, while the total number of trajectory nodes in normal operation was 24.33, which is consistent with the situation where there was no abnormal prompt in the early stage of normal operation (before the 9th sequence).

Although certain abnormal behaviors of drilling operations are less likely to cause offshore drilling accidents, abnormal behaviors are an accurate reflection of the current operating status of operators and a significant cause of offshore drilling accidents. As such, using the trajectory node corresponding method to prompt about the abnormal operation status of drilling operators can effectively avoid the continuation of the abnormal status of operators, thereby preventing the occurrence of severe accidents. Because the abnormal state of operation does not necessarily lead to accidents, an early-warning method for accidents caused by offshore drilling operators' unsafe behavior is needed for prevention purposes.

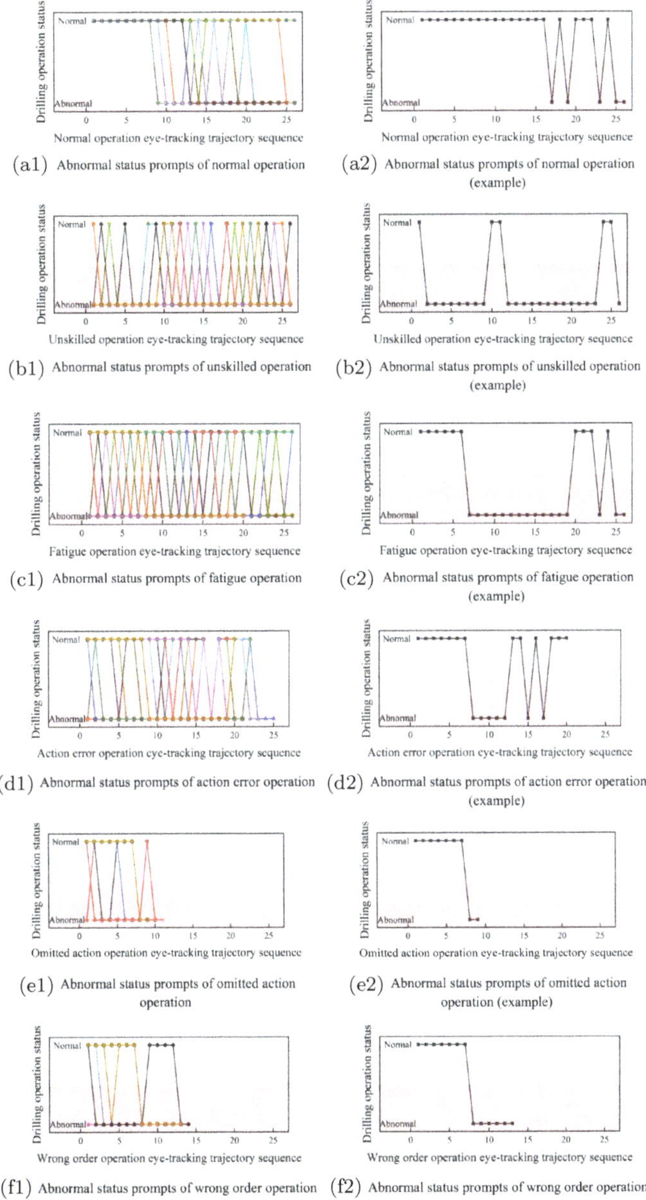

Fig. 5.42. The abnormal status prompting of different offshore drilling operations based on the trajectory node correspondence method.

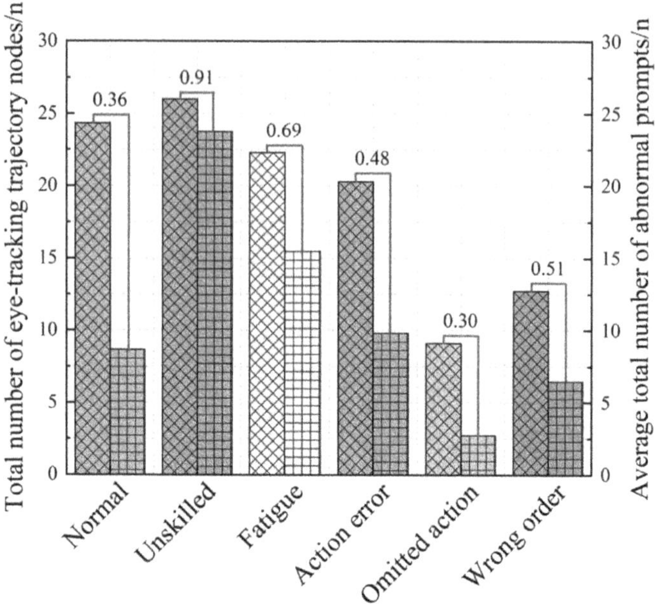

Fig. 5.43. The total number of trajectory nodes and the average number of abnormal prompts of different offshore drilling operations.

5.4.3.4 *Early warning of accidents caused by offshore drilling operators' unsafe behavior based on the trajectory node coupling method*

In view of the problem that the abnormal state can only be prompted about in the process of operation, but early warning of offshore drilling accidents cannot be provided, an early-warning method for accidents caused by offshore drilling operators' unsafe behavior based on trajectory node coupling was proposed. DLD similarity calculation was conducted using the method of node coupling between the serialized results of the eye-tracking trajectory of the drilling operators collected in real time and the standard trajectory sequence. Ultimately, coupling similarity results of the unsafe behavior of the operators were obtained. If the coupling similarity of unsafe behavior among drilling operators falls below a specified threshold, timely action should be taken to either halt the risky operation of the operators or address potential accidents, so as to prevent significant losses resulting from unsafe human behavior. The average trajectory node

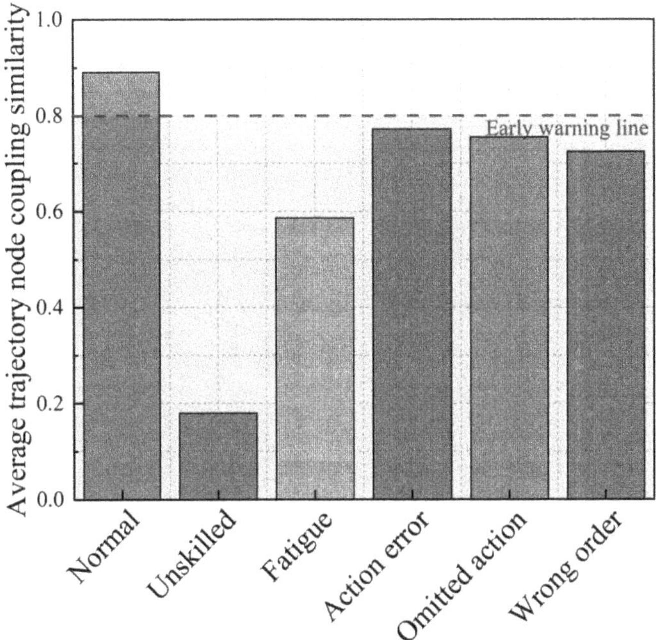

Fig. 5.44. The average trajectory node coupling similarity of offshore different drilling operations.

coupling similarity of different offshore drilling operations is shown in Fig. 5.44.

By calculating the coupling similarity of the collected 35 trajectories of the normal drilling operator's eye-tracking trajectory sequence (removing 7 trajectories that were significantly different from the eye-tracking trajectory of the normal drilling operation), the average trajectory node coupling similarity of the normal drilling operation was found to be 0.89, and the coupling similarity of each normal drilling operation trajectory node was greater than 0.8. Therefore, the early-warning threshold of accidents caused by drilling operators' unsafe behavior was set to 0.8.

According to the calculation method of the coupling similarity of the eye-tracking trajectory nodes, if there is an error when the trajectory node is 1, it is easy to lead to the generation of early-warning information. Obviously, the early-warning information at this time cannot correctly reflect the actual operating status of the operator. Therefore, after determining the warning threshold, it is

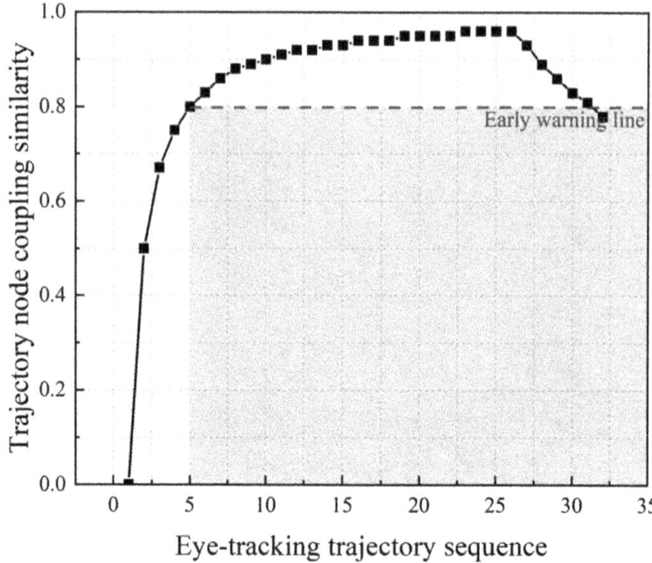

Fig. 5.45. Trajectory node coupling similarity in the case of nodes 1 and 26 error.

necessary to determine the warning eye-tracking trajectory sequence interval. When errors occur at nodes 1 and 26, the coupling similarity of trajectory nodes is shown in Fig. 5.45.

As shown in Fig. 5.45, if the drilling operator did not place the sight in the correct position at trajectory node 1, the coupling similarity could reach over 0.8 as long as the following four sequences were correct. As shown in Section 5.3, the standard eye-tracking trajectory sequence has 26 nodes. Therefore, the warning area was set as nodes 5–26. When the standard drilling process of an offshore drilling operation exceeds 26 steps, this model can still realize the early warning of unsafe behavior accidents. In order to ensure the safety of the first five steps of offshore drilling operations, an abnormal status prompts method based on the eye-tracking trajectory was proposed. As a supplement to the early-warning method of unsafe behavior, this method can meet the needs of prompting and early warning of unsafe behavior of different types of drilling operators. The early warning of accidents caused by offshore drilling operators' unsafe behavior based on the trajectory node coupling method is shown in Fig. 5.46.

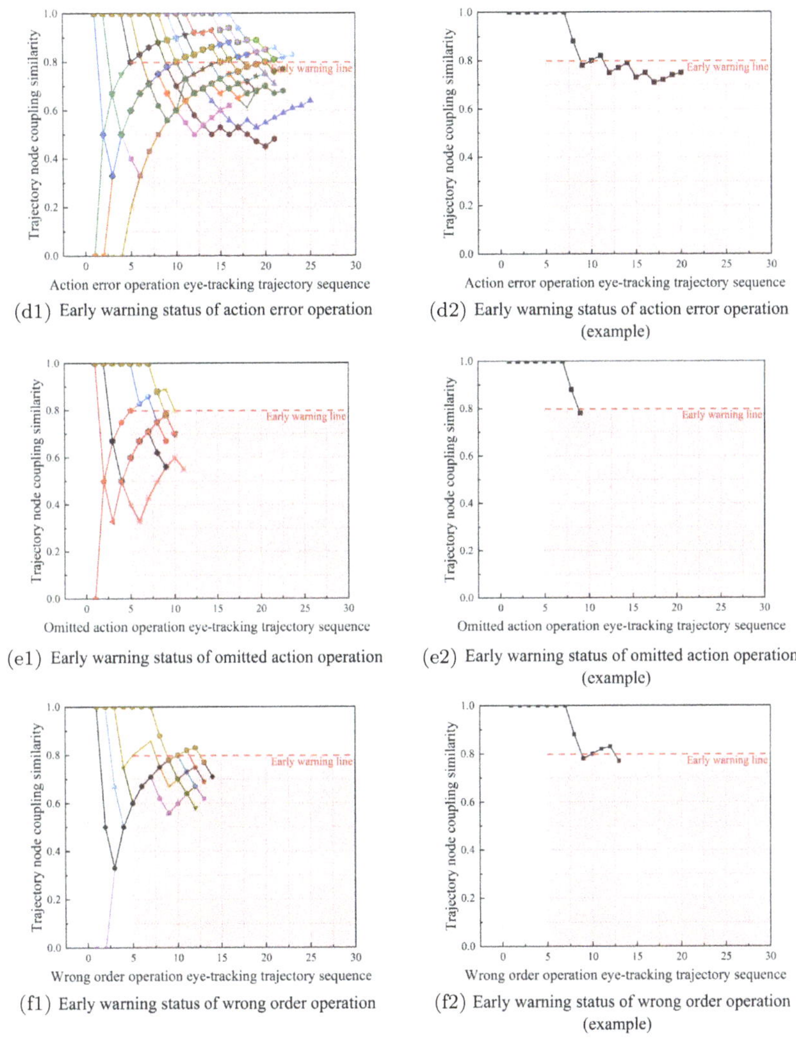

Fig. 5.46. The early warning of accidents caused by offshore drilling operators' unsafe behavior is based on the trajectory node coupling method.

As can be seen from Fig. 5.46, the early-warning area could be used to realize the early warning of accidents caused by offshore drilling operators' unsafe behavior. Taking Fig. 5.46 (a2) as an example, although the coupling similarity of normal drilling operations decreased to a certain extent during the operation, with the timely

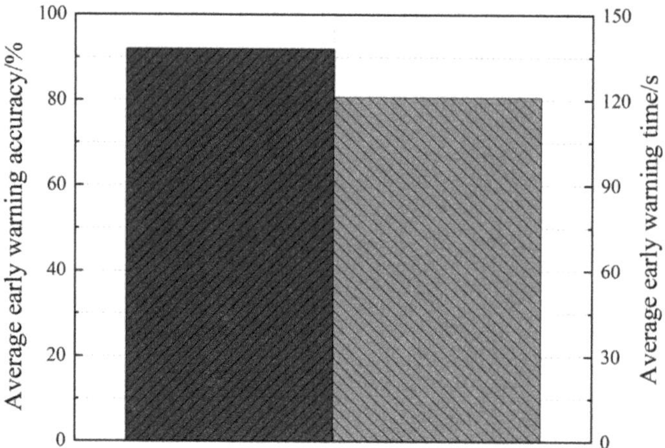

Fig. 5.47. The average early-warning accuracy and average early-warning time of unsafe behavior accidents of offshore drilling operators.

correction of operators, the coupling similarity continued to rise and did not reach the warning threshold. Taking Fig. 5.46 (c2) as an example, when the eye-tracking trajectory node was 13, a timely warning was provided about the operation behavior, and the operator needed to stop the operation behavior in time and conduct emergency treatment. Taking Fig. 5.46 (d2) as an example, an early warning occurred when the trajectory node was 9. Due to timely adjustment by operators, the coupling similarity increased to the safe area above the early-warning line. When the trajectory node was 12, the warning occurred again. The occurrence of the first warning signal was utilized as the standard for the operator to cease their operational behavior, and the corresponding warning node and time were recorded. The average early-warning accuracy and average early warning time of unsafe behavior of offshore drilling operators are shown in Figs. 5.47–5.49.

In the present study, a total of 210 unsafe behavior eye-tracking trajectories of offshore drilling operators were selected, including 42 unskilled operations, 42 fatigue operations, 42 action errors, 42 omitted actions, and 42 wrong order operations. As can be seen from Fig. 5.40, the IETTSM-DLD model was adopted for the early warning of accidents caused by offshore drilling operators' unsafe

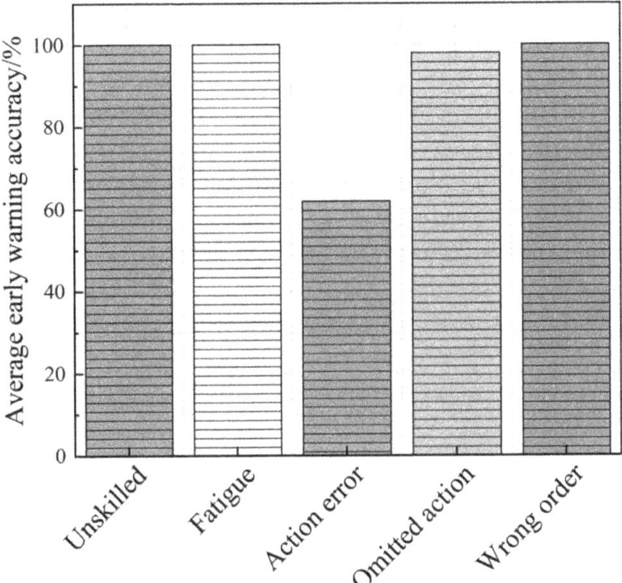

Fig. 5.48. The average early-warning accuracy of offshore drilling operators under different operating statuses.

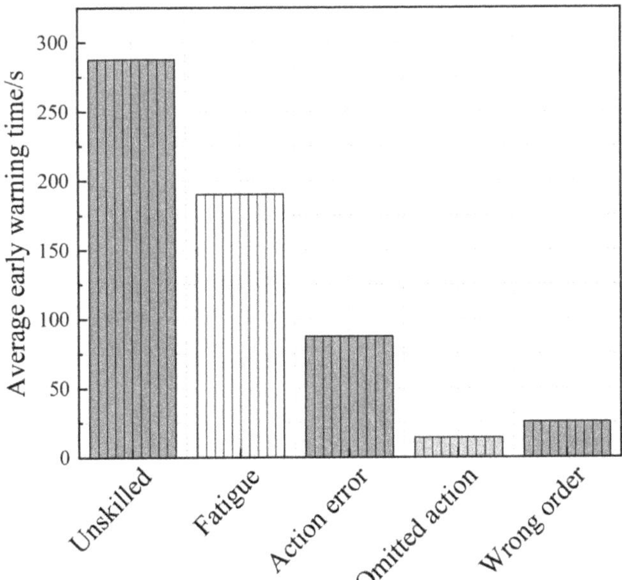

Fig. 5.49. The average early-warning time of offshore drilling operators under different operating statuses.

behavior. The average early-warning accuracy was 91.9% and the average early-warning time was 120.97 s.

As shown in Fig. 5.41, through the analysis of the average early-warning accuracy of accidents caused by offshore drilling operators' unsafe behavior, the average early-warning accuracy of unskilled operation, fatigue operation, and wrong order operation could reach 100%. The average early-warning accuracy of omitted action was 97.62%, while the early-warning accuracy of error behavior was only 61.9% (26 times).

As shown in Fig. 5.42, through the analysis of the average early-warning time of different offshore drilling operators' unsafe behavior, the average early-warning times of unskilled operation, fatigue operation, and action error were relatively long, being 287.62 s, 190.11 s, and 87.42 s, respectively. The early-warning times of omitted action and wrong order were relatively short, being 14.12 s and 25.57 s, respectively.

From the aforementioned calculation results, the IETTSM-DLD model can realize the early warning of accidents caused by offshore drilling operators' unsafe behavior, and the early-warning time can meet the needs of drilling operators for emergency stop operations or emergency treatment.

References

[1] Hu J, Hu J, Zhang X. Evaluation of training effect of 3D simulation fire fighting and rescue based on line of sight tracking. *Safety & Security*, 2019, 40(7): 58–62. (In Chinese)

[2] Hu J, Zhang L, Hu J. Study on human error recognition of process operators based on eye tracking technology. *Journal of Safety Science and Technology*, 2019, 15(5): 142–147. (In Chinese)

[3] Hu J. Research on the human error recognition of oil and gas production process operators based on eye tracking. PhD Thesis, *China University of Petroleum*, Beijing, 2020. (In Chinese)

[4] Fei X, Zhu H, Cai L, *et al.* Research on the advancement of reducing human errors standard of American Petroleum Industry. *Total Corrosion Control*, 2015, 29(12): 17–20. (In Chinese)

[5] Selected Accident Cases of PetroChina 2003–2005 Editorial Committee. *Selected Accident Cases of PetroChina in 2003–2005*. Petroleum Industry Press, Beijing, 2006. (In Chinese)

[6]   Hermens F, Flin R, Ahmed I. Eye movements in surgery: A literature review. *Journal of Eye Movement Research*, 2013, 6(4): 1–11.

[7]   Raney GE, Campbell SJ, Bovee JC. Using eye movements to evaluate the cognitive processes involved in text comprehension. *Journal of Visualized Experiments*, 2014, 83: e50780.

[8]   Recarte MA, Nunes LM. Effects of verbal and spatial-imagery tasks on eye fixations while driving. *Journal of Experimental Psychology Applied*, 2000, 6(1): 31–43.

[9]   Stasi LD, Contreras D, Cándido A. Behavioral and eye-movement measures to track improvements in driving skills of vulnerable road users: First-time motorcycle riders. *Transportation Research*, 2011, 14(1): 26–35.

[10]   Pradhan AK, Hammel KR, Deramus R. Using eye movements to evaluate effects of driver age on risk perception in a driving simulator. *Human Factors*, 2005, 47(4): 840–852.

[11]   Kilingaru K, Tweedale JW, Thatcher S, *et al.* Monitoring pilot situation awareness. *Journal of Intelligent & Fuzzy Systems*, 2013, 4(3): 457–466.

[12]   Fletcher L, Zelinsky A. Driver inattention detection based on eye gaze-road event correlation. *The International Journal of Robotics Research*, 2009, 28(6): 774–801.

[13]   Kodappully M, Srinivasan B, Srinivasan R. Towards predicting human error: Eye gaze analysis for identification of cognitive steps performed by control room operators. *Journal of Loss Prevention in the Process Industries*, 2015, 42: 35–46.

[14]   Wang L. Eye tracking technology application in the human-computer interaction. PhD Thesis, *Northwestern Polytechnical University*, 2016. (In Chinese)

[15]   Chen C, Hu J, Zhang L, *et al.* Identification method for safety hazard behavior in offshore drilling operators. *Ocean Engineering*, 2024, 301(1): 117447.

[16]   Chen C, Hu J, Zhang L, *et al.* Early warning method of unsafe behavior accidents for offshore drilling operators based on eye-tracking trajectory. *Process Safety and Environmental Protection*, 2023, 177: 1506–1522.

Chapter 6

Typical Examples of Safety Alerts for Shale Gas Fracturing Systems

6.1 Real-Time Diagnosis and Early Warning of Downhole Accidents during Fracturing Based on Qualitative Trend Analysis

6.1.1 *Basic theories*

6.1.1.1 *Qualitative trend analysis (QTA)*

A qualitative trend is the most natural representation of a feature and has been utilized for fault diagnosis in chemical processes. To extract the qualitative trend from measured process variables, a qualitative trend analysis is proposed. There are two main steps involved in such an analysis: (a) determining the primitives to represent the qualitative trend and (b) extracting and identifying the qualitative trend. A brief discussion follows.

(1) Determining primitives to represent the qualitative trend

Primitives are fundamental elements to describe the qualitative trend of process variables (QTPV). Seven primitives, i.e., $A(0,0)$, $B(+,+)$, $C(+,0)$, $D(+,-)$, $E(-,+)$, $F(-,0)$, and $G(-,-)$, where the signs are of the first and second derivatives are shown in Fig. 6.1. For practical application, several of them can be selected to represent the trend feature according to the fitting function as listed in Table 6.1. To reduce computation complexity, first-order polynomial fitting is adopted throughout this chapter.

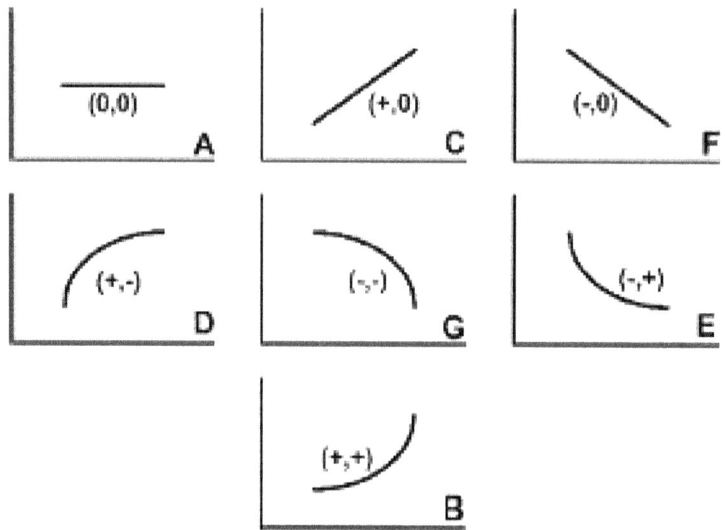

Fig. 6.1. Fundamental language: primitives.

Table 6.1. Rules of selecting primitives according to the fitting function.

Fitting function	Primitives
First-order polynomial	A$(0,0)$, C$(+,0)$, and F$(-,0)$
Second-order polynomial	A$(0,0)$, B$(+,+)$, C$(+,0)$, D$(+,-)$, E$(-,+)$, F$(-,0)$, G$(-,-)$

(2) Extracting and identifying the qualitative trend

Before extracting the QTPV within a sliding window with a fixed length β, the sliding window is discretized into a series of segments of equal length (χ), as shown in Fig. 6.2, and then the first polynomial in each segment can be obtained as Eq. (6.1):

$$y_\Delta(t) = pt + y_0 \quad t \in [t_s, t_e], \tag{6.1}$$

where p is the first derivative, y_0 is the intercept, and t_s and t_e are the start and end times of each segment, respectively.

Once the first polynomial is estimated, a primitive is assigned based on the sign of the first derivative, as listed in Table 6.2, where p_0 is a non-negative threshold value that allows the first derivative to fluctuate around zero due to the existence of noise.

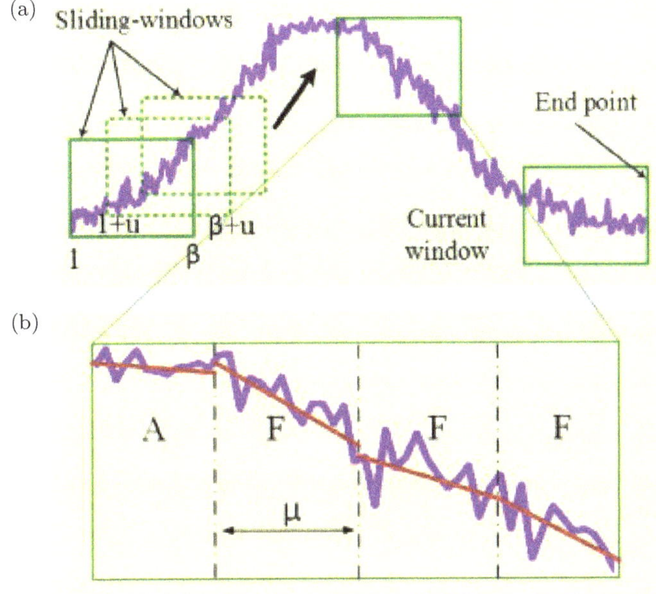

Fig. 6.2. Illustration of qualitative trend extraction using a sliding window: (a) the manner of sliding window; (b) linear fitting of each segment in one window.

Table 6.2. Rules of identifying primitives.

Primitive	The sign of the first derivative	First derivative
A	0	$-p_0 \leq p \leq p_0$
C	+	$p_0 < p$
F	−	$p < -p_0$

6.1.1.2 *Support vector machine (SVM)*

The basic SVM deals with two-class problems, but it can be developed for multi-class classification. In practice, multi-class classification is transformed into several two-class problems. Thus, strategies like "one-versus-all (OVA)" or "one-versus-one (OVO)" are put forward to solve multi-class problems. In order to achieve multi-incident diagnosis, the OVO strategy is applied throughout this section due to its superior convergence speed and accuracy.

(1) Binary support vector machine (BSVM)

The BSVM transforms the original feature space into a higher-dimensional space to decide an optimum separating hyperplane that maximizes the margin between the two nearest datapoints belonging to two separate classes. Given an input training dataset, (x_i, y_i), $i = 1, 2, \ldots, s, x \in R^d, y \in \{-1, +1\}$ can be separated by the hyperplane; then, a hyperplane can be formulated as $\omega^T \phi(x) + b = 0$ where ϕ is the transformation function, and ω and b are the weight vector and the bias, respectively. This hyperplane is determined to maximize the distance D between the nearest classes using the optimization in Eq. (6.2):

$$\max_{\omega \in R^n, b \in R} D \quad \text{subject to } y_i(\omega^T \phi(x) + b) \geq D, \ \forall i. \tag{6.2}$$

Considering that input data are accompanied by a high noise level, one can obtain Eq. (6.3) as follows:

$$\min_{\omega \in R^n, b \in R} \left\{ \frac{1}{2}\omega^2 + C \sum_{i=1}^{s} \xi_i \right\}, \tag{6.3}$$

subject to $\xi_i \geq 0, \ y_i(\omega^T \phi(x) + b) \geq 1 - \xi_i, \ \xi_i \geq 0, \ \forall i,$

where ξ_i are non-negative slack variables and C is the penalty parameter that controls the trade-off between the complexity of the decision function and the number of misclassified training examples.

Then, the optimal decision function is given by Eq. (6.4):

$$f(x) = \text{sgn}\left(\sum_{i,j=1}^{s} y_i\alpha_i(\phi^T(x_i)\phi(x_j)) + b \right), \tag{6.4}$$

where α_i are Lagrange multipliers. The hyperplane function can be decided by kernel function $K(x_i, x_j) = \phi^T(x_i)\phi(x_j)$ by calculating the inner products without specifying the explicit form of the transformation function. Several types of kernel functions can be formulated, such as linear, polynomial, and the radial basis function (RBF). After a kernel function is selected, the decision function becomes Eq. (6.5):

$$f(x) = \text{sgn}\left(\sum_{i,j=1}^{s} y_i\alpha_i K(x_i, x_j) + b \right). \tag{6.5}$$

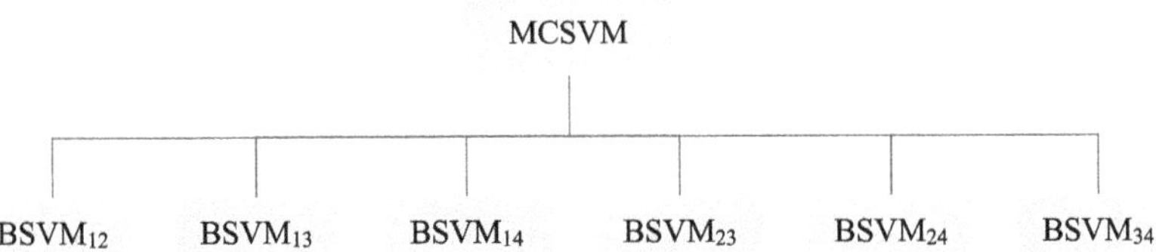

Fig. 6.3. MCSVM classifier for a four-class classification problem.

(2) Multi-class support vector machine (MCSVM)

In order to learn a K-class classification problem, the MCSVM classifier can be built by training $K(K-1)/2$ binary SVM according to the OVO strategy. Each BSVM is trained to separate one class from another. When classifying a new example, each BSVM produces a class label and the one with the highest confidence is selected as the end result. Figure 6.3 displays the structure of the MCSVM for the four-class classification problem.

6.1.2 *Steps for real-time diagnosis and early warning of downhole incidents during the fracturing process*

The proposed real-time diagnosis and alarm methodology in this section is divided into two stages. The first stage is establishing an offline multi-incident classifier for the shale gas well fracturing process. Specifically, the qualitative trend of process variables extracted by QTA is utilized for developing the MCSVM classifier. The second stage is building the real-time diagnosis and alarm procedure of downhole incidents by means of real-time process data coupled with the offline multi-incident classifier.

6.1.2.1 *Offline multi-incident classifier development*

Figure 6.4 illustrates the general procedure for constructing an offline multi-incident classifier. The procedure is described in detail as follows:

(1) Step 1: Data acquisition

Collecting process data is the basis for both extracting the QTPV and further training the multi-incident classifier. Thus, n original

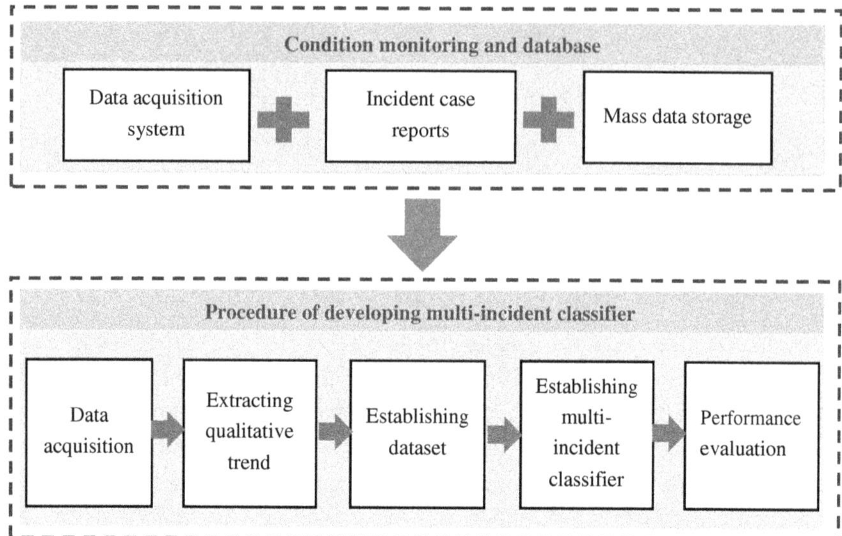

Fig. 6.4. Procedure for establishing the offline multi-incident classifier.

datasets are manually selected from the data acquisition system for each type of incident. Each original dataset is composed of three process variables, i.e., wellhead pressure, casing pressure, and displacement. Notice that each original dataset should comprise the tire process data throughout the development of the downhole incident. The acquired original datasets will be used for extracting the qualitative trend in the next step.

(2) Step 2: Extracting qualitative trend

First of all, the length of the sliding window (β) and that of each segment (μ) are determined by trial and error. The first polynomial (Eq. (6.1)) of each segment is then fitted using the linear least-squares method. Finally, the primitives in a sliding window are identified according to the rules of identifying primitives (Table 6.2).

Suppose the number of samples of the θth original dataset is l_θ and $l_\theta - \beta + 1$ sliding windows are obtained using the sliding-window approach. Let $R_k = \sum_{i=1}^{n} (l_i - \beta + 1)$ be the number of all sliding windows of the kth type of incident. Thus, the primitives in the rth sliding window are expressed as the input vector in Eq. (6.6) (also called the qualitative trend vector):

$$L_k^r = (\alpha_{11}, \ldots, \alpha_{1U}, \alpha_{21}, \ldots, \alpha_{2u}, \ldots, \alpha_{2U}, \ldots, \alpha_{3U})$$

$$k = 1, 2, \ldots, K \, r = 1, 2, \ldots, R_k \, u = 1, 2, \ldots, U, \tag{6.6}$$

where α_{1u}, α_{2u}, and α_{3u} represent the uth primitive of wellhead pressure, casing pressure, and displacement, respectively. U is the number of segments in a sliding window and K is the number of types of incidents.

(3) Step 3: Establishing the dataset

All obtained input vectors in Step 2 are arranged in a matrix in which each row represents one qualitative trend vector of process variables and its label in each sliding window. For K various downhole incidents, a collection of $R_{\text{sum}} = \sum_{k=1}^{K} R_k$ input vectors is recorded, where ϕ_k indicates the class label of the kth type of incident in Eq. (6.7):

$$T = \begin{bmatrix} \boldsymbol{L}_1^1 & \varphi_1 \\ \boldsymbol{L}_1^2 & \varphi_1 \\ \vdots & \vdots \\ \boldsymbol{L}_k^r & \varphi_k \\ \vdots & \vdots \\ \boldsymbol{L}_K^{R_K} & \varphi_K \end{bmatrix}. \tag{6.7}$$

Next, the matrix $\boldsymbol{T}$ is randomly split into the training matrix ($\boldsymbol{T}_{\text{train}}$) and the test matrix ($\boldsymbol{T}_{\text{test}}$). The matrices are responsible for training the offline multi-incident classifier and evaluating the performance of the classifier, respectively. As stated earlier, $K(K-1)/2$ binary SVM models need to be constructed for a K-class problem. Therefore, BSVM $\varepsilon\sigma$ is used to represent each binary classifier and distinguish between different events labeled as $\varepsilon(\varepsilon = 1, 2, \ldots, K-1)$ and labeled as $\sigma(\sigma = 2, 3, \ldots, K)$. In this way, the corresponding sub-matrix for training BSVM$_{\varepsilon\sigma}$ is selected from $\boldsymbol{T}_{\text{train}}$.

(4) Step 4: Establishing the multi-incident classifier

The greatest problem encountered in setting up the MCSVM model is how to determine the penalty parameter (C) and kernel function

Table 6.3. Confusion matrix.

			Happening incident (true label)			
		1	2	$\cdots$	$K-1$	K
Classified incident	1	$v_{1,1}$	$v_{2,1}$	$\cdots$	$v_{K-1,1}$	$v_{K,1}$
(calculated label)	2	$v_{1,2}$	$v_{2,2}$	$\cdots$	$v_{K-1,2}$	$v_{K,2}$
	$\cdots$	$\cdots$	$\cdots$	$\cdots$	$\cdots$	$\cdots$
	$K-1$	$v_{1,K-1}$	$v_{2,K-1}$	$\cdots$	$v_{K-1,K-1}$	$v_{K,K-1}$
	K	$v_{1,K}$	$v_{2,K}$	$\cdots$	$v_{K-1,K}$	$v_{K,K}$

parameters, such as the value of gamma (γ) for the RBF kernel. Thus, particle swarm optimization is adopted to optimize these parameters for each $\text{BSVM}_{\varepsilon\sigma}$. Then, the multi-incident classifier is built using the training dataset according to the OVO strategy.

(5) Step 5: Performance evaluation

During the testing phase, the vectors in $\boldsymbol{T}_{\text{test}}$ are first fed into the established multi-incident classifier to infer the classified incidents. The performance evaluation of the developed multi-incident classifier is then implemented by comparing classified incidents and true incidents for all test samples. In this order, classified outcomes are arranged in a confusion matrix as presented in Table 6.3 where $v_{i,j}(1 \leq i,j \leq K)$ represents the number of event samples labeled as i in reality that are misclassified as labeled j after being diagnosed by the classifier.

Afterward, several performance indexes, including global accuracy (GA), false alarm rate (FAR), and missing alarm rate (MAR), are calculated and their definitions and mathematical forms are shown in the following.

GA is defined as the percentage of all samples classified correctly and the whole test sample in Eq. (6.8):

$$GA = \frac{\sum_{i=1}^{K} v_{i,i}}{\sum_{i=1}^{K} \sum_{j=1}^{K} v_{i,j}}. \tag{6.8}$$

FAR is defined as the ratio of samples classified as event samples to all non event samples, i.e., the false alarm rate, as shown in formula

(6.9), where no actual events occur.

$$FAR = \frac{\sum_{j=2}^{K} v_{1,j}}{\sum_{j=1}^{K} v_{1,j}}.$$ (6.9)

MAR is defined as the ratio of samples that actually occurred but were classified as non event samples to all samples classified as events, i.e., the false positive rate, as shown in formula (6.10).

$$MAR = \frac{\sum_{i=2}^{K} v_{i,1}}{\sum_{i=2}^{K} \sum_{j=1}^{K} v_{i,j}}.$$ (6.10)

6.1.2.2 *Real-time diagnosis and alarm for downhole incidents*

In this stage, real-time process data are used to diagnose downhole incidents in real time. The automatic diagnosis and alarm procedure for downhole incidents are illustrated in Fig. 6.5. This real-time diagnosis process is achieved using the sliding-window approach. Specifically, when the latest samples of process variables are observed, the endpoint of the current sliding window should coincide with the latest samples, as shown in Fig. 6.2. The updated qualitative trend vector is then calculated and transferred into the developed multi-incident classifier. Subsequently, real-time diagnosis and a hierarchy alarm are implemented.

Providing that an incident has been identified, an alert signal will be sent to operators in the control center. In the following, a hierarchy alarm mode containing two levels of alarm is introduced and described:

(i) Each time an incident is detected, an alarm (one-level alarm or two-level alarm) is initiated.

(ii) If an incident is diagnosed for the first time, a one-level alarm is activated.

(iii) In the case of a continuous alarm, the one-level alarm is upgraded to a two-level alarm if the one-level alarm lasts more than 0.5 min.

(iv) In the case of an intermittent alarm, if the total time of the one-level alarm within one minute before the current alarm time exceeds 0.5 min, the one-level alarm escalates into a two-level alarm, otherwise remaining a one-level alarm.

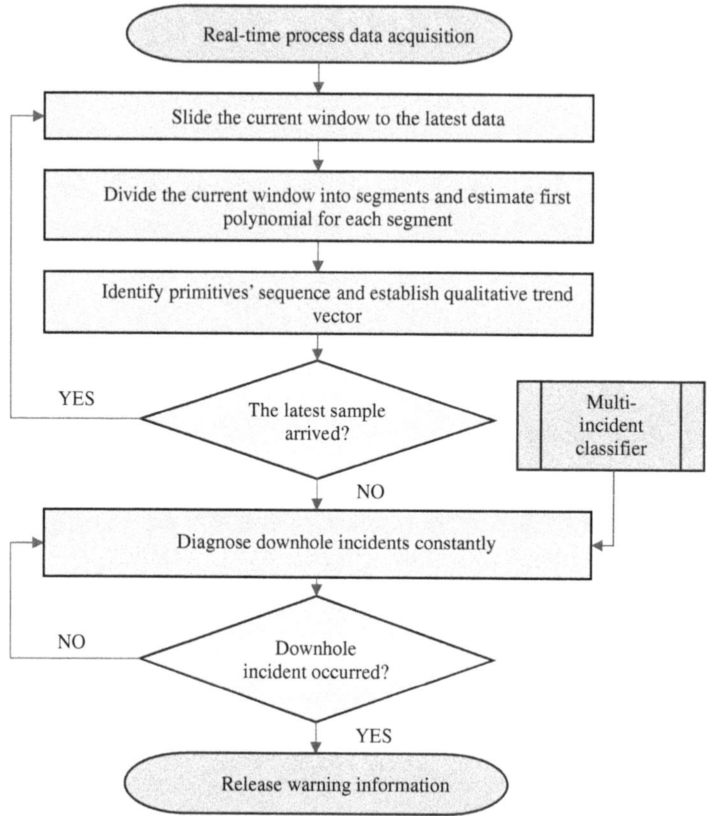

Fig. 6.5.　Workflow of real-time diagnosis and alarm.

(v) Similarly, if the cumulative time of the two-level alarm within one minute before the current alarm time is no more than 0.5 min, the two-level alarm is downgraded into a one-level alarm, otherwise remaining a two-level alarm.

6.1.3　*Case study*

6.1.3.1　*Offline multi-incident classifier development*

(1) Step 1: Data acquisition

Table 6.4 provides the basic information of the original datasets for each type of incident. As provided in Table 6.4, $K = 4$ different conditions, including the non-incident and three downhole incidents,

Table 6.4. Basic information of the original datasets for each type of incident.

Downhole condition	ϕ	No.	Length of dataset (minute)	l_θ	$l_\theta - \beta + 1$	R_k	Source of each dataset
Non-incident	1	Case 1	30	360	325	975	Dingye 1HF
		Case 2	30	360	325		Dingye 2HF
		Case 3	30	360	325		Dingye 3HF
Cracks forming in the strata	2	Case 4	21	252	217	819	Dingye 4HF
		Case 5	27	324	289		Dingye 5HF
		Case 6	29	348	313		Dingye 6HF
Channeling near wellbore area	3	Case 7	19	228	193	699	Dingye 7HF
		Case 8	22	264	229		Dingye 8HF
		Case 9	26	312	277		Dingye 9HF
Sand plug	4	Case 10	12	144	109	411	Dingye 10HF
		Case 11	13	156	121		Dingye 11HF
		Case 12	18	216	181		Dingye 12HF

as studied in the previous section, and $n = 3$ datasets, with different lengths selected for each kind of condition. The original process data of all datasets are shown in Fig. 6A.1. It should be noted that a non-incident condition is necessary when training the classifier. For the sake of convenience, the non-incident is also called one incident in the following.

(2) Step 2: Extracting the qualitative trend

In this step, we consider the case where the length of the sliding window and that of each segment are 3.0 min and 1.0 min, respectively. Then, the related parameters are $\beta = 36$ and $\mu = 12$. Based on these data, the number of sliding windows for each dataset $(l_\theta - \beta + 1)$ and the number of qualitative trend vectors (R_k) of each type of incident are listed in the sixth and seventh columns in Table 6.4, respectively.

(3) Step 3: Establishing the dataset

As observed in Table 6.4, a total of $R_{\text{sum}} = 2904$ qualitative trend vectors are obtained for $K = 4$ incidents. Furthermore, two-thirds of the qualitative trend vectors are then used for training the MCSVM classifier and the rest for testing. Equation (6.11) shows the partial qualitative trend vectors extracted at different moments 9:

$$
T_{\text{par}} = \begin{array}{c} \begin{matrix} \alpha_{11} & \alpha_{12} & \alpha_{13} & \alpha_{14} & \alpha_{15} & \alpha_{16} & \alpha_{17} & \alpha_{18} & \alpha_{19} & \varphi_k \end{matrix} \\ \begin{bmatrix} A & A & F & A & A & A & C & A & A & 1 \\ C & F & A & A & A & A & A & F & A & 1 \\ A & C & C & A & A & F & A & C & C & 2 \\ C & A & C & A & C & A & C & C & C & 2 \\ F & F & A & C & C & C & A & F & A & 3 \\ A & F & F & C & C & A & C & A & A & 3 \\ C & C & C & C & A & A & A & F & F & 4 \\ C & C & C & A & A & C & A & A & A & 4 \end{bmatrix} \end{array} . \quad (6.11)
$$

(4) Step 4: Establishing a multi-incident classifier

Based on the obtained training dataset, three MCSVM classifiers with different kernels, namely, linear, polynomial, and radial basis kernel function (RBF), are considered with the aim of determining the best kernel function. These classifiers are, respectively, termed the MCSVM-linear, MCSVM-poly, and MCSVM-RBF. For the polynomial kernel, the degree d of this kernel is varied in the range of [2, 5] so as to span polynomials with low and high flexibility. The associated parameters C and _ for the RBF kernel are varied in the ranges of [1, 1000] and [0, 1] for the purpose of covering high and low regularization of the classification model and fat as well as thin kernels, respectively. Table 6.5 provides the optimal parameters C and γ for the RBF kernel optimized by particle swarm optimization.

(5) Step 5: Performance evaluation

For comparison, the original process data are directly applied to establish the aforesaid classifiers in Step 4. Figure 6.6 compares the performance of the MCSVM-linear, MCSVM-poly, and MCSVM-RBF classifiers developed by the QTPV and original process data.

Table 6.5. Optimal parameters for MCSVM – RBF.

(MCSVM – RBF)	BSVM_{12}	BSVM_{13}	BSVM_{14}	BSVM_{23}	BSVM_{24}	BSVM_{34}
C	32.556	48.361	15.214	73.265	40.210	39.448
γ	0.365	0.589	0.165	0.067	0.464	0.105

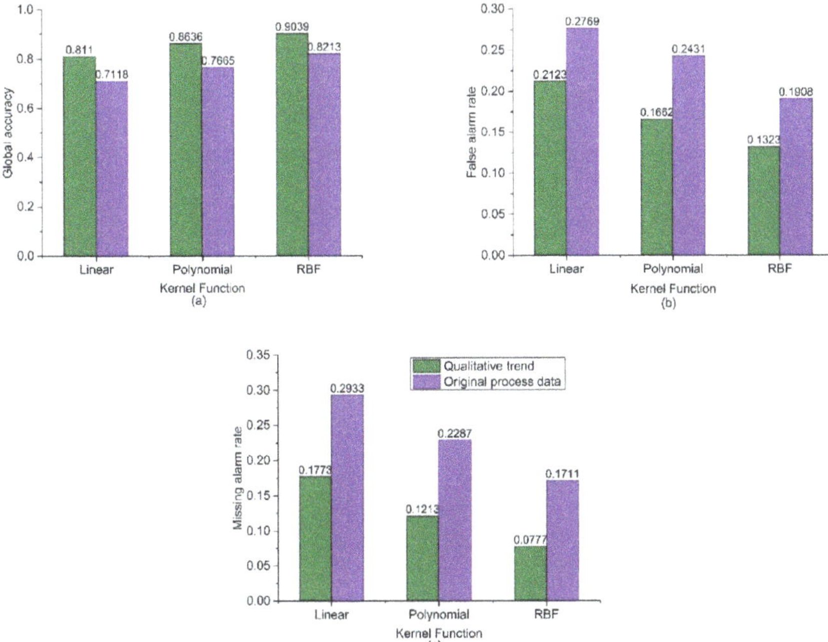

Fig. 6.6.   Performance comparison of MCSVM developed by different modeling methods.

It is clear from Fig. 6.6 that the MCSVM classifiers trained by the QTPV provide higher global accuracy as well as a lower false alarm rate and missing alarm rate when compared to those trained by the original process data, no matter which kernel function is adopted. This implies that the extracted QTPV better captures significant features of downhole conditions despite a limited number of incident cases.

However, more importantly, the MCSVM-RBF classifier provides the best results (0.9039 for global accuracy, 0.1323 for false alarm rate, and 0.0777 for missing alarm rate), indicating that the MCSVM-RBF model is the most suitable choice for real-time diagnosis of downhole incidents. This phenomenon can be attributed to the fact that MCSVM-RBF captures the basic problem characteristics while not introducing redundant noise that could have lowered the performance of the MCSVM.

For further validation, AIS and ANN classifiers are applied for incident classification. The same training and testing procedure is

Table 6.6. Performance comparison of different classifiers.

Classifier	Global accuracy	False alarm rate	Missing alarm rate
MCSVM-RBF	0.9039	0.1323	0.0777
AIS	0.8491	0.1015	0.1757
ANN	0.8584	0.1508	0.1369

implemented for both the classifiers using the obtained training dataset and test dataset in Step 3. Table 6.6 lists the performance indexes for each classifier. Comparatively, the MCSVM-RBF classifier outperformed the two other classifiers. The MCSVM-RBF classifier achieved the optimal results in terms of global accuracy (0.9039) and missing alarm rate (0.0777), while the AIS classifier achieved the optimal false alarm rate (0.1015). Therefore, the MCSVM RBF classifier is adopted for real-time diagnosis of incidents.

6.2 Dynamic Prediction of Remaining Life of Fracturing Pump System Based on Dynamic Object-Oriented Bayesian Network

Pump injection systems have four characteristics: they are large, interconnected, dynamic, and uncertain. Such systems are usually composed of similar or identical subsystems; for example, a pump injection system may include sixteen fracturing pump subsystems. Also, different subsystems interact with each other through various flows; for example, an oil supply subsystem provides power-end subsystems with a certain flow of lubricating oil. In addition, system degradation trends are dynamic and uncertain due to the constantly changing internal conditions and exterior environment factors. Extensive research has been conducted on residual life estimation for simple systems, such as bearings and gearboxes. However, few studies take into account the correlation among subsystems within a complex system or the dynamic and uncertain nature of system degradation.

When an abnormal condition occurs, traditional safety assessment methods often provide many possible reasons for it, especially for complex systems. However, in the real world, not all hazard causes would happen at the same time because many of them have a very low probability of occurrence. To find the true fault, maintenance personnel often must exclude false causes one by one, based on the identified possible reasons; this process requires a great deal of effort, and it is possible to make wrong maintenance decisions, which may lead to the best maintenance time being missed.

With regard to the prognosis of residual life or system performance, the existing research only considers failure-related data and sensor data to forecast future system performance and does not consider the current system status that can be analyzed by safety assessment or fault diagnosis. Excluding such considerations would make prognoses inaccurate.

Taking into account the interactions within a system, the uncertainty of system degradation, and the impact of a system's current state on future performance, a comprehensive safety management method for a complex pump injection system is proposed based on the Object-Oriented Bayesian Network (OOBN) and the Dynamic Object-Oriented Bayesian Network (DOOBN). This model is developed by taking advantage of the system structure, failure-related data available from the construction units, offline monitoring data, and expert experience. When abnormal events occur, the proposed method can implement fault diagnosis by using the abnormal online monitoring parameters as an input item and also predict the remaining useful life of the system together with the future trend of system health indicators, starting from the current system state.

6.2.1 *Overview of basic methodology*

6.2.1.1 *Object-oriented Bayesian network*

A Bayesian network (BN) involves directed acyclic graphs for visual representation and probability inference of causal processes with uncertainty, and it has been extensively applied in various engineering and technical fields. Practical application cases suggest that BN is accurate only in relatively static and small-scaled systems. For large mechanical systems in industry, such as fracking pump systems, modeling BN directly is a challenging task. A large BN model weakens the visualization of the model and increases the computational effort of the parameter update process. The Object-Oriented

Bayesian Network (OOBN) has gained attention in the field of risk assessment of large systems by incorporating an "object-oriented" idea into the BN modeling process, which improves the efficiency of modeling and reasoning.

The OOBN can be regarded as a meta-network composed of a series of special BN fragments. The basic element of the OOBN is a class, i.e., a network fragment with input nodes, internal nodes, and output nodes. Input nodes and output nodes are the interfaces of a class, through which information is exchanged between different classes or objects, and the nodes contained in a class must satisfy four conditions:

(1) The input node accepts the probability distribution of the immediate parent node and has no parent node within the class.
(2) The input node is a reference node, i.e., a mapping of upstream nodes.
(3) The internal node has no parent or child nodes outside this class.
(4) An output node passes probabilities to other input nodes and has no child nodes within this class.

Network fragments obtained by the instantiation of classes are called objects, which can also be encapsulated in other classes as basic elements, and encapsulation allows the attributes of network fragments to be transferred to each other. The OOBN allows hierarchical modeling of a complex system, increasing the accuracy and readability of the model through different levels of abstraction.

6.2.1.2 *Dynamic Bayesian network*

In the real world, some time-dependent factors make systems behave dynamically; therefore, to capture the dynamic stochastic process during their lifecycle, BN is extended into Dynamic Bayesian Networks (DBNs) by introducing related temporal dependencies to analyze the evolution of probability distributions over time-dependent domains. The DBN contains dynamic nodes whose states are hidden and cannot be observed directly (e.g., bearing wear) and static nodes whose states can be monitored by multiple sensors or people (e.g., temperature of lubricating oil). The symbol X_t is used to denote the set of state variables at time t, the symbol Y_t denotes the set of observed nodes, and the observation is denoted as $Y_t = y_t$, where y_t is some set of values of the variable.

The DBN includes three types of information: prior probability distribution of state nodes $P(X_0)$, state transfer probability distribution $P(X_{t+1}|X_{1:t})$, and causality model $P(Y_t|X_t)$. The DBN model satisfies the Markov and observation assumptions.

(1) **Markov assumption.** It is assumed that the current state of the state variable is only related to the previous state and not related to all states prior to the previous state, denoted as Eq. (6.12), and therefore, the state transfer probability can be expressed as $P(X_{t+1}|X_{1:t})$:

$$P(X_{t+1}|X_{1:t}) = P(X_{t+1}|X_t). \tag{6.12}$$

(2) **Observation assumption.** It is assumed that the variable values of the observed variables depend only on the current state and are independent of the state at other moments, expressed as Eq. (6.13), and thus the causality model can be expressed as $P(Y_t|X_t)$:

$$P(Y_t|X_{0:t}Y_{0:t-1}) = P(Y_t|X_t). \tag{6.13}$$

Based on this, the forward inference formula Eq. (6.14) can be used to predict the state probability distribution of a state node or an observed node at any moment:

$$P(X_{0:t+1}) = P(X_0) \prod_{t=0}^{T} P(X_{t+1}|X_t). \tag{6.14}$$

6.2.1.3 *Dynamic object-oriented Bayesian network*

A DOOBN has certain properties that are analogous to object-oriented modeling. A DOOBN can also be viewed as a meta-network linking a number of DBNs with added features that make them reusable as part of a larger DBN. The typical feature of a DOOBN is its internal nodes together with interface nodes: input and output nodes that import and export information across multiple DBN fragments. A DOOBN follows three fundamentals of object-oriented modeling: abstraction, inheritance, and encapsulation. To use object-oriented programming terminology, each DBN fragment acts as a class, while fragments resulting from instantiating these classes are termed as objects. Also, an instantiated class can be used to represent different DBN classes within other fragments. The notion of encapsulation allows the transmission of all properties of the fragment. The reasons why this section proposes the use of OOBN- and DOOBN-based approaches for online fault diagnosis and early warning studies of fracturing pump systems are as follows:

(1) Application of a modular approach can greatly simplify program modifications by decomposing a large-scale complex system into several smaller subsystems.
(2) The inference process is particularly efficient because the standard nodes inside an instance are conditionally independent of the rest of the DOOBN.
(3) Compared to standard BN or DBN, DOOBN has a less complex structure and can communicate and explain better.

6.2.2 *Dynamic prediction of remaining life of fracturing pump systems*

In this section, the OOBN and DOOBN are combined to establish a dynamic prediction method for the remaining life of fracturing pump systems. Firstly, a system function model is established by decomposing the system function, and the causal relationship between the abnormal state of the subsystem internal parts and the symptoms is analyzed. The OOBN-based Reasoning Model for Abnormal State of Components (RMASC) is also established, and multi-source diagnostic knowledge (supervisory information) acquired in real time is used as the input for the RMASC. The DOOBN-based Degradation Trend Prediction Model (DTPM) is updated according to the results of the RMASC to predict the degradation trend of the system and then estimate the development trend of the number of remaining fracturing segments and functional indicators of the components. The overall modeling process is described as follows.

6.2.2.1 *System function modeling*

Fracturing pump systems are physically complex and structurally large, and a system function model is used to express their internal system associations. The system function model is a hierarchical, function-based modeling approach that expresses the knowledge of the entire system from the perspective of transport flows (i.e.,

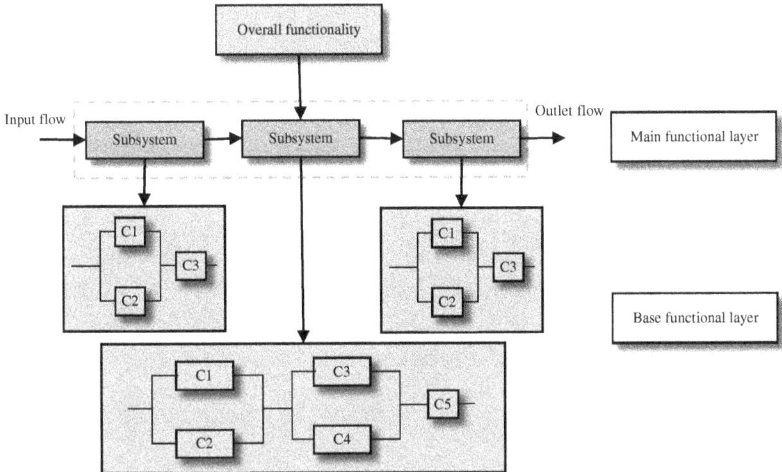

Fig. 6.7. Modeling flow of the fracturing pump system functional model.

material, energy, and information flows) and describes the interaction processes between subsystems within the system (Fig. 6.7).

The specific realization steps are as follows:

(1) **Define the overall function of the system:** For the fracturing pump system, its overall function is to pressurize the fracturing fluid from the sand mixing truck, generate high-pressure fracturing fluid that meets the requirements, and pump it into the high-pressure piping system in the discharge area of the pumping truck.

(2) **Decomposing the overall function of the system from the main function level:** The overall function of the system is realized through functional coupling between the internal subsystems. Through the analysis of the functional coupling relationship, the overall function of the system is divided into a number of main function modules, each of which corresponds to a physical subsystem; the input and output flows of each main function module are determined, and oriented arrows are used to indicate the direction of the transmission flow so as to construct the system function model at the main function level.

(3) **Decompose each main function module from the basic function level:** Each main function module corresponds to a physical subsystem, and each subsystem is further decomposed at the component level to determine the basic function module

corresponding to each component; at the same time, the transmission flow of each basic function module is analyzed, and a directed arrow is used to indicate the direction of the transmission flow so as to construct the main function model at the basic function level. For any complex system, the system function is realized through the coupling of a series of basic functions and main functions, so the system performance can be assessed by quantitatively evaluating the system's basic functions or main functions. The functional model of the fracturing pump system will provide the basis for the subsequent development of the causal model, RMASC model, and DTPM model.

6.2.2.2 *Causality modeling*

The primary role of the frac pump system causality model is to do the following: (1) It can identify observed variables and their deviations that characterize the health of the frac pump system, e.g., lubricant temperature. The attribute values of these variables can be obtained through data acquisition systems, testing techniques, or manual observation at the shale gas fracturing site. (2) It can analyze fault variables and the potential consequences associated with the observed variables. The behavior of the failure variables can be determined by inspection techniques or time series data of the observed variables, e.g., the fouling state of the filter and the wear state of the crankshaft. (3) It can reveal uncertainty in the correspondence between observed variables and fault variables. As shown in Fig. 6.19, since the subsystems interact with each other through transport flow, the potential consequences of the former subsystem may be the cause of the anomalies of the latter subsystem, and based on this relationship, the fault propagation process within the fracturing pump system can also be described. In this section, Hazard and Operability (HAZOP) analysis is used to identify the observed variables and deviations of the system, determine the fault variables and behaviors, and reveal the causal relationships between the variables at a qualitative level.

6.2.2.3 *Component abnormal state reasoning model*

In this section, the OOBN is used as the theoretical basis to integrate the qualitative causal relationships in the system causal model, quantify the uncertainty of the causal relationships, and build the RMASC model of the fracturing pump system. The following four steps are required to build the RMASC model:

Step 1: Determine the nodes and state spaces of the BN category based on the observed variables and fault variables in the system causal model.

Step 2: Construct the network structure of the BN class by creating directed edges based on the uncertain causal relationships between the nodes.

Step 3: Under the condition of determining the network structure of a BN class, estimate the parameters of the BN class, i.e., the conditional probability distribution between faulty nodes and symptomatic nodes.

Step 4: Establish the OOBN model, i.e., the RMASC model of the fracturing pump system, based on the transmission flow in the system function model.

(1) Determine the nodes of a BN class

The BN class describes the fault causality of the subsystem, and the nodes of the BN class are divided into fault nodes and symptom nodes; the symptom nodes include the input nodes and output nodes of the class. The flow of determining the nodes is shown in Fig. 6.8, and the main rules are expressed as follows:

The physical components of each subsystem in the system functional model are selected as fault nodes, and the states of the components are extracted from the "probable cause" item of the HAZOP analysis report to establish the state space of the fault nodes.

The observed variables of each subsystem in the system functional model are selected as symptom nodes, and the state space of the symptom nodes is extracted from the HAZOP analysis report under the item "Guiding words".

According to the transmission flow between subsystems, the symptom nodes that characterize the transmission flow are converted into input/output nodes of the BN class. Since the neighboring subsystems are connected by the transport flow, the symptom node characterizing the transport flow is both the output node of the BN class

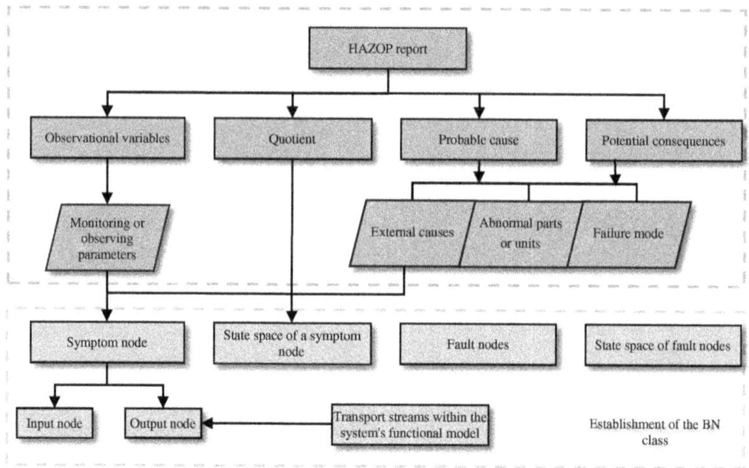

Fig. 6.8. Flow of determining nodes.

of the former subsystem and the input node of the BN class of the latter subsystem.

(2) Constructing the network structure of a BN class

There are two kinds of action relationships in the system causal model: (1) the effect of fault variables on observed variables, which indicates the effect of the state of fault nodes on the state of symptom nodes, e.g., the clogging of outlet valves leads to the low flow rate of outlet manifolds, and (2) the mutual influence between observed variables, which indicates the effect of symptom nodes on another symptom node, e.g., the pressure difference of the gas filter is too large, causing the gas pressure at the inlet of the pneumatic motor to be too low, which may affect the normal operation of the pneumatic motor. According to the above causal relationship, a directed arrow is used to connect the nodes of each BN class.

(3) Quantify the parameters of the BN class

The parameters of the BN class include the a priori probability distribution of the faulty nodes and the conditional probability distribution between the nodes. The a priori probability describes the probability of occurrence of each state of the faulty node inferred from existing knowledge. In practice, the BN network utilizes conditional probability to update the a priori probability of the faulty node when new evidence is obtained for posterior probability, and then, based on the size of the a posteriori probability, the most probable faulty node is determined; therefore, the a priori probability of each state within each faulty node is assumed to be the same to highlight the important role of the conditional probability in the fault inference process. For example, if the faulty node includes M states, the a priori probability of each state is $1/M$.

Conditional probability quantifies the dependency between the symptomatic node and the faulty node. There are usually two methods for determining the conditional probability: (1) expert knowledge method, which is especially suitable for the lack of historical monitoring or testing data, e.g., equipment without a data acquisition system installed; (2) parameter learning algorithms are suitable for systems with a large amount of fault sample data. Common representative algorithms include Maximum Likelihood Estimation (MLE) and Maximum A Posteriori Estimation (MAP). However, since the available fault sample data are very limited, the expert knowledge method is usually used to determine the conditional probability distribution of the BN class.

(4) Instantiating BN Classes

Based on the transmission flow in the system function model, the BN classes are connected together to build an abnormal state inference model for components based on the OOBN model.

6.2.2.4 *System degradation trend prediction model*

Based on the DOOBN model, adopt the inverse modeling method to map the system function model to the DTPM model.

(1) Establish the DBN base class at the component level

From Fig. 6.7, it can be seen that the system functional model consists of three functional layers from the bottom to the top: the base functional layer, the main functional layer, and the overall functional layer. The base functional layer depicts the functions of physical components and their input/output streams, and the realization of each base function depends on the joint action of physical components and various input streams. When establishing the DBN class of a physical component, its input and output streams are treated as the input and output nodes of the DBN class, respectively; the physical component is converted into a dynamic discrete node of the DBN class due to its degradation behavior, and the base function box is converted into a function node (a class of intermediate nodes) of the DBN class. Take Fig. 6.9 as an example: The left side shows the basic functional model of the lubricant filter, and its input stream is "Input: lubricant with impurities", which is converted into the input node "Input_node" of the DBN class. The right side shows the output stream and "Output: filtered lubricant" is converted into the output node "Output_node" of the DBN class. The physical component "Com: lubricant filter" is converted into the physical component "Com: lubricant filter". The physical component "Com: Lubricant Filter" is converted into the dynamic discrete node Com t of DBN class. The basic function box is converted into the function node "Func" of the DBN class.

The conditional probability of a DBN class quantitatively describes the logical relationship between a functional node and its connected nodes. A definite logical relationship between a functional node and its connected nodes exists if each node includes two states and there is no spare unit for each physical part or unit. Taking Fig. 6.9 as an example, assuming that the state space of Input_node, Com $t+1$, and Func is {Normal, Abnormal}, the conditional probability distributions between the three are shown in Table 6.7.

If there exists a dynamic node that includes three or more states, there is an indeterminate logical relationship between the functional node and its parent node. Taking Fig. 6.9 as an example, assuming that the state space of the lubricant filter Com is {Normal, Degradation, Abnormal} and the rest of the nodes are in binary states, the

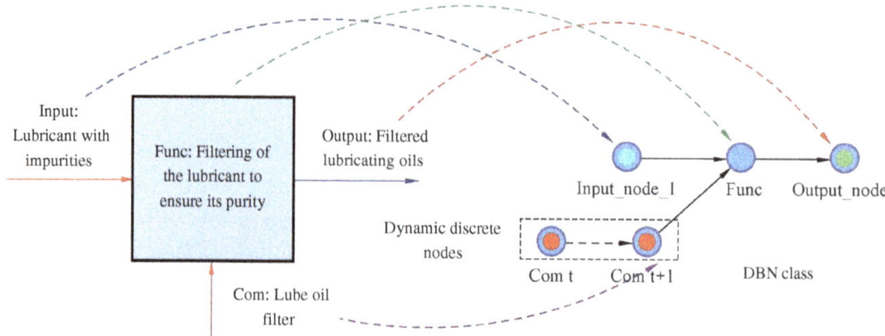

Fig. 6.9. Schematic diagram of the rules for converting a base functional layer to a DBN class.

Table 6.7. A conditional probability distribution table based on determined logical relationship conditions.

Input_node		Normal		Abnormal	
Com $t+1$		Normal	Abnormal	Normal	Abnormal
Func	Normal	1	0	0	0
	Abnormal	0	1	1	1

Table 6.8. Conditional probability distributions under uncertain logistic relationships.

Input_node		Normal			Abnormal		
Com $t+1$		Normal	Degradation	Abnormal	Normal	Degradation	Abnormal
Func	Normal	1	$0 < p < 1$	0	0	0	0
	Abnormal	0	$1 - p$	1	1	1	1

conditional probability distributions between Input_node, Com $t+1$, and Func are shown in Table 6.8, where p denotes the probability that the functional node Func is in the normal state.

If a physical component has a spare unit or multiple components of the same type operate together to realize a certain basic function, the relationship can be described by using parallel and series structures. Figure 6.10 shows the basic function of the lubrication pump and its corresponding two DBN base classes, in which the physical components include the main lubrication pump Com_1 and the standby lubrication pump Com_2, and the standby lubrication pump is automatically turned on when the main lubrication pump fails. In the second DBN base class, a simplification of the conditional probability table is realized by adding the logic node Operator (or gate).

The prior probabilities of the DBN class include the prior probabilities of the input nodes and the state transfer probability distributions of the dynamic nodes. The a priori probabilities of the input nodes are determined by expert knowledge or obtained from statistical history. The transfer probabilities of the dynamic nodes quantify the transfer relationships between different states of the same component. Figure 6.11 shows the transfer relationship between the

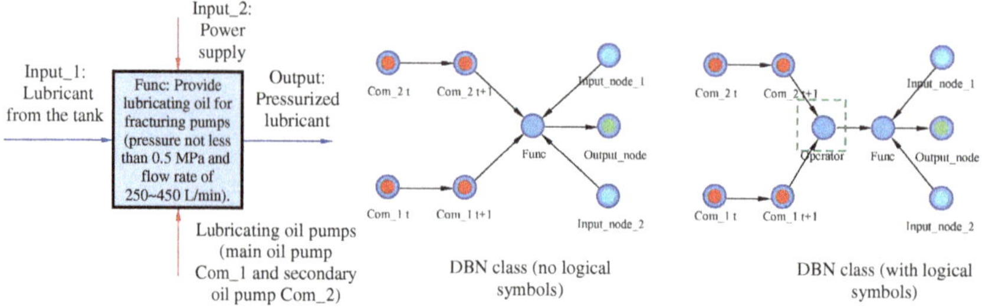

Fig. 6.10. Two DBN base classes.

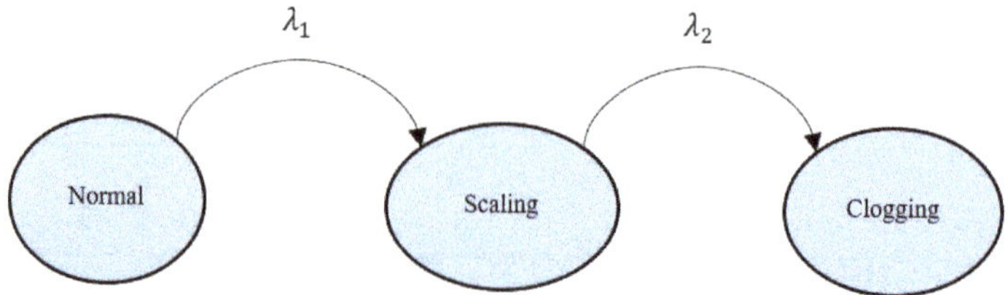

Fig. 6.11. State transfer diagram.

states of the lubricant filter. The life distribution of physical components follows an exponential distribution, and its failure rate is a constant λ. Methods for determining the failure rate λ include the expert knowledge method and the parameter learning method. For irreparable components with a small failure rate, due to the limited accumulation of life samples, on-site maintenance personnel are invited to estimate the Mean Time to Failure ($MTTF$), and the failure rate is calculated using Eq. (6.15). "Failure" is a relative concept. In Fig. 6.11, "fouling" is a failure state relative to "normal", and "clogging" is a failure state; "clogging" is also a failure state relative to "scaling".

$$\lambda_2 = \frac{1}{MTTF}.\tag{6.15}$$

For components with large failure rates, which have accumulated more life data, Bayesian estimation is used to determine their failure rates. The specific inference process can be described as follows:

The conjugate prior of the exponential model is the Gamma distribution, so Gamma distribution is used to describe the uncertainty of the failure rate λ. The Gamma prior distribution consists of the shape parameter α_{prior} and the scale parameter β_{prior}. In the engineering application, α_{prior} denotes the number of observed failed components and β_{prior} denotes the total service time of the α_{prior} components before they fail. Fracturing equipment maintenance personnel are first asked to estimate α_{prior} and β_{prior} empirically.

Gamma distribution is the conjugate prior of the exponential distribution, so the posterior distribution of the failure rate λ is still the Gamma distribution. n denotes the number of failed parts collected; $t_{\text{total}} = MTTF_1 + MTTF_2 + \cdots + MTTF_n$ denotes the total running time of the n failed components. The shape parameter of the posterior distribution is $\alpha_{\text{post}} = \alpha_{\text{post}} + n$, and the scale parameter is $\beta_{\text{post}} = \beta_{\text{prior}} + t_{\text{total}}$.

In Eq. (6.16), the a posteriori mean value, i.e., the failure rate of the component, λ_{post}, is calculated:

$$\lambda_{\text{post}} = \frac{\alpha_{\text{post}}}{\beta_{\text{post}}} = \frac{\alpha_{\text{prior}} + n}{\beta_{\text{prior}} + t_{\text{total}}}. \tag{6.16}$$

(2) Establish DBN intermediate classes at the subsystem level

A subsystem is composed of a series of physical components and transport streams. Each physical component corresponds to its basic function, and according to the correlation between the basic functions, the corresponding DBN classes are instantiated and connected; the output stream (output node) of the previous class may be the input stream (input node) of the next DBN class, so as to build the DBN intermediate class at the subsystem level. Since the parameters of DBN classes at the component level have been determined, this step does not involve parameter adjustment.

(3) Establish the DOOBN model at the overall system level

Instantiate all the DBN classes at the subsystem level according to the input and output flows between subsystems, and connect the instantiated classes together to build a DOOBN-based DTPM model.

6.2.2.5 *Dynamic remaining life prediction process*

The process for dynamic prediction of remaining life is shown in Fig. 6.12.

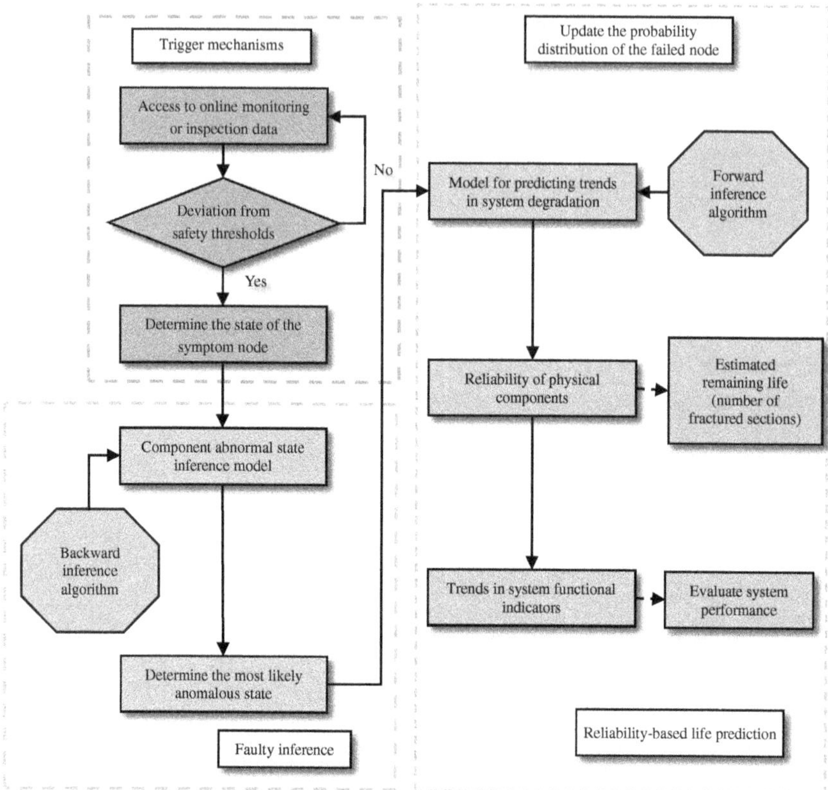

Fig. 6.12. Dynamic residual life prediction process.

It is specifically described as follows:

(1) Obtain the online monitoring data or test data of the symptomatic node and determine whether the safety threshold is exceeded. If the safety threshold is exceeded, the state of the symptomatic node is determined and used as an input for the component abnormal state inference model.

(2) Component abnormal state inference involves taking the state of the symptomatic node as the input for the RMASC, updating the a priori probability distribution of the faulty node by using the backward inference algorithm, and inferring the most probable abnormal component based on the a posteriori probability distribution.

Assumption $C_{1:N}^t$ denotes N faulty nodes at time t, and $C_k^t(k = 1, 2, \ldots, N)$ denote the state values of the faulty nodes; $S_{1:M}^t$ denotes M symptomatic nodes (multiple monitoring variables), and $S_r^t(r = 1, 2, \ldots, j, \ldots, M)$ denote the state values of the S_r^t symptomatic nodes.

The probability of the fault state $C_n^t = i$ is $\varphi_t(i) = P(C_n^t = i | S_m^t = j)$; see Eq. (6.17).

$$P(C_n^t = i | S_m^t = j) = \frac{P(C_n^t = i, S_m^t = j)}{P(S_m^t = j)}. \tag{6.17}$$

(3) Based on the inference in (2), predict the remaining lifespan. Based on the abnormal state of the component inferred by the RMASC, the state probability distribution of the corresponding node in the DTPM is updated, and the degradation trend of the dynamic discrete node in the future is inferred from the current state of the component using Eq. (6.14). The degradation trend of the "normal state" of the dynamic discrete node can be used as a reliability indicator of the component. If the probability of a "normal state" of a physical component is within the reliability threshold, the physical component is considered safe and no maintenance activities are taken. It is also possible to evaluate system performance by predicting the trend of basic and main function indicators at the subsystem or whole system level according to Eq. (6.14).

In summary, by firstly establishing the functional model and causal model of the fracturing pump system, and then establishing the RMASC based on the OOBN and the DTPM based on the DOOBN, the RMASC model can reason out the abnormal status of the current system components and provide a reference basis for the maintenance personnel to formulate preventive maintenance. The RMASC model can reason out the current abnormal state of the system components and provide a reference basis for maintenance personnel to troubleshoot. The DTPM model can provide an early warning on the degradation trend of components and the development trend of system function indicators, which is helpful for maintenance personnel to formulate preventive maintenance strategies.

6.2.3 *Case study*

6.2.3.1 *Description of pump injection system*

A pump injection system used for shale gas well fracturing first pressurizes the fracturing fluid from the sand blenders to a certain pressure, then pumps the pressurized liquid into secondary manifolds, and finally transports the collected fluid into the primary manifold connected with the wellhead device. A pump injection system may comprise 4–16 fracturing pumps, two sets of secondary manifolds, and a primary manifold. The number of fracturing pumps in a pump injection system depends on various factors, including rock density, rock porosity, and formation pressure. In this chapter, the proposed approach is applied to a real-world pump injection system with eight fracturing pumps to demonstrate fault diagnosis and residual life prediction. The configuration of the pump injection system under study (PISS) is given in Fig. 6.13; this shows that every four fracturing pumps are linked with one secondary manifold. The basic information for the PISS is provided in Table 6.9.

6.2.3.2 *System function model development*

The PISS is divided into a series of subsystems, including one-level subsystems and two-level subsystems, as shown in Fig. 6.14. The SFM for these subsystems is built and illustrated in Fig. 6.15. A single fracturing pump subsystem is equipped with a specific pneumatic transmission system and a specific lubricating oil system. Therefore, the PISS contains eight identical fracturing pump subsystems, eight identical pneumatic transmission subsystems, eight identical lubricating oil subsystems, and two identical secondary manifolds.

6.2.3.3 *Causal model development*

To achieve an accurate result, HAZOP analysis is performed for each subsystem, with the assistance of an equipment operator and two maintenance technicians on the shale gas well fracturing site. Taking the hydraulic end subsystem as an example, the structures of DBN classes for the repeatable subsystems are shown in Fig. 6.16.

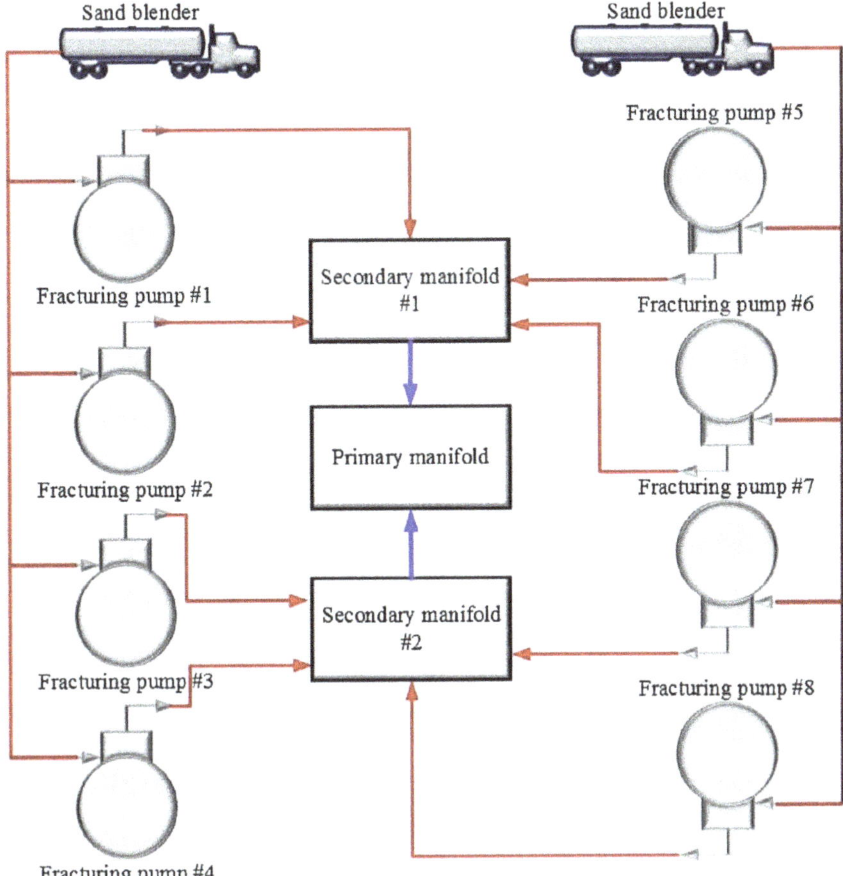

Fig. 6.13.   Configuration of the current pump injection system.

6.2.3.4 *System behavior model development*

The continuous probability distributions of dynamic nodes are determined on the basis of the available lifetime data of multiple components. Using an outlet valve that has two states (normal and failure) as an example, the information for such a component that is obtained from the PISS is listed in Table 6.10. For example, the outlet valve with serial number 1 lived through 55 fracturing stages before failure, and the accumulated operating time was about

Table 6.9. Basic information of the PISS.

Fracturing pump subsystem		Manifold subsystem		Lubricating oil subsystem	
Model	5ZB-2800 plunger pump	Model	1502 type	Maximum outlet power	0.85 MPa
Maximum input power	2080 kW (2800 hp)	Maximum working pressure	105 MPa	Rotating speed	940 r/min
Maximum pressure	108 MPa	Maximum outlet flow	21 m^3/min	Maximum outlet flow	0.3 m/min
Maximum flow	3.5 m^3/min	Pump configurations	8		
Plunger diameter	3.7″	Nominal bore size	8 in		
Gears	8				
Rotating speed	1450 rpm				

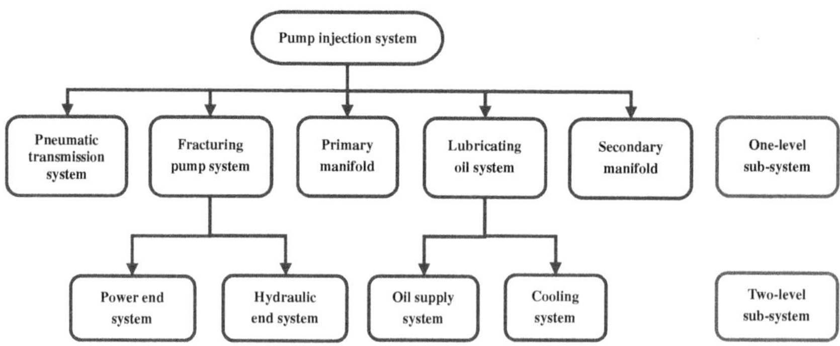

Fig. 6.14. Main subsystems of the current pump injection system.

388.5 h. As shown in Table 6.10, the number of failure components (22) is more than eight; therefore, it is considered that sufficient failure data have been obtained. Furthermore, non-information prior distribution is adopted to estimate the Weibull parameters by means of Bayesian inference. The resulting Weibull parameters are provided in Table 6.11 and the conditional probability distribution for such an outlet valve is shown in Table 6.12.

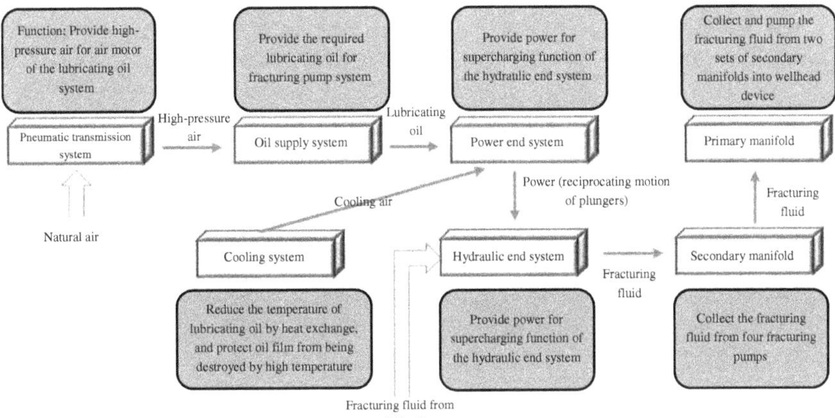

Fig. 6.15. System function model for the pump injection system.

Because maintenance records may be incomplete or the service time of the PISS may be relatively short, collected failure data are scarce for some components. For example, the number of crankshaft failures from the PISS is six (see Table 6A.1), which is less than eight; hence, the failure data from the other two PISs (PIS#13 and PIS#27) are also collected to make the estimated failure rate function more accurate. All the collected lifetime data of crankshafts are presented in Table 6A.1. A crankshaft has three states: normal, wear, and failure. If a crankshaft becomes worn and has little influence on the fracturing pump, then it will be treated using grinding technology. This treatment is regarded as minimal maintenance, and the processed crankshaft is still in its degraded state. If failure occurs, then the failed crankshaft would be replaced at once and the replaced part would stay in its normal state. Based on this, the Weibull parameters and the corresponding conditional probability distribution at time t for such a crankshaft are computed and shown in Tables 6.13 and 6.14, respectively.

Regarding the conditional probability distributions among nodes, three nodes (Y3_6, Y3_7, X3_10) that form a local DBN fragment are used as an example to demonstrate the process of parameter learning by searching the historical inspection records, maintenance records, and abnormal alarms of surface temperature of cylinders. Fifty samples are displayed in Fig. 6.17, with states classified as follows: for Y3_6, 1 indicates "normal" state, 2 is "wear", and 3 is "failure"; for

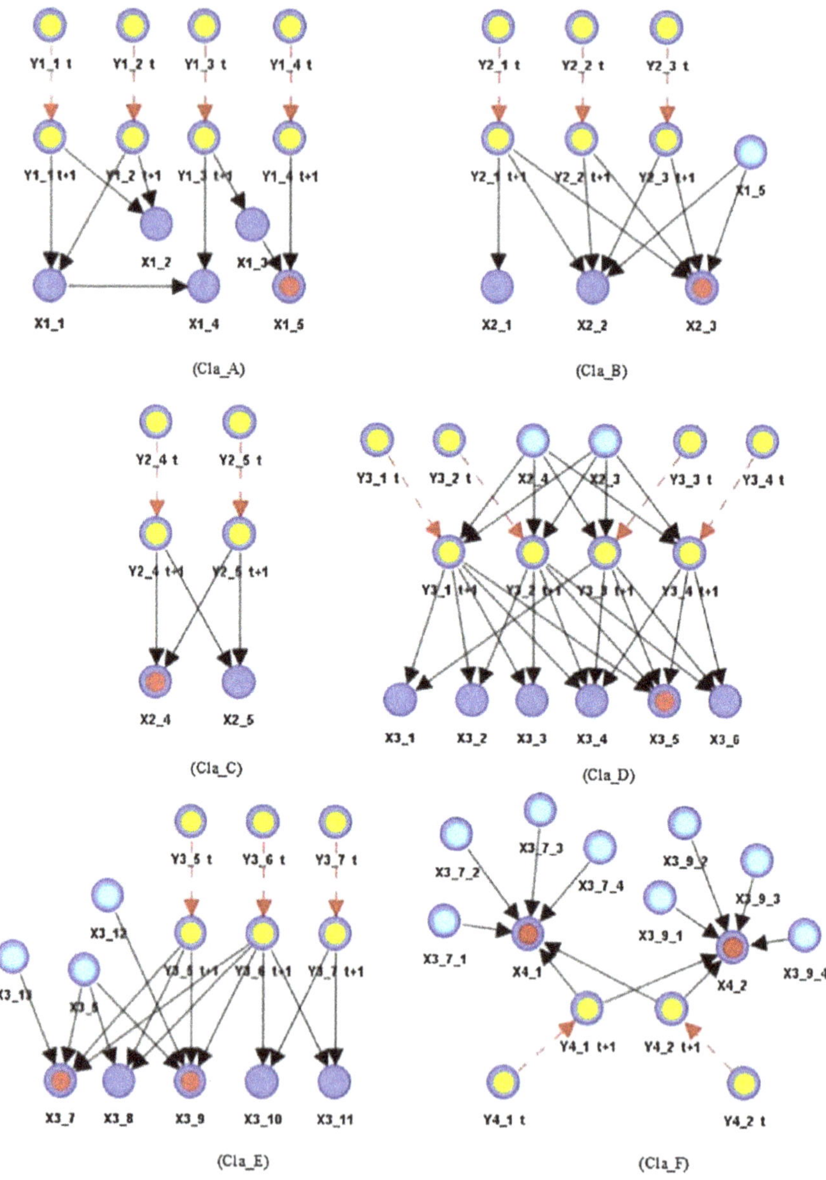

Fig. 6.16. Six DBN classes for subsystems: (1) Cla_A: pneumatic transmission system; (2) Cla_B: oil supply system; (3) Cla_C: cooling system; (4) Cla_D: power end system; (5) Cla_E: hydraulic end system; and (6) Cla_F: secondary manifold.

Table 6.10.   Life data of the outlet valves.

No.	Fracturing stages before failure	Maintenance mode	Accumulated running time
1	55	Replacement	388.5
2	73	Replacement	518
3	78	Replacement	554.4
4	63	Replacement	444.6
5	87	Replacement	616.5
6	98	Replacement	700.3
7	116	Replacement	779.5
8	106	Replacement	747.1
9	104	Replacement	726.6
10	101	Replacement	707.8
11	113	Replacement	750.2
12	76	Replacement	547.7
13	119	Replacement	838.6
14	97	Replacement	692.5
15	106	Replacement	748.9
16	135	Replacement	923.1
17	47	Replacement	250.3
18	84	Replacement	560.9
19	116	Replacement	808.1
20	118	Replacement	821.1
21	109	Replacement	765.5
22	62	Replacement	410.3

Table 6.11.   Weibull parameters for outlet valve.

Component	Node	Parameters	Mean value	Standard deviation
Outlet valve	Y3.5	Shape	4.78	0.85
		Scale	711.84	85.36

Y3_7, 1 is "normal" and 2 is "failure"; and for X3_10, 1 is "normal" and 2 is "high". The results obtained when using the MLE algorithm are given in Table 6.15.

The pneumatic transmission subsystem is used as an example to illustrate the specific verification process. Firstly, assume that all dynamic nodes of Cla_A in Fig. 6.16 have a priori failure probability

Table 6.12. Conditional probability distribution of outlet valve.

Y3_5 at $t+1$	Y3_5 at t	
	Normal	Failure
Normal	$e^{\frac{t^{4.78}-(t+1)^{4.78}}{711.84^{4.78}}}$	$1 - e^{\frac{t^{4.78}-(t+1)^{4.78}}{711.84^{4.78}}}$
Failure	0	1

Table 6.13. Weibull parameters for crankshaft.

Component	Node	State transition	Parameters	Prior mean value	Posterior information Mean value	Posterior information Standard deviation
Crankshaft	Y3_3	From normal to wear	Shape	34.52	10.58	1.86
			Scale	695.78	673.78	62.75
		From wear to failure	Shape	9.91	11.68	1.67
			Scale	545.31	556.28	70.36
		From normal to failure	Shape	10.91	10.78	1.72
			Scale	1171.54	1183.67	112.59

Table 6.14. Conditional probability distribution of crankshaft.

Y3_3 at $t+1$	Y3_3 at t		
	Normal	Wear	Failure
Normal	$e^{\frac{t^{10.58}-(t+1)^{10.58}}{673.78^{10.58}}} + e^{\frac{t^{10.91}-(t+1)^{10.91}}{1171.54^{10.91}}} - 1$	$1 - e^{\frac{t^{10.58}-(t+1)^{10.58}}{673.78^{10.58}}}$	$1 - e^{\frac{t^{10.91}-(t+1)^{10.91}}{1171.54^{10.91}}}$
Wear	0	$e^{\frac{t^{11.68}-(t+1)^{11.68}}{556.28^{11.68}}}$	$1 - e^{\frac{t^{11.68}-(t+1)^{11.68}}{556.28^{11.68}}}$
Failure	0	0	1

of 0.01; then, set the Y1_1 scaling probability and failure probability, and the Y1_2 wear probability and failure probability, to be 1.00 in turns. The probability of an abnormal state of X1_1 (the sum of the probabilities of low-pressure difference and high-pressure difference) increases from 0.0445 to 0.8647, 0.9345, 0.9653, and 1.00 one

Table 6.15. Conditional probability distribution of crankshaft.

		X3_10	
X3_6	X3_7	Normal	Failure
Normal	Normal	0.994	0.006
	Failure	0.126	0.874
Wear	Normal	0.203	0.797
	Failure	0.045	0.955
Failure	Normal	0.114	0.886
	Failure	0.007	0.993

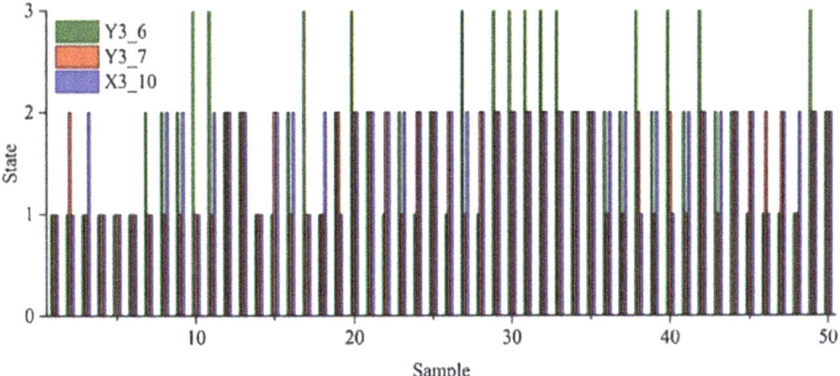

Fig. 6.17. Historical data for parameter learning (states: Y3_6: 1 - "normal", 2 - "wear", 3 - "failure"; Y3_7: 1 - "normal", 2 - "failure"; X3_10: 1 - "normal", 2 - "high").

by one. Similarly, the probability updating process for the remaining observation variable nodes can be realized by using the abovementioned process. In conclusion, such a process is satisfied by the "three-axiom" for all static nodes; therefore, the accuracy of such DBN classes is partially verified.

Because each fracturing pump subsystem is equipped with a lubricating oil subsystem and a pneumatic transmission subsystem, and to make the structure of the SBM more compact and more expressive, five DBN classes (Cla_A, Cla_B, Cla_C, Cla_D, and Cla_E) of the three subsystems are instantiated and linked together according to the flows described in Fig. 6.15. They are further encapsulated into a

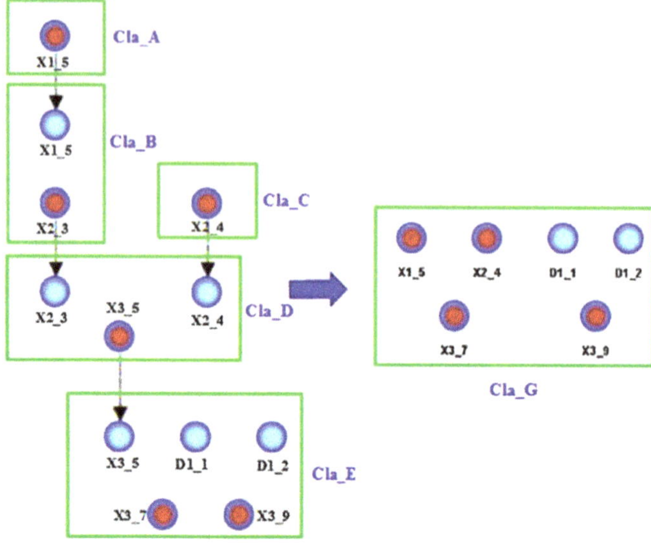

Fig. 6.18. Structure of the new class Cla_G.

high-level class (Cla_G) as displayed in Fig. 6.18. For the PISS with eight fracturing pump subsystems, the new class Cla_G is instantiated eight times to create eight corresponding objects (Obj_G1, Obj_G2, Obj_G3, Obj_G4, Obj_G5, Obj_G6, Obj_G7, and Obj_G8) as shown in Fig. 6.19. In the same vein, the DBN class of the secondary manifold is instantiated twice to build two objects (Obj F1 and Obj F2). In addition, the DBN fragment of the primary manifold is constructed. Ultimately, given the flows in Fig. 6.15, all the built objects are connected to develop the SBM as illustrated in Fig. 6.19.

The following abnormal condition is considered in this case study: During the online monitoring of the PISS in late December 2017, the crankcase temperature of fracturing pump #6 (X3_3) was 80.4°C by minute 186 of the 47th fracturing stage (see Fig. 6.20), and the motor current (X3_4) increased to 765.3 A after 2 min (see Fig. 6.21). These are beyond the upper limits of safety intervals; other monitoring parameters remained in a normal state.

Initially, the possible root causes for such an abnormal condition were determined from the HAZOP report as provided in Table 6.16. However, there is no quantitative information to indicate the most likely reason, which increases the difficulty of finding the true

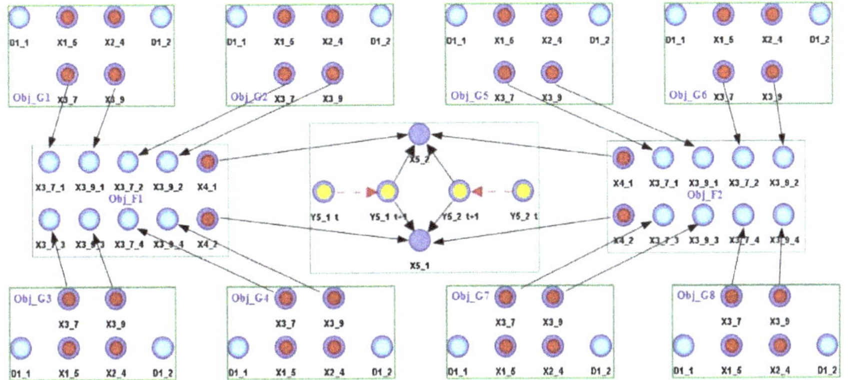

Fig. 6.19. System function model based on the DOOBN.

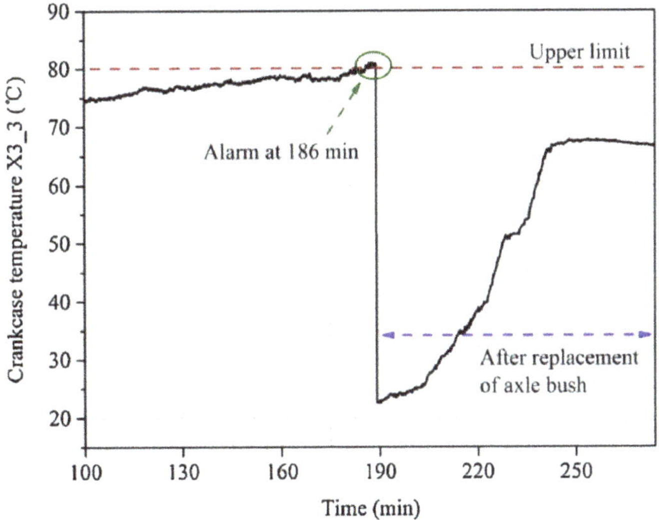

Fig. 6.20. Crankcase temperature of fracturing pump #6 (X3_3).

fault. If several abnormal monitoring parameters occur simultaneously, then many potential reasons would be listed in the HAZOP report; therefore, the maintenance team will spend a lot of time troubleshooting.

Next, the states of static nodes (i.e., the "high" state of X3_3, the "large" state of X3_4, and the "normal" state of other nodes) are taken as inputs for the SBM, and the first four nodes with larger

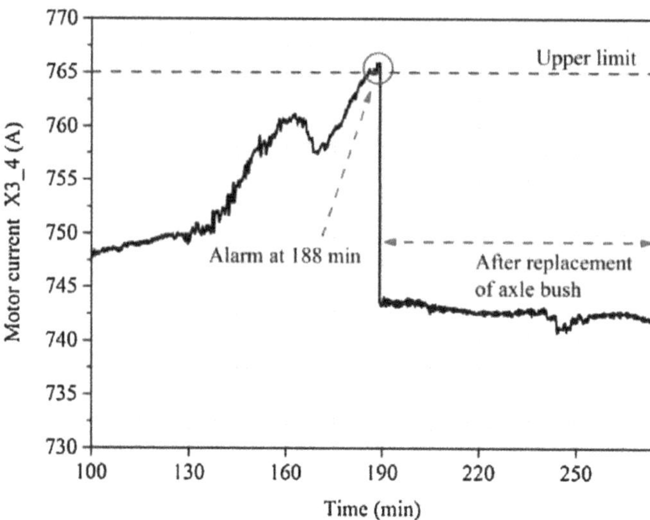

Fig. 6.21. Motor current of fracturing pump #6 (X3_4).

Table 6.16. Results of HAZOP analysis.

Possible causes	Safety measurements
High temperature of lubricating oil at the inlet pipe of crankcase	Improve the power of the cooling system
A large amount of heat produced by the wear of axle bush	Replace the worn axle bush
Excessive heat produced by crankshaft wear	Maintain or replace the worn crankshaft
Wear or failure of thrust bearing	Replace the thrust bearing
Wear of the crosshead	Spray or replace the crosshead
Metamorphism of lubricating oil	Replace the lubricating oil
Less lubricating oil in crankcase	Add lubricating oil

posterior probabilities of abnormal situations, in descending order, are Y3_2, Y3_4, Y3_1, and Y3_3 (see Fig. 6.22). For the dynamic node Y3_2, the state space is {normal; wear; failure} and the corresponding posterior probability is {0.1342; 0.8104; 0.0554}. The results indicate that the most probable root cause is that the axle bushes became worn. To keep the fracturing pump in motion with high efficiency, it was necessary to stop the pump and replace the worn or damaged

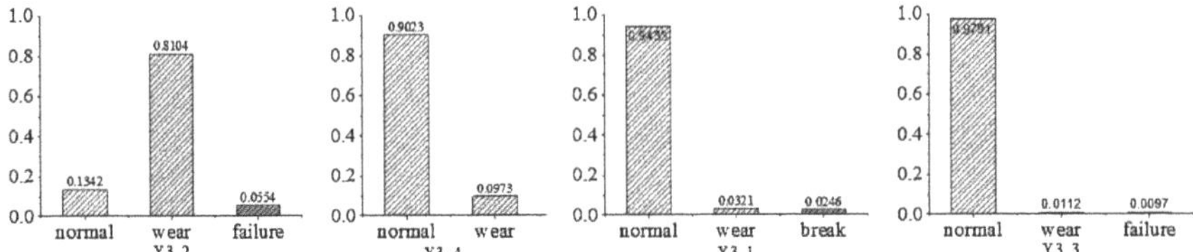

Fig. 6.22. Posterior probability distribution of the first four dynamic nodes.

axle bushes. Because the oil supply flow (X2_3) and oil temperature at the outlet (X2_4) were in a normal state, it was indicated that the wear of the axle bushes may have resulted from the oxidation of the lubricating oil or excessive impurity content in the lubricating oil.

Using the results of the online fault diagnosis model, the on-site maintenance personnel inspected the axle bushes of the power end system of fracturing pump #6 and found that an axle bush was being subjected to abrasive wear. This was because a large quantity of impurities entered the clearance between the axle bush and the journal, which meant that the oil area of the axle bush was reduced and further abrasion occurred. The maintenance staff changed the worn axle bush and updated the lubricating oil. After restarting fracturing pump #6, the crankcase temperature and the motor current returned to normal levels (see Figs. 6.20 and 6.21).

If the traditional HAZOP method for fault analysis was applied in this situation, all possible parallel fault reasons would come up without any priority rating, so field engineers would not know which safety-related action to implement first (see Table 6.16). In contrast, the online fault diagnosis model proposed in this chapter gave the possible fault reasons with priority ratings, which reduced the time and cost for troubleshooting, enabling a targeted maintenance plan. Hence, the proposed approach successfully overcame limitations of the traditional HAZOP method. In this way, the efficiency of safety inspection and maintenance is improved, and measures to avoid accidents or catastrophic events can be implemented as soon as possible.

Figure 6.22 shows that the probabilities of the normal states of Y3_1, Y3_3, and Y3_4, which can be considered as the reliability indexes of those components, are 0.9433, 0.9791, and 0.9023, respectively. Furthermore, starting from the current degraded state of the power end subsystem of fracturing pump #6, the future reliability trends of these three components are predicted. The remaining life

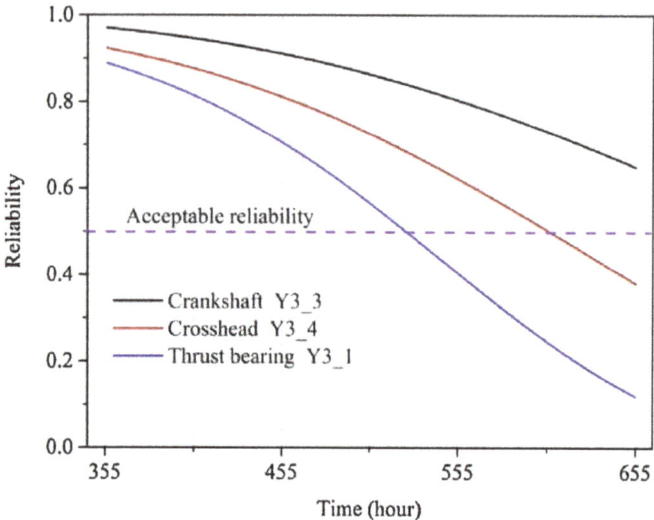

Fig. 6.23. Future reliability trends of the degraded power end system.

of Y3_1 and Y3_4 is about 170 h and 252 h, respectively (Fig. 6.23), if the acceptable reliability in industrial practice of 0.50 is used as a criterion for replacement; that is, Y3_1 and Y3_4 can still be used for about 23 and 34 fracturing stages, respectively. Although crankshaft Y3_3 is currently highly reliable, its degradation rate will increase after 200 h; therefore, to prolong the service life of the crankshaft, it is suggested that timely maintenance measures should be undertaken at about 555 h.

The reciprocating motion of the plunger (X3-5) is the process of transferring energy from the power terminal system to the hydraulic terminal system, and its predicted probability of being in a normal state in the future is provided in Fig. 6.24. This indicates that the probability of plunger Y3_6 keeping a normal reciprocating motion would reduce rapidly in the case of a worn axle bush, although it will still be greater than 0.50 in the next 136 h. In consideration of the interaction between the power end subsystem and the hydraulic end subsystem, the trends of performance indexes of the hydraulic end subsystem (the pressure of outlet pipe X3_7 and the flow at outlet pipe X3_9), are calculated and shown in Fig. 6.24. The wear of the axle bush will accelerate the speed at which X3_7 and X3_9 deviate from their normal intervals (Fig. 6.25). This shows that the degraded

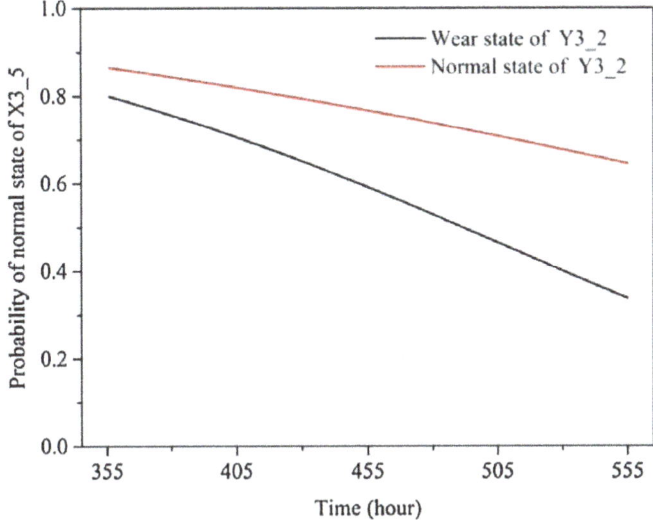

Fig. 6.24. Future performance trend of power end system.

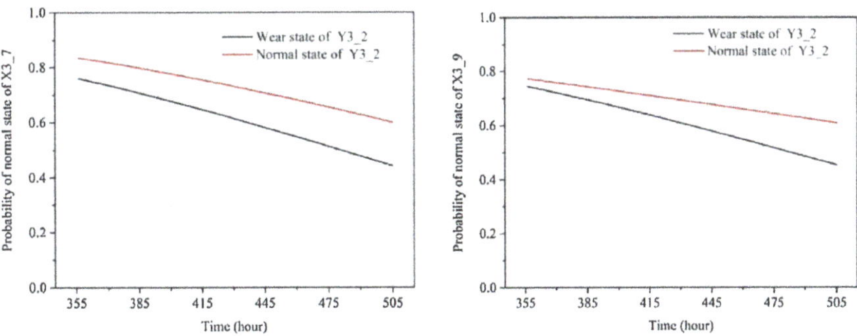

Fig. 6.25.   Future performance trend of the hydraulic end subsystem.

states of the axle bush in the power end subsystem would reduce the performance of the hydraulic end subsystem because of the flows between them.

The following abnormal condition is considered in this case study: By minute 228 of the 53rd fracturing stage (the cumulative operation time of the PISS is now about 392 h), the field instrument system sent an alarm that the oil supply flow (X2_3) at the thrust bearing (X3_1) of the power end subsystem in fracturing pump #8 was low and beyond its lower limit (see Fig. 6.26) and that the pressure difference

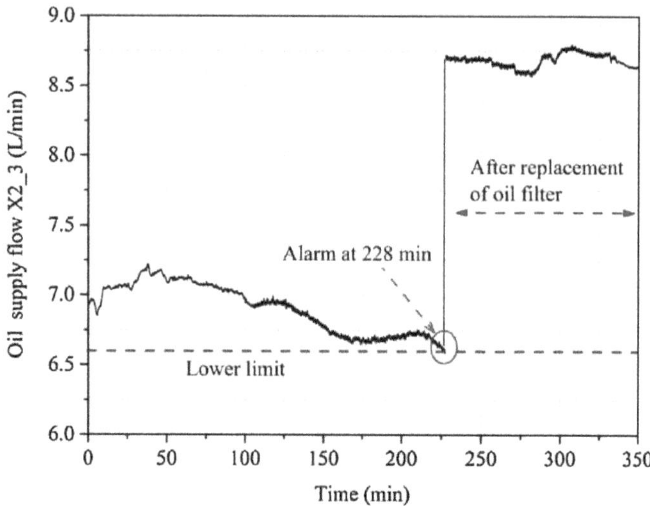

Fig. 6.26.   Oil supply flow of the power end subsystem in fracturing pump #8 (X2_3).

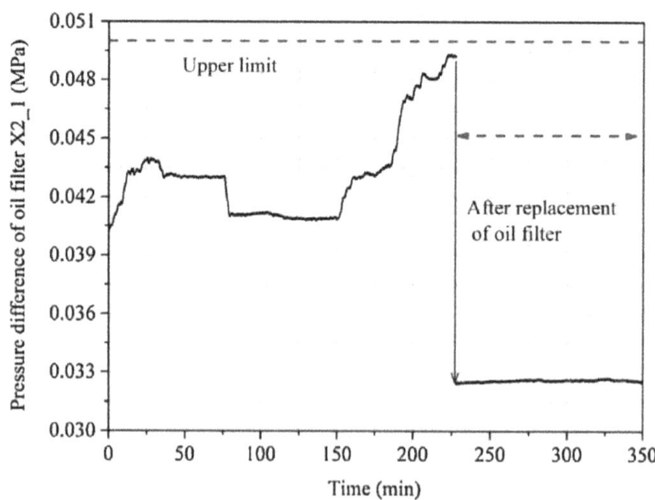

Fig. 6.27.   Pressure difference of oil filter in fracturing pump #8 (X2_1).

of the oil filter (X2_1) was high but not beyond its upper limit (see Fig. 6.27).

To avoid damage from a lack of lubricating oil, fracturing pump #8 was shut down immediately. Although X2_1 was not beyond the upper limit of its safety range, it has a tendency to exceed it. Therefore, the states of static nodes (i.e., the "low" state of X2_3, the "high" state of X2_1, and the "normal" state of other nodes) were taken as the inputs of the SBM. The posterior probability distributions of three dynamic nodes (oil filter Y2_1, oil pump Y2_2, and oil tube Y2_3) were obtained. For Y2_1, its state space is normal, blockage, and breakdown and the corresponding posterior probability distribution is 0.0756, 0.9212, and 0.0032. For Y2_3, its state space is normal, scaling, and rupture and the corresponding posterior probability distribution is 0.6768, 0.3185, and 0.0047. For Y2_2, its state space is normal, degraded, and blockage and the corresponding posterior probability distribution is 0.7612, 0.2334, and 0.0054. It can be seen that the most likely reason is the blockage of Y2_1, followed by the scaling of Y2_3. After inspection by on-site personnel, it was found that the filter core of Y2_1 was covered with impurities. The blocked filter core was removed and a new one was installed. After restarting fracturing pump #8, the pressure difference of the oil filter (X2_1) and the oil supply flow (X2_3) returned to within their respective safety ranges, as shown in Figs. 6.28 and 6.29.

Furthermore, starting from the current state of the lubricating oil subsystem, the reliability trends of Y2_2 and Y2_3 are predicted and

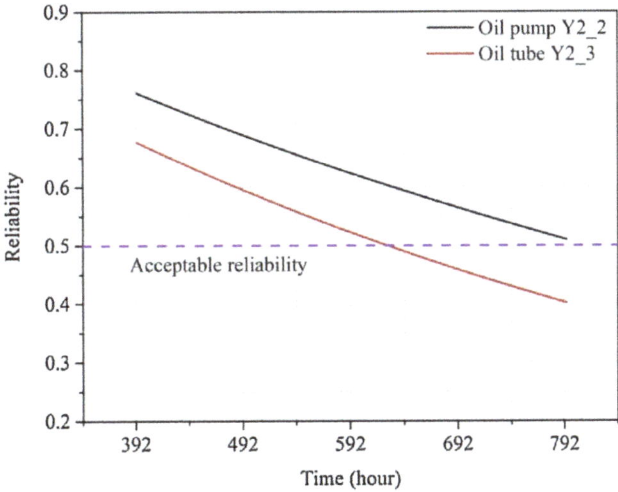

Fig. 6.28. Reliability trends of the oil pump (Y2_2) and oil tube (Y2_3).

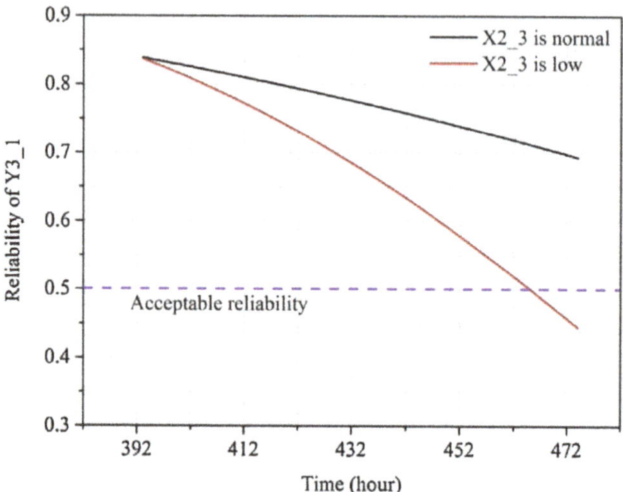

Fig. 6.29. Reliability trends of the thrust bearing of fracturing pump #8 (Y3_1).

shown in Fig. 6.28. The oil pump (Y2_2) can continue in service for about 400 h, which is another 53 fracturing stages; the oil tube (Y2_3) can continue to be used for 233 h, which is another 31 fracturing stages (Fig. 6.28).

As observed from Fig. 6.29, the lubricating oil subsystem provides the necessary oil for the power end subsystem; therefore, if the flow of the lubricating oil (X2_3) is inadequate, the degradation of the power end subsystem would be accelerated. Using the residual life prediction algorithm, the degradation trends of critical parts (thrust bearing Y3_1, crankshaft Y3_3, and crosshead Y3_4) when X2_3 is low are derived (see Figs. 6.29–6.31, respectively). The remaining life of Y3_1, Y3_3, and Y3_4 is reduced to varying extents: If the flow of lubricating oil X2_3 is low, the residual life will be about 73, 115, and 163 h, respectively (Figs. 6.29–6.31).

Safety management is very important for improving the safety and reliability of the pump injection system used in the shale gas well fracturing process. Such a system is of a large scale and complicated, including multiple similar or identical subsystems and a number of monitoring parameters. Within the system, subsystems or components interact closely with each other through various flows, resulting

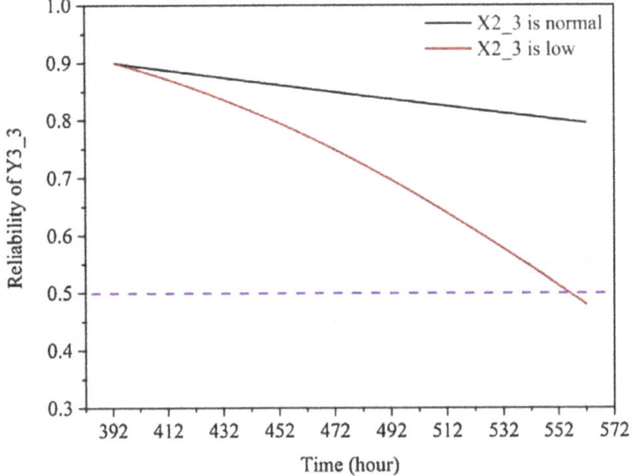

Fig. 6.30.   Reliability trends of the crankshaft of fracturing pump #8 (Y3_3).

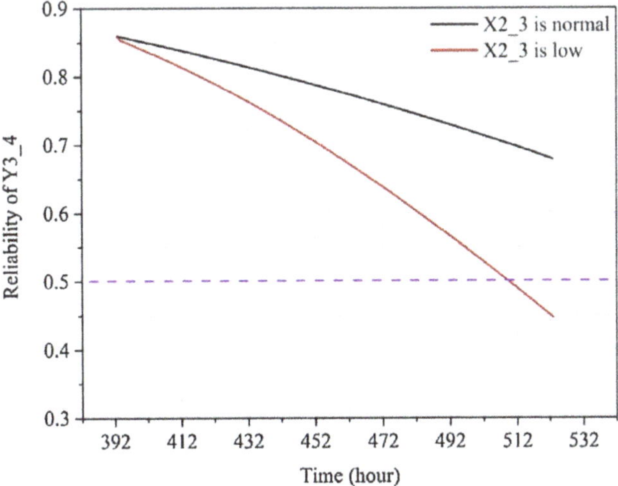

Fig. 6.31.   Reliability trends of the crosshead of fracturing pump #8 (Y3_4).

in uncertainty in system degradation. This chapter describes the limitations of conventional methods for safety assessment and prognosis in these complex systems, given the interactions and dependencies between entities. To ensure safety and increase the operational performance of pump injection systems, a novel integrated model based on a DOOBN methodology is proposed to implement fault diagnosis of and remaining life prediction for the target system. Using a DOOBN method is advantageous because it breaks a large-scale complex system into small reusable units using a modular approach. The proposed safety pre-warning methodology is unique because it has the following distinct features:

(1) The approach clearly represents the causal relationships between monitoring parameter deviations and component or subsystem failure through hierarchical and component-by-component analysis.
(2) The potential root causes are diagnosed with priority ratings when one or several symptoms (abnormal monitoring parameters) arise.
(3) The remaining useful life of components and health indexes representing the state of the system can be predicted, starting from the current state of the pump injection system.

The effectiveness and validation of this approach have been demonstrated through two case studies of a real-world pump injection system. In the first case, fault diagnosis was conducted using the proposed model and the HAZOP method for comparison. The new approach provided the most probable root cause under the conditions of several symptoms. Also, in both cases, the residual life of components was predicted based on the current system state; for example, when the oil filter in the lubricating oil subsystem is blocked, the oil tube could continue to be used for about 31 fracturing stages. These results indicate that this proposed model is a reasonable starting point for the safety management of pump injection systems and could be integrated into safety monitoring devices for field application during fracturing operations.

Appendix

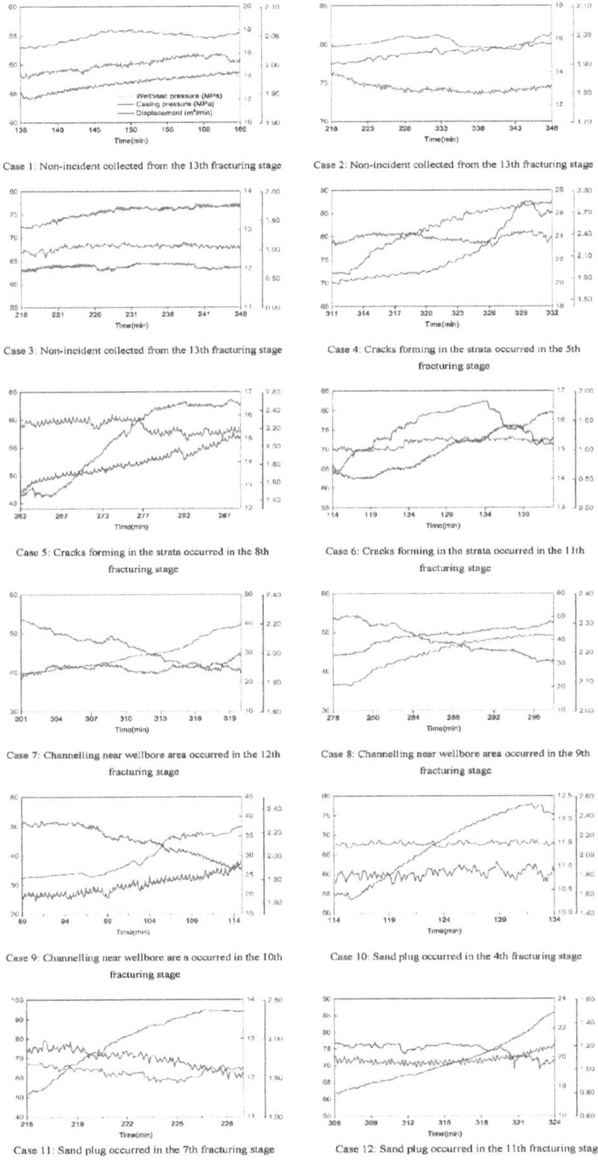

Case 1: Non-incident collected from the 13th fracturing stage

Case 2: Non-incident collected from the 13th fracturing stage

Case 3: Non-incident collected from the 13th fracturing stage

Case 4: Cracks forming in the strata occurred in the 5th fracturing stage

Case 5: Cracks forming in the strata occurred in the 8th fracturing stage

Case 6: Cracks forming in the strata occurred in the 11th fracturing stage

Case 7: Channelling near wellbore area occurred in the 12th fracturing stage

Case 8: Channelling near wellbore area occurred in the 9th fracturing stage

Case 9: Channelling near wellbore are a occurred in the 10th fracturing stage

Case 10: Sand plug occurred in the 4th fracturing stage

Case 11: Sand plug occurred in the 7th fracturing stage

Case 12: Sand plug occurred in the 11th fracturing stage

Fig. 6A.1. Wellhead pressure, casing pressure, and displacement of each case in Table 6.4.

Table 6A.1. Failure data of crankshafts.

System	No.	State before maintenance	Fracturing stages before failure	Accumulated running time	Maintenance mode	State after maintenance
PISS	1	Wear	86	651.4	Grinding treatment	Degraded
		Failure	64	483.3	Replacement	Normal
	2	Failure	129	967.5	Replacement	Normal
	3	Failure	138	1039.8	Replacement	Normal
	4	Wear	95	712.1	Grinding treatment	Degraded
		Failure	57	427.5	Replacement	Normal
	5	Wear	89	671.5	Grinding treatment	Degraded
		Failure	80	583.6	Replacement	Normal
	6	Failure	144	1145.7	Replacement	Normal
PIS #13	7	Failure	151	1154.5	Replacement	Normal
	8	Failure	160	1209.2	Replacement	Normal
	9	Failure	134	1031.8	Replacement	Normal
	10	Failure	175	1312.7	Replacement	Normal
	11	Wear	95	701.2	Grinding treatment	Degraded
		Failure	78	573.6	Replacement	Normal
PIS #27	12	Failure	157	1169.5	Replacement	Normal
	13	Failure	142	1057.4	Replacement	Normal
	14	Failure	161	1176.9	Replacement	Normal
	15	Failure	130	955.7	Replacement	Normal
	16	Wear	61	450.5	Grinding treatment	Degraded
		Failure	73	549.8	Replacement	Normal
	17	Failure	175	1315.5	Replacement	Normal

References

[1]   Laibin Zhang, Xin Zhang, Jinqiu Hu, *et al.* A comprehensive method for safety management of a complex pump injection system used for shale-gas well fracturing. *Process Safety and Environmental Protection*, 2018, 120: 370–387.

[2]   Xin Zhang, Laibin Zhang, and Jinqiu Hu. Real-time risk assessment of a fracturing manifold system used for shale-gas well hydraulic fracturing activity based on a hybrid Bayesian network. *Journal of Natural Gas Science and Engineering*, 2019, 62: 79–91.

[3]   Xin Zhang, Laibin Zhang, and Jinqiu Hu. Real-time diagnosis and alarm of down-hole incidents in shale-gas well fracturing process. *Process Safety and Environmental Protection*, 2018, 116: 243–253.

[4]   Xin Zhang, Laibin Zhang, and Jinqiu Hu. A fault inference method under uncertainty: Case study on crankshafts in fracturing pumps. *IOP Conference Series Materials Science and Engineering*, 2019, 5: 12010.

[5]   Zhang Xin, Hu Jinqiu, Zhang Laibin, *et al.* High-accuracy fault diagnosis of chemical processes based on RS and SVM. *Acta Petrolei Sinica* (Petroleum Processing Section), 2017, 33(04): 777–784 (in Chinese).

[6]   Xin Zhang, Jinqiu Hu, and Labin Zhang. Performance assessment of fault classifier of chemical plant based on support vector machine. In *12th International Conference on Natural Computation, Fuzzy Systems and Knowledge Discovery*, 2016.

[7]   Zhang Xin. Research on real-time risk assessment method for shale gas fracturing operation and key equipment based on Bayesian network. China University of Petroleum, Beijing, 2019 (in Chinese).

[8]   Wei Jia, and Xin Yongliang. Research on the causes and strategies of sand plugging in shale gas fracturing construction. *China Petrochem*, 2017(09): 95–96 (in Chinese).

[9]   Zeng Yuyun, Liu Jingquan, Yang Chunzhen, *et al.* A machine learning based system performance prediction model for small reactors. *Nuclear Power Engineering*, 2018, 39(01): 117–121 (in Chinese).

[10]   Kim KO, Zuo MJ. General model for the risk priority number in failure mode and effects analysis. *Reliability Engineering and System Safety*, 2018, 169: 321–329.

[11]   Khakzad N, Khan F, Amyotte P. Dynamic safety analysis of process systems by mapping bow-tie into Bayesian network. *Process Safety and Environmental Protection*, 2013, 91(1–2): 46–53.

[12]   Li X, Chen G, Zhu H. Quantitative risk analysis on leakage failure of submarine oil and gas pipelines using Bayesian network. *Process Safety and Environmental Protection*, 2016, 103: 163–173.

[13] Ayodeji A, Liu Y. Support vector ensemble for incipient fault diagnosis in nuclear plant components. *Nuclear Engineering and Technology*, 2018, 50(8): 1306–1313.

[14] Biagetti T, Sciubba E. Automatic diagnostics and prognostics of energy conversion processes via knowledge-based systems. *Energy*, 2004, 29: 2553–2572.

[15] Feng E, Yang H, Gao M. Fuzzy expert system for real-time process condition monitoring and incident prevention. *Expert Systems with Applications*, 1998, 15(3–4): 383–390.

[16] Abimbola M, Khan F. Resilience modeling of engineering systems using dynamic object-oriented Bayesian network approach. *Computers & Industrial Engineering*, 2019, 130: 108–118.

[17] Cai B, Liu Y, Zhang Y, *et al.* Dynamic Bayesian networks based performance evaluation of subsea blowout preventers in presence of imperfect repair. *Expert Systems with Applications*, 2013, 40(18): 7544–7554.

[18] Cai B, Liu Y, Ma Y, *et al.* Real-time reliability evaluation methodology based on dynamic Bayesian networks: A case study of a subsea pipe ram BOP system. *ISA Transactions*, 2015, 8: 595–604.

[19] Song G, Khan F, Yang M. Probabilistic assessment of integrated safety and security related abnormal events: A case of chemical plants. *Safety Science*, 2018, 113: 115–125.

Chapter 7

Typical Cases of Multi-Level Correlation Warning for Refining and Chemical Equipment

7.1 Optimization of Process Alarm Systems Based on Alarm Clustering

As the complexity of industrial alarm systems and the number of alarms grow exponentially, taking effective measures for alarm management helps operators find critical alarms and make correct decisions in time when abnormal working conditions occur. In order to prevent the occurrence of correlated alarms, too many jitter alarms, or even alarm flooding, this section proposes a process alarm system optimization method based on alarm clustering to address the lack of representative binary alarm sequences and adaptive challenges of existing alarm correlation analysis methods. Based on the word embedding method, we adaptively learn the vector representation of each alarm variable. An integrated clustering method is used to adaptively group the correlated alarms, and the clustering results are visualized through multi-dimensional scaling. Finally, we propose an optimization strategy for the alarm system, which helps detect and eliminate redundant and chattering alarms, perform causal analysis and locate the root causes of alarms, and give operators the opportunity to quickly target and solve problems, thereby avoiding potential faults and accidents.

7.1.1 *Process alarm optimization issues and difficulties*

An alarm system is a type of system designed to direct the operator's attention to abnormal process conditions. A well-designed alarm system should meet the requirements of existing standards and guidelines such as EEMUA-191 and ANSI/ISA-18.2 As technology continues to evolve, the configuration of process alarms is becoming easier. In this case, operators are overwhelmed by the number of alarms that far exceed their ability to handle them, thus contributing significantly to industrial accidents. Nowadays, the increasing emphasis on process safety has drawn extensive attention from industry and academia to advanced alarm management techniques, of which alarm correlation and clustering analysis are a top priority for effective alarm management.

There are often many interactions in industrial processes and some alarms may be caused by the same initial abnormal event. These closely related alarms often convey similar process information and are often referred to as sequence or correlated alarms. The literature has defined, "Sequence alarms are those in which one alarm occurs first under the same activation conditions, followed by related alarms in no particular order." Both sequential and correlated alarms are referred to as associated alarms. Recognizing correlated alarms allows operators to eliminate redundant alarms and find the root cause of alarms in a timely manner, preventing the process from deteriorating and thus ensuring process safety. Today, the main difficulties in alarm correlation and clustering analysis are as follows:

(1) How to obtain representative alarm sequences is the key to correlation and clustering analysis. Most of the existing methods are based on the binary alarm sequences of each process variable, however, some variables usually do not have alarms frequently, so it is difficult to obtain representative binary alarm sequences.

(2) The alarm data in the alarm log is a set of text information generated by the DCS system, usually with the following basic attributes: variable name, alarm level, and time stamp information. Alarm levels usually include "HI (High)", "HH (High-High)", "LO (Low)", "LL (LowLow)" and so on. In this section, the variable name and alarm level are defined together as an alarm variable type, and the timestamp is the time when the

alarm occurs. An alarm log consists of a series of alarms that occur in chronological order. The core problem that needs to be solved for alarm correlation and clustering analysis based on alarm logs is how to quantify the alarm data and preserve the correlation among the alarm variables.

Aiming at the above challenges, combined with the rich multivariate alarm information in the alarm log, this section proposes a process alarm system optimization method based on alarm clustering, which employs the Word Embedding method "Word2Vec" to map each alarm variable into a corresponding real-valued vector. As a Natural Language Processing (NLP) technique, the Word Embedding method can capture different degrees of similarity between words, and the words that frequently appear one after another in the text will be very close to each other in the vector space, so the distance between the alarm vectors can indicate the degree of correlation between the alarm variables. In addition, in order to overcome the immutability of hierarchical clustering and the influence of the initial clustering center on the k-mean algorithm, this section adopts an integrated clustering method to group the associated alarms and proposes an alarm system optimization strategy to eliminate redundant and jittery alarms, and to analyze the correlation relationship between alarm variables within each category, which can help to identify the root cause of the alarms in order to avoid the occurrence of alarm flooding.

7.1.2 *Fundamental principle*

The basic methods used for alarm clustering and visualization of clustering results mainly include word embedding, cluster analysis, and multi-dimensional scaling methods.

7.1.2.1 *Word embedding*

Word Embeddings are a class of word representations in NLP that use real-valued vectors to represent words in order to use machine learning algorithms for various NLP tasks, such as syntactic analysis and sentiment analysis. Word2Vec is a two-layer neural network for generating word vectors, which takes the corpus as input and generates n-dimensional vectors (n is the dimensionality of the embedding space) for each word in the corpus. In the embedding space, the corresponding vectors of the words that usually appear one after another in the corpus are very similar in distance, so the word vectors maintain the similarity of the words in the context.

By comparing various Word Embedding methods, the literature concludes that Word2Vec yields the best representation of word vectors in a low-dimensional semantic space. Word2Vec can compute vector representations of words using two model structures: the Continuous Bag-of-Words (CBOW) and Skip-Gram models. The CBOW model predicts a target word based on the contextual words of that word and assumes that the order of the contextual words has no effect on the prediction result. In contrast, the Skip-Gram model uses each current word as an input to predict its contextual word, and words closer to the current word are given more weight. In contrast, CBOW takes less time, but the Skip-Gram model expresses less frequent words better. The training goal of the Skip-Gram model is to find word vector representations that help to predict the contextual words in a sentence or document, and given the training words $m_1, m_2, \ldots, m_N$, the Skip-Gram aims at maximizing the mean log probability P_m, see Eq. (7.1):

$$P_m = \frac{1}{N} \sum_{n=1}^{N} \sum_{-c \leq i \leq c, i \neq 0} \lg p(m_{n+i} \mid m_n), \tag{7.1}$$

where c is the width of the context window for each current word. the larger c is, the more training samples are needed, which improves the training accuracy at the cost of more training time. The basic Skip-Gram uses softmax function to define $p(m_{n+i} \mid m_n)$, see Eq. (7.2):

$$p(m_O \mid m_I) = \frac{\exp(\nu'^T_{m_O} \nu_{m_I})}{\sum_{m=1}^{M} \exp(\nu'^T_{m} \nu_{m_I})}, \tag{7.2}$$

where ν'_m and ν_m are the vector representations of the output and input of the word m, respectively, and M is the size of the vocabulary.

In order to reduce the computational effort, Word2Vec models can be trained using Hierarchical Softmax (HS) or Negative Sampling (NS) methods. The basic idea of Word2Vec is to maximize the similarity between the corresponding vectors of frequently occurring words in the text. To this end, the HS method uses Huffman trees to reduce the computational effort, while the NS method handles this maximization problem by minimizing the log-likelihood of sampling negative instances. Literature studies have concluded that the Skip-Gram with Negative Sampling (SGNS) model outperforms other Word Embedding methods on similarity tasks.

Therefore, in this section, the SGNS model is used to learn the vector representation of each alarm variable, the alarm variable is regarded as a word in the text, the alarm log is regarded as a text, and the generated alarm vectors retain the correlation information among the alarm variables. The basic principle of the SGNS method is as follows:

For a central word m (positive example), the $2c$ words of its context are remembered as $context(m)$. The purpose of negative sampling is to get neg center words m_i ($i = 1, 2, \ldots, neg$) that are different from m, constituting neg negative examples $[context(m)$ with $m_i]$ that do not really exist.

Assuming that the size of the vocabulary list is M, a line segment of length 1 is divided into M parts, which correspond to M words in the vocabulary list one by one. The line segments are of different lengths, with long line segments corresponding to high-frequency words and short line segments corresponding to low-frequency words, and the length of the line segment for each word m is computed in the Word2Vec model by using Eq. (7.3):

$$len(m) = \frac{count(m)^{3/4}}{\sum_{word \in vocabulary} count(word)^{3/4}}. \tag{7.3}$$

In order to obtain neg negative examples by negative sampling method, the segment of length 1 is first divided into W equal parts (W is much larger than M, in Word2Vec, $W = 10^8$) to ensure that the segment corresponding to each word m will be divided into corresponding chunks. In negative sampling, we only need to sample the corresponding line segments in neg positions out of the line segments in W positions, and the word to which it belongs is the negative example word.

Negative sampling computes the model parameters through logistic regression, noting the positive example (center word) as m_0 and the negative sampling instance $[context(m_0), m_i]$, where $i = 1, 2, \ldots, neg$. In logistic regression, for the positive and negative instances, Eqs. (7.4) and (7.5) are expected to be satisfied, respectively, where $\sigma(\ldots)$ is the sigmoid function and ν_{m_0} is the word vector corresponding to m_0:

$$P[context(m_0), m_i] = \sigma(\nu_{m_0}^T \theta^{m_i}) \quad (y_i = 1, i = 0), \tag{7.4}$$

$$P[context(m_0), m_i] = 1 - \sigma(\nu_{m_0}^T \theta^{m_i}) \quad (y_i = 0, i = 1, 2, \ldots, neg). \tag{7.5}$$

Based on the Stochastic Gradient Ascent, the SGNS algorithm flow can be obtained as follows:

The inputs are Skip-Gram-based vocabulary training samples with context size $2c$ and step size η, and the outputs are model parameters θ and word **vectors** ν_m corresponding to each word.

Step 1: Initialize all parameters θ and word **vectors** ν_m randomly.

Step 2: For each training sample $[context(m_0), m_0]$, sample neg negative example center words m_i, where $i = 1, 2, \ldots, neg$.

Step 3: Perform a gradient ascent iteration as shown in Fig. 7.1 for each training sample $[context(m_0), m_0, m_1, \ldots, m_{neg}]$ as follows:

(1) *for $j = 1$ to $2c$*:

   ① $\varepsilon = 0$

   ② *for $i = 0$ to neg* :

$$f = \sigma(v_{m_{0j}}^T \theta^{m_i})$$

$$g = (y_i - f)\eta$$

$$\varepsilon = \varepsilon + g\theta^{m_i}$$

$$\theta^{m_i} = \theta^{m_i} + gv_{m_{0j}}$$

   ③ Updating word vectors $v_{m_{0j}} = v_{m_{0j}} + \varepsilon$

(2) If the gradient converges, terminate the iteration and the algorithm ends, otherwise go to step (1)

Fig. 7.1. SGNS algorithm flow.

7.1.2.2 *Cluster analysis*

Cluster analysis is a class of data mining tools for statistical data analysis, which is an unsupervised classification method that divides similar objects within the same class, while objects within different classes have great dissimilarity. Among them, Agglomerative Hierarchical Clustering and k-means clustering methods are two commonly used cluster analysis methods.

(1) Agglomerative Hierarchical Clustering methods

Agglomerative Hierarchical Clustering (AHC) is a bottom-up clustering method in which the algorithm initializes each object (each alarm variable type) in a separate class, and based on some similarity criterion, merges the two closest classes at each step until only one class remains. The degree of dissimilarity is expressed by calculating the distance d_{ij} between alarm variable types i and j, see Eq. (7.6):

$$d_{ij} = \sqrt{\sum_{k=1}^{n}(x_{ik} - x_{jk})^2}, \tag{7.6}$$

where n is the dimension of the alarm vector, i.e., the dimension of the embedding space; x_{ik} and x_{jk} are the kth component of the vector corresponding to alarm variable types i and j, respectively.

In this section, average linkage is used to compute the similarity between different classes. Average linkage defines the distance between two classes p and q as the average of the two-by-two distances d_{ij} $(i \subset p, j \subset q)$ of all points in the two classes. The system tree diagram graphically displays the clustering results and the distances between the classes. The AHC method does not require a pre-set number of clusters, but once two classes have been merged, it cannot be undone.

(2) k-means clustering method

The k-means algorithm is a class of distance-based non-hierarchical clustering methods that uses distance as an evaluation metric for similarity, with the goal of making all distances between objects within a class close enough and all distances between objects between classes as far as possible. The algorithm first randomly selects k objects (i.e., alarm vectors) as the initial clustering centers, and assigns each of the remaining objects to the closest class based on the magnitude of the distance to each clustering center. The algorithm recalculates the average value $\bar{e}_h$ of the distance e_{ih} from object i to the clustering center within each class h as the new clustering center in each iteration, and reassigns each object to the closest class until the clustering center does not change anymore, which indicates that the algorithm has converged, as shown in Eq. (7.7):

$$\bar{e}_h = \frac{1}{n_h} \sum_{i \subset h} e_{ih},\tag{7.7}$$

where n_h is the number of objects in class h.

The k-means clustering algorithm is easy to implement and can produce tighter classes than hierarchical clustering. However, as a heuristic algorithm, the method is susceptible to the initial clustering center.

7.1.2.3 *Multi-dimensional scaling*

Multi-dimensional Scaling (MDS) allows for the visualization of alarm vectors in 2D space. The method can obtain a good low-dimensional representation of the distances between the original high-dimensional data and thus visualize the alarm clustering results. The following section focuses on the MDS method for visualizing n-dimensional alarm vectors.

For Sn-dimensional alarm vectors $\boldsymbol{x}_i$ ($i = 1, 2, \ldots, S$), the Euclidean distance δ_{ij} ($i, j = 1, 2, \ldots, S$) between two vectors can be expressed as the following matrix form $\boldsymbol{D}_{Eu}$ in Eq. (7.8):

$$\boldsymbol{D}_{Eu} = \begin{bmatrix} \delta_{11} & \delta_{12} & \cdots & \delta_{1S} \\ \delta_{21} & \delta_{22} & \cdots & \delta_{2S} \\ \vdots & \vdots & \ddots & \vdots \\ \delta_{S1} & \delta_{S2} & \cdots \delta_{SS} \end{bmatrix}.\tag{7.8}$$

The basic principle of MDS is to embed these S vectors into a low-dimensional (2D) space in order to keep the similarity (distance) between them as constant as possible. Each alarm vector $\boldsymbol{x}_i$ is first initialized as a two-dimensional vector $\boldsymbol{y}_i$, and the optimal $\boldsymbol{y}_i'$ value can be obtained by minimizing the objective function O [Eq. (7.9)] to represent each alarm vector $\boldsymbol{x}_i$:

$$O = \sum_{i=1}^{S} \sum_{j=1}^{S} (\delta_{ij} - \lambda_{ij})^2, \tag{7.9}$$

where λ_{ij} is the Euclidean distance between $\boldsymbol{y}_i$ and $\boldsymbol{y}_j$.

Finally, each vector $\boldsymbol{y}_i'$ is represented on a 2D scatter plot for the purpose of visualizing each high-dimensional alarm vector $\boldsymbol{x}_i$.

7.1.3 *Alarm system optimization method based on alarm clustering*

Based on the Word Embedding approach, process alarms can be represented in vector form. In the embedding space, the corresponding vectors of the alarm variables that frequently appear one after another in the alarm log are also at similar distances, thus preserving the correlation information among the alarm variables. In this section, an integrated clustering method is used to cluster the correlated alarm vectors, and the clustering results are visualized on a 2D scatter plot using the MDS method. By performing a correlation analysis of alarm variables within each category, an alarm system optimization strategy is proposed for eliminating redundant and jitter alarms and suppressing correlated alarms. The proposed method will help to optimize the management of process alarms, such as analyzing the root causes of alarms and controlling alarm flooding. Figure 7.2 shows the flow of the proposed method.

7.1.3.1 *Alarm data preprocessing*

An alarm log usually contains several months or even a full year's worth of alarms. If the time interval between two neighboring alarms is too long, it does not make much sense to consider the correlation between the two alarms. Therefore, it is necessary to set a time threshold to divide the alarm log. If the time interval between two neighboring alarms exceeds the set threshold, the latter will be used as the first alarm variable in another alarm sequence.

Jitter alarms are a class of alarms that occur frequently within a short period of time, and alarms that occur three or more times within a minute are usually considered the most serious jitter alarms.

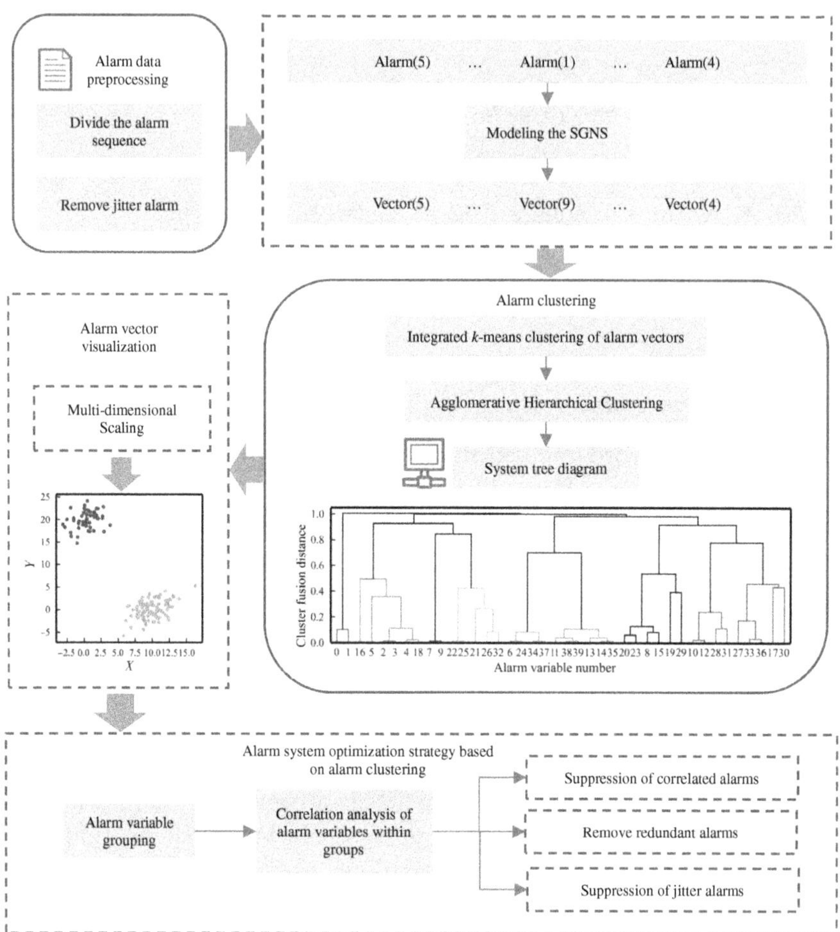

Fig. 7.2. Process alarm system optimization method flow based on alarm clustering.

Most of the jitter alarms are redundant and affect the effectiveness of Word Embedding methods to produce high-quality alarm vectors. In this section, alarms that occur more than three times in a minute are fused into one alarm to eliminate jitter alarms.

7.1.3.2 *Generate alarm vector*

The SGNS model is used to learn the vector representation of each alarm variable, and the SGNS model is trained by a powerful Python

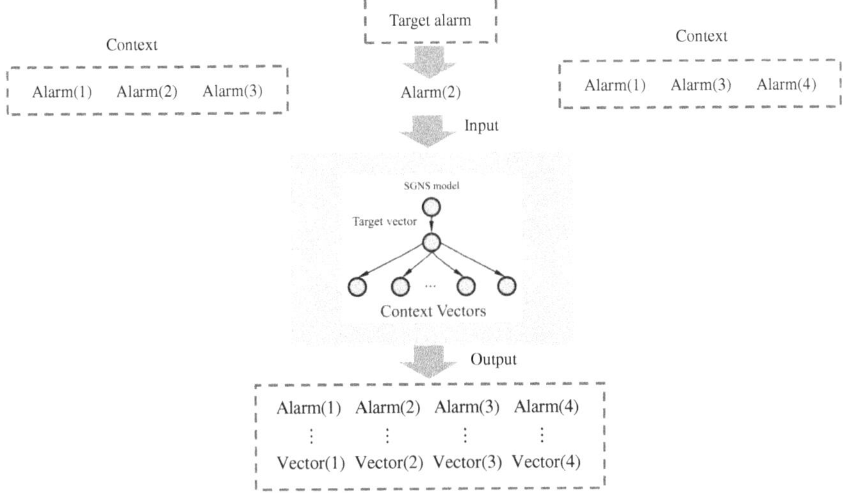

Fig. 7.3. Alarm vector generation process.

library Gensim to obtain the required alarm vectors. Gensim requires that the inputs must be in the form of sequences, so each alarm sequence is taken as the input to the model, and each sequence consists of alarm variables arranged in chronological order. The SGNS model adaptively maps each alarm variable into a corresponding n-dimensional vector to obtain the embedding matrix $S \times n$, where S is the number of alarm variable types and n is the dimension of the embedding space.

Figure 7.3 gives the generation process of alarm vectors, where the inputs and outputs of the SGNS model are the vectors corresponding to the target alarms and their Context alarm vectors, respectively.

7.1.3.3 *Alarm clustering and visualization*

The S alarm vectors generated by the Word Embedding method are grouped by an integrated k-means clustering method.

Considering the severe impact of the initial clustering centers on the k-means clustering results, here a collection of solutions is integrated into a single hierarchical classification result. The solutions from the rth run of the k-means algorithm are first recorded in a binary matrix $\boldsymbol{R}^{(r)}$, and the matrix element $R_{ih}^{(r)}$ is written as 1 if the alarm variable i is assigned to class h at the rth run, and 0 otherwise. The matrix $\boldsymbol{R}$ is obtained by concatenating the solutions $R^{(r)}$ obtained from the Mth run, see Eq. (7.10):

$$\boldsymbol{R} = [R^{(1)}, R^{(2)}, \ldots, R^{(M)}]. \tag{7.10}$$

The integration distance matrix $\boldsymbol{D}$ can be calculated by Eq. (7.11), where 1 denotes the 1 matrix:

$$\boldsymbol{D} = 1 - \frac{1}{M} RR^{T}. \tag{7.11}$$

Here, the matrix $\boldsymbol{D}$ is a dissimilarity measure that can be used as an input to the AHC algorithm to generate a system tree diagram for the user to decide on the final clustering result. In order to adaptively determine the optimal value of k, the number of clusters is set sequentially from 2 to k_{max} and the k-means algorithm is run. The value of k_{max} can be determined by comparing the results obtained with different experimental values of k'_{max}. Increasing k'_{max} from 2, the k-means algorithm is randomly run M_{runs} times with $k = k'_{\mathrm{max}}$ as the number of clusters, respectively, to cluster the alarm vectors, and the integrated distance matrix $\boldsymbol{D}(k'_{\mathrm{max}})$ is computed according to Eqs. (7.10) and (7.11). The difference between the distance matrices obtained using k'_{max} and $k'_{\mathrm{max}} + 1$ as the clustering numbers can be evaluated by the Sum of Squared Errors (SSE) of the errors between their one-to-one corresponding elements, denoted as $\varepsilon(k'_{\mathrm{max}})$, see Eq. (7.12):

$$\varepsilon(k'_{\mathrm{max}}) = \sum_{i<j} [D_{ij}(k'_{\mathrm{max}} + 1) - D_{ij}(k'_{\mathrm{max}})]^2. \tag{7.12}$$

The parameter k_{max} is the minimum value of k'_{max} that brings $\varepsilon(k'_{\mathrm{max}})$ to convergence. The converged distance matrix $\boldsymbol{D}(k_{\mathrm{max}})$ is used as an input to the AHC algorithm to generate the final system tree diagram from which the appropriate alarm clustering results are selected. The method can adaptively determine the parameter k_{max}, avoiding the immutability of the heuristic solution obtained by AHC and the subjective estimation of the number of clusters by the k-means method. In order to visualize the clustering results, the MDS method is used to generate a 2D scatter plot of each alarm vector, in which different colors are used to indicate different categories.

For Sn-dimensional alarm vectors x_i $(i = 1, 2, \ldots, S)$, calculate the Euclidean distance δ_{ij} $(i, j = 1, 2, \ldots, S)$ between two vectors to obtain the distance matrix $\boldsymbol{D_{Eu}}$ [Eq. (7.8)]. In order to embed the S alarm vectors into a two-dimensional space, each alarm vector $\boldsymbol{x_i}$ is initialized as a two-dimensional vector $\boldsymbol{y_i}$, and the optimal $\boldsymbol{y_i'}$ value can be obtained by minimizing the objective function O [Eq. (7.9)], which is used to represent each alarm vector $\boldsymbol{x_i}$. Finally, the vector $\boldsymbol{y_i'}$ is represented on a two-dimensional scatter plot for the purpose of visualizing each high-dimensional alarm vector x_i.

7.1.3.4 *Alarm system optimization strategy based on alarm clustering*

Based on the alarm clustering results, each alarm variable is grouped, and the correlation between the alarm variables within each category is analyzed by combining process knowledge and expert experience, in order to optimize the alarm system and avoid alarm flooding by suppressing correlated and jitter alarms and eliminating redundant alarms. The main flow of this alarm system optimization strategy is shown in Fig. 7.4.

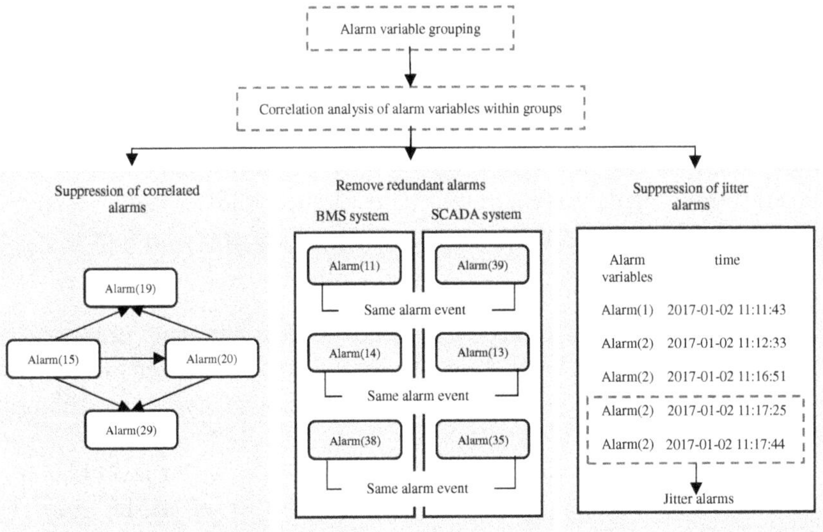

Fig. 7.4. Alarm system optimization process.

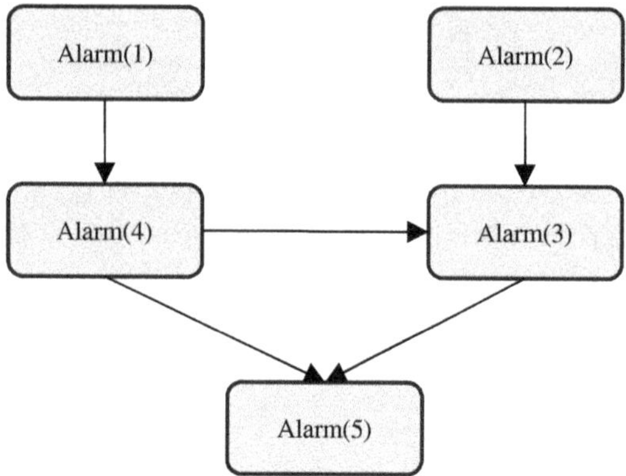

Fig. 7.5.   Example of a process alarm cause and effect diagram.

(1) Suppression of correlated alarms

Process alarm clustering is used to group and monitor the associated alarms. For the alarm variables with causal relationships within each class, the causal network is constructed by combining the process knowledge as shown in Fig. 7.5. The causal graph consists of nodes and directed arrows, where the nodes are used to represent the alarm variables, and the directed arrows are used to represent the causal relationships between the two variables. By analyzing the correlation of alarm variables within a class, if Alarm(i) can cause Alarm(j) to alarm, the directed arrow will point from Alarm(i) to Alarm(j).

When an alarm occurs, it suppresses related alarms that have a causal relationship and assists in locating the root cause of the alarm through the alarm cause and effect diagram, avoiding the flooding of alarms and the occurrence of potential failures and accidents.

(2) Remove redundant alarms

Alarm correlation analysis is used to identify the presence of redundant alarms in the system, where redundant alarms are defined as redundant alarms indicating the same alarm event in different monitoring systems.

As shown in Fig. 7.6, for example, a boiler low water level alarm is generated by both the Burner Management System (BMS) and the SCADA system. The existence of redundant alarm messages may lead to the phenomenon of alarm flooding, which prevents operators

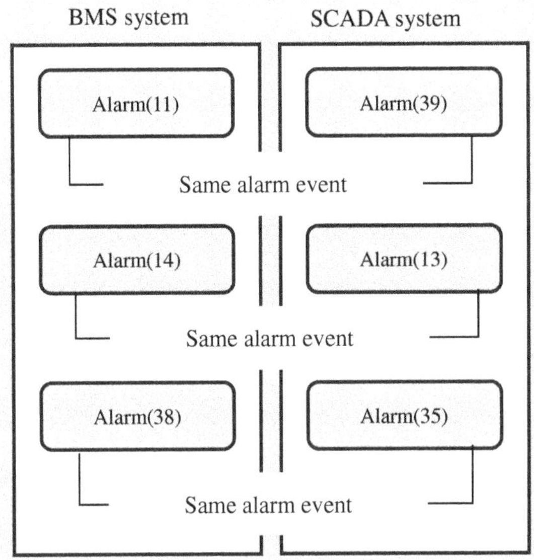

Fig. 7.6. Redundant alarm example.

from detecting critical alarms from numerous alarm messages in time, thus leading to accidents. For this reason, based on the results of cluster analysis, it is possible to identify whether there are redundant alarms in the alarm logs that indicate the same alarm event generated by different monitoring systems and to streamline the alarm system by eliminating redundant alarm variables to prevent the occurrence of alarm flooding.

(3) Suppression of jitter alarms

A jitter alarm refers to a type of alarm that repeatedly transitions between an alarm state and a normal state within a short period of time. Usually, the same alarm that occurs three or more times in one minute is regarded as the most serious jitter alarm. It is stipulated here that if the time interval between two alarms is less than 20 s, an alarm suppression strategy will be adopted for the latter.

As shown in Fig. 7.7, the time interval between two alarms occurring in Alarm (2) is 18 s, so the second Alarm (2) alarm is suppressed.

This alarm suppression strategy prevents jitter and jitter alarms from occurring frequently, interfering with the operator's ability to

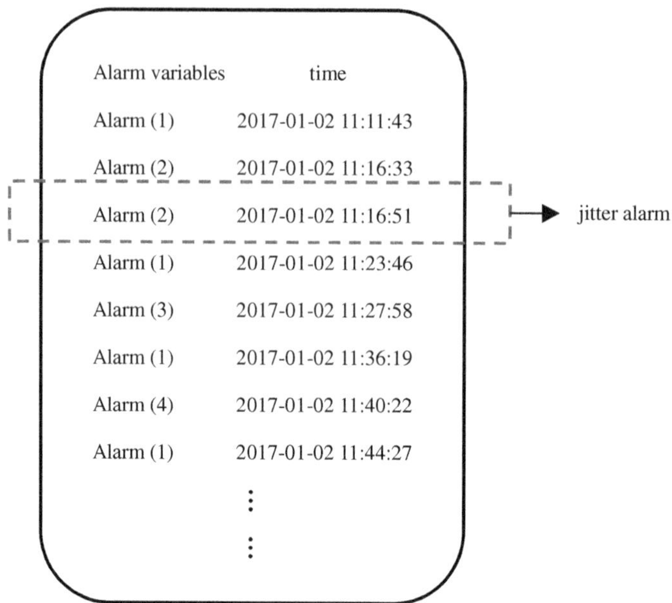

Fig. 7.7.   Example of jitter alarm.

correctly judge the current process state and thus ignoring critical alarm information.

By suppressing correlated alarms and jitter alarms and eliminating redundant alarms through the above method, the number of alarms in the system can be effectively reduced and the frequent occurrence of alarm flooding can be prevented. In addition, this method can help locate the root cause of alarms by analyzing the cause and effect of alarm variables through correlated alarm clustering, so that operators can find the causes of faults in time to ensure the safety of industrial processes.

7.1.4   *Case study*

This section presents a case study of the Central Heating and Cooling Plant (CHCP) at the University of California, Davis. The plant, which is primarily fueled by petroleum, consists of two parts: a steam generation and distribution system and a cooling water generation, storage, and distribution system, and is designed to meet most of

(a) Cooling tower and condenser pumps　　　(b) Chillers and cooling water distribution pumps

(c) Boiler No. 4　　　(d) Thermal energy storage tanks

Fig. 7.8.　Site photographs of major equipment at the CHCP plant.

the heating and cooling needs of the campus buildings. The main equipment of the CHCP plant is shown in Fig. 7.8.

The steam production system consists of four boilers with a combined steam production capacity of 192.8 t/h. The system also includes make-up water treatment, condensate collection, deaeration, and air emission control facilities as necessary. The cooling water system consists of eight motorized centrifugal cooling units and eight distribution pumps with a cooling capacity of 18,000 t. A storage tank of 18,927 m^3 holds approximately 70,000 t/h of cooling water. The cooling units are equipped with associated cooling water pumps, condenser water pumps, evaporative cooling towers, and water treatment systems.

It is understood from CHCP's engineers that they are primarily concerned with process alarm information rather than process data and that many alarms do not have process data, so they have never

Table 7.1. Alarm log form.

Time	Alarm variable name	Variable description
4/21/2017 22:24:22	CHCP_CES#TT8233_HiHiAlm.AD.Msg	Hot Water Heat Exchanger 101 Outlet Temp High High Alarm
4/21/2017 22:24:45	CHCP_CES#TT8233_HiHiAlm.AD.Msg	Hot Water Heat Exchanger 101 Outlet Temp High High Alarm
4/21/2017 22:26:29	CHCP_CES#TT8233_HiHiAlm.AD.Msg	Hot Water Heat Exchanger 101 Outlet Temp High High Alarm
4/21/2017 22:42:47	CHCP_CES#TT8233_HiAlm.AD.Msg	Hot Water Heat Exchanger 101 Outlet Temp High Alarm
4/21/2017 22:43:46	CHCP_CES#TT8233_HiAlm.AD.Msg	Hot Water Heat Exchanger 101 Outlet Temp High Alarm
4/21/2017 22:53:04	CHCP_CES#TT8233_HiAlm.AD.Msg	Hot Water Heat Exchanger 101 Outlet Temp High Alarm
4/21/2017 23:17:23	CHCP_CES#DPT8270.ALHi	Tercero 3 Hot Water Differential Pressure high alarm
4/21/2017 23:19:00	CHCP_CES#DPT8270.ALHi	Tercero 3 Hot Water Differential Pressure high alarm

extracted process data for analysis. In this section, the proposed method is demonstrated using the alarm logs of CHCP in 2017, in which there are 703 alarm variable types and 216,991 alarm records, and the names of each variable are recorded in the following form. For example, "CHCP_STM4#AlmCCSnoxpvhi.ALAlarm" indicates the alarm of High (HI) measurement value of Nitrogen Oxides (NOx) in boiler #4 of CHCP, and the form of the alarm records in the alarm log is shown in Table 7.1.

In reality, the steam production system and the cooling water system are independent of each other, and as a result, alarms in the two systems are completely unrelated. This section focuses on a case study of alarm data from the steam production system at the CHCP plant in 2017.

7.1.4.1 *CHCP alarm preprocessing and alarm vector generation*

Alarm data in the steam production system of the CHCP plant in 2017 were preprocessed based on the methodology described in Section 7.1.2. By consulting with the CHCP engineers, the alarm data were divided into discrete sequences using a time threshold of 15 min. If the time interval between two neighboring alarms exceeds this threshold, the latter will be the first alarm of the next alarm sequence. In order to further eliminate jitter alarms, alarms occurring more than three times in one minute are fused into one alarm here. If the sequence length is too short, its available information will be too little, so sequences with less than five alarms are discarded. In addition, there are alarm types that occurred less than five times in the year-long dataset, and these alarms will not be considered in this section.

The SGNS model was trained using Gensim (see Section 7.1.3), and the preprocessed sequences of each alarm were used as model inputs to map each alarm variable type to a vector with a corresponding dimension of 30. Considering that the number of alarm types is much less than the vocabulary in NLP, an embedding dimension of 30 is chosen. These alarm vectors will be used as inputs to the clustering algorithm.

7.1.4.2 *CHCP alarm clustering and visualization*

The 40 alarm types (more than 10 alarms of each type) listed in Table 7A.1 were selected for cluster analysis.

(1) Cluster the alarm vectors corresponding to these 40 alarms

Starting from 2, increase the number of clusters $k'_{\max}$. For each $k'_{\max}$, randomly run the k-means algorithm $M_{\mathrm{runs}} = 100$ times and calculate its corresponding distance matrix $D(k'_{max})$. Figure 7.9 shows that with the increase of $k'_{\max}$, $\varepsilon(k'_{\max})$ decreases gradually and converges to $k'_{\max} = 10$, so let $k_{\max} = 10$ and calculate the converged distance matrix $D(k'_{max})$ as the input of the AHC algorithm using average linkage to get the system tree diagram shown in Fig. 7.10. The system tree diagram shown, in which nine distinct clusters are represented by different colors respectively, it can be seen that the resulting classification is good.

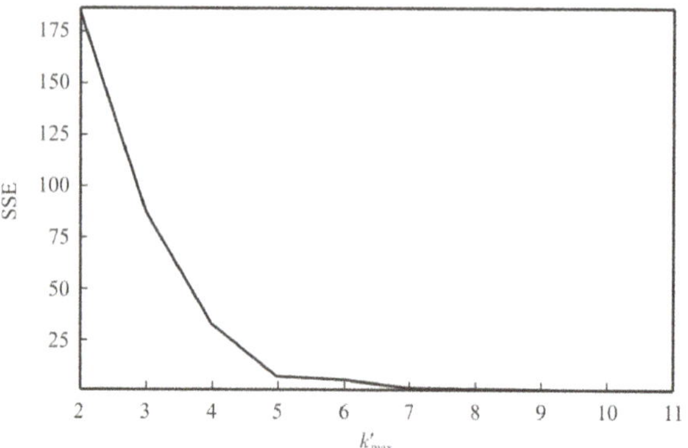

Fig. 7.9. $\varepsilon(k'_{\max})$ curve.

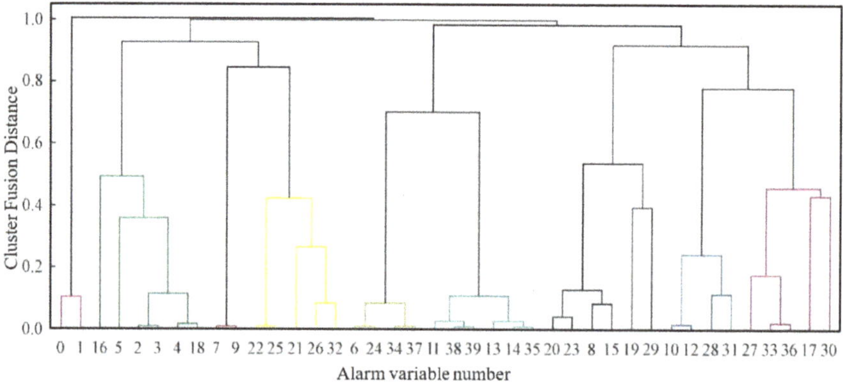

Fig. 7.10. System tree diagram of integrated clustering results.

(2) Adopt the MDS method to realize the visualization of alarm vectors in two-dimensional space.

As shown in Fig. 7.11, different colors indicate the alarm vectors in different classes (the colors correspond to Fig. 7.10 one by one). From the scatter plot, it can be seen that the clustering results are good, thus verifying the feasibility and effectiveness of the alarm clustering method. In Fig. 7.11, the very close Alarm (0) [here each alarm variable is abbreviated as Alarm (i), and i is used to distinguish between different alarm types] and Alarm (1) also have low fusion distances in Fig. 7.10, indicating that vectors with similar distances are clustered into one class. It can also be seen that Alarm (7)

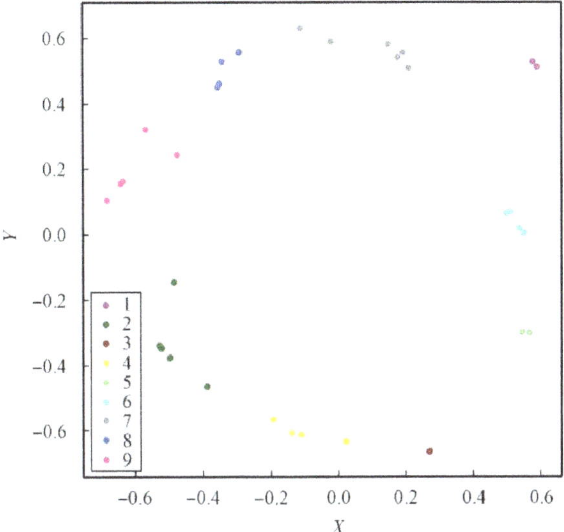

Fig. 7.11. Alarm vector MDS scatterplot.

and Alarm (9), which are almost coincident in Fig. 7.11, have much shorter fusion distances (Fig. 7.10) than Alarm (0) and Alarm (1), indicating that the shorter the distance between vectors, the shorter the fusion distance for clustering.

(3) Alarm system optimization strategy based on alarm clustering

(i) Alarm correlation analysis:

The CHCP alarm clustering results are shown in Table 7.2.

Combining process knowledge and expert experience, the correlation between alarm variables within each category is analyzed as follows:

(a) In class 1, Alarm (0) (Continuous Emission Monitoring System (CEMS) fault) is associated with Alarm (1) (CEMS Cabinet Temperature Excessive) and may be caused by an alarm occurring in Alarm (1).

(b) In class 2, the fusion distance between Alarm (2) and Alarm (3) is extremely short, when the hot water economizer inlet flow rate (FIT-8110) decreases, Alarm (3) should occur first, and if that flow rate continues to decrease to reach the low flow rate limit, Alarm (2) will follow. the fusion distance between Alarm (4)

Table 7.2. Serial number of each alarm type.

Class number	Alarm variable serial number
1	0, 1
2	2, 3, 4, 5, 16, 18
3	7, 9
4	21, 22, 25, 26, 32
5	6, 24, 34, 37
6	11, 13, 14, 35, 38, 39
7	8, 15, 19, 20, 23, 29
8	10, 12, 28, 31
9	17, 27, 30, 33, 36

and Alarm (18) is similarly short. Make-up water energy saver inlet flow (FIT-8100) low alarm Alarm (4) should occur first, and if that flow continues to decrease to reach the low flow limit, Alarm (18) will follow. Alarm (5) energy saver Fan Variable Frequency Drives (VFD) Fault with Alarm (16) energy saver fan VFD trip alarm is associated with the Alarm (16) energy saver fan VFD trip alarm. Although there is no obvious causal relationship between the alarm types in this category, they are all part of the same Condensing Energy Saving System (CES) and associated with the same heat exchanger, and they always appear one after the other in the alarm log.

(c) In class 3, Alarm (7) (Boiler No. 3 Trouble) shall occur before Alarm (9) (Boiler No. 3 Shutdown).

(d) In class 4, Alarm (22) may occur at the same time as Alarm (21), or Alarm (25), and Alarm (26) and Alarm (32) may also occur at the same time, especially when Alarm (25) occurs with a high probability of occurring at the same time. Alarms in this category may be caused by high outlet temperature of hot water exchanger.

(e) In class 5, each alarm variable is associated with the CEMS system, and these alarms may occur when the CEMS system malfunctions.

(f) In class 6, a low water level alarm (Alarm (13) or Alarm (14)) may result in a low water cutoff if the operator does not respond in a timely manner (the other four alarm variables in this class). In addition, it was found that Alarm (11) and Alarm (39) in this category refer to the same alarm event, Alarm (14) is the same as Alarm (13) and Alarm (38) and Alarm (35) are also the same alarm event, with one of the alarm types being generated by the Burner Management System (BMS), and the other coming from the SCADA system. This helps the operator to remove redundant alarms and clean up the alarm system.

(g) In class 7, Alarm (15) should occur before Alarm (8), and Alarm (20) should occur before Alarm (23). Alarm (20) and Alarm (23) are clearly caused by Alarm (8) and Alarm (15). These four alarms can cause Alarm (19) and Alarm (29) to alarm.

(h) In class 8, the alarm types are interrelated, are from the Combustion Control System (CCS), and appear one after the other in the alarm log.

(i) In class 9, each alarm type belongs to the CCS system, Alarm (17) is associated with Alarm (30), and Alarm (27), Alarm (33) and Alarm (36) are also associated with each other.

In summary, the proposed alarm clustering method recognizes the correlation between the types of alarm variables, and alarm types that occur frequently in succession are grouped into the same class. This method can be used to group related alarms that occur frequently due to process disturbances and system failures.

(ii) Alarm system optimization strategy based on clustering results

(a) Suppress associated alarms: Group and monitor associated alarms through process alarm clustering, and for alarm variables with causal relationships within each category, construct their causal network diagrams by combining process knowledge. Taking class 7 as an example, Alarm (15) and Alarm (8), Alarm (20) and Alarm (23) belong to different levels of alarms (high or ultra-high alarms) occurring in the same process variable, so in order to simplify the causal network, only Alarm (15) and Alarm (20) are retained here. By analyzing the association of alarm variables within the class, Alarm (20) can be caused by the occurrence of alarms in Alarm (15), and at the same time, these two alarms can lead to the occurrence of alarms in Alarm (19) and Alarm (29). The constructed process alarm causal diagram is shown in Fig. 7.12.

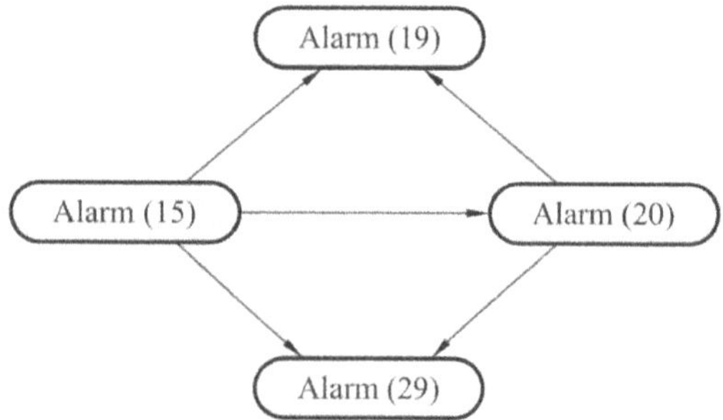

Fig. 7.12. Process alarm cause and effect diagram.

When an alarm occurs, it suppresses related alarms that have a cause-and-effect relationship, and accurately locates the root cause of the alarm through the alarm cause-and-effect diagram, so as to avoid the flooding of alarms and the occurrence of potential malfunctions and accidents.

(b) Eliminate redundant alarms: Through the above analysis of alarm correlation, three pairs of redundant alarms can be identified in the system: Alarm (11) and Alarm (39) belong to the same alarm event, Alarm (14) and Alarm (13), Alarm (38), and Alarm (35) belong to the same alarm event, of which Alarm (11), Alarm (14), and Alarm (38) are generated by the BMS system, and Alarm (39), Alarm (13), and Alarm (35) come from the SCADA system. Alarm (11), Alarm (14) and Alarm (38) are generated by the BMS system, while Alarm (39), Alarm (13) and Alarm (35) are from the SCADA system. In order to prevent redundant alarms and alarm flooding, the three redundant alarms Alarm (11), Alarm (14), and Alarm (38) generated by the BMS system are excluded.

(c) Suppressing jitter alarms: If the time interval between two alarms is less than 20s, the alarm suppression strategy is adopted for the latter. For example, if Alarm (2) occurs at 13:14:04 on September 4, 2017 and 13:14:21 on September 4, 2017, the second Alarm (2) alarm is suppressed.

Through the above method, the alarm system of the steam production system of the CHCP plant is optimized based on the clustering results, and taking the alarm data of the steam production system

Table 7.3. Comparison of the number of alarms by month before and after system optimization.

Statistical time	Before optimization	After optimization	Percentage of reduction in the number of alarms
September 2017	1135	811	28.5%
October 2017	663	389	41.3%
November 2017	5079	2662	47.6%
December 2017	20, 372	6701	67.1%
September through December 2017	27, 249	10, 563	61.2%

of the CHCP plant from September to December 2017 as an example, the number of alarms in each month before and after the optimization of the system is counted separately as shown in Table 7.3. It can be seen that the number of system alarms from September to December 2017 is effectively reduced by suppressing correlated alarms and jitter alarms and eliminating redundant alarms, and the number of alarms is reduced by a total of 61.2% compared with the threshold-based alarm system before optimization, thus avoiding the frequent occurrence of alarm flooding to a great extent. In addition, the method carries out causal analysis of alarm variables through correlated alarm clustering, which helps to further locate the root causes of alarms so that operators can discover the causes of failures in a timely manner to ensure the safe and reliable operation of the process.

7.2 Multi-Round Coupled Alarm Optimization

7.2.1 *Implementation of multi-round coupled alert optimization methods*

The multi-round coupled alarm optimization method contains five alarm optimization strategies, which are individual sensor alarm optimization strategy, alarm optimization strategy of sensor group, coupled alarm strategy based on Bayesian network, sensor's own abnormality judgment strategy, and redundant alarm suppression

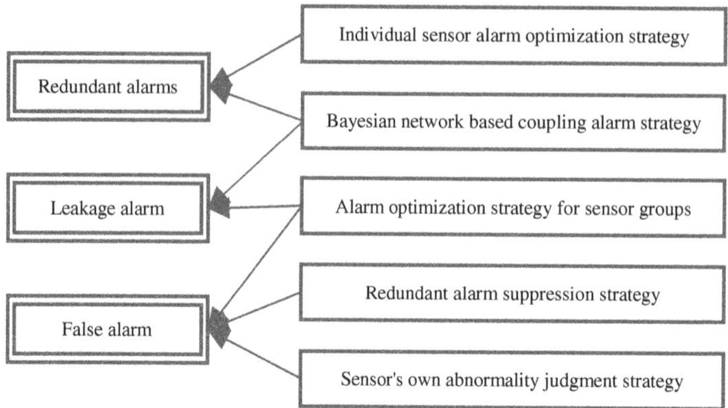

Fig. 7.13. Multi-round coupled alarm optimization mechanism of action.

strategy, whose comprehensive optimization mechanism is shown in Fig. 7.13 and the method flow is shown in Fig. 7.14.

7.2.1.1 *Individual sensor alarm optimization strategy*

Sensors are installed at key equipment and nodes in the refining and chemical production process to measure parameters such as temperature, pressure, and liquid level of the plant unit. Alarm thresholds are set according to the attributes of the production process monitored by each sensor, and alarms are triggered when the sensor measurement data reaches the alarm thresholds.

When the alarm is triggered, the device is in a dangerous state, and when and only when the sensor parameters are lower than the upper alarm limit or higher than the lower alarm limit by 5%, the normal state is restored, and the next alarm judgment is carried out, and the judgment logic structure is shown in Fig. 7.15.

Through this optimization method, a large number of redundant alarms can be avoided due to the oscillation of the sensor monitoring parameters around the alarm limit for reasons such as operation adjustment, and the workload of the operator in dealing with redundant alarms can be reduced.

7.2.1.2 *Alarm optimization strategy for sensor groups*

For a group of sensor parameters installed in the same unit of the device (or the same functional block) for processing, when the same

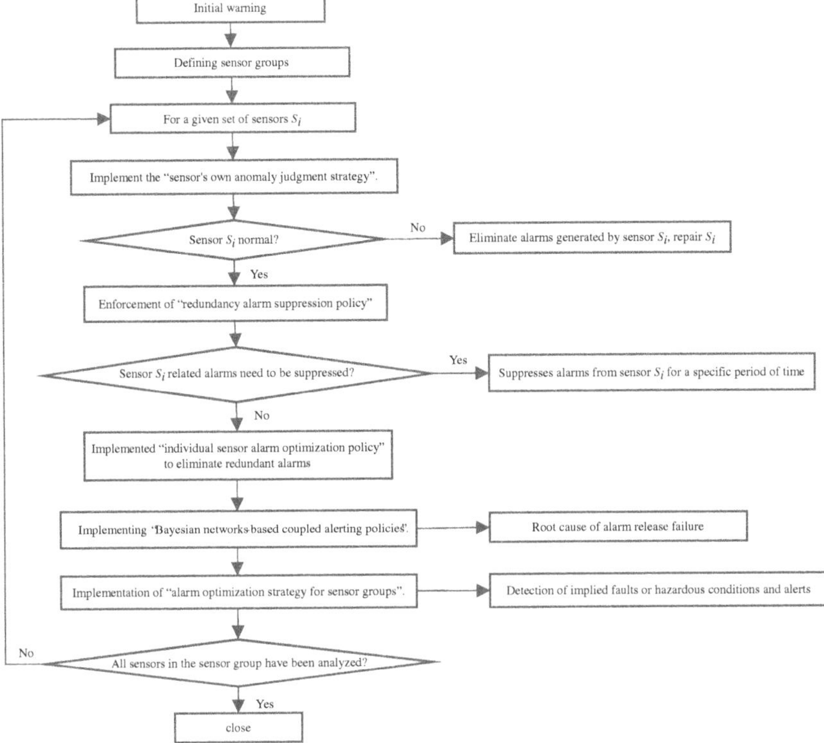

Fig. 7.14. Flow of implementation of the multi-round coupled alarm optimization approach.

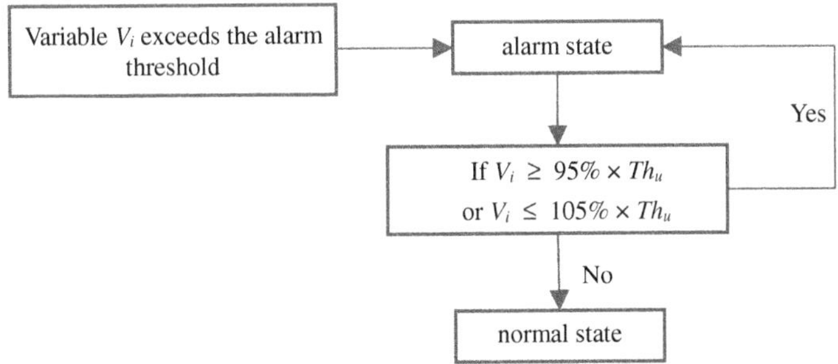

Fig. 7.15. Logical structure of the optimization method for individual sensor alarms.

group of sensor data at the same time close to the upper limit or lower limit of the alarm, but did not trigger any alarm, the device may also be in a dangerous state, it should be alarmed. The sensor group-based alarm optimization method comprehensively analyzes a group of sensor monitoring parameters within the same unit (or the same functional block) of the device, and gives the following alarm conditions. see Eqs. (7.13) and (7.14).

Define the ratio between the parameters of the sensors and the upper limit of the alarm as the upper limit judgment factor α and the lower limit judgment factor β:

$$\alpha_i = \frac{Q_i}{A_i}, \tag{7.13}$$

where α_i is the alarm limit judgment factor of the ith sensor; Q_i is the parameter of the ith sensor; A_i is the upper alarm value of the ith sensor.

$$\beta_i = \frac{Q_i}{B_i}, \tag{7.14}$$

where β_i is the judgment factor of the lower alarm limit of the ith sensor; Q_i is the parameter of the ith sensor; B_i is the lower alarm value of the ith sensor.

The alarm trigger condition for the judgment factor is

$$\sum_i = \alpha_i \geq \partial \quad \text{or} \quad \sum_i = \beta_i \geq \partial,$$

where ∂ is the alarm value of the judgment factor.

When the triggering conditions are met, the alarm is conducted, and the logical structure is shown in Fig. 7.16.

As a supplement to the single sensor alarm, the alarm optimization method for the sensor group can be used to warn of special dangerous states and warn the abnormal equipment earlier, so as to reserve more time for the operator to take measures.

7.2.1.3 *Bayesian network-based coupled alarm strategy*

For the alarm mode for a single node, multiple nodes will trigger alarms at the same time when their parameters are abnormal due to a root cause, and a large number of alarm messages will appear in

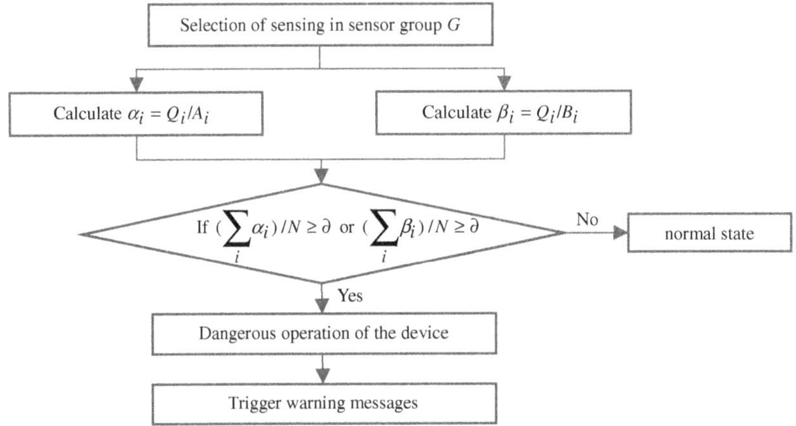

Fig. 7.16. Logical structure of the sensor group alarm optimization approach.

a short period of time, and redundant alarm messages will increase the information processing workload of the security maintenance personnel and affect their trust, alertness and decision-making power to formulate security measures for the alarms.

According to the established association warning model and fault inference method based on a dynamic Bayesian network (see Chapter 6 for details of dynamic Bayesian network modeling), the association warning model can be further utilized in the alarm optimization process, and the coupling relationship between alarm nodes can be found out through the computation of conditional probability of Bayesian network, so that redundant alarms can be eliminated by digging out the root cause of faults, which triggers different signs of faults and cascading faults. By mining the root cause of the fault, the redundant alarm problem caused by different fault signs and cascading faults triggered by the same fault root cause is eliminated.

When the parameters of multiple nodes meet the alarm thresholds of a hazardous scenario, the coupled alarm optimization mechanism is triggered, and the root cause of the hazardous scenario is inferred according to the Bayesian network graph of the correlation warning model and its inference method, so that the alarms are issued for the root cause, and the safety recommendations are given at the same time, and the logical structure is shown in Fig. 7.17.

This alarm optimization strategy reduces a large number of redundant alarm messages generated by non-root faulty nodes in the case

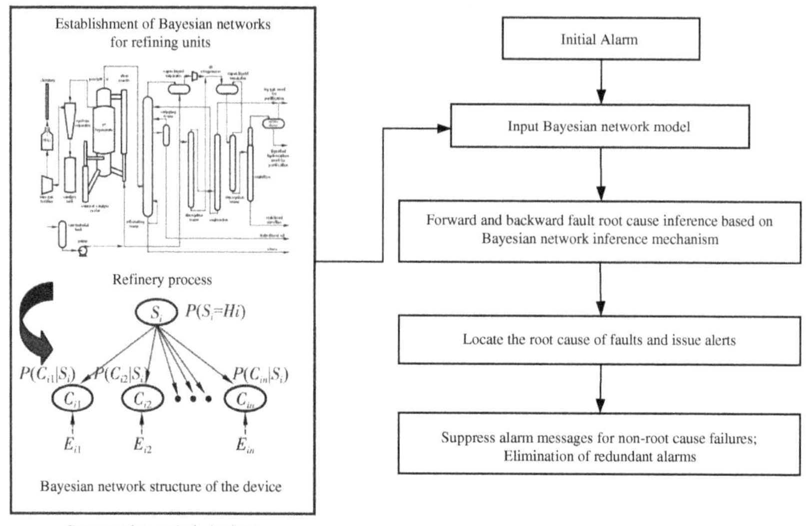

Fig. 7.17. Logical structure of Bayesian network-based coupled alerting approach.

of multi-node anomalies and improves the alarm efficiency and the speed of the safety maintenance personnel to process the information and take measures.

7.2.1.4 *Sensor itself abnormal judgment strategy*

When the sensor cannot carry out the normal collection function due to power exhaustion, the external environment, its own loss, and other factors, false alarms or missed alarms often occur. In such a situation, the sensor's own state of judgment, when the sensor failure is judged, the sensor will be removed from the alarm system.

In the continuous time t, when the sensor acquisition parameters are 0 or beyond the environment of reasonable parameter values (such as the tower's extreme maximum temperature value), the judgment that the sensor is damaged will be isolated from the alarm system after the alarm step, and prompted to replace the sensor, the logical structure as shown in Fig. 7.18.

7.2.1.5 *Redundant alarm suppression strategy*

When a parameter collected exceeds the alarm limit, the judgment of alarm suppression is carried out. If the first four collections are

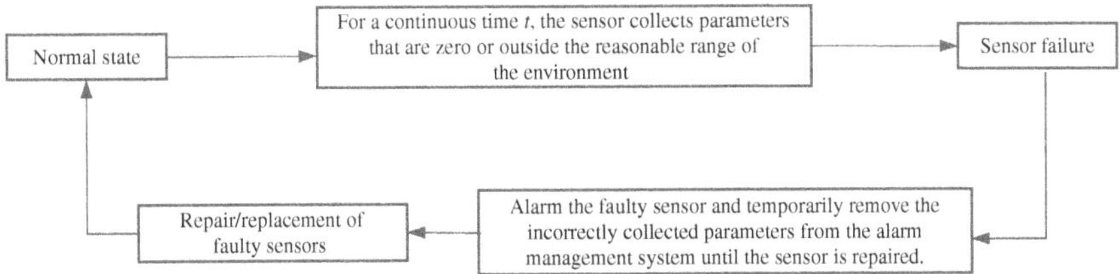

Fig. 7.18. Logical structure of the sensor's own abnormality judgment method.

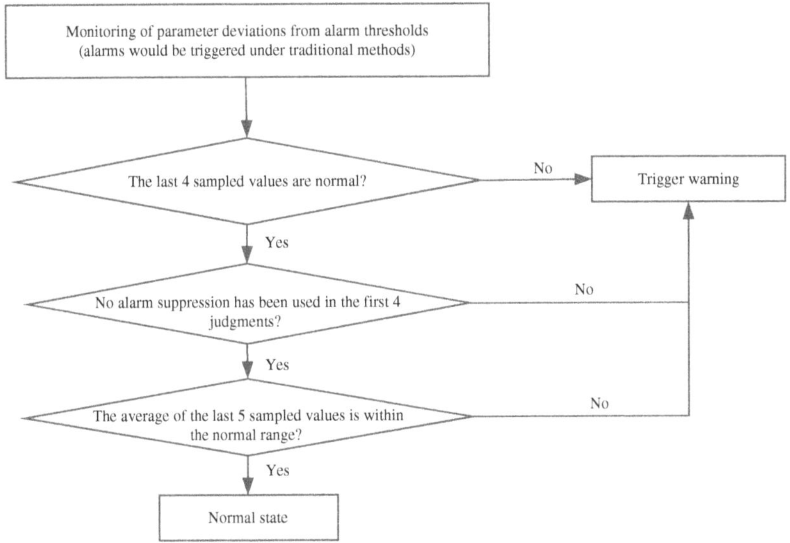

Fig. 7.19. Logical structure of redundant alarm suppression policy.

within the normal threshold and there is no alarm suppression, the average value of the parameter of the five times is taken to carry out the judgment of alarm again, and if the second judgment does not exceed the alarm limit, alarm suppression is carried out, and vice versa, the alarm is normal. The logical structure is shown in Fig. 7.19.

7.2.2 *Hazard scenario creation and case study*

In accordance with the application process of the alarm optimization method, a Bayesian network model is established based on the flow of the reaction regeneration part of the catalytic cracking process

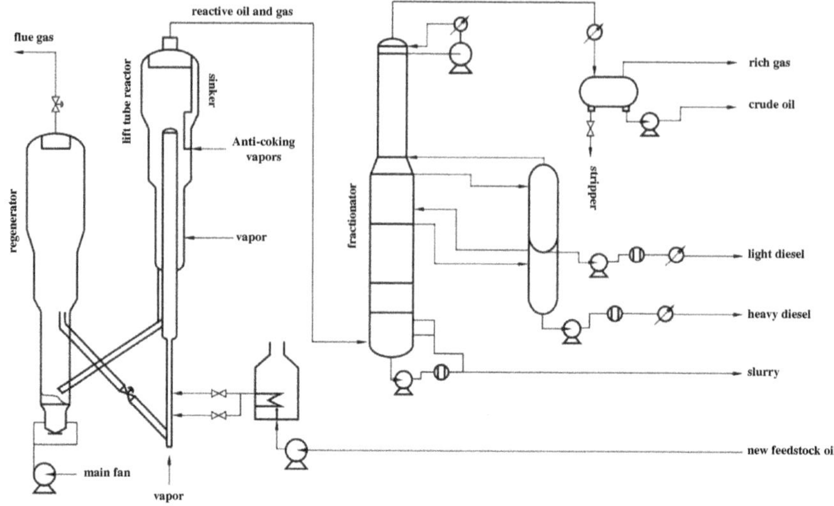

Fig. 7.20. Catalytic cracking reaction regeneration and fractionation partial flow.

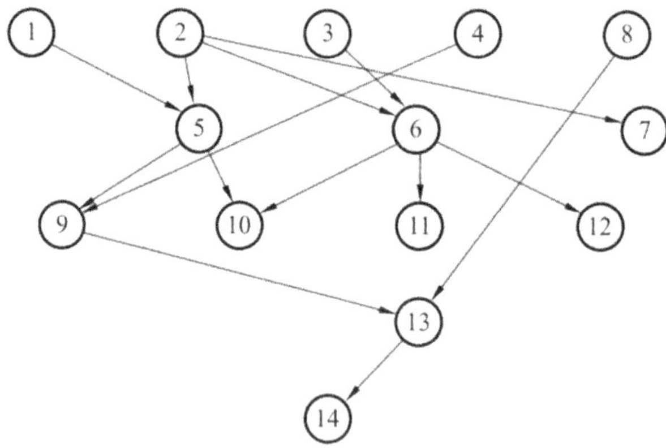

Fig. 7.21. Reaction regeneration process Bayesian network structure.

(Fig. 7.20), as shown in Fig. 7.21, in which the events represented by the numbers in the Bayesian network structure diagram are shown in Table 7.4. The four hazardous scenarios in the reaction regeneration process are further modeled by the Bayesian network as shown in Table 7.5.

Table 7.4. Reaction regeneration process Bayesian networks list of basic events.

Serial number	Basic event	Serial number	Basic event
1	Lifting wind gauge malfunction	8	Steam with water
2	External heat extraction kit leakage	9	High lift tube reactor temperatures
3	Double-acting slide valve malfunction fully open	10	Low oxygen content in regenerator flue gas
4	High steam flow rate of the lifting tube	11	High flue gas turbine pressure
5	High regenerator temperature	12	Regenerator main air volume fluctuation
6	High regenerator pressure	13	High pressure in the sinker
7	Low level of externally heated packages	14	High speed of air compressor

7.2.2.1 *Case 1: comparison of optimization results of alerts for the same failure signs with different root causes*

(1) Lift the wind instrument failure fault alarm analysis

Figure 7.22 shows the status monitoring data of the alarm parameters related to the lifting air instrument failure fault, whose monitoring parameters are: lifting pipe steam flow rate, regenerator temperature, regenerator pressure, external heat-take packet level, lifting pipe reactor temperature, regenerator flue gas oxygen content, flue gas turbine pressure, regenerator main airflow, settler internal pressure, and pneumatic compressor rotational speed.

Taking 600 min as a sample time unit, the collection frequency is once per minute. In the figure, it can be seen that the first 300 min is the normal state, and at the 300th min, the lifting wind instrument failure fault occurs, which leads to changes in parameters such as regenerator temperature, flue gas turbine pressure, internal pressure of the settler, and speed of the pneumatic compressor, so the last 300 min is the state after the fault occurs.

Table 7.5. Hazardous scenarios for reaction regeneration processes.

Serial number	Root cause	Alarm nodes and status
1	Lifting wind gauge malfunction	High regenerator temperature, high lift tube reactor temperature, low regenerator flue gas oxygen content, high settler internal pressure, high pneumatic compressor speed
2	External heat extraction kit leakage	High regenerator temperature, high regenerator pressure, low level of the external heat extraction package, high temperature of the lift tube reactor, high internal pressure in the settler, high speed of the pneumatic compressor, low oxygen content of the flue gas in the regenerator, high pressure of the flue gas turbine, fluctuations in the main airflow of the regenerator
3	Double-acting slide valve malfunction fully open	High regenerator pressure, low regenerator flue gas oxygen content, high flue gas turbine pressure, fluctuating regenerator mains air volume
4	Steam with water	High internal pressure in the sinker and high speed of the pneumatic compressor

Twenty samples were selected for analysis after the fault occurred, each containing 10min of data, and the average values of the ten monitoring parameters for each group of samples are shown in Table 7A.2. Under the traditional alarm mechanism, if the monitoring parameters exceed the limit, an alarm is generated, and the results of the alarm analysis are shown in Table 7.6. Applying the research result "multi-round coupling alarm optimization method" in this section, for this abnormal event, the root cause of the alarm is given as "lifting wind instrument failure fault", and the alarms of other fault signs will be suppressed, and the results of the alarm analysis are shown in Table 7.6.

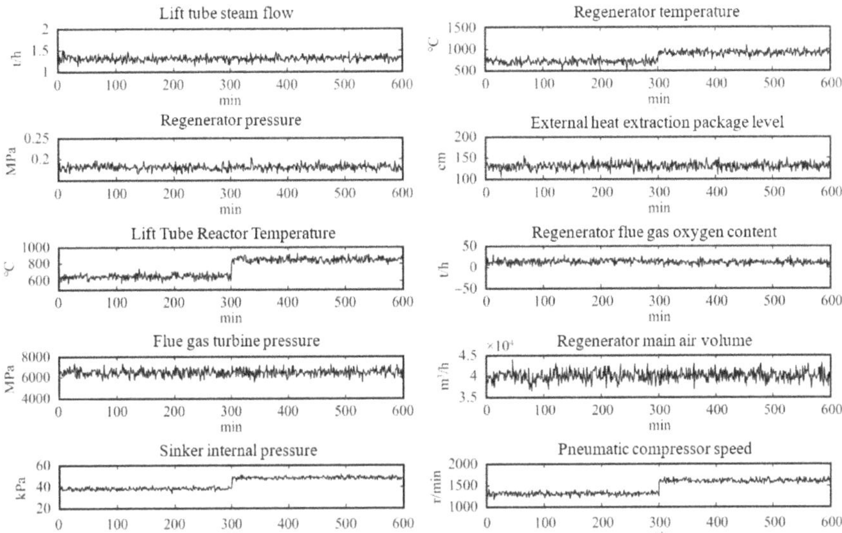

Fig. 7.22. Condition monitoring data of node parameters for lift the wind instrumentation failure.

The results of optimizing the integrated alarms of 20 samples and the traditional over-limit alarms are compared and counted, and the results are shown in Table 7.7. The results are shown in Table 7.7. Correct alarm rate = number of correct alarms/total number of faults that should be alarmed; missing alarm rate = number of missing alarms/total number of faults that should be alarmed.

Case study: In the multi-round coupled alarm optimization method, the optimization method of alarm suppression reduces the false alarm rate in the traditional single alarm method, such as the 1st parameter of the 8th sample in Table 7A.2, which is judged to be an abnormal value in the traditional over-limit alarm method, while in the multi-round coupled alarm optimization method, this alarm is suppressed by judging that the parameters of the first four samples are in the range of normal thresholds, and then by judging that the average value of parameters of the 4–8 samples 1.294 is also within the normal threshold range, thus suppressing the alarm and eliminating a false alarm.

Through the multi-round coupling alarm optimization method, the root cause of the fault is reported directly, eliminating redundant

Table 7.6. Alarm results and types of alarm results for lift wind instrument failure.

Serial number	Optimize alert results	Alarm result type	Traditional overrun alarm results	Alarm result type
1	Failure to lift the wind instrumentation	Correct alarms	Parameter alarms 2, 9, 10	4 correct alarms, 3 redundancy alarms, 1 false alarm
2	Failure to lift the wind instrumentation	Correct alarms	Parameter alarms 2, 5, 9, 10	3 correct alarms, 2 redundancy alarms, 1 false alarm
3	Failure to lift the wind instrumentation	Correct alarms	Parameter alarms 2, 5, 7, 9, 10	4 correct alarms, 3 redundancy alarms
4	Failure to lift the wind instrumentation	Correct alarms	Parameter alarms 2, 5, 7, 9, 10	4 correct alarms, 3 redundancy alarms
5	Failure to lift the wind instrumentation	Correct alarms	Parameter alarms 2, 5, 9	3 correct alarms, 2 redundancy alarms, 1 false alarm
6	Failure to lift the wind instrumentation	Correct alarms	Parameter alarms 2, 5, 7, 9, 10	4 correct alarms, 3 redundancy alarms
7	Failure to lift the wind instrumentation	Correct alarms	Parameter alarms 2, 5, 9, 10	4 correct alarms, 3 redundancy alarms
8	Failure to lift the wind instrumentation	Correct alarms	Parameter alarms 1, 2, 5, 9, 10	4 correct alarms, 3 redundancy alarms, 1 false alarm
9	Failure to lift the wind instrumentation	Correct alarms	Parameter alarms 2, 5, 9, 10	4 correct alarms, 3 redundancy alarms
10	Failure to lift the wind instrumentation	Correct alarms	Parameter alarms 2, 5, 9, 10	4 correct alarms, 3 redundancy alarms

(Continued)

Table 7.6. (*Continued.*)

Serial number	Optimize alert results	Alarm result type	Traditional overrun alarm results	Alarm result type
11	Failure to lift the wind instrumentation	Correct alarms	Parameter alarms 2, 5, 9	3 correct alarms, 2 redundancy alarms, 1 false alarm
12	Failure to lift the wind instrumentation	Correct alarms	Parameter alarms 2, 5, 9, 10	4 correct alarms, 3 redundancy alarms
13	Failure to lift the wind instrumentation	Correct alarms	Parameter alarms 2, 5, 9, 10	4 correct alarms, 3 redundancy alarms
14	Failure to lift the wind instrumentation	Correct alarms	Parameter alarms 2, 5, 9, 10	4 correct alarms, 3 redundancy alarms
15	Failure to lift the wind instrumentation	Correct alarms	Parameter alarms 2, 5, 9, 10	4 correct alarms, 3 redundancy alarms
16	Failure to lift the wind instrumentation	Correct alarms	Parameter alarms 2, 5, 9, 10	4 correct alarms, 3 redundancy alarms
17	Failure to lift the wind instrumentation	Correct alarms	Parameter alarms 2, 5, 9, 10	4 correct alarms, 3 redundancy alarms
18	Failure to lift the wind instrumentation	Correct alarms	Parameter alarms 2, 4, 5, 9, 10	4 correct alarms, 3 redundancy alarms, 1 false alarm
19	Failure to lift the wind instrumentation	Correct alarms	Parameter alarms 2, 9	2 correct alarms, 1 redundancy alarms, 2 false alarms
20	Failure to lift the wind instrumentation	Correct alarms	Parameter alarms 2, 5, 9, 10	4 correct alarms, 3 redundancy alarms

Table 7.7. Comprehensive alarm optimization approach for boosting wind instrument failure vs. traditional single alarm approach.

Alarm mode	Correct alarm rate	Missed alarm rate	Number of false alarms	Number of redundant alarms
Traditional overrun alarm methods	93.75%	6.25%	3	55
Alarm optimization method	100%	0	0	0

alarms, whereas the traditional alarm method not only warns of the direct parameter changes under the root cause but also warns of the deviation of other parameters, such as the 5th, 9th, and 10th parameters, which not only cannot find the root cause directly but also reduces the processing efficiency of the safety operators.

The sensor group alarm optimization method in the multi-round coupled alarm optimization method reduces the leakage alarm rate, e.g., parameter 10 in Sample 11 is close to the upper alarm limit but still in the normal threshold range, which would be judged as a normal value in the traditional alarm method, whereas the sensor group alarm optimization strategy reduces the leakage alarm rate by comprehensively judging the abnormal changes of the associated parameters in the hazardous scenario. If the whole meets the alarm condition $\sum_i \alpha_i \geq \partial$, an alarm will be issued for this hazardous scene, eliminating missed alarms.

(2) Steam with water fault alarm analysis

The steam with water fault also causes the same two-node alarm states of high pressure inside the settler and high speed of the pneumatic compressor, and the values of the ten monitoring parameters under this root cause of the fault are shown in Fig. 7.23. Twenty samples were selected for analysis after the fault occurred, each containing 10 min of data, and the average values of the ten monitoring parameters for each set of samples are shown in Table 7A.3. Under the traditional alarm mechanism, if the monitoring parameter exceeds the limit, an alarm is issued, and the results of the alarm analysis are shown in Table 7.8. Applying the research result

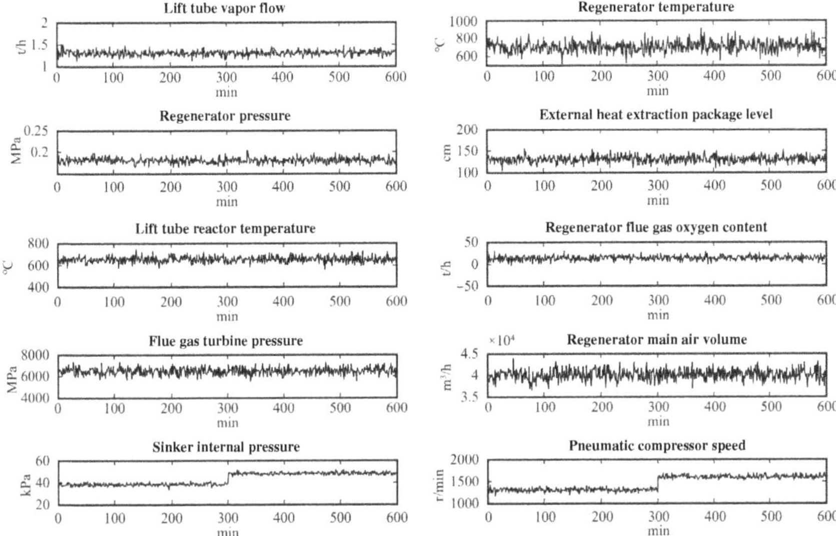

Fig. 7.23. Nodal parameter state values for steam with water.

"multi-round coupling alarm optimization method" in this section, for this abnormal event, the root cause of the alarm is given as "steam with water", and the alarms of other fault signs will be suppressed, and the results of the alarm analysis are shown in Table 7.8.

The results of the integrated alarm optimization results of the 20 samples and the traditional over-limit alarm results are compared and counted, and the results are shown in Table 7.9.

The 2nd parameter of the 17th sample in Table 7A.3 is judged to be an abnormal value in the single alarm method, while in the multi-round coupled alarm optimization method, the alarm is suppressed by judging that the parameters of the first four samples are all within the normal threshold range, and then by judging that the average values of the parameters of the 4–8 samples are also within the normal threshold range, thus eliminating a false alarm. In the 18th sample in the same table, another false alarm occurs, and at this time, according to the judgment of alarm suppression, the alarm cannot be suppressed, therefore, both methods have a false alarm. The missed alarm rate and the number of redundant alarms are similarly reduced by the multi-round coupled alarm optimization method.

Table 7.8. Alarm results and types of alarm results for lift wind instrument failure.

Serial number	Optimize alert results	Alarm result type	Traditional overrun alarm results	Alarm result type
1	Steam with water	Correct alarms	Parameter alarms 7, 9, 10	2 correct alarms, 1 redundancy alarm, 1 false alarm
2	Steam with water	Correct alarms	Parameter alarms 9, 10	2 correct alarms, 1 redundancy alarm
3	Steam with water	Correct alarms	Parameter alarms 9, 10	2 correct alarms, 1 redundancy alarm
4	Steam with water	Correct alarms	Parameter alarms 4, 9, 10	2 correct alarms, 1 redundancy alarm, 1 false alarm
5	Steam with water	Correct alarms	Parameter alarms 9, 10	1 correct alarm, 0 redundancy alarms, 1 false alarm
6	Steam with water	Correct alarms	Parameter alarms 9, 10	2 correct alarms, 1 redundancy alarm
7	Steam with water	Correct alarms	Parameter alarms 9, 10	2 correct alarms, 1 redundancy alarm
8	Steam with water	Correct alarms	Parameter alarms 9, 10	2 correct alarms, 1 redundancy alarm
9	Steam with water	Correct alarms	Parameter alarms 9, 10	2 correct alarms, 1 redundancy alarm
10	Steam with water	Correct alarms	Parameter alarms 9, 10	2 correct alarms, 1 redundancy alarm

(Continued)

Table 7.8. (*Continued.*)

Serial number	Optimize alert results	Alarm result type	Traditional overrun alarm results	Alarm result type
11	Steam with water	Correct alarms	Parameter alarms 9, 10	1 correct alarm, 0 redundancy alarms, 1 false alarm
12	Steam with water	Correct alarms	Parameter alarms 9, 10	1 correct alarm, 0 redundancy alarms, 1 false alarm
13	Steam with water	Correct alarms	Parameter alarms 10	1 correct alarm, 1 redundancy alarm, 1 false alarm
14	Steam with water	Correct alarms	Parameter alarms 9, 10	2 correct alarms, 1 redundancy alarm
15	Steam with water	Correct alarms	Parameter alarms 9, 10	2 correct alarms, 1 redundancy alarm
16	Steam with water	Correct alarms	Parameter alarms 9, 10	2 correct alarms, 1 redundancy alarm
17	Steam with water	Correct alarms	Parameter alarms 2, 9, 10	2 correct alarms, 1 redundancy alarm, 1 false alarm
18	Steam with water	Correct alarms	Parameter alarms 4, 9, 10	2 correct alarms, 1 redundancy alarm, 1 false alarm
19	Steam with water	Correct alarms	Parameter alarms 9, 10	1 correct alarm, 0 redundancy alarms, 1 false alarm
20	Steam with water	Correct alarms	Parameter alarms 9, 10	2 correct alarms, 1 redundancy alarm

Table 7.9.　Comprehensive alarm optimization approach for Steam with water failure vs. traditional single alarm approach.

Alarm mode	Correct alarm rate	Missed alarm rate	Number of false alarms	Number of redundant alarms
Traditional overrun alarm methods	75%	25%	4	16
Alarm optimization method	100%	0	1	0

7.2.2.2　*Case 2: comparison of alarm optimization results for correlated faults*

The lifting air meter failure and the double-acting slide valve failure fully open have the same nodal fault state under both root causes: low regenerator flue gas oxygen content, which is an associated fault. The values of the ten monitored parameters under this root cause of failure are shown in Fig. 7.24. Twenty samples are selected for analysis after the fault occurs, each sample contains 10 min of data, and the average values of the ten monitoring parameters for each group of samples are shown in Table 7A.4. Under the traditional alarm mechanism, if the monitoring parameter exceeds the limit, an alarm is issued, and the results of the alarm analysis are shown in Table 7.10. Applying the research result "multi-round coupling alarm optimization method" in this section, for this abnormal event, the root cause of the alarm is given as "lifting wind instrument malfunction, double-acting slide valve failure fully open", and the alarm of other fault signs will be suppressed, and the results of the alarm analysis are shown in Table 7.10.

The integrated alarm optimization results of the 20 samples and the traditional overrun alarm results are compared and counted, and the results are shown in Table 7.11.

The second parameter of the 17th sample in Table 7A.4 is judged to be an abnormal value in the traditional over-limit alarm method, while in the multi-round coupled alarm optimization method, the alarm is suppressed by judging that the parameters of the first four samples are within the normal threshold range, and then by judging that the average value of the parameters of the 4–8 samples is also

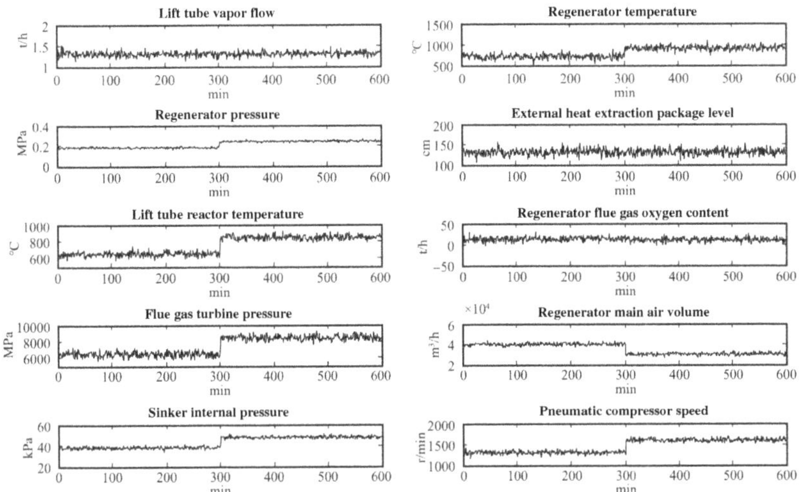

Fig. 7.24. Nodal parameter states for double-acting slide valve failure full open and hoisting wind instrumentation failure.

within the normal threshold range, thus eliminating a false alarm. The 18th sample in the same table has another false alarm, and at this time, according to the judgment of alarm suppression, the alarm cannot be suppressed, therefore, both methods have false alarms. The missed alarm rate and the number of redundant alarms are similarly reduced by the multi-round coupled alarm optimization method.

Through the above two case studies, the multi-round coupled alarm optimization method improves the correct alarm rate of the traditional over-limit alarm method (the correct alarm rate is maximally increased by 33.3%), reduces the rate of false alarms and missed alarms, and reduces the number of redundant alarms. Therefore, the multi-round coupled alarm optimization method established in this section can be applied to the alarm management of refining and chemical plants during the long production cycle, which can improve the positioning, processing efficiency and relevance of plant faults, and avoid secondary accidents and economic losses due to improper alarm processing.

7.2.2.3 *Analysis and summary*

(1) In this section, based on the research of correlation warning model and inference mechanism, for the problems of omission,

Table 7.10. Alarm results and types of alarm results for double-acting slide valve failure to fully open and lift air instrumentation failure.

Serial number	Optimize alert results	Alarm result type	Traditional overrun alarm results	Alarm result type
1	Double-acting slide valve malfunction fully open, lifting wind instrumentation failure	Correct alarms	Parameter alarms 2, 3, 5, 6, 8, 9, 10	8 correct alarms, 5 redundancy alarms, 1 false alarm
2	Double-acting slide valve malfunction fully open, steam with water	Correct alarm 1 time, false alarm 1 time, missed alarm 1 time	Parameter alarms 3, 6, 7, 8, 9, 10	6 correct alarms, 4 redundancy alarms, 2 false alarms
3	Double-acting slide valve malfunction fully open, lifting wind instrumentation failure	Correct alarms	Parameter alarms 2, 3, 5, 6, 7, 8, 9, 10	8 correct alarms, 6 redundancy alarms
4	Double-acting slide valve malfunction fully open, lifting wind instrumentation failure	Correct alarms	Parameter alarms 2, 3, 4, 5, 6, 7, 8, 9, 10	8 correct alarms, 6 redundancy alarms, 1 false alarm
5	Double-acting slide valve malfunction fully open, lifting wind instrumentation failure	Correct alarms	Parameter alarms 2, 3, 6, 7, 8, 9, 10	7 correct alarms, 5 redundancy alarms, 1 false alarm
6	Double-acting slide valve malfunction fully open, lifting wind instrumentation failure	Correct alarms	Parameter alarms 2, 3, 5, 6, 7, 8, 9, 10	8 correct alarms, 6 redundancy alarms
7	Double-acting slide valve malfunction fully open, lifting wind instrumentation failure	Correct alarms	Parameter alarms 2, 5, 6, 7, 8, 9, 10	7 correct alarms, 6 redundancy alarms, 1 false alarm

(Continued)

Table 7.10. (*Continued.*)

Serial number	Optimize alert results	Alarm result type	Traditional overrun alarm results	Alarm result type
8	Double-acting slide valve malfunction fully open, lifting wind instrumentation failure	Correct alarms	Parameter alarms 2, 3, 5, 6, 7, 8, 9, 10	8 correct alarms, 6 redundancy alarms
9	Double-acting slide valve malfunction fully open, lifting wind instrumentation failure	Correct alarms	Parameter alarms 2, 3, 5, 6, 7, 8, 9, 10	8 correct alarms, 6 redundancy alarms
10	Double-acting slide valve malfunction fully open, lifting wind instrumentation failure	Correct alarms	Parameter alarms 2, 3, 5, 6, 7, 8, 9, 10	8 correct alarms, 6 redundancy alarms
11	Double-acting slide valve malfunction fully open, lifting wind instrumentation failure	Correct alarms	Parameter alarms 2, 3, 5, 6, 7, 8, 9, 10	8 correct alarms, 6 redundancy alarms
12	Double-acting slide valve malfunction fully open, lifting wind instrumentation failure	Correct alarms	Parameter alarms 2, 3, 5, 6, 7, 8, 9, 10	8 correct alarms, 6 redundancy alarms
13	Double-acting slide valve malfunction fully open, lifting wind instrumentation failure	Correct alarms	Parameter alarms 2, 3, 5, 6, 7, 8, 9, 10	8 correct alarms, 6 redundancy alarms
14	Double-acting slide valve malfunction fully open, lifting wind instrumentation failure	Correct alarms	Parameter alarms 2, 3, 5, 6, 7, 8, 9, 10	8 correct alarms, 6 redundancy alarms

(*Continued*)

Table 7.10. (*Continued.*)

Serial number	Optimize alert results	Alarm result type	Traditional overrun alarm results	Alarm result type
15	Double-acting slide valve malfunction fully open, lifting wind instrumentation failure	Correct alarms	Parameter alarms 2, 3, 5, 6, 7, 8, 9, 10	8 correct alarms, 6 redundancy alarms
16	Double-acting slide valve malfunction fully open, lifting wind instrumentation failure	Correct alarms	Parameter alarms 2, 3, 5, 6, 7, 8, 9, 10	8 correct alarms, 6 redundancy alarms
17	Double-acting slide valve malfunction fully open, lifting wind instrumentation failure	Correct alarms	Parameter alarms 2, 3, 5, 6, 7, 8, 9, 10	8 correct alarms, 6 redundancy alarms
18	Double-acting slide valve malfunction fully open, lifting air meter failure, steam with water, external heat extraction Packet leakage	Correct alarm 1 times, false alarm 2 times	Parameter alarms 1, 2, 3, 4, 5, 6, 7, 8, 9, 10	8 correct alarms, 6 redundancy alarms, 2 false alarms
19	Double-acting slide valve malfunction fully open, lifting wind instrumentation failure	Correct alarms	Parameter alarms 2, 3, 6, 7, 8, 9, 10	7 correct alarms, 5 redundancy alarms, 1 false alarm
20	Double-acting slide valve malfunction fully open, lifting wind instrumentation failure	Correct alarms	Parameter alarms 2, 3, 5, 6, 7, 8, 9, 10	8 correct alarms, 6 redundancy alarms

Table 7.11. Coupled alarm optimization method for double-acting slide valve failure full opening and hoisting wind instrument failure vs. traditional overrun alarm method.

Alarm mode	Correct alarm rate	Missed alarm rate	Number of false alarms	Number of redundant alarms
Traditional overrun alarm methods	96.25%	3.75%	3	155
Alarm optimization methods	98.125%	1.875%	3	0

false alarm, and redundant alarms existing in the production site, combined with the results of correlation warning inference and fusion decision-making, we propose the optimization method of multiple rounds of coupled alarms, which on the one hand, generalize and aggregate the low-level alarms, and remove the redundant alarms; On the other hand, considering multi-hazardous factors and multi-failure scenarios, the alarm messages triggered by deviation sequences are optimized to give and publish the root causes of the failure signs and enhance the directionality of the alarms.

(2) The multi-round coupled alarm optimization method contains five alarm optimization strategies, which are individual sensor alarm optimization strategy, alarm optimization strategy of sensor group, coupled alarm strategy based on Bayesian network, sensor's own abnormality judgment strategy, and redundant alarm suppression strategy.

(3) Verified by a number of case studies, the multi-round coupling alarm optimization method alarm correct rate of more than 98%, compared with the traditional DCS overrun alarms, it will be correct to maximize the alarm rate of 33.3%, and completely remove redundant alarms.

7.3 Early Warning of Real-Time Fault Correlation Prediction for Refining and Chemical Equipment

Hazardous factors objectively exist in refining and chemical production, and these factors exist in various forms within the system, which are interrelated, influenced, and eventually transformed into accidents under certain conditions. Therefore, it is necessary to make qualitative and quantitative predictions on whether the process parameters will deviate and the failure trend after the deviation, so as to fully grasp the trend of hazardous factors in a timely manner before the occurrence of accidents, to predict the deviation state of process parameters and to estimate the consequences of their impact, so as to take measures in advance to inhibit, eliminate the danger of the first signs, and to prevent the expansion and development of the hazardous factors as well as their transformation into major accidents.

In this section, we first predict the fault development trend from a qualitative perspective and propose a fault trend prediction method based on the feature data segmentation method. Then, from the quantitative point of view, the prediction results are further refined, and a grey correlation analysis-based Kalman filter correlation prediction method is proposed for "safety behavior-state". The quantitative prediction of future deviation values of parameters and the early alarm can help to inhibit the transition from abnormal to accidental conditions and reduce the number of unplanned shutdowns in a timely manner.

7.3.1　*Fault trend prediction based on feature data segmentation methods*

Trend prediction of process parameters requires the extraction of characteristic fragments from a large number of process parameters and the prediction of future trends through the integration of certain laws. The trend of process parameters can reflect the development speed and tendency of important parameters of the process operation state. The method of trend extraction has been improved many times at home and abroad: In 1991, Venkatasubramanian and Janusz of Purdue University in the United States defined seven kinds of primitives to describe the trend of parameters through the positive and negative values of first-order derivatives and second-order derivatives, but the method is not applicable to trend extraction of noisy data: In 2005, Charbonnier *et al.* proposed a method to judge the trend based on the combination of three simple primitives, which is more concise and suitable for online parameter trend analysis: In 2010,

Song Zhenghui *et al.* introduced the derivative confidence interval on the basis of the two, which improved the robustness and adaptability of the trend recognition algorithm: In 2011, Chen Junping *et al.* proposed an improved trend analysis method for step signals, and the trend information obtained through segmentation and feature extraction of process parameters is widely used in the fields of adaptive control, industrial process diagnosis, machinery fault diagnosis, and stock price trend prediction. The above method only relies on the numerical model of time series for analysis and prediction, and the prediction data are relatively isolated, with low accuracy and applicability.

Considering the continuity of the state of matter in the refining process, the parameter deviations affect each other, and there are certain intrinsic correlations in the chemical process model, which provide a great deal of information about the future development of the fault. The intrinsic correlation of the process parameters of the refining process should be fully explored, and the future trend of the predicted object should be predicted by taking into account the development law of the predicted object itself and the development law of other parameters associated with the predicted object, i.e., to cognize the development law of the things from the systematic point of view and to improve the accuracy of the fault prediction.

Under this idea, this section establishes a fault trend prediction method based on the feature data segmentation method by studying the intrinsic correlation of the process parameters of the refining and chemical device, predicts the fault trend through the feature data segmentation, and verifies the fault trend prediction results through the correlation data, and at the same time obtains the credibility of the prediction results. Through the integration of parameter deviation and trend analysis, a more accurate fault prediction method for process parameters is provided, which can predict the fault trend and is conducive to realizing the over-advanced defense of refining system faults.

7.3.1.1 *Fault trend prediction methods*

(1) Feature data segmentation methods

The traditional data segmentation method takes a specific fixed amount of data as the division unit, which destroys the integrity of the trend unit and affects the accuracy of trend prediction. In the refining process, the deviation of process parameters is mainly divided into step deviation and cumulative deviation, so the method of data segmentation is proposed by finding feature points and feature line segments to retain the complete trend characteristics. The feature points are divided for the step parameters, and the feature line segments are divided for the trend changes caused by the gradual accumulation of deviations in the process parameters.

Let t be a judgment moment in the parameter, it is the value of the parameter at the moment t. Let $I_{\max}$ be the maximum value of the parameter and $I_{\min}$ be the minimum value of the parameter. Let $I_h = I_{\max} - I_{\min}$, $k_1 = I_2 - I_1$, $k_2 = I_3 - I_2$, $k_{i-1} = I_i - I_{i-1}, \ldots$. Definitions for feature points and feature line segments are as follows:

Characteristic line segment: when $|\sum_1^{i-1} k_{i-1}| > 0.2 I_h$, the line segment $I_1 I_i$ is a characteristic line segment.

Characteristic point: a point I_t is characteristic if $|I_t - I_{t-1}| > 0.1 I_h$.

After finding the corresponding feature points and feature line segments through the parameter features, let $n = 0$, I_n represents the sequence value of the parameter in each feature cell, and k is the minimum number of parameters in the set feature cell, and the data segmentation is carried out through Fig. 7.25.

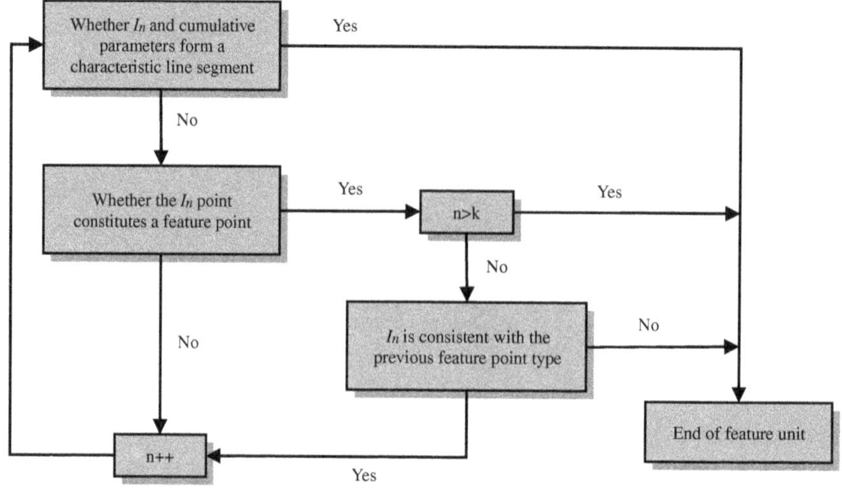

Fig. 7.25. Feature data segmentation methods.

(2) Trend fitting methods

In statistics, linear regression is a type of regression analysis that uses the least square function of a linear regression equation to model the relationship between one or more independent and dependent variables. This function is a linear combination of one or more models called regression coefficients. The case of only one independent variable is called simple regression and the case of more than one independent variable is called multiple regression.

In linear regression, the data are modeled using a linear predictive function, and the unknown model parameters are estimated from the data, these models are called linear models. The most commonly used linear regression model is one in which the conditional mean of Y, given a value of X, is an affine function of X. The median or some other median given a value of X can also be used. A linear model can also be a median or some other quantile of the conditional distribution of Y given X expressed as a linear function of X.

The practical uses of linear regression fall into the following two broad categories:

(1) If the goal is prediction or mapping, linear regression can be used to fit a predictive model to the observed dataset and X values. When such a model is completed, a Y value can be predicted from this fitted model for an added X value, given no Y to match it.
(2) Given a variable Y and a number of variables $X_1, \ldots, X_p$ that are potentially correlated with Y, linear regression analysis can be used to quantify the strength of the correlation between Y and X, to evaluate the X_j that are the least correlated with Y, and to identify which subsets of X_j contain redundant information about Y.

Linear regression models are often fitted with least squares approximation, in order to improve the efficiency of the fit and thus achieve the purpose of online fitting, after the segmentation of the feature data, each segmented part is fitted separately using a single linear fitting function, and the coefficients a and b in the fitted linear equation $y = ax + b$ are calculated from Eqs. (7.15) and (7.16):

$$a = \frac{n \sum_{k=0}^{n-1} x_k y_k - \sum_{k=0}^{n-1} x_k \sum_{k=0}^{n-1} y_k}{n \sum_{k=0}^{n-1} x_k^2 - \sum_{k=0}^{n-1} x_k \sum_{k=0}^{n-1} x_k}, \tag{7.15}$$

$$b = \frac{\sum_{k=0}^{n-1} y_k - a \sum_{k=0}^{n-1} x_k}{n}. \tag{7.16}$$

Table 7.12. Trend combination patterns.

Late stage	Pre-production		
	Up	Invariant	Down
Up	Up	Up	Bottom-up transient
Invariant	Goose-step	Invariant	Stride forward
Down	Up-down transient	Down	Down

The fitted function a is compared with a predetermined judgmental slope $a_1(a_1 > 0)$ to obtain a simple trend for each segmented segment. If $a > a_1$, it is an ascending segment; if $a_1 > a > -a_1$, it is a stabilizing segment; if $a < -a_1$, it is a descending segment.

(3) Trend forecasting methods

The fitted trends are combined sequentially according to the forward and backward order, and combined according to the combining rule in Table 7.12 until two neighboring trends cannot be combined. By combining the trends, seven types of trends are obtained: rising, falling, unchanged, positive step, negative step, up-down transient, and down-up transient. Based on the trend combination results of the last two segments, the future trends of the process parameters can be predicted.

(4) Failure trend analysis

For the low alarm threshold $l_{\min}$ and the high alarm threshold $l_{\max}$, let $k = l_{\max} - l_{\min}$, when the parameter l meets the danger range of $l_{\min} < l < l_{\min} + 3\%k$ or $l_{\max} - 3\%k < l < l_{\max}$ at a certain point in time, then l is a dangerous parameter.

When $l_{\min} < l < l_{\min} + 3\%k$, and the predicted trend is up or down-up transient, the high threshold alarm is triggered; when $l_{\max} - 3\%k < l < l_{\max}$, and the predicted trend is down or up-down transient, the low threshold alarm is triggered.

7.3.1.2 *Process parameter correlation validation*

(1) Process parameter correlation

In the process of trend prediction, only single-parameter time-series relationships are used for analysis, while in the production process of refining and chemical plant equipment units and operations associated with a high degree of strong coupling of hazardous factors and has the characteristics of the whole emergence, etc., you can establish a correlation model to fit the relationship between the parameters, and mutual verification of the results of the trend prediction of the faults, and correlation between the parameters of the relationship between the fitted as shown in Table 7.13.

By fitting the relationships of the process parameters, the relevant parameters conforming to Relationships 1 and 2 can be verified in the failure trend prediction accordingly, e.g., the stabilizer tower reflux pressure and the reflux tank level.

(2) Associated parameter validation

Fault trend prediction results are validated by correlation-consistent process parameters to verify the reliability of the results, and the validation steps are shown in Fig. 7.26.

The credibility of the original fault prediction trend results is obtained through the validation results of the associated parameters, thus providing a basis for the field operator's judgment of the overall fault status.

7.3.1.3 *Example validation*

(1) Field data validation

In the petroleum refining process, the first step of crude oil processing is distillation, and the atmospheric decompression process is the key process to fractionate crude oil into different components and provide raw materials for secondary processing afterward. The safe and smooth operation of the atmospheric decompression unit is directly related to the production efficiency of the whole refinery. The stabilizer tower is the key device in the NRP process. Crude oil is dewatered and warmed up in the heat exchanger and then enters the stabilizer tower, where the crude oil is partially gasified inside the stabilizer tower, and the gasified portion is distilled in the upper part. The carbon five component of the gas in the top of the tower is cooled and separated by circulating water, and part of it is sent back to the tower as the top reflux, and the carbon one to carbon four is output in a gaseous state, and the stabilized crude oil flows out from the bottom of the stabilized tower into the storage tanks or

Table 7.13. Process parameter fitting relationships.

Serial number	Depiction	Concrete relation	
1	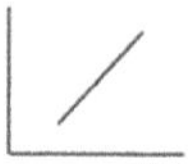	Descriptive	Represents a linear relationship where the cause variable is proportional to the effect variable
		Example	Temperature and pressure in ideal gases
		Formula	$y = ax + b(a > 0)$
2		Descriptive	Represents a linear relationship where the cause variable is inversely related to the effect variable
		Example	Volume and pressure in an ideal gas
		Formula	$y = -ax + b(a > 0)$
3		Descriptive	As the cause variable increases, the consequence variable increases slowly and then increases faster
		Example	The inlet flow rate is greater than the outlet flow rate and is increasing, level vs. inlet flow rate
		Formula	$dy/dt = x_1 - x_2$
4	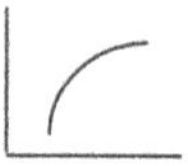	Descriptive	As the cause variable increases, the consequence variable increases rapidly and then increases more slowly
		Example	Outlet flow rate is less than inlet flow rate and is increasing, level vs. outlet flow rate
		Formula	$dy/dt = x_1 - x_2$
5	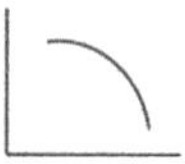	Descriptive	As the cause variable increases, the consequence variable slowly decreases and then decreases more rapidly
		Example	The outlet flow rate is greater than the inlet flow rate and is increasing, level vs. outlet flow rate
		Formula	$dy/dt = x_1 - x_2$
6	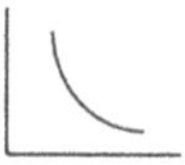	Descriptive	As the cause variable increases, the consequence variable decreases rapidly and then increases at a slower rate
		Example	The inlet flow rate is less than the outlet flow rate and is increasing, level vs. inlet flow rate
		Formula	$dy/dt = x_1 - x_2$

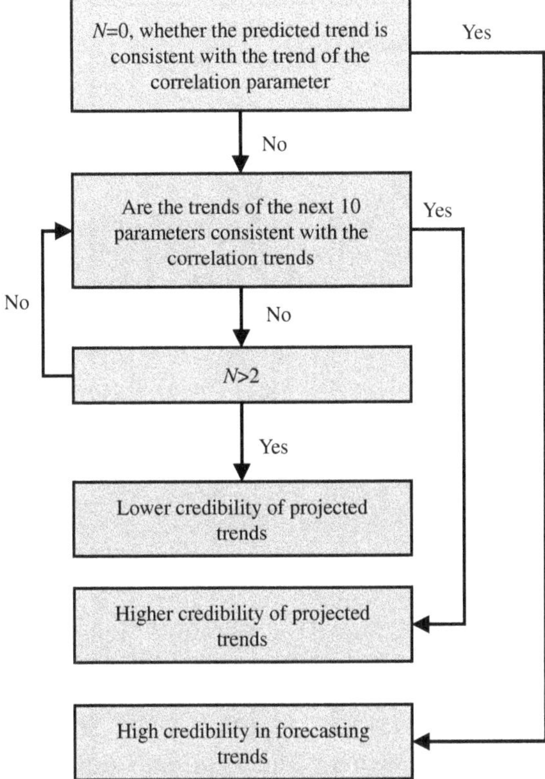

Fig. 7.26. Associative parameter validation steps.

the external transport. Field data from the stabilizer tower are used for trend prediction.

The stabilized tower top reflux tank level parameters for a certain period of time in this refinery are shown in Fig. 7.27, with a high alarm threshold of 32%, and the alarm was triggered in the 32nds due to the high level of the reflux tank caused by the imbalance of the incoming and outgoing materials and then restored to normal after system adjustments. Through the parameters of reflux tank level from 0–30 s to predict the failure trend from 30–35 s, and through the associated parameters to stabilize the reflux pressure at the top of the tower to verify the credibility of the prediction results, and finally use the real failure trend of reflux tank level from 30–35 s to check the prediction accuracy.

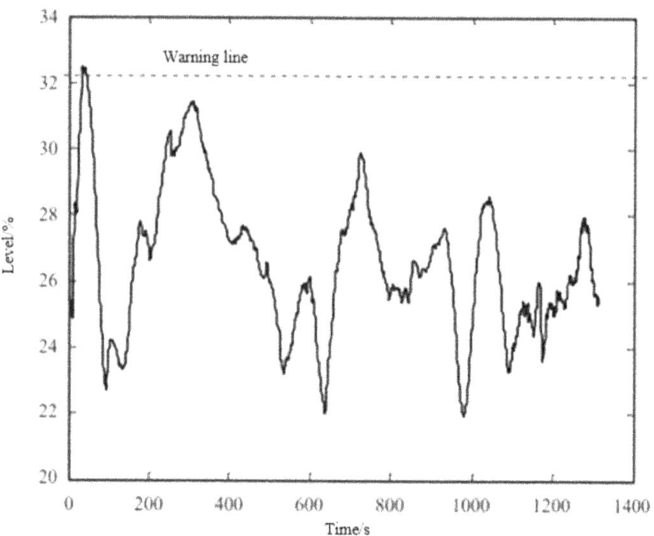

Fig. 7.27. Stabilizing tower top reflux tank levels.

According to the feature data segmentation algorithm to find the feature points and feature line segments, the first 30 s parameters of the tower top temperature are divided into 6 feature segments: 0–10 s, 10–11 s, 11–12 s, 12–16 s, 16–21 s, and 21–30 s. The segmented feature segments are shown in Fig. 7.28, and different symbols are used to distinguish different segments.

Each segmentation part is then fitted separately using a one-time linear fitting function to obtain the fitted line segments for the six segments of segmented data as shown in Fig. 7.29, and the expression for the fitted line segments is as follows:

$$y_1 = -0.0844x + 25.7093$$
$$y_2 = 1.8810x + 6.0980$$
$$y_3 = -0.1710x + 28.6700$$
$$y_4 = 0.4127x + 21.7682$$
$$y_5 = -0.0441x + 29.0081$$
$$y_6 = 0.3478x + 21.0038$$

By judging the trend of the fitted line segments, the trends of the six segmented segments are: y_1 down, y_2 up, y_3 down, y_4 up, y_5 unchanged, y_6 up, and according to the trend combining law in

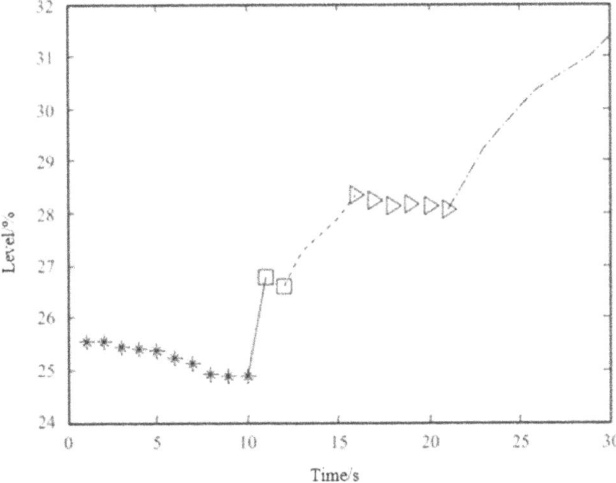

Fig. 7.28. Reflux tank level data segmentation.

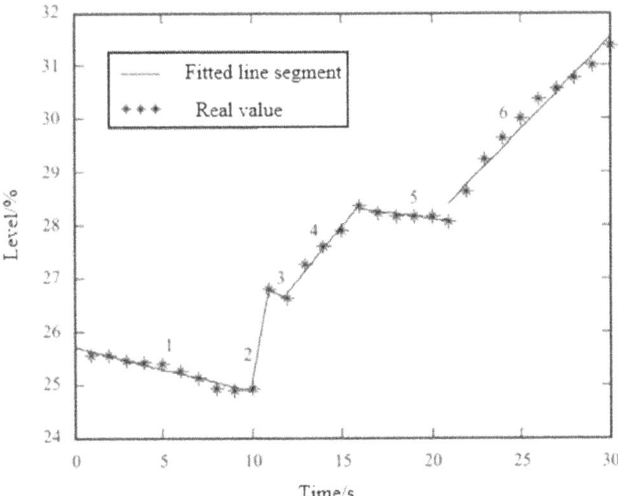

Fig. 7.29. Reflux tank level 0–30 s fitted line segment.

Table 7.12, the total trend of the last two segments is up. Meanwhile, considering that the parameter value of the 30ths is 31.38, which is within the dangerous parameter range, it is predicted that the reflux tank level will exceed the high-value warning line and the fault trend will rise.

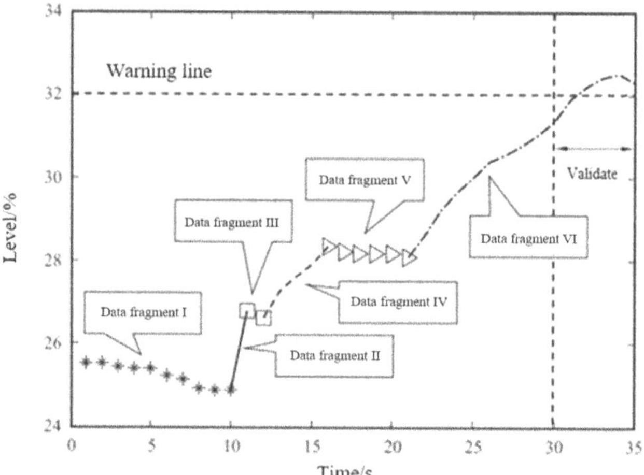

Fig. 7.30. Segmentation results of reflux tank level data (0–35 s).

The reliability of the failure prediction results was verified by analyzing the trend of the return tank pressure parameters. The results show that the predicted trend is consistent with the measured data. Based on the standard judgment process assessment, the confidence level of the fault prediction results for the return tank level within 30–35 seconds is high.

To verify the accuracy of the trend prediction from 0–30 s with the real trend of the reflux tank level from 30–35 s, Figs. 7.30 and 7.31 show the segmentation results of the characteristic data and the fitted line segments of the reflux tank level from 0–35 s, respectively. According to the data segmentation and line fitting of the reflux tank level from 0–35 s, the last characteristic line segment is from 21–35 s, and the equation of the fitted line segment is $y = 0.3207x + 21.7831$, and the alarm line is indeed triggered at 32 s, and the fault trend is rising, which is consistent with the predicted results.

(2) Simulation data validation

The primary distillation tower is also an important device in the atmospheric decompression process. The device carries out preliminary gas-liquid separation of mixed oil and gas, and the crude oil enters the primary distillation column after heat exchange to about 220–240°C. The liquid phase flowing into the bottom of the column is partially sent to the atmospheric column, and the gas phase rises to the top of the column, where the reformed raw material or light

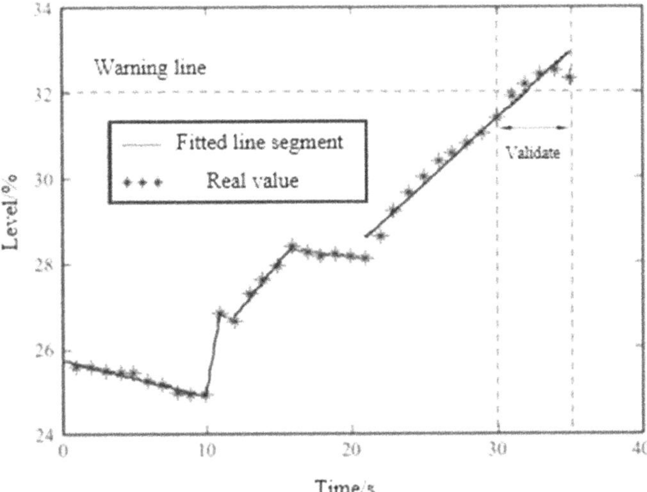

Fig. 7.31. Reflux tank level fitting results (0–35 s).

gasoline is fractionated out at the top of the primary distillation column. The process model of the primary distillation column is established in gPROMS software by modeling the gas-liquid separation process.

By simulating the heating anomaly state, the heating anomaly occurs at 60 s due to an electrical fault, and the parameters of the gas-phase yield of the primary distillation column are shown in Fig. 7.32. The failure trend is predicted based on the parameters of gas-phase production from 0–63 s, and the prediction results are verified using the correlation parameter (i.e., the temperature of the top of the primary distillation column) at the same time to get the credibility of the failure prediction results, and finally the accuracy of the prediction results is checked based on the real data values of gas-phase production from 63–70 s.

According to the feature data segmentation algorithm to find the feature points and feature line segments, the parameters of the first 63 s of the gas-phase yield are divided into three feature segments: 0–10 s, 10.1–60.4 s, and 60.5–63 s. The divided feature segments are shown in Fig. 7.33, and different symbols are used to distinguish the different segments.

Each segmentation part is then fitted separately using a one-time linear fitting function to obtain the fitted line segments for the three

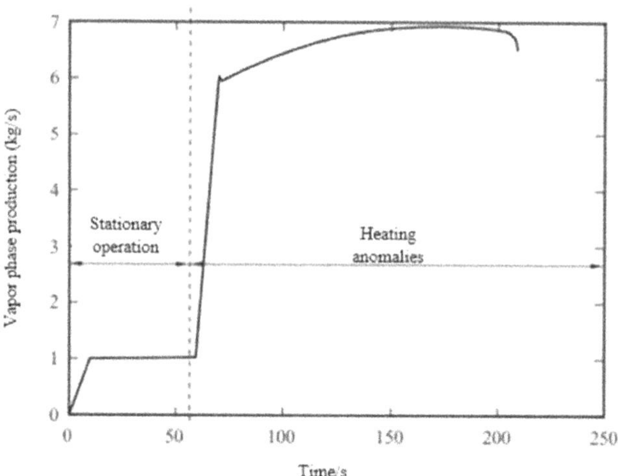

Fig. 7.32.   Variation of gas-phase production under heating anomalies.

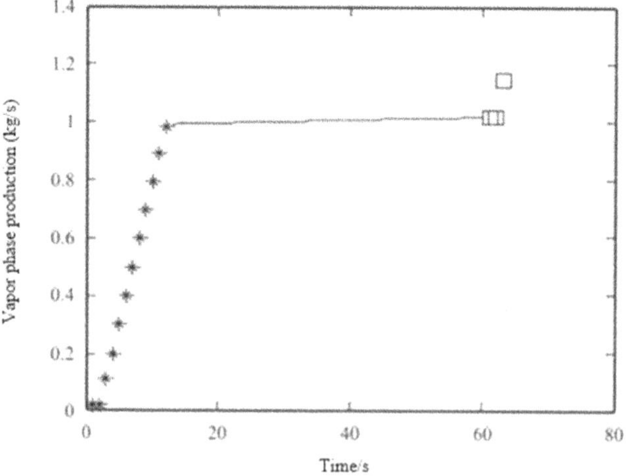

Fig. 7.33.   Segmentation of gas-phase production data.

segmented data as shown in Fig. 7.34, and the expression for the fitted line segments is as follows:

$$y_1 = 0.0990x - 0.0100$$

$$y_2 = 0.0006x + 0.9851$$

$$y_3 = 0.4879x - 28.3133$$

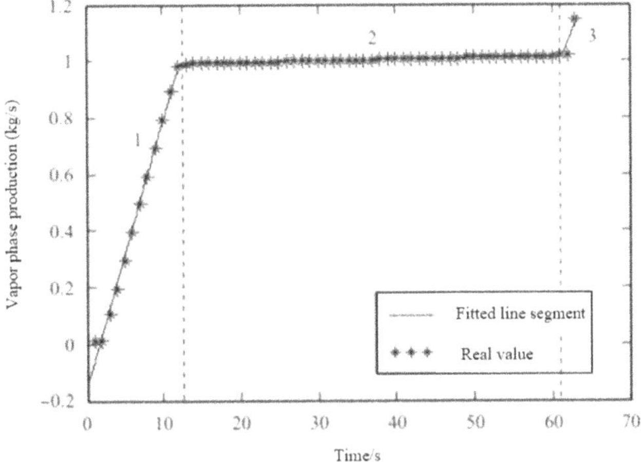

Fig. 7.34. Gas-phase yield 0–63 s fitted line segment.

By judging the trend of the fitted line segments, the trends of the three segmented segments were obtained as follows: y_1 rising, y_2 unchanged, and y_3 rising, and according to the trend combining law in Table 7.12, the general trend of the latter two trends was rising, and the parameter value of the 63rd second was 1.149, which was in the hazardous range, and therefore it was predicted that the gas-phase yield would exceed the high warning line and the failure trend was rising.

The prediction results are verified by the associated parameter primary distillation column temperature, which is consistent with its trend, and are verified according to the judgment step, and the confidence level of the trend results of the failure prediction of the gas-phase production rate from 63–66 s is judged to be: high confidence.

The accuracy of the trend prediction of 0–30 s is verified with the real trend of gas-phase yield 63–66 s parameters. Figures 7.35 and 7.36 show the segmentation results of the characteristic data and the fitted line segments of the gas-phase yield 0–66 s parameters, respectively. According to the data segmentation and line segment fitting for the gas-phase yield 0–35 s parameter, the last characteristic line segment is 60–66 s, and the alarm line is indeed triggered at 64 s, and the fault trend is rising, which is consistent with the prediction results.

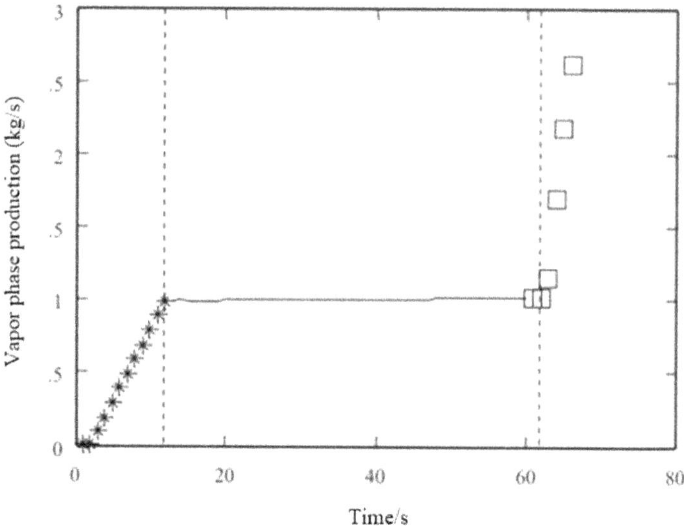

Fig. 7.35.   Segmentation of gas-phase yield 0–66 s data.

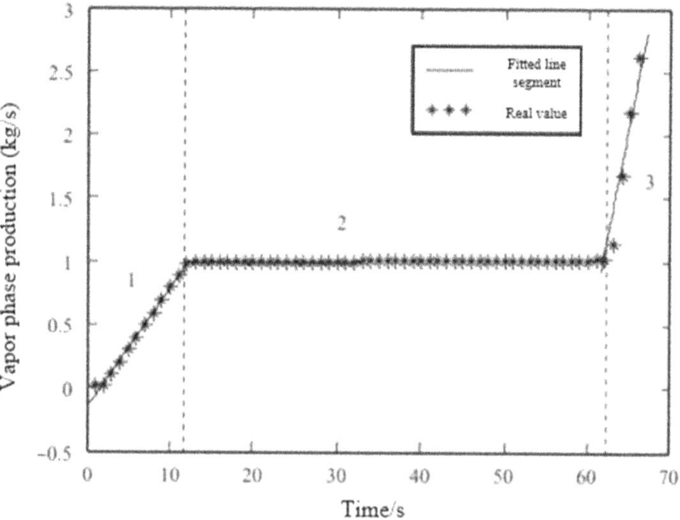

Fig. 7.36.   Linear fit for gas phase yield 0–66 s.

7.3.2 Grey correlation analysis based "safety behavior state" Kalman filter correlation prediction method

7.3.2.1 Grey correlation analysis of influencing factors

The refining and chemical production process is a nonlinear and complex system, and each unit is not only subject to the constraints of manual operation and the inherent properties of the unit during operation, but also affected by inflow, outflow and various random factors. In the case of the temperature at the top of the primary distillation column, it is not only affected by the production state in the primary distillation column, but also closely related to the production state in the neighboring unit.

The measure of the magnitude of the correlation between two systems with respect to factors that vary over time or with different objects is called the degree of correlation. During the development of the system, if the trend of change of the two factors is consistent, i.e., the degree of synchronous change is high, i.e., the degree of correlation between the two is high; conversely, it is low. Therefore, grey correlation analysis is a method to measure the degree of correlation between factors based on the degree of similarity or dissimilarity in the trends of the factors, i.e., "grey correlation".

The grey system theory proposes the concept of grey correlation analysis of subsystems, with the intention of seeking numerical relationships between subsystems (or factors) in a system through certain methods. Therefore, grey correlation analysis provides a quantitative measure of a system's developmental changes and is suitable for dynamic history analysis. Grey correlation analysis has the advantages of requiring fewer samples, not needing to calculate statistical eigenvalues and being easy to compute, and has been widely applied to measure and analyze the degree of correlation between factors in an uncertain system.

Considering a given sequence of parameters x_0 of a research object in the production process of a refining unit and a sequence of process parameters x_i of the production state of the unit and the adjacent units, according to the idea of grey equilibrium proximity correlation analysis, the following calculation steps are designed to determine the main influencing factors of the temperature of the top of the atmospheric tower and their degree of influence.

(1) Identify the reference series that characterize the behavior of the system and the sub-series that influence the behavior of the system

A sequence of data that reflects the behavioral characteristics of a system is called a reference series. The data series consisting of factors affecting the behavior of the system is called the sub-series. For the original data series, the $\{x_i(j)\}$ $(i = 1, \ldots, l; j = 1, \ldots, n)$, where l is the number of subsequences and n is the length of the subsequence. Determine its reference sequence x_0 and subsequence x_i and calculate the sequence difference at each moment point $\Delta_i(j) = |x_0(j) - x_i(j)|$, e maximum and minimum values of $\Delta_{\max}$ and $\Delta_{\min}$.

(2) Dimensionless processing of reference and comparison series

Due to the different physical significance of the factors in the system, the resulting data may not always have the same dimension, which is not easy to compare, or it is difficult to get the correct conclusion when comparing. Therefore, in the grey correlation analysis, generally have to carry out the dimensionless data processing.

(3) Calculate the grey correlation coefficient $r_i(j)$ between the reference sequence x_0 and each subsequence x_i at each moment point

The degree of correlation is essentially the degree of difference in geometry between the curves. Therefore, the size of the difference between the curves can be used as a measure of the degree of association. For a reference series X_0 there are several comparison series $X_1, X_2, \ldots, X_n$. The correlation coefficient $r_i(j)$ between each comparison series and the reference series at various moments (i.e., points in the curve) is calculated in Eq. (7.17) where ρ is the resolution coefficient, generally 0–1, usually taken as 0.5:

$$r_i(j) = (\Delta_{\max} + \rho\Delta_{\min})/[\Delta_i(j) + \rho\Delta_{\max}]. \qquad (7.17)$$

Calculate the grey correlation coefficient distribution mapping $p_i(j)$ for the subsequence in Eq. (7.18):

$$p_i(j) = r_i(j) \left/ \sum_{j=1}^{n} r_i(j) \right. . \qquad (7.18)$$

Calculate the correlation coefficient entropy $H(R_i)$ for the ith grey correlation coefficient sequence $R_i = \{r_i(j)\}$ in Eq. (7.19):

$$H(R_i) = -\sum_{j=1}^{n} p_i(j) \ln p_i(j). \tag{7.19}$$

The magnitude of correlation coefficient entropy reflects the degree of equilibrium of grey correlation coefficients.

Calculate the degree of balance B_i of the grey correlation coefficient sequence R_i in Eq. (7.20):

$$B_i = H(R_i)/H_{\max}(R_i), \tag{7.20}$$

where $H_{\max}(R_i) = \text{In } n$ is the maximum entropy value of R_i.

Equilibrium proximity $B_a(i)$ is obtained from grey correlation and equilibrium in Eq. (7.21):

$$B_a(i) = B_i \times V_{oi}$$
$$V_{oi} = \frac{1}{n} \sum_{j=1}^{n} r_i(j), \tag{7.21}$$

where V_{oi} is the grey correlation degree.

All the calculated equilibrium proximity is ranked, the greater the equilibrium proximity, the stronger the correlation between the factor and the research object parameter, and vice versa, the weaker the correlation. A large number of experiments and practical applications show that the balanced proximity method has greater advantages than other grey correlation analysis methods, so this method is chosen to select the main influence factors of the research object parameters.

7.3.2.2 *Kalman filter-based correlation prediction algorithm*

The prediction method of Kalman filter has the characteristics of fewer model parameters and convenient calculation, which is suitable for online fast prediction, but its conventional model is based on the historical data of the sampling time series for the parameter prediction of the next time period, which leads to the defects of the prediction method of Kalman filter in terms of adaptability and prediction accuracy due to the lack of consideration of the influencing factors of the whole device process.

In order to introduce the influencing factors of other process parameters within a given refining unit into the Kalman filter model, it is necessary to select the main factors from a large number of possible influencing factors by means of correlation analysis, so as to establish a multivariate relational model of a certain parameter to be predicted in the refining unit. Combined with the characteristics of Kalman filter theory, based on grey correlation analysis and Kalman filtering to establish a comprehensive chemical correlation data model, the prediction algorithm for process parameters is as follows.

Step 1: Calculate the equilibrium proximity between a parameter to be predicted in the refining unit and the relevant influencing factors according to the grey theory, select m variables closely related to it, and establish the following regression equation:

$$\begin{cases} x_0(k+1) = b_{00}x_0(k) + b_{01}x_1(k) + b_{02}x_2(k) + \cdots + b_{0m}x_m(k) + \varepsilon_0 \\ x_1(k+1) = b_{10}x_0(k) + b_{11}x_1(k) + b_{12}x_2(k) + \cdots + b_{1m}x_m(k) + \varepsilon_1 \\ \vdots \\ x_m(k+1) = b_{m0}x_0(k) + b_{m1}x_1(k) + b_{m2}x_2(k) + \cdots \\ \qquad\qquad + b_{mm}x_m(k) + \varepsilon_m \end{cases},$$

$$(7.22)$$

where parameters $b_{00}, b_{01}, \ldots, b_{mm}$ and $\varepsilon_0, \varepsilon_1, \ldots, \varepsilon_m$ are regression coefficients, which can be obtained by least squares.

$$\begin{bmatrix} \varepsilon_0 & \varepsilon_1 & \cdots & \varepsilon_m \\ b_{00} & b_{10} & \cdots & b_{m0} \\ b_{01} & b_{11} & \cdots & b_{m1} \\ \vdots & \vdots & \vdots & \vdots \\ b_{0m} & b_{1m} & \cdots & b_{mm} \end{bmatrix}$$

$$= [T' \cdot T]^{-1} T' \cdot \begin{bmatrix} x_0(2) & x_1(2) & x_2(2) & \cdots & x_m(2) \\ x_0(3) & x_1(3) & x_2(3) & \cdots & x_m(3) \\ \vdots & \vdots & \vdots & & \vdots \\ x_0(n) & x_1(n) & x_2(n) & \cdots & x_m(n) \end{bmatrix}.$$

$$(7.23)$$

Among them,

$$
T = \begin{bmatrix}
1 & x_0(1) & x_1(1) & \cdots & x_m(1) \\
1 & x_0(2) & x_1(2) & \cdots & x_m(2) \\
\vdots & \vdots & \vdots & & \vdots \\
1 & x_0(n-1) & x_1(n-1) & \cdots & x_m(n-1)
\end{bmatrix}.
$$

Step 2: Establish the equation of state such that $X(k) = [x_0(k)\ x_1(k)\ \cdots\ x_m(k)]'$:

$$
\begin{cases}
X(k+1) = B(k)X(k) + w(k) \\
y(k+1) = A(k) \cdot [x_0(k+1)\ x_1(k+1)\ \cdots\ x_m(k+1)]' + v(k)
\end{cases},
\tag{7.24}
$$

among others,

$$
B(k) = \begin{bmatrix}
b_{00} & b_{01} & \cdots & b_{0n} \\
b_{10} & b_{11} & \cdots & b_{1n} \\
\vdots & \vdots & & \vdots \\
b_{m0} & b_{m1} & \cdots & b_{mn}
\end{bmatrix},
$$

$$
A(k) = [1\ 0\ \cdots\ 0], \quad w(k) = [\varepsilon_0\ \varepsilon_1\ \cdots\ \varepsilon_m]',
$$

$v(k)$ is the observation noise at time k.

Step 3: Initialize the filtered variance array $P(0)$ and observation $X(0)$.

Step 4: Recursive calculation:

$$
P(k|k-1) = B(k)P(k-1)B'(k) + Q(k-1),
\tag{7.25}
$$

where $Q(k-1)$ is a symmetric non-negative definite matrix.

Step 5: Calculate the Kalman filter coefficients:

$$
K(k) = P(k|k-1)A'(k)[A(k) \cdot P(k|k-1)A'(k) + R(k)]^{-1}.
\tag{7.26}
$$

Step 6: Perform the status update according to the following formula.

$$
\hat{x}(k) = B(k)\hat{x}(k-1) + K(t) \cdot [y(k) - A(k)B(k)\hat{x}(k-1)],
\tag{7.27}
$$

$$
P(k) = [1 - K(k)A(k)]P(k|k-1),
\tag{7.28}
$$

where $y(k) = x_0(k)$.

Step 7: Let $k = k + 1$ and go back to step 4 for iterative calculation until the termination condition.

Step 8: Calculate the predicted value of the temperature at the top of the atmospheric tower $y(k) = A(k) \cdot X(k)$, and evaluate the prediction.

7.3.2.3   *Field examples*

(1) Case 1: predicting and judging the dangerous state of the liquid level in the reflux tank at the top of the stabilization tower

The petroleum refining process is selected as the case study object of the normally reduced pressure device. Stabilization tower is a key device in the NRP process. Crude oil is dewatered, warmed up in the heat exchanger, and then enters the stabilization tower, where the crude oil is partially gasified inside the stabilization tower, and the gasified portion is distilled in the upper part of the tower. The carbon five component of the gas at the top of the tower is cooled and separated by circulating water, and part of it is sent back to the tower as the top reflux, and the carbon one to carbon four is output in a gaseous state, and the stabilized crude oil flows out from the bottom of the stabilization tower into the storage tanks or external transport. Trend prediction using field data from the stabilizer tower.

The role of the reflux tank at the top of the stabilizing tower is to adjust the purity of the product at the top of the tower. The greater the reflux ratio, the fewer theoretical plates are required, and the higher the quality of the product achieved by production with the number of plates remaining constant. The reflux tank can improve the cold reflux on the tower plate, take away the excess heat, and maintain the heat balance in the tower; the gas and liquid phases are in reverse contact on the tower plate, and the recombined components in the upward gas condense, and the light components in the downward liquid absorbs the heat and vaporizes, and the recurring condensation-vaporization effect further increases the precision of product separation. During the production process, it is necessary to maintain the level of the reflux tank to continuously meet the reflux requirements.

A refinery's top reflux tank of a stabilization tower had an over high warning level of 33.0% and an under low warning level of 24.7%. Using the Kalman filter modeling method based on grey correlation model, the time series of the reflux tank liquid level $\{x(k)\}(k = 1, 2, \ldots, 60)$. The short-term prediction experiment was conducted to determine whether the 60 level parameters within 60 min of a certain period of time were within the normal operating range.

According to the grey correlation analysis method, a grey correlation model is established based on the sequence of reflux tank level parameters x_0 at the top of the stabilizer tower and the sequence of production process parameters x_i in the stabilizer tower and the adjacent units, and the influencing factors of the reflux tank level and their grey correlation results are calculated as shown in Table 7.14.

According to the results of the grey balanced proximity analysis method, the two influencing factors with the largest correlation and the average value greater than 0.3, namely, the feed temperature of the stabilizer tower and the lower temperature of the stabilizer tower, are taken as covariates of the reflux tank liquid level regression model. According to equation (7.20), the time series regression model of reflux tank level was established, and the parameters of the regression model were obtained by the least squares method. Given the time observation values of the reflux tank level within 60 min in the prediction interval, the prediction was carried out according to the conventional Kalman filtering prediction method and the Kalman correlation regression model, respectively, and the results are shown in Figs. 7.37 and 7.38.

From Figs. 7.38 and 7.39, it can be seen that some parameters are close to the warning line within 60 min, which is a dangerous state but not exceeding the warning line, and they are all within the normal operation range. Analyzing the prediction results of the Kalman correlation model, the predicted state is always consistent with the real value, and the error is very small. According to the prediction results of the conventional Kalman filtering method, at 9 min and 34 min, the predicted state is always consistent with the real value, and the error is very small. Exceeding the high/low alarm line once does not correspond to the actual state of the real value and is a false alarm.

Table 7.14. Grey correlation of influencing factors of reflux tank level.

Serial number	Factor	Grey correlation
1	Stabilization tower feed temperature	0.46291
2	Stabilizer return tank pressure	0.039491
3	Stabilization of tower cooling return flow	0.210494
4	Stabilization tower reboiler level	0.14422
5	Stabilization tower t304 lower temperature	0.359296
6	Fractionator bottom level	0.142829
7	Fractionator stirring vapor flow	0.010703
8	Circulating oil flow rate at the top of the fractional distillation column	0.189645
9	Fractionator middle oil flow	0.267718
10	Fractionator top pressure	−0.05013

In order to test the sample sequence $\{x(t)\}$ and the sequence of predicted values $\{x'(t)\}(t = 1, 2, \ldots, 60)$, the following error indicators are introduced:

$$\text{Average relative error } e_r = \frac{1}{60} \sum_{i=1}^{60} \frac{x'(t) - x(t)}{x(t)}.$$

$$\text{Maximum relative error}: \ e_{r,\max} = \max \left| \frac{x'(t) - x(t)}{x(t)} \right|.$$

Relative error squared and root mean square e'_r

$$= \sqrt{\frac{1}{60} \sum_{t=1}^{60} \left[\frac{x'(t) - x(t)}{x(t)} \right]^2}.$$

The results of the return tank level prediction of the conventional Kalman filter prediction method and the Kalman correlation regression model prediction method were examined, and a comparison of the error analyses is shown in Table 7.15. From the prediction error analysis in Table 7.15, it can be seen that the Kalman correlation regression model prediction method is able to track the time series

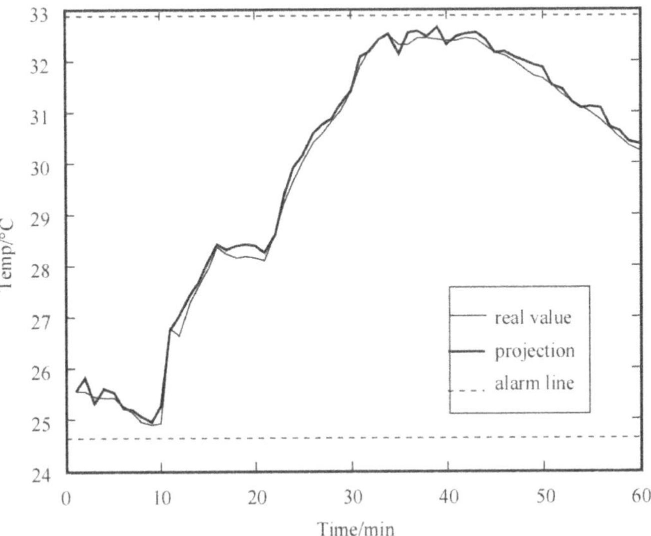

Fig. 7.37. Predictions from the Kalman correlation regression model.

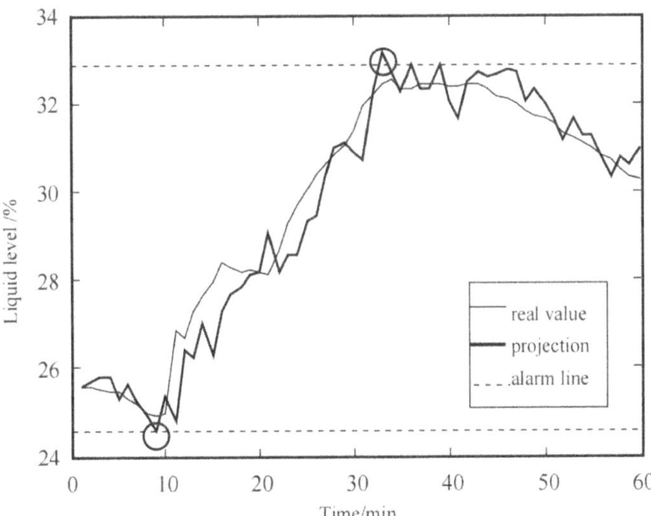

Fig. 7.38. Conventional Kalman filter prediction results.

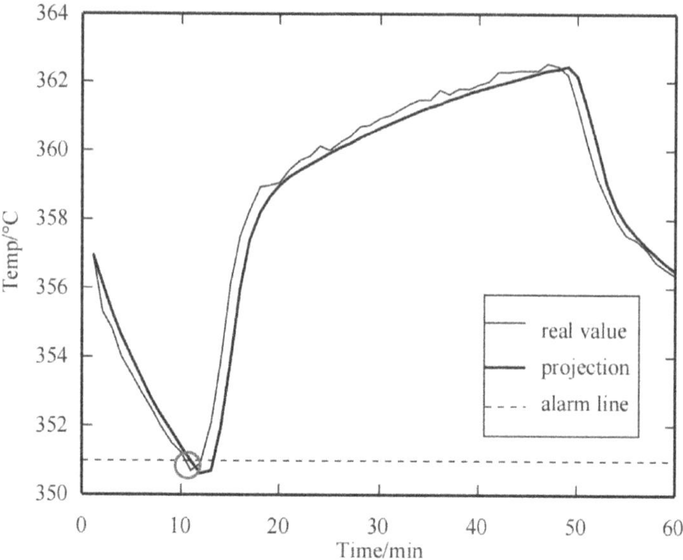

Fig. 7.39. Predictions from the Kalman correlation regression model.

Table 7.15. Comparison of prediction error test metrics for the two methods.

Error indicator	Average relative error	Maximum relative error	Relative error squared and root mean square
Conventional Kalman filter prediction methods	0.36%	2.48%	0.16
Kalman correlation regression model prediction method	0.042%	1.55%	0.10

trend of the reflux tank level more quickly and has a smaller relative error maximum and average relative error than the conventional Kalman filter prediction method because the influencing factors of the process parameters of the reflux tank device and the neighboring devices are taken into account in the prediction of the reflux tank level. In the prediction process, the two influencing factors with the highest degree of correlation with the research object are selected, and the predicted values of these two influencing factors are directly used for recursive calculation, and the prediction accuracy is significantly improved.

(2) Case 2: prediction of ultra-low temperature failure of temperature parameter at the top of atmospheric pressure tower

The atmospheric pressure tower is the core device of the atmospheric decompression device, and the distillation products are mainly obtained from the atmospheric pressure tower. The top of the atmospheric tower separates the lighter naphtha component, the bottom of the tower produces heavy oil, and the side lines produce diesel or wax oil components in between. Atmospheric columns are generally set up 3–5 side lines, the number of side lines is mainly based on the number of product types to determine the number of products equal to the atmospheric tower product types N minus the top and bottom of these two products, that is, $N - 2$.

A refinery's atmospheric tower top temperature has an ultra-high alarm temperature of $369°C$ and an ultra-low alarm temperature of $351°C$. Using the Kalman filter modeling method based on grey correlation model, the time series of atmospheric pressure tower top temperature $\{x(k)\}(k = 1, 2, \ldots, 60)$ the short-term prediction experiment is conducted to determine whether the 60 temperature parameters within $60\,\mathrm{min}$ of a certain period of time are within the normal operating range.

Firstly, according to the grey correlation analysis method, a grey correlation model is established based on the sequence of temperature parameters x_0 at the top of the atmospheric pressure tower, the sequence of process parameters x_i at the atmospheric pressure tower, and the production state in the adjacent plant to calculate the influencing factors of the temperature at the top of the atmospheric pressure tower and their grey correlations, and the results are shown in Table 7.16.

According to the results of the grey equilibrium proximity analysis method, the two influencing factors with the largest correlation degree, the top pressure of the atmospheric pressure tower and the bottom temperature of the atmospheric pressure tower, were taken as the covariates of the regression model of the top temperature of atmospheric pressure tower. Accordingly, a time-series

Table 7.16. Grey correlation of factors influencing the temperature at the top of the atmospheric tower.

Serial number	Factor	Grey correlation
1	Pressure at the top of an atmospheric tower	0.706435
2	Liquid level at the bottom of an atmospheric tower	0.395016
3	Bottom temperature of atmospheric tower	0.782185
4	Constant top circulation flow	0.292439
5	Return flow of the first intermediate section of the atmospheric pressure tower	−0.164750
6	Return flow rate of the second intermediate section of the atmospheric pressure tower	0.194884
7	Atmospheric furnace east feed flow	0.341415
8	Atmospheric furnace west road feed flow	0.280297
9	Primary distillation column reflux tank outlets	0.107107
10	Primary distillation tower top temperature	0.081779
11	Primary distillation tower top pressure	0.196804
12	Primary distillation tower bottom level	0.294884

regression model of atmospheric pressure tower top temperature was established, and the parameters of the regression model were obtained by the least-squares method. Given the observed values of atmospheric pressure tower top temperature within 60 min in the prediction interval, the prediction was carried out according to the conventional Kalman filtering prediction method and the Kalman correlation regression model, respectively, and the results are shown in Figs. 7.39 and 7.40.

From Figs. 7.39 and 7.40, it can be seen that the temperature parameter of the atmospheric pressure tower is below the alarm line at 12 min and 13 min, which belongs to the operation of non-normal state, and the alarm information should be generated and adjusted in time. Analyzing the prediction results of Kalman correlation regression model, the predicted state is always consistent with the real value, and according to the prediction results at 11 min, the temperature parameter at 12 min will be lower than the alarm line, and the warning is carried out 1 min in advance, and the warning

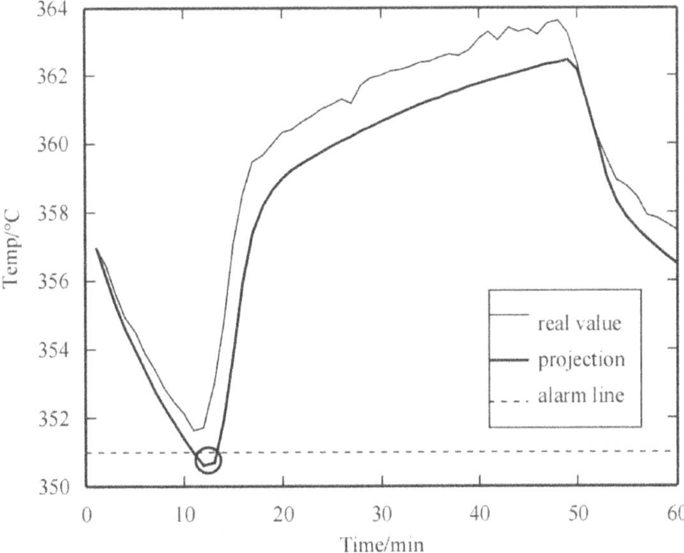

Fig. 7.40. Conventional Kalman filter prediction results.

results are accurate. According to the prediction results of the conventional Kalman filtering method, it fails to accurately predict the non-normal state below the alarm line, which is not consistent with the actual state of the real value and belongs to the omission of the alarm.

The predicted results of the reflux tank level of the conventional Kalman filter prediction method and the Kalman correlation regression model prediction method are examined separately, and the error analysis comparison is shown in Table 7.17.

From the prediction error analysis in Table 7.17, it can be seen that the Kalman correlation regression model prediction method is able to track the change trend of the time series of the atmospheric pressure tower temperature more quickly, and has a smaller relative error maximum and average relative error than the conventional Kalman filter prediction method, because the influencing factors of the process parameters in the atmospheric pressure tower device and the neighboring devices are considered in the prediction process of the atmospheric pressure tower temperature. In the prediction process, the two influencing factors with the highest degree of correlation with

Table 7.17. Comparison of prediction error test metrics for the two methods.

Error indicator	Average relative error	Maximum relative error	Relative error squared and root mean square
Conventional Kalman filter prediction methods	0.31%	0.63%	0.027
Kalman correlation regression model prediction method	0.027%	0.38%	0.014

the research object are selected, and the predicted values of these two influencing factors are directly used for recursive calculation, and the prediction accuracy is significantly improved.

7.3.2.4　*Analysis and summary*

(1) The idea of using the correlation between the parameters of the refining process to accurately predict the fault development trend is proposed, and the fault trend prediction method based on the feature data segmentation method is established from the perspective of qualitative prediction of the fault development trend. The systematic correlation influence of the refining process is analyzed to get the fitting relationship of the process parameters, the special data segmentation method is used to carry out reasonable segmentation and feature extraction of the data, the prediction trend is obtained through the combination law of the three simple trend primitives, the hazardous parameters and trend are analyzed in combination with the alarm information of the system, so as to predict the failure and the failure trend of the system and the credibility is obtained through the validation of the correlation parameters.

(2) The main feature of this method is that the data division for the characteristics is more reasonable and accurate, and the system failure is predicted in advance, while the failure development trend and credibility information are obtained at the same time. The field data and simulation experimental data show that the

method can accurately predict the failure trend of refining process, which is conducive to the realization of the refining system failure in advance of the defense.

(3) From the perspective of quantitative prediction of fault development trend, a Kalman filter correlation prediction method based on grey correlation analysis is proposed for "safety behavior state". The influencing factors of the reference sequence are ranked by grey correlation analysis, the main influencing factors are extracted, and the regression model and Kalman filter prediction are established on the basis of the grey correlation analysis. After the on-site data comparison and verification, this method takes into account the influencing factors of process parameters of neighboring units, so the adaptability of the prediction model is better than that of the conventional Kalman filtering method, and it can accurately predict the operating status of the unit and anticipate the failure status in advance, which is conducive to the accurate release of early-warning information of the refining and chemical production process.

7.4　Text Mining Based Process Alarm Prediction

Alarm systems play a very critical role in modern process safety management. Effective prediction of process alarms helps field operators to respond in time, improves the pre-control timeliness of the alarm system, and avoids the alarm flooding phenomenon and the occurrence of abnormal working conditions. Aiming at the limitations of the existing prediction methods based on process data, we propose a process alarm prediction method based on text mining by using the rich information of alarm logs and combining deep learning and Natural language processing (NLP) technology. The vector representation of each alarm variable is constructed adaptively based on the Word Embedding method, and the resulting alarm vectors are used as inputs to a long and short-term memory neural network for model training and prediction of possible alarms. Mentioned in this section is not only a new approach in the field of alarm management but also a new application of NLP and deep learning methods, which helps to improve the effectiveness of the alarm system and the timeliness of pre-control.

7.4.1 *Problems and difficulties*

With the continuous advancement of modern technology, alarm configuration of process variables has become easier, leading to the generation of redundant and jitter alarms, which ultimately reduces the effectiveness of the alarm system. In order to improve the performance of alarm systems, alarm management has attracted a lot of attention from many companies and researchers. Nowadays, there are many standards and guidelines for alarm management in the industry, such as EEMUA-191 and ANSI/ISA-18.2. In particular, EEMUA-191 proposes a five-level model for assessing the performance of alarm systems. From the lowest level of "overload" to "active", "stable" and "robust" to the highest level of "predictability". Process alarm predictability is the optimal level of alarm system performance. From an engineering point of view, the main difficulties in process alarm prediction are as follows:

(1) Existing prediction methods are mostly based on process data, while some process variables usually do not have frequent alarms or do not have process data, such as numerical alarms indicating specific events or situations, which are only ON/OFF ("1" or "0") values, it is difficult to obtain a representative time series, thus increasing the limitations of existing prediction methods.

(2) Most machine learning methods (e.g., regression prediction, neural networks, support vector machines) must have numerical data as inputs, while process alarm logs record textual information. Therefore, how to convert process alarm information into real-valued data is a prerequisite and a major difficulty for further alarm prediction. In addition, due to the complexity of industrial processes and the connectivity of equipment, there are often different degrees of correlation between process variables, and if a process variable has an alarm, it often causes its associated variables to have an alarm. Therefore, how to retain the correlation information between the alarm variables while converting the data is another difficulty to be solved.

(3) Since there is a certain degree of correlation between alarm variables, the proposed prediction method needs to use historical process alarm information to explore the relationship between alarm variables, so as to predict the upcoming alarms. The literature proposes a probabilistic prediction method based on the N-gram model, which predicts the probability that the next

alarm may occur when the previous N alarms are known to have occurred. The N-gram model is a probabilistic language model for predicting the next possible occurrence of a word in a text sentence. The model is based on a Markov process, which assumes that the occurrence of the Nth word is only related to the previous $N - 1$ words. The N-gram model has been widely used in the field of NLP and bio sequence analysis; however, the model cannot take into account the long-time interactions between alarm variables.

However, this model is unable to consider the long-term interactions between alarm variables. Therefore, considering the correlation and long-term interactions between alarm variables, we study the numerical expression of alarm variables and alarm prediction, and combine Word Embedding method and Long Short-Term Memory (LSTM) neural network, to propose a process alarm prediction method based on text mining, which breaks through the limitations of the existing prediction methods based on process data, and improves the accuracy and alarm prediction. It improves the accuracy of alarm prediction and the timeliness of alarm system pre-control, which is conducive to the operators to formulate pre-control strategies in advance to prevent further development of risks and possible accidents.

7.4.2 *LSTM*

Recurrent Neural Networks (RNN) models are promising in the field of NLP, which are a class of neural networks with feedback links, in which the output of each unit not only provides information for the next layer of units but also feeds certain information back to itself. The LSTM models are the most widely used class of RNN models, which can better represent long-time dependencies compared to standard LSTM models are the most widely used class of RNN models today, and they can better represent long-time dependencies than standard RNN models. Figure 7.41 shows the structure of a typical LSTM network with one input layer, one output layer, and two hidden layers that are expanded in five steps in the time dimension.

A typical application of LSTM for NLP is to predict the next word in the text. As shown in Fig. 7.42, for each word, the next word after

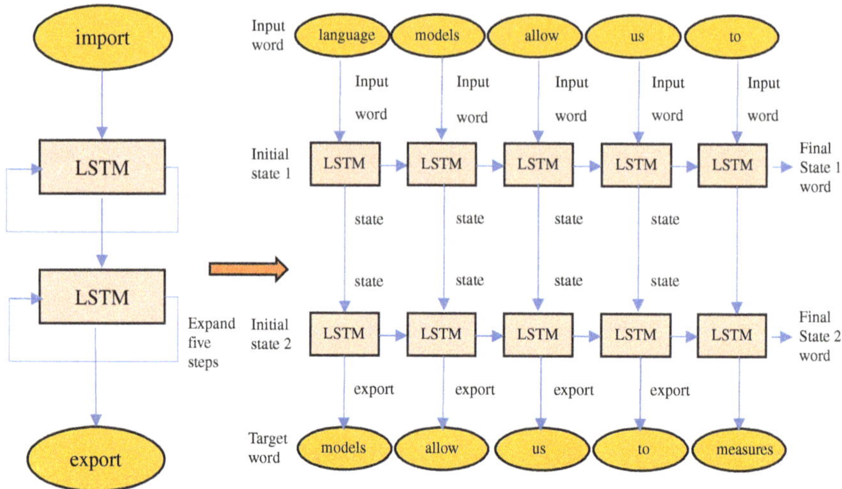

Fig. 7.41. Double hidden layer LSTM model with time step 5.

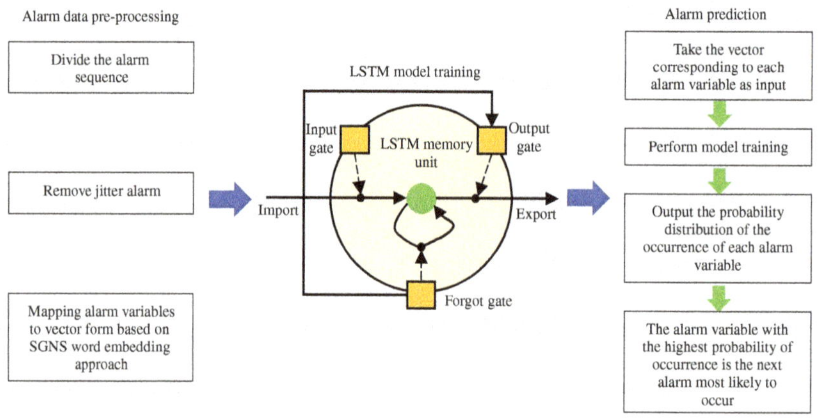

Fig. 7.42. Process alarm prediction method based on LSTM and SGNS.

it is the desired target. For example, given the input "to" in Fig. 7.42, the expected output of the prediction would be "measure". LSTM makes full use of the information in the previous words ("language", "models", "allow", "us") to improve the accuracy of the prediction result.

The only difference between LSTM and standard RNN models is the use of LSTM cells in the hidden layers. Equations (7.29)–(7.34) list the definitions of LSTM cells:

$$f_t = \sigma(W_f \cdot [h_{t-1}, x_t] + b_f), \tag{7.29}$$

$$i_t = \sigma(W_i \cdot [h_{t-1}, x_t] + b_i), \tag{7.30}$$

$$\overline{c_t} = \tanh(W_c \cdot [h_{t-1}, x_t] + b_c), \tag{7.31}$$

$$c_t = f_t \times c_{t-1} + i_t \times \overline{c_t}, \tag{7.32}$$

$$o_t = \sigma(W_o \cdot [h_{t-1}, x_t] + b_o), \tag{7.33}$$

$$h_t = o_t \cdot \tanh(c_t). \tag{7.34}$$

In the above-mentioned equation, c, x, h, f, i, o are vectors in n-dimensional (R^n) space. The weights W and bias b are trainable parameters. The number of LSTM cells per layer, i.e., dimension n, can be set as appropriate, and $[\cdots]$ denotes the vector cascade. The functions $\sigma(\cdots)$ and $\tanh(\cdots)$ are denoted as follows:

$$\sigma(x) = \frac{1}{1 + e^{-x}}, \tag{7.35}$$

$$\tanh(x) = \frac{e^x - e^{-x}}{e^x + e^{-x}}. \tag{7.36}$$

An LSTM cell passes states through three gate computations (i.e., forgot gate, input gate, and output gate). The forgetting gate is responsible for deciding how much of the previous moment's cell state is to be retained in the current moment's cell state, and the input gate is responsible for deciding how much of the current moment's input is to be retained in the current moment's cell state, and the output gate is responsible for deciding how much of the current moment's cell state information is to be output. The forgetting gate f_t [Eq. (7.29)] determines how much of the previous moment's state information needs to be forgotten. Equations (7.35) and (7.36) determine how the cell state is controlled and updated. The input gate i_t [Eq. (7.30)] controls the amount of new information input to the current cell. The new information $\overline{c_t}$ generated by the current input is calculated by Eq. (7.31) and combined with c_{t-1} to get the new cell state c_t [Eq. (7.32)]. Finally, the output gate o_t [Eq. (7.33)] is used to obtain the final output state h_t [Eq. (7.34)].

Many scholars have proposed improved versions of the LSTM, but experimental results obtained in the literature show that the traditional LSTM structure performs very well on a wide range of datasets, and the use of other variants does not significantly improve the effectiveness of the LSTM.

LSTM network models have been successfully applied in many areas of NLP, such as machine translation, speech recognition, and handwriting recognition. Based on its successful application in NLP (especially in predicting the next word in a sentence), in this section, the LSTM model will be used to predict process alarms.

7.4.3 *Process alarm prediction method based on deep learning and word embedding*

By considering the chronological alarm records as text and analogizing each alarm variable as a word, a text mining-based process alarm prediction method is proposed to predict the next alarm that may occur in an industrial process based on LSTM and SGNS word embedding methods. The implementation flow of the proposed method is shown in Fig. 7.42, which mainly includes three parts: data preparation, method implementation, and alarm prediction.

7.4.3.1 *Data preparation*

Data preparation consists of two main parts: alarm sequence preprocessing and generating alarm vectors. First, the alarm data in the alarm log is preprocessed. The alarm sequences are segmented based on time thresholds to maintain the correlation between each alarm variable and to reduce redundant and meaningless information by eliminating jitter alarms. Subsequently, a word embedding method is used to obtain the vector representation corresponding to each alarm variable.

(1) Alarm sequence preprocessing

Alarm data is a series of text messages generated by the DCS system. When a process variable exceeds a preset alarm threshold, the corresponding alarm is triggered and stored in the alarm log.

Alarm information usually has the following basic attributes: variable name, alarm level, and time stamp information. Alarm levels usually include "HI (High)", "LO (Low)", "HH (HighHigh)", "LL (LowLow)" and so on. In this section, the variable name and alarm level are defined together as an alarm variable type, and the timestamp is the time when the alarm occurs. Analogous to NLP, an alarm variable is equivalent to a word, an alarm log is equivalent to a text, and the number of alarm variable types is equivalent to the size of the vocabulary in the text.

Typically, an alarm log contains several months to a full year's worth of recorded alarm information. If the time interval between two neighboring alarms is too long, it is not meaningful to consider the correlation between the two variables, so it is necessary to set a time threshold to divide the different alarm sequences. If the time interval between two neighboring alarms exceeds this threshold, the latter will be considered as the first alarm of another sequence.

As shown in Fig. 7.43, the alarm type is noted as Alarm (i) ("i" denotes an alarm type). Here, the time threshold is set to 30 min, based on which the sequence is divided into Sequence 1 and Sequence 2. If the sequence length is too short, it will result in too little sequence information available for predicting the subsequent alarms, so alarm sequences with less than five alarms are not considered in this section.

It is worth noting that the length of an alarm sequence may be much longer than the length of a typical sentence in natural language, and the number of alarm variables is much smaller than the size of the textual vocabulary. These two different features allow alarm prediction to have fewer kinds of prediction targets (i.e., number of alarm categories) and more information about long sequences, potentially giving the method better prediction results than NLP.

Recurring jitter alarms over a short period of time are a typical class of nuisance alarms. Predicting jitter alarms is not a practical guide and may reduce the accuracy of the prediction, so in this section, more than three alarms in a minute are considered jitter alarms and are eliminated from the alarm sequence. For example, in Fig. 7.44, Alarm (2) has three alarms in one minute, so only one alarm record for Alarm (2) is kept at that location.

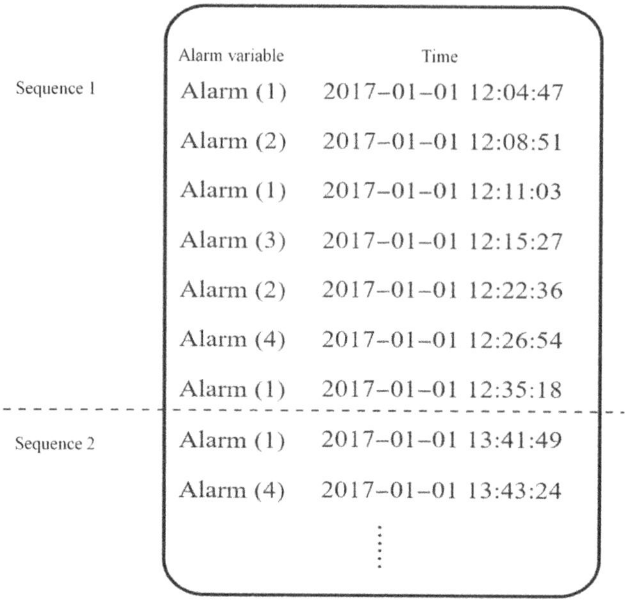

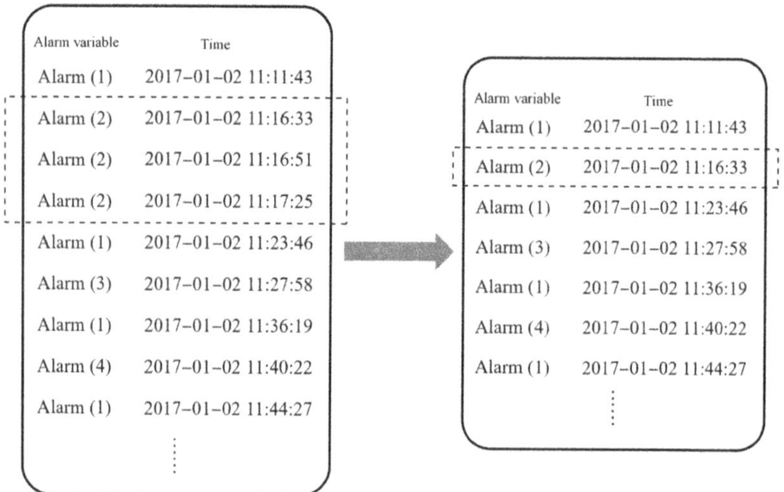

Fig. 7.43. Sequence segmentation based on time thresholding.

Fig. 7.44. Example of vibration alarm filtering.

(2) Generate alarm vectors

The input to many machine learning methods (including LSTM) must be in numerical form, so words in the text need to be transformed into real-valued data. One-hot encoding is one of the encoding methods that transforms words into vectors whose dimension is the size of the vocabulary, so this encoding method often produces a large number of sparse vectors. In contrast, the vectors generated by word embedding can have lower dimensions. The word embedding method maps the words in the vocabulary to an n-dimensional vector space by constructing an embedding matrix $(M \times n)$, where M is the size of the vocabulary and n is the dimension of the embedding space, i.e., the dimension of the selected word vector. The resulting word vectors adaptively preserve the similarity of words in terms of contextual relationships, and words that occur frequently one after another in the text will be very close in the vector space. These word vectors can be used as input x_t to the LSTM in Eqs. (7.29)–(7.31), whose output is a probability distribution of all word occurrences, and the word with the highest probability is used as the next word predicted.

Literature studies have concluded that Skip-Gram with Negative Sampling (SGNS) based on Negative Sampling (NS) outperforms other word embedding methods in the performance similarity task, so in this section, the SGNS model is used to learn the vector representation of each alarm variable.

Gensim is an open-source Python toolkit for topic modeling, document indexing, and similarity retrieval of large corpora. In this section, the Gensim package is used to train the SGNS model to obtain the vector representation of each alarm variable. Gensim requires that the input to the model must be a sentence composed of words, so the preprocessed alarm sequences are used as model inputs, and each sequence includes several alarms. An $S \times n$-dimensional embedding matrix can be obtained from the SGNS model, which maps each alarm variable to a vector in the corresponding n-dimensional space, where S is the total number of alarm types, i.e., the number of corresponding alarm vectors. Finally, each alarm vector is used as an input to the LSTM model.

7.4.3.2 *Methodology implementation*

Tensorflow is a powerful software package for deep learning. It provides a Python application programming interface and many predefined functions for constructing different neural networks. In this section, we use the Python programming language to build a two-hidden layer LSTM network model based on Tensorflow, and the network structure is shown in Fig. 7.41. The model parameters need to be set as appropriate, especially the time step of the unfolding and the dimension of the embedding space. Since the inputs to the LSTM model must be in real number form, each alarm vector is used as the training data for the model. In this model, the corresponding vectors of the first few alarms (whose number is the time step in the network) are used as inputs for each training, and their next alarm in the sequence is used as the target alarm variable.

The dimension of the embedding space can be set based on the number of alarm types. The increase in dimension may improve the predictive performance of the model, but it also increases the computational burden. The effect of this parameter on the predictive performance of the model is discussed for the alarm sequences with the length of the time step used as a single model input, and their state information is retained through network training. The longer the step length, the more previous alarm information can be considered, but it also increases the computational burden. This section addresses the effect of the time step on the predictive performance of the model, noting that the other parameters have a much smaller effect on model performance than the above two parameters and that their effects on model performance are practically independent.

The proposed method trains the model in batches. For example, there are 2000 samples in an alarm log, if the Batch size is 20, it takes 100 iterations to train all the samples once, i.e., to complete an Epoch. Dropout is an effective method used in this section to prevent the model from overfitting, which randomly discards a certain LSTM cell during the training process according to a certain probability.

7.4.3.3 *Alarm prediction*

With the previous alarms known (the number of alarms is the time step of the LSTM network), the next alarm that may occur during the process can be predicted by using its corresponding alarm vector as a network input to the resulting LSTM model that was trained.

The output of the model is a probability of occurrence distribution over all alarm types, with the alarm variable with the highest probability of occurrence being the next most likely alarm. Figure 7.45 briefly depicts the prediction process. Assuming that the time step of the LSTM network is 5, the first five alarms are first mapped to their corresponding vectors by word embedding, and these vectors are used as inputs to the LSTM model. The output of the model is an occurrence probability distribution over all S alarm types, and the alarm variable Alarm (5) with the highest occurrence probability is the predicted next alarm. This method helps the operator to detect anomalies and prevent correlated alarms before they occur or before the process operation deteriorates.

7.4.4 *Case study*

A case study of the Central Heating and Cooling Plant (CHCP) at the University of California (Davis). The plant, which is primarily oil-fueled, consists of a steam generation and distribution system and a cooling water generation, storage, and distribution system, and is designed to meet most of the heating and cooling needs of the campus buildings.

The proposed methodology was demonstrated using the alarm logs of the 2017 CHCP, in which there are 703 alarm variable types and 216,991 process alarm records, and the names of the variables are recorded in the alarm system in the following form. For example, "CHCP_STM4#AlmCCSnoxpvhi.ALAlarm" indicates a High (HI) alarm for the measured value of Nitrogen oxides (NOx) in boiler CHCP#4.

7.4.4.1 *CHCP data preprocessing*

According to the alarm prediction method described in Section 7.4.3, first, an alarm sequence is divided and jitter alarms are removed using 30 min as the time threshold. If an alarm sequence has less than 5 alarms, the alarm sequence is removed. Figure 7.46 gives the number of different types of alarms frequent in each month of 2017 after

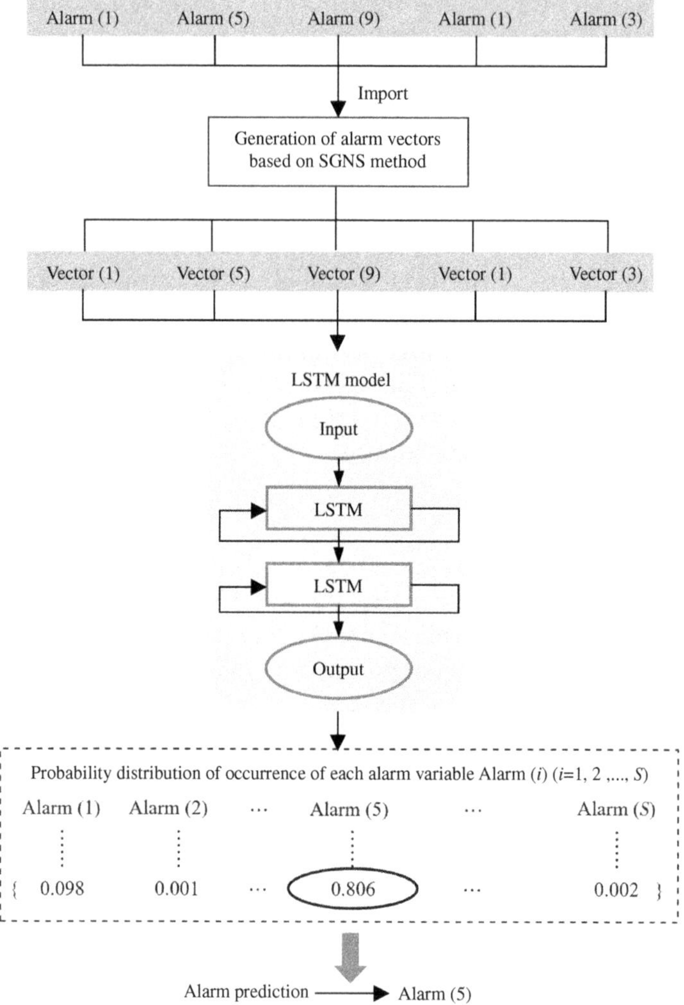

Fig. 7.45. Example of alarm prediction process.

preprocessing. The variable for each alarm is abbreviated as $\text{Alarm}(i)$ ("i" is used to distinguish between different types of alarms). Figure 7.46 shows that some of the alarms show a clear seasonal pattern.

Gensim is used to train the SGNS model to obtain the vector representation of each alarm variable, map each alarm to a vector

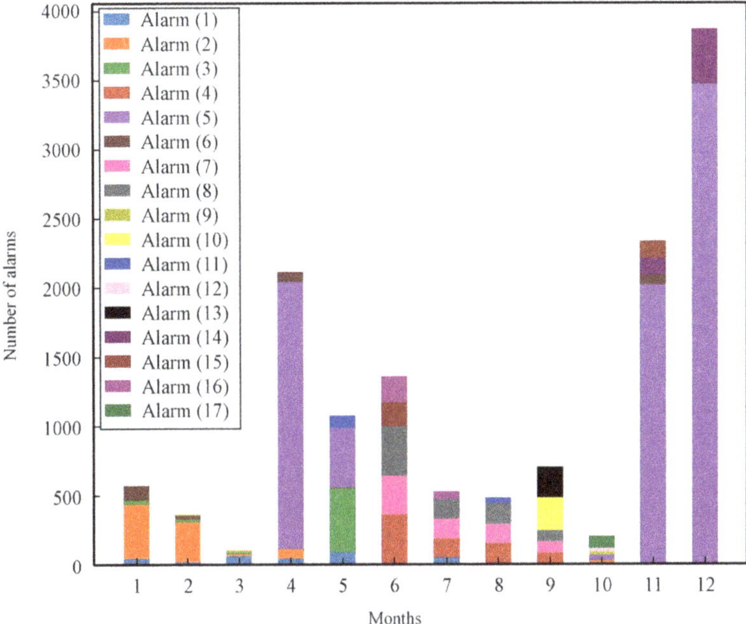

Fig. 7.46. Number of frequent and different alarms by month in 2017.

space of dimension 80, and finally use the vector corresponding to each alarm as the model input for LSTM.

7.4.4.2 *Model training*

A two-hidden layer LSTM network model is built based on Tensorflow, and the network structure is shown in Fig. 7.41. In this model, the corresponding vectors of the first five alarms (whose number is the time step in the network) are used as inputs for each training, and their next alarm in the sequence is used as the target alarm variable.

The dimension of the embedding space can be set based on the number of alarm types. The increase in dimensionality may improve the predictive performance of the model, but it also increases the computational burden. Alarm sequences of the length of the time step used are used as a single model input, and their state information is retained through network training. The longer the step, the more previous alarm information can be considered.

Table 7.18.   LSTM model parameters.

Model parameters	Parameter value
Batch size	20
Epoch number	100
Base learning rate	0.0001
Learning rate decay	0.9
Dropout probability	0.2
LSTM forgot bias	0.1

The proposed method trains the model in batches with a batch size of 20. Dropout is an effective method used in this section to prevent overfitting of the model, which randomly discards a particular LSTM cell with a certain probability during the training process. The parameters of the modeled LSTM network are shown in Table 7.18.

Select the alarm data from February to October in the alarm log to train the LSTM model with 100 iterations of an epoch. The network generalization ability of the model is evaluated by 10-fold cross-validation, and the prediction accuracy is defined in this section as the ratio of the number of correctly predicted alarms to the total number of predicted alarms. Figure 7.47 shows the training accuracy and validation accuracy (averaged over the 10-fold cross-validation) for each epoch.

From Fig. 7.47, 100 epochs are enough to get the optimal and stable prediction results, and the cross-validation accuracy is similar to and slightly lower than the training accuracy. From Fig. 7.47, it can be seen that the accuracy rate of the first 20 epochs shows a significant upward trend, and then the model gradually converges.

7.4.4.3   *Alarm prediction*

The time step of the LSTM network is 5, so the first 5 alarms are known, and the trained LSTM model predicts the next possible alarm (an example of the prediction process is shown in Fig. 7.45).

The model was tested on 500 alarms for the month of January and the resulting prediction accuracy (the ratio of correctly predicted alarms to the total number of predicted alarms) was 78.00%. Figure 7.48 summarizes the total number of alarms and the number of correctly predicted alarms for each type of alarm in the test

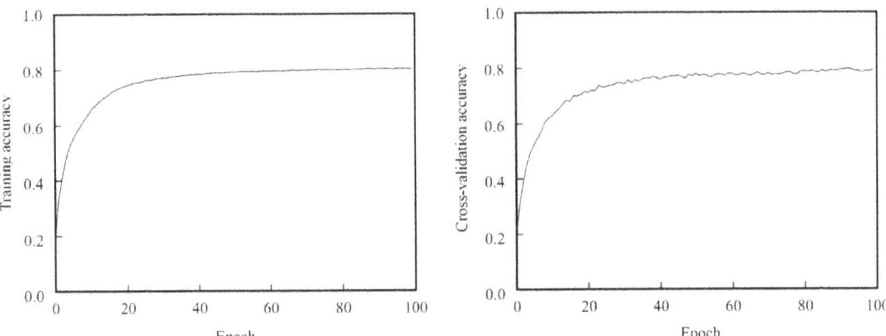

Fig. 7.47. Training accuracy and validation accuracy for each training cycle.

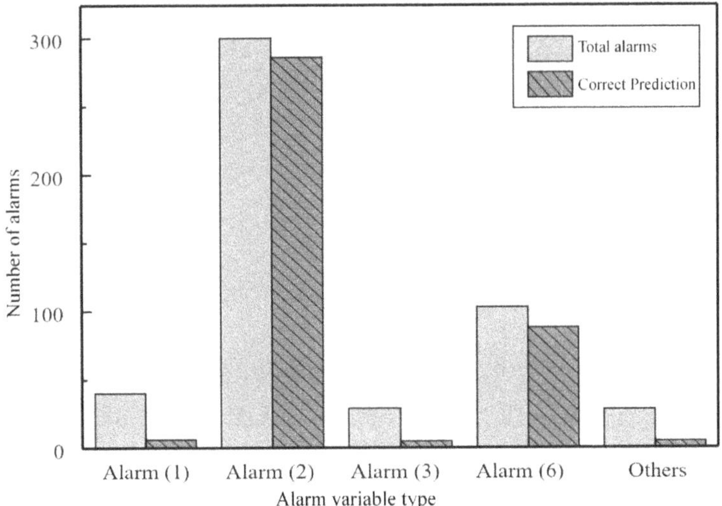

Fig. 7.48. Total number of alarms of each type in the test set and prediction accuracy.

dataset. From Fig. 7.48, it can be seen that Alarm (2) and Alarm (6) were predicted with higher accuracy, which may be related to the fact that these alarms occurred more frequently in the test dataset. The less frequent alarm types are more likely to be misidentified as Alarm (2) or Alarm (6). Figure 7.49 gives the distribution of the lead time of each correctly predicted alarm compared to its actual alarm time (the interval is 30 s). It can be seen that the predicted alarm lead time is mainly concentrated in the range of 0–10 min, with the

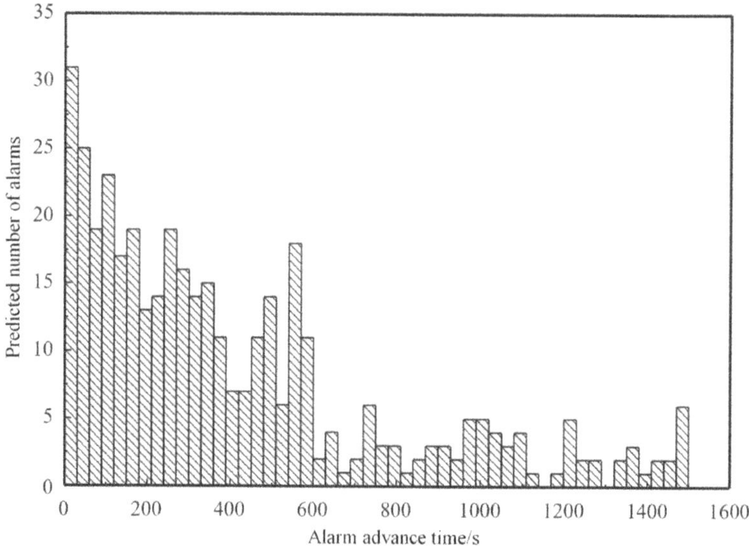

Fig. 7.49. Distribution of predicted alarm advance times.

longest alarm lead time being 24 min 59 s, and the average alarm lead time is calculated to be 6 min 49 s.

7.4.4.4 *The effect of embedding spatial dimensions*

In order to analyze the effect of the embedding dimension on the prediction performance, the LSTM model is trained with 320, 160, 40, 20, and 10 embedding dimensions, respectively, keeping the other parameters unchanged, since the model has basically converged to a stable accuracy value after 50 epoch, the mean values of the training and validation accuracies of the last 50 epoch are shown in Fig. 7.50 as the training and validation accuracies under each embedding dimension.

As shown in Fig. 7.50, as the embedding dimension decreases from 80 to 40, 20 to 10, the training and validation accuracy decreases. However, when increasing the dimensionality to 160 and 320, the increase in embedding dimension no longer has a significant effect on the training and validation accuracy.

Considering the accuracy rate and computational burden, the benchmark dimension chosen in this section is 80, and as can be seen

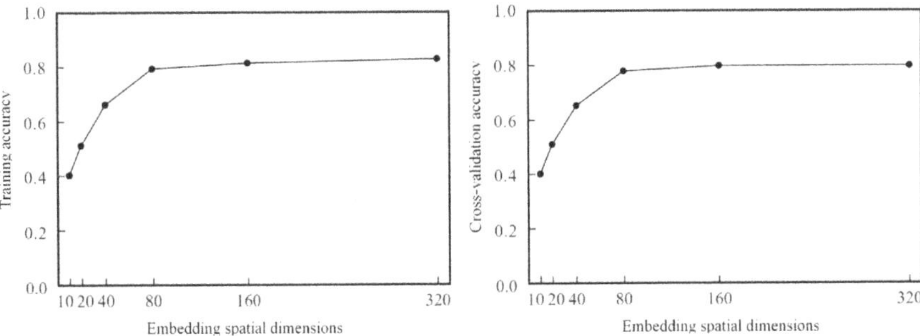

Fig. 7.50. Training and verification accuracy for different embedding dimensions.

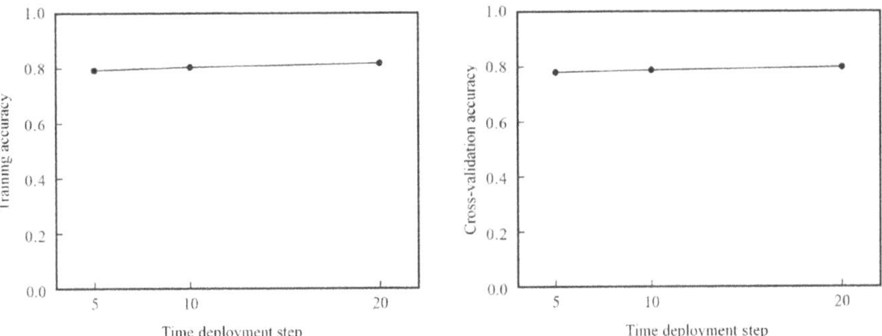

Fig. 7.51. Training and validation accuracy for different time step lengths.

from Fig. 7.51, the accuracy rate of training and validation under this dimension has basically reached the optimal level.

7.4.4.5 *Effect of time step*

In order to analyze the effect of the time step of network expansion on the prediction performance, other parameters are kept constant, and the LSTM model is trained with 20 and 10 as the time steps, respectively, because the model has basically converged to a stable accuracy value after 50 epoch, so Fig. 7.51 shows the average of training and validation accuracy under each embedding dimension since the model has basically converged to a stable accuracy value after 50 epoch, the mean values of the training and validation accuracies after 50 epoch are shown in Fig. 7.51 as the training and validation accuracies under each embedding dimension.

As shown in Fig. 7.51, when the time step is 5, the training and verification accuracy is close to the optimal, and increasing the time step has no significant effect on the training and verification accuracy, so the benchmark time step chosen in this section is 5.

7.4.4.6 *Impact of data features*

There were 10,945 alarm messages and 210 alarm variable types in the February through October alarm logs. For February through August, there are 9703 alarms and 164 alarm variable types. From Fig. 7.46, it can be seen that the number of Alarm (5) alarms occurs much higher than the other alarms in the months of November to December. These different data characteristics may affect the accuracy of the prediction results to a certain extent, so other parameters are kept unchanged, and the LSTM model is trained with the alarm data from February to August and the alarm data from December, respectively, and the resulting training and validation accuracies for each training cycle are shown in Fig. 7.52.

As can be seen in Fig. 7.52, the model trained using the December alarm data converges the fastest and achieves a higher level of accuracy. Compared with the other two training datasets, December has fewer alarms and alarm types, and one alarm type, Alarm (5), occurs much more often than the others, which is typical of an unbalanced dataset. The model trained using the February through October alarm data was the slowest to converge and had the lowest training and validation accuracy. In contrast, the February through August training data had fewer alarm types.

The LSTM model trained with the December data is tested on 500 alarms from November, and the prediction accuracy (the ratio

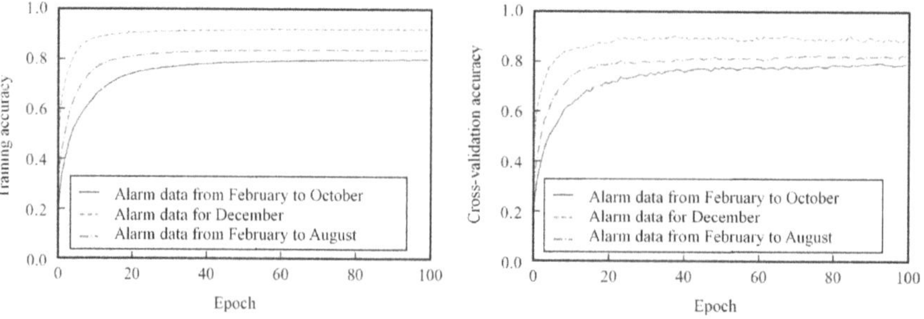

Fig. 7.52. Training and validation accuracy for different training data.

Table 7.19. Forecast of test data for November.

Alarm variable type	Total	Number of accurate predictions
Alarm (5)	458	432
Other alarms	42	7

Table 7.20. Comparison of prediction results for different training sets.

Training data	Prediction accuracy	Total	Number of alarm types
February to August	81.40%	9703	164
February to October	78.00%	10945	210

of the number of correctly predicted alarms to the total number of predicted alarms) is 87.8%. As shown in Table 7.19, almost all Alarms (5) can be predicted accurately, while the number of accurate predictions for other types of alarms is very low, thus indicating that the model obtained using the unbalanced training dataset has poor generalization ability.

The LSTM model trained with data from February to August is tested with 500 alarms, as shown in Table 7.20, and the resulting prediction accuracy is 81.40%, which is higher than that of the LSTM model trained with data from February to October. Normally, more training data may give better prediction results, however, the results obtained are contrary to expectations, which indicates that the number of alarm types has an impact on the prediction results. From Fig. 7.47, it can be seen that the alarm types that occur in January frequently occur in February to May, but rarely occur in September to October, indicating that reducing the number of alarm types that differ from the test data improves the accuracy of the prediction results. Therefore, the fewer alarm types, the better the prediction accuracy may be.

7.4.4.7 *Comparison of methods*

The N-gram model is a probabilistic language model that, given the first $N - 1$ words, the method predicts the next word in the text. The N-gram ($N = 5$) model based on the Kneser-Ney smoothing method is trained using the alarm data from February to October

Table 7.21. Prediction accuracy of the proposed method and N-gram model.

Training data	Proposed method	N-gram model
February to August	81.40%	73.20%
February to October	78.00%	71.40%

and from February to August, respectively, and the accuracy of the model is tested using the same data as in Section 7.1.4, as shown in Table 7.21, and the prediction accuracies of the proposed methods are higher than that of the N-gram model.

For the two different training datasets in Table 7.21, the proposed method improves the prediction accuracy by 8.20% and 6.6%, respectively, with an average improvement of 7.4% compared to the N-gram model. The N-gram model is simple and easy to implement, but it is inferior to the LSTM model in dealing with long-time dependencies. Compared with the N-gram model, the proposed method has better performance in process alarm prediction.

7.4.4.8 *Analysis and summary*

(1) Effective prediction of process alarms improves the timeliness of the alarm system's pre-control and prevents the occurrence of related alarms or even alarm flooding. Existing prediction methods are mostly based on process data, while some process variables usually do not have frequent alarms or do not have process data (e.g., numerical alarms indicating specific events or situations), so it is difficult to obtain a representative time series, making the existing prediction methods have certain limitations. For this reason, this section proposes a text mining-based process alarm prediction method based on the rich information content of alarm logs and different degrees of correlation between alarm variables, combined with NLP word embedding methods and LSTM neural networks.

Since the inputs to many machine learning methods (including LSTM) must be in numerical form, this section adaptively constructs vector representations of each alarm variable based on word embedding methods in order to adequately express the varying degrees of correlation levels among the alarm variables and uses the resulting alarm vectors as the inputs to train a neural network model to make predictions about possible correlated alarms. This method is both a new approach in the field of alarm management and a new application of NLP and deep learning methods.

(2) A case study of the alarm log data of CHCP in 2017 was conducted to establish an alarm prediction model based on the word embedding method and LSTM neural network, and by testing the constructed model, it was obtained that the predicted alarm lead times were mainly concentrated in the range of 0–10 min, and the average alarm lead time was 6 min 49 s. Compared with the N-gram model, the prediction accuracy of the proposed method is improved by 7.4% on average.

In summary, this section of the process alarm prediction method based on text mining overcomes the limitations of the existing prediction methods based on process data, and improves the accuracy of the alarm prediction and the timeliness of the alarm system's pre-control, and the resulting prediction results can provide a reasonable reference for the operator to formulate the pre-control decision, so as to ensure the safety and reliability of the industrial process.

Appendix

Table 7A.1. Alarm types and descriptions.

Serial number	Alarm variable type	Description
0	CHCP_STM4#CEMS. STFault	Continuous emission monitoring system (CEMS) malfunction
1	CHCP_STM4# AlmCEMShighsheltertemp. ALAlarm	CEMS cabinet temperature too high
2	CHCP_CES# FIT8110.ALLoFlowLimit	Hot water energy saver inlet flow (FIT-8110) low limit alarm
3	CHCP_CES# FIT8110.ALLoFlow	Hot water energy saver inlet flow (FIT-8110) low alarm
4	CHCP_CES# FIT8100.ALLoFlow	Make-up water energy saver inlet flow (FIT-8100) low alarm
5	CHCP_CES#Fan.ALVFD	Variable Frequency Drives (VFD) malfunction

(Continued)

Table 7A.1.   (*Continued.*)

Serial number	Alarm variable type	Description
6	CHCP_STM4#AlmCEM Sanycalinprogres.ALAlarm	CEMS system calibration stopped
7	CHCP_STM1#YA1300A.ALAlarm	No. 3 boiler malfunction
8	CHCP_CES#TT8227_ HiHiAlm.AD.Msg	102 Heat Exchanger Outlet Temperature Excess Alarm
9	CHCP_STM1#YA1300.ALAlarm	Boiler No. 3 shutdown
10	CHCP_STM4#AlmCCSoxpvsrcfail. ALAlarm	Oxygen quantity regulator connection interrupted
11	CHCP_STM4#AlmBMSfoalwco. ALAlarm	Backup equipment low water level cut off
12	CHCP_STM4#AlmCCSairoll. ALAlarm	Gas flow output low limit alarm
13	CHCP_STM4#LSLBoiler.STSwitch	Boiler water level low alarm
14	CHCP_STM4#AlmBMSlwa. ALAlarm	Boiler water level low alarm
15	CHCP_CES#TT8227_HiAlm. AD.Msg	102 heat exchanger outlet temperature high alarm
16	CHCP_CES#Fan.ALTrip	Energy saver fan VFD trip alarm
17	CHCP_STM4#AlmCCSfdwoll. ALAlarm	Boiler feed water outlet flow low limit alarm
18	CHCP_CES#FIT8100. ALLoFlowLimit	Make-up water energy saver inlet flow (FIT-8100) low limit alarm
19	CHCP_CES#TT8291.ALHi	Apartment 4 hot water supply temperature high alarm
20	CHCP_CES#TT8238_HiAlm. AD.Msg	Hot water supply temperature high alarm
21	CHCP_CES#TT8271.ALHi	High temperature alarm for hot water supply in apartment 3
22	CHCP_CES#DPT8270.ALHi	High differential pressure alarm for hot water supply in apartment 3
23	CHCP_CES#TT8238_ HiHiAlm.AD.Msg	Hot water supply temperature high alarm

(*Continued*)

Table 7A.1. (*Continued.*)

Serial number	Alarm variable type	Description
24	CHCP_STM4#AlmCEMSback? shinprgrs.ALAlarm	CEMS backwash mode stop
25	CHCP_CES#DPT8280.ALHi	Differential pressure high alarm for hot water in apartment cafeteria
26	CHCP_CES#TT8281.ALHi	High hot water supply temperature alarm in apartment cafeteria
27	CHCP_STM4#AlmCCSnoxolh. ALAlarm	High NOx output alarm
28	CHCP_STM4#AlmCCScombair? waif.ALAlarm	Abnormal combustion gas flow alarm
29	CHCP_CES#TT8292.ALHi	Apartment 4 hot water return temperature high alarm
30	CHCP_STM4#AlmCCSdrmoll. ALAlarm	Boiler No. 4 boiler ladle water level output low limit alarm
31	CHCP_STM4#AlmCCSairpvsrcfail. ALAlarm	Gas flow measurement connection interrupted
32	CHCP_CES#TT8282.ALHi	High hot water return temperature alarm in apartment cafeteria
33	CHCP_STM4#AlmCCSnoxpvhi. ALAlarm	High alarm for NOx concentration measurement
34	CHCP_STM4#AlmCEM Soutofservice.ALAlarm	CEMS deactivation
35	CHCP_STM4#LSLLBoiler. STSwitch	Low water level disconnection
36	CHCP_STM4#AlmCCSnoxerrpos. ALAlarm	High alarm for deviation of NOx concentration measured value from set value
37	CHCP_STM4#AlmCEMSanycalfail. ALAlarm	CEMS calibration failure
38	CHCP_STM4#AlmBMSsolwco. ALAlarm	Low water level cutoff
39	CHCP_STM4#LSLLBoilerAux. STSwitch	Auxiliary equipment low water level cutoff

Table 7A.2. Ten parameter values for lifting wind instrumentation failure.

Serial number	Lift tube steam flow t/min	Regenerator tempera- ture °	Regenerator pressure MPa	External heat extraction package level cm	Lift tube reactor temper- ature °	Regenerator flue gas oxygen content kg/h	Flue gas turbine pressure kPa	Regenerator main air volume m³/h	Sinker internal pressure kPa	Pneumatic compressor speed r/min
1	1.30	756	0.189	134	680.1	13.0	6366	40991	477.4	1332
2	1.25	754	0.182	125	656.7	12.9	6437	40496	476.1	1338
3	1.25	755	0.191	125	682.4	13.3	6382	38902	482.7	1360
4	1.30	763	0.187	117	663.3	12.9	6576	40648	480.1	1341
5	1.27	757	0.181	138	658.1	12.9	6484	38235	478.5	1297
6	1.33	768	0.186	131	676.1	12.8	6491	40325	475.1	1365
7	1.34	762	0.181	138	680.0	12.9	6412	40674	476.3	1362
8	1.19	761	0.178	142	683.4	12.9	6360	40153	483.9	1359
9	1.31	770	0.182	120	680.8	12.5	6478	39035	480.0	1336
10	1.37	758	0.180	135	676.5	13.1	6406	40253	478.8	1376
11	1.26	761	0.178	130	686.8	12.9	6589	39788	475.2	1312
12	1.32	763	0.191	137	683.7	13.3	6413	40201	482.9	1343
13	1.33	756	0.181	139	666.8	13.2	6370	39684	475.0	1359
14	1.33	760	0.198	136	671.4	12.7	6428	40513	480.7	1366
15	1.28	752	0.178	123	661.2	13.1	6558	39498	477.3	1348
16	1.29	765	0.184	128	668.5	12.7	6393	38700	477.5	1327
17	1.29	769	0.190	132	677.9	12.7	6440	40909	473.8	1331
18	1.30	761	0.198	155	674.1	12.9	6302	40976	476.1	1353
19	1.26	749	0.182	140	659.8	12.7	6462	40839	475.6	1318
20	1.39	753	0.175	148	662.5	12.7	6411	38397	477.4	1407
Thresholds	1.20–1.40	700–720	1.75–2.00	120–150	640–660	12.5–13.5	6300–6600	38000–41000	375–382	1280–1330

Table 7A.3. Ten parameter values for steam with water.

Serial number	Lift tube steam flow t/min	Regenerator tempera-ture °	Regenerator pressure MPa	External heat extraction package level cm	Lift tube reactor tempera-ture °	Regenerator flue gas oxygen content kg/h	Flue gas turbine pressure kPa	Regenerator main air volume m^3/h	Sinker internal pressure kPa	Pneumatic compressor speed r/min
1	1.30	713	0.189	134	654	13.0	6266	40991	477.4	1332
2	1.25	714	0.182	125	642	12.9	6437	40496	476.1	1338
3	1.25	715	0.191	125	643	13.3	6382	38902	482.7	1360
4	1.30	718	0.187	117	657	12.9	6576	40648	480.1	1341
5	1.27	707	0.181	138	647	12.9	6484	38235	478.5	1297
6	1.33	709	0.186	131	646	12.8	6491	40325	475.1	1365
7	1.34	711	0.181	138	659	12.9	6412	40674	476.3	1362
8	1.19	717	0.178	142	654	12.9	6360	40153	483.9	1359
9	1.31	713	0.182	120	657	12.5	6478	39035	480.0	1336
10	1.37	706	0.180	135	646	13.1	6406	40253	478.8	1376
11	1.26	716	0.178	130	656	12.9	6589	39788	475.2	1312
12	1.32	711	0.191	137	648	13.3	6413	40201	482.9	1343
13	1.33	702	0.181	139	657	13.2	6370	39684	475.0	1359
14	1.33	710	0.198	136	659	12.7	6428	40513	480.7	1366
15	1.28	702	0.178	123	650	13.1	6558	39498	477.3	1348
16	1.29	703	0.184	128	646	12.7	6393	38700	477.5	1327
17	1.29	695	0.190	132	650	12.7	6440	40909	473.8	1331
18	1.30	714	0.198	155	647	12.9	6302	40976	476.1	1353
19	1.26	704	0.182	140	648	12.7	6462	40839	475.6	1318
20	1.39	724	0.175	148	654	12.7	6411	38397	477.4	1407
Thresholds	1.20–1.40	700–720	1.75–2.00	120–150	640–660	12.5–13.5	6300–6600	38000–41000	375–382	1280–1330

Table 7A.4. Ten parameter values for double-acting slide valve failure full opening and hoisting wind instrumentation failure.

Serial number	Lift tube steam flow t/min	Regenerator tempera-ture °	Regenerator pressure MPa	External heat extraction package level cm	Lift tube reactor tempera-ture °	Regenerator flue gas oxygen content kg/h	Flue gas turbine pressure kPa	Regenerator main air volume m^3/h	Sinker internal pressure kPa	Pneumatic compressor speed r/min
1	1.30	756	0.219	134	680.1	14.0	6566	37991	475.2	1288
2	1.25	754	0.212	125	656.7	13.9	6737	37496	482.9	1279
3	1.25	715	0.191	125	682.4	13.3	6382	35902	475.0	1296
4	1.30	718	0.187	117	663.3	13.9	6576	37648	480.7	1325
5	1.27	707	0.181	138	658.1	13.9	6784	35235	477.3	1301
6	1.33	709	0.186	131	676.1	13.8	6791	37325	477.5	1284
7	1.34	711	0.181	138	680.0	13.9	6712	37674	473.8	1294
8	1.19	717	0.178	142	683.4	14.9	6860	37153	476.1	1299
9	1.31	713	0.182	120	680.8	13.5	6778	36035	475.6	1307
10	1.37	706	0.180	135	676.5	14.1	6706	37253	477.4	1283
11	1.26	716	0.178	130	686.8	13.9	6889	36788	477.4	1302
12	1.32	711	0.191	137	683.7	14.3	6713	37201	476.1	1338
13	1.33	702	0.181	139	666.8	14.2	6670	36684	482.7	1297
14	1.33	710	0.198	136	671.4	14.7	6728	37513	480.1	1300
15	1.28	702	0.178	123	661.2	14.1	6858	36498	478.5	1295
16	1.29	703	0.184	128	668.5	13.7	6693	35700	475.1	1286
17	1.29	695	0.190	132	677.9	13.7	6740	37909	476.3	1307
18	1.30	714	0.198	155	674.1	13.9	6602	37976	483.9	1306
19	1.26	704	0.182	140	659.8	13.7	6762	37839	480.0	1297
20	1.39	724	0.175	148	662.5	13.7	6711	35397	478.8	1304
Thresholds	1.20–1.40	700–720	1.75–2.00	120–150	640–660	12.5–13.5	6300–6600	38000–41000	375–382	1280–1330

References

[1] Hu J, Zhang L, Yi Y, *et al.* Real-time fault associated early warning for petrochemical process equipment under abnormal condition. *China Safety Science Journal*, 2016, 26(9): 140–145. (In Chinese)

[2] Cai S. Adaptive Correlation Analysis and Prediction Method for Industrial Process Alarm based on Text Mining. PhD Thesis, *China University of Petroleum (Beijing)*, 2019. (In Chinese)

[3] Hu J, Zhang L, Yi Y, *et al.* Research on dynamic alarm management method of refining device based on Bayesian estimation. *Journal of Safety Science and Technology*, 2016, 12(10): 81–85. (In Chinese)

[4] Cai S, Hu J, Zhang L, *et al.* Modeling study of main wind turbine alarm analysis based on SDG and MFM. *Plant Maintenance Engineering*, 2015(S2): 320–324. (In Chinese)

[5] Yi Y, Hu J, Zhang L, *et al.* Study on refining equipment fault trend prediction method based on correlation between technological parameters. *China Safety Science Journal*, 2014, 24(12): 70–75. (In Chinese)

[6] Hu J, Zhang L, Liang W. Opportunistic predictive maintenance for complex multi-component systems based on DBN-HAZOP model. *Process Safety and Environmental Protection*, 2012, 90(5): 376–388.

[7] Hu J, Zhang L, Wang Y. A systematic modeling of fault interdependencies in petroleum process system for early warning. In: *The 5th World Conference of Safety of Oil and Gas Industry*, June 8–11, 2014.

[8] Hu J, Zhang L, Wang Y. A systematic modeling of fault interdependencies in petroleum process system for early warning. *Journal of Chemical Engineering of Japan*, 2015, 48(8): 678–683.

[9] Hu J, Yi Y. A two-level intelligent alarm management framework for process safety. *Safety Science*, 2016, 82: 432–444.

[10] Hu J, Zhang L, Wang A. Quantitative safety early warning method of fault propagation for petrochemical plants. *CIESC Journal*, 2016, 67(7): 3091–3100. (In Chinese)

[11] Hu J, Zhang L, Wang A, Wan F. Fault propagation analysis and causal fault diagnosis for petrochemical plant based on granger causality test. In: *12th Global Congress on Process Safety*, April 10–14, 2016, Houston, USA.

[12] Hu J, Zhang L, Tian W, *et al.* DBN based failure prognosis method considering the response of protective layers for the complex industrial systems. *Engineering Failure Analysis*, 2017, 79: 504–519.

Chapter 8

Typical Cases of Early Warning of Abnormal Events in Deepwater Oil and Gas Extraction

8.1 Real-Time Drilling Well Leakage Early-Warning Method Based on Multi-Source Information Fusion

Early real-time diagnosis and early warning of drilling downhole abnormal events (well leakage, well surge) can gain valuable time for reestablishing the pressure balance at the bottom of the well and can also reduce the possibility of occurrence, spreading, propagation, and escalation of blowout and other accidents. The traditional method of monitoring downhole events suffers from insufficient accuracy. In the traditional method, drilling fluid first passes through the pressure-controlled drilling throttling manifold system and then enters the drilling fluid pool through the liquid–gas separation system, but does not accurately respond to the actual flow rate change in the wellbore. The parameter values obtained through theoretical derivation are subject to errors due to the influence of other factors, such as mechanical efficiency and water upstream efficiency. The trend of process parameter data (e.g., drilling fluid pool) is not obvious or is delayed (when the wellbore flow rate changes, it takes a certain amount of time for the drilling fluid to flow into the drilling fluid pool through the outlet conduit), and the nonlinearity of the change rule of drilling parameters is another drawback.

Therefore, the establishment of a real-time-based drilling process parameter abnormality model is of great practical engineering significance for early warning of the occurrence of drilling abnormality events. Using this mode, one can determine the warning indicators that characterize various types of drilling anomalies, extract the indicator feature vectors, analyze the trend of the indicator parameters and feature values, and determine whether the anomalies occur by comparing the measured values of the feature quantities with the dynamic safety thresholds. Considering the possibility of false alarms and missed alarms, a multi-indicator fusion early-warning model based on Bayesian estimation is established to realize early real-time diagnosis and early warning of drilling downhole abnormal events.

8.1.1   *Early-warning indicators of drilling abnormal events*

Drilling downhole anomalies can be characterized by abnormal changes in drilling process parameters. Based on the drilling process parameters, field experience, and existing literature, analyzing the changing law of process parameters, early warning of abnormalities and taking measures can effectively avoid the occurrence of abnormalities and inhibit the escalation of events. In this section, process parameters are taken as early-warning indicators of drilling anomalies, and the corresponding early-warning indicators are analyzed when different anomalies occur. Some downhole anomalies and early-warning indicators are shown in Table 8.1.

Well leakage drilling process parameters are characterized by a decrease in riser pressure; a decrease in drilling fluid outlet flow rate or even a zero-outlet flow rate; a decrease in inlet flow rate; a slow decrease in total pool volume; a bottomhole pressure greater than the formation pore, leakage, or fracture pressures; and a continuous decrease in bottomhole pressure. The signs corresponding to the emergence of well surge are an increase in relative export flow rate, increase in drilling fluid pool volume, increase in standing pressure, and increase in gas content of drilling fluid or a change in chloride content of drilling fluid, which manifests as a decrease in the density of drilling fluid. Among them, the relative outlet flow rate of drilling

Table 8.1. Early-warning indicators of drilling anomalies and their changing patterns.

Downhole anomalies	Main early-warning indicators		Secondary early-warning indicators	
	Indicator	Change	Indicator	Change
Well surge	Relative outlet flow	↑	Riser pressure	↑
	Pit volume	↑	Drilling fluid density	↓
Well leakage	Relative outlet flow	↓	Hook weigh	↑
	Pit volume	↓	Pump stroke	↑
	Riser pressure	↓	Drilling hook lowering speed	↑
	Inlet flow	↓		
Drill tool puncture leakage	Riser pressure	↓	Relative outlet flow	↑
			Pump stroke	↑
Drilling tool breakage	Hook load	↓	Pump stroke	↑
	Riser pressure	↓	Relative outlet flow	↑
	Torque	↓	Pit volume	↑
Stuck drill	Hook load	↑	Riser pressure	↑
	Torque	↑	Pump stroke	↓
Hydrogen sulfide leakage	Hydrogen sulfide content	↑	Riser pressure	↑
Plugging nozzle	Riser pressure	↑	Relative outlet flow	↓
			Pump stroke	↓

fluid and the volume of the drilling fluid pool are commonly used early-warning indicators of well surge in offshore drilling operations. Relative export flow rate monitoring can provide earlier warning of overflow, and online drilling fluid volume monitoring can avoid the possibility of missing alarms in flow rate monitoring. Therefore, a combination of two or more indicators is used commonly to provide early warning of well surges in the field.

Vectors that characterize the process parameters (warning indicators) and parameter data changes of well surges, well leaks, and other events during the drilling warning process are the objects of focus. Vectors must be able to characterize the abnormal trend of data changes to ensure the accuracy of downhole anomaly detection. The extraction of feature vectors should meet the principles of sensitivity, stability, reliability, and real-time performance. The eigenvectors selected in this section include dynamic mean, standard deviation, relative standard deviation, rate of change and autocorrelation coefficient. The extraction of feature vectors is relatively complex, and the data change characteristics exhibited by various types of drilling downhole events are different. For drilling events such as well leakage and well surge, the dynamic trend and characteristics of sensor data over a period of time need to be analyzed to determine whether an abnormality occurs; for downhole events such as a sudden breakage of the drilling tool or stabbing leakage, a small amount of data over a short period of time needs to be analyzed and judged. Therefore, according to different types of downhole events, the time period of the data taken can be determined and the characteristics of data (long-term or short-term data) can be analyzed.

In the drilling downhole anomaly warning process, the long-term data LI is taken, and LI is equally divided into n short-term data SI_i, i.e., $LI = \{SI_n, SI_{n-1}, \ldots, SI_i, \ldots, SI_1\}$. At the same time, the short-term data SI is equally divided into m time intervals ITi, i.e., $SI = \{IT_m, IT_{m-1}, \ldots, IT_i, \ldots, IT_1\}$. Record the feature quantity (μ, σ, σ_R, v, ρ) of IT_i sequentially and get the dynamic feature quantity of short-term SI as (μ_i, σ_i, σ_{Ri}, v_i, ρ_i), then we have

$$\mu_{SI} = \{\mu_m, \mu_{m-1}, \ldots, \mu_i, \ldots, \mu_1\}, \tag{8.1}$$

$$\sigma_{SI} = \{\sigma_m, \sigma_{m-1}, \ldots, \sigma_i, \ldots, \sigma_1\}, \tag{8.2}$$

$$\sigma_{RSI} = \{\sigma_{Rm}, \sigma_{R(m-1)}, \ldots, \sigma_{Ri}, \ldots, \sigma_{R1}\}, \tag{8.3}$$

$$\nu_{SI} = \{\nu_m, \nu_{m-1}, \ldots, \nu_i, \ldots, \nu_1\}, \tag{8.4}$$

$$\rho_{SI} = \{\rho_m, \rho_{m-1}, \ldots, \rho_i, \ldots, \rho_1\}. \tag{8.5}$$

Similarly, the dynamic eigenvolume within each SI_i of the LI is obtained, and the mean value of the eigenvolume within SI_i is $\left(\frac{\sum_{i=1}^{m} \mu_i}{m}, \frac{\sum_{i=1}^{m} \sigma_i}{m}, \frac{\sum_{i=1}^{m} \sigma_{Ri}}{m}, \frac{\sum_{i=1}^{m} \nu_i}{m}, \frac{\sum_{i=1}^{m} \rho_i}{m} \right)$. Obtain the dynamic eigenvolume of the long-term LI $\left(\left(\frac{\sum_{i=1}^{m} \mu_i}{m} \right)_j, \left(\frac{\sum_{i=1}^{m} \sigma_i}{m} \right)_j, \right.$

$\left(\left(\frac{\sum_{i=1}^{m} \sigma_{Ri}}{m} \right)_j, \left(\frac{\sum_{i=1}^{m} \nu_i}{m} \right)_j, \left(\frac{\sum_{i=1}^{m} \rho_i}{m} \right)_j \right)$, then we have

$$\mu_{\mathrm{LI}} = \left\{ \left(\frac{\sum_{i=1}^{m} \mu_i}{m} \right)_n, \left(\frac{\sum_{i=1}^{m} \mu_i}{m} \right)_{n-1}, \ldots, \right.$$
$$\left. \times \left(\frac{\sum_{i=1}^{m} \mu_i}{m} \right)_j, \ldots, \left(\frac{\sum_{i=1}^{m} \mu_i}{m} \right)_1 \right\}, \qquad (8.6)$$

$$\sigma_{\mathrm{LI}} = \left\{ \left(\frac{\sum_{i=1}^{m} \sigma_i}{m} \right)_n, \left(\frac{\sum_{i=1}^{m} \sigma_i}{m} \right)_{n-1}, \ldots, \right.$$
$$\left. \times \left(\frac{\sum_{i=1}^{m} \sigma_i}{m} \right)_j, \ldots, \left(\frac{\sum_{i=1}^{m} \sigma_i}{m} \right)_1 \right\}, \qquad (8.7)$$

$$\sigma_{\mathrm{RLI}} = \left\{ \left(\frac{\sum_{i=1}^{m} \sigma_{Ri}}{m} \right)_n, \left(\frac{\sum_{i=1}^{m} \sigma_{Ri}}{m} \right)_{n-1}, \ldots, \right.$$
$$\left. \times \left(\frac{\sum_{i=1}^{m} \sigma_{Ri}}{m} \right)_j, \ldots, \left(\frac{\sum_{i=1}^{m} \sigma_{Ri}}{m} \right)_1 \right\}, \qquad (8.8)$$

$$\nu_{\mathrm{LI}} = \left\{ \left(\frac{\sum_{i=1}^{m} \nu_i}{m} \right)_n, \left(\frac{\sum_{i=1}^{m} \nu_i}{m} \right)_{n-1}, \ldots, \right.$$
$$\left. \times \left(\frac{\sum_{i=1}^{m} \nu_i}{m} \right)_j, \ldots, \left(\frac{\sum_{i=1}^{m} \nu_i}{m} \right)_1 \right\}. \qquad (8.9)$$

Through statistical and eigenvalue analysis of long-term and short-term data, one can avoid the influence of errors in data processing and also identify the current operating status and development trend of the drilling process, so as to carry out real-time and accurate diagnosis and analysis of downhole anomalies.

8.1.2 *Dynamic safety threshold discrimination model*

8.1.2.1 *Determination of early-warning indicator thresholds*

In order to be able to make early real-time warnings of drilling down-hole events, it is necessary to comprehensively analyze the range of risk reference values of various drilling anomaly indicators in the study area, i.e., the safety warning threshold. Once several or a

Table 8.2. Magnitude of change in drilling warning indicators.

Drilling indicators	Trends	Amplitude of change
Bit pressure	Fluctuation or sudden increase	$\geq 100\,\mathrm{kN}$
Large hook load	Sudden increase or decrease	$100\text{--}200\,\mathrm{kN}$
Turntable torque	Continuous increase or decrease	$10\%\text{--}20\%$
Standpipe pressure	Gradual decrease	$0.5\text{--}1\,\mathrm{MPa}$
	Sudden increase or decrease	$\geq 2\,\mathrm{MPa}$
Volume of slurry tank	Gradual decrease	$0.5\text{--}2\,\mathrm{m}^3$

section of parameter data deviate from the parameter risk reference value range in a row, or exceed the warning threshold, drilling anomalies may occur. The safety warning thresholds vary for different formations and well sections. The magnitude of changes in some of the drilling warning indicators suggested by the relevant literature is shown in Table 8.2.

8.1.2.2 *Dynamic safety threshold judgment*

Judging the type of drilling abnormality based on empirical observations of whether the change of relevant drilling parameters exceeds the safety threshold cannot guarantee the timeliness and accuracy of the early warning. The dynamic threshold method can avoid these shortcomings and can solve the defects of different judgment standards due to different geological and well depth conditions through the change of dynamic thresholds. The method calculates the dynamic safety thresholds of the parameters according to the warning thresholds given by the actual geology and well depth conditions. The so-called parameter abnormality means that the characteristic value of the parameter exceeds the acceptable range compared with the normal value, i.e., the threshold value, including the upper threshold value and the lower threshold value. Eigenvalues of the same parameter greater than the upper threshold or less than the lower threshold predict a downhole anomaly.

The available literature provides predictive values for drilling parameters. Upward and downward fluctuations exceeding 10% of the warning base value indicate anomalies; anything other than those are within acceptable safety thresholds. The disadvantage of this method is the lack of real-time data. Therefore, this book proposes a dynamic safety threshold judgment function for the drilling process, with the upper limit value R_{DU} and lower limit value R_{DL} of the dynamic safety threshold calculated in Eqs. (8.10) and (8.11) as shown:

$$R_{\mathrm{DU}} = \frac{\sum_{i=1}^{n} R(t_i)}{n} \left(1 + \frac{R_U}{100} \right), \tag{8.10}$$

$$R_{\mathrm{DL}} = \frac{\sum_{i=1}^{n} R(t_i)}{n} \left(1 - \frac{R_L}{100} \right), \tag{8.11}$$

where $R(t_i)$ is the drilling parameter or variable at the moment, R_U is the upper threshold limit of the variable, and R_L is the lower threshold limit of the variable.

The judgment function of the upper and lower limits of the dynamic safety threshold of the variable is shown in Eqs. (8.12) and (8.13):

$$F(R_{\mathrm{DU}}) = R_{\mathrm{DU}} - R(t_i), \tag{8.12}$$

$$F(R_{\mathrm{DL}}) = R_{\mathrm{DL}} - R(t_i). \tag{8.13}$$

If $F(R_{\mathrm{DU}}) < 0$, the judgment variable exceeds the upper limit abnormality. If $F(R_{\mathrm{DL}}) < 0$, the judgment variable exceeds the lower limit abnormality.

The judgment results based on the dynamic threshold method may have false alarms or omissions in the judgment of downhole anomalies. Therefore, this section proposes an early-warning model for downhole anomalous events based on Bayes parameter estimation with multi-indicator fusion. It is known from historical and practical experience that the drilling process parameters approximately obey Gaussian distribution, and the parameter estimation and analysis are carried out using the Bayesian inference method. In order to improve the accuracy, the prior distribution and likelihood function of the relevant sign parameters are analyzed, the posterior distribution of the anomaly is determined, and multi-indicator fusion warning is carried out by introducing weighting coefficients and Gaussian distribution density function.

8.1.3 *Real-time early-warning model with multi-source information fusion*

The probability density functions of commonly used random variables are uniform distribution, Gaussian distribution (normal distribution), lognormal distribution, etc. Considering the uncertainty and randomness of the geological conditions of the drilling process, the distribution of drilling anomaly warning indicator variables in this book is mainly Gaussian distribution.

Gaussian distribution (also known as normal distribution) is the distribution law of continuous random variables describing many things. If the random variable X obeys Gaussian distribution, the expression of its probability density function is

$$f(x) = \frac{1}{\sqrt{2\pi\sigma^2}} e^{-\frac{(x-\mu)^2}{2\sigma^2}}, \quad -\infty < x < +\infty, \tag{8.14}$$

where μ, σ are constants, X is said to obey the normal distribution with parameters μ, σ, denoted as X–N (μ, σ^2), μ is the mean value of X, and σ is the standard deviation of X.

A random variable obeying a normal distribution has a probability density function $f(x)$ whose curve is symmetric about $x = \mu$, obtains a maximum at $x = \mu$, and has an inflection point at $x = \mu + \sigma$. The smaller σ, the steeper the curve, and the greater the probability that X falls near μ. Conversely, the flatter the curve, the more dispersed the distribution of X.

Let there be two states of events in the drilling process: the normal condition N and the abnormal condition A. The probability that the normal condition occurs is $p(A)$ and the probability that the abnormal condition occurs is $p(N) = 1 - p(A)$.

Assuming that the drilling parameter R_i normal and abnormal states conform to a Gaussian distribution, and that the drilling normal case, $p(R_i|N)$, satisfies the $N(R_i|\mu, \sigma^2)$ distribution by Bayesian inference, the likelihood estimation of the dataset is shown in Eq. (8.15):

$$N(R_i|\mu, \sigma^2) = (2\pi\sigma^2)^{-1/2} \exp[-(R_i - \mu)^2/2\sigma^2]. \tag{8.15}$$

The drilling anomaly $p(R_i|A)$ satisfies the $N(R_i|\mu + 3\sigma, \sigma^2)$ distribution, and the likelihood estimation of the dataset is shown in Eq. (8.16):

$$N(R_i|\mu, \sigma^2) = (2\pi\sigma^2)^{-1/2} \exp[-(R_i - \mu - 3\sigma)^2/2\sigma^2]. \tag{8.16}$$

The probabilistic risk prediction model for the sign parameter R_i is shown in Eq. (8.17):

$$p(A|R_i) = \frac{p(R_i|A)p(A)}{\Sigma p(R_i|A)p(A)}. \tag{8.17}$$

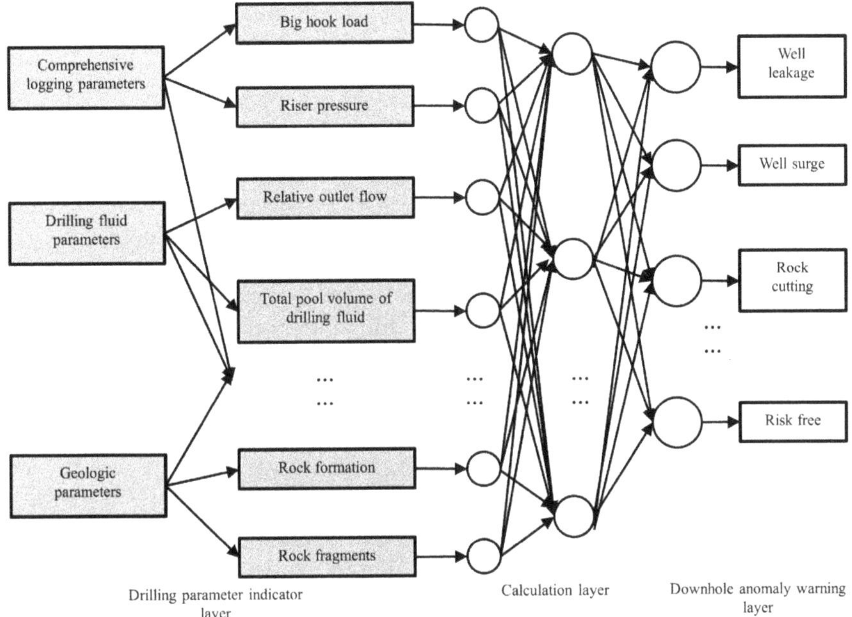

Fig. 8.1. Multi-indicator, multi-level fusion early-warning model.

The principle of multi-indicator fusion is to achieve system design, data information processing, coordination and optimization, decision-making, and estimation, which mainly includes the fusion of data layer, feature layer, and decision-making layer. Firstly, the data need to be preprocessed to eliminate coarse error points; secondly, the processed data need to be characterized by feature calculations, and the fusion of data and features is carried out to finally obtain the decision, as shown in Fig. 8.1.

The real-time warning of downhole events is performed by extracting the feature quantity and fusion calculation of drilling indicators. Data fusion is to synthesize data information from multiple parameters to obtain decision-making results, and then to judge and warn about downhole anomalies.

A nonlinearly distributed continuous variable can be a superposition of multiple probability distributions, e.g., approximated by the linear superposition of multiple Gaussian distributions in a certain proportion. Given that a variable can be expressed as a weighted sum of m sub-Gaussian distributions with mean μ_m, variance σ_m, weights

ω_m, and anomaly correction factor φ_m, the Gaussian decomposition of the probability distribution of a nonlinear continuous variable is expressed as shown in Eq. (8.18):

$$p(A|R_i) = \varphi_m \sum_{m=1}^{M} \omega_m N(R_i; \mu_m, \sigma_m). \qquad (8.18)$$

8.1.4   *Application*

In this section, a well leakage event that occurred during the drilling operation of a production well in an oil and gas field is taken as an example to analyze the abnormal rule of change of the comprehensive logging parameters and drilling fluid parameters that characterize the well leakage event. The production well had a well leak at a depth of 2875–2880 m in the 81/2 in well section. The design density of drilling fluid was 16.0 ppg, the pump displacement increased from 150 gpm to 170 gpm and then decreased to 50 gpm, and the pump pressure increased from 180 psi to 370 psi and then decreased to 86 psi when the dynamic leakage rate of 10 bbl/h developed with no return. In this example, we select the drilling process parameter data information of the drilling time period from 19:00 to 01:00 of the next day, and set the dynamic leakage rate of the well leakage to be greater than 10 bbl/h to judge it as a well leakage, and analyze it as an early warning. Firstly, standardized preprocessing is carried out for long and short cycle monitoring data, then key parameters are extracted and their eigenvalues are calculated. Based on the fusion technology of multi-source indicators, combined with dynamic safety threshold determination and Bayesian probabilistic early warning model, the quantitative assessment of well leakage risk is realized, and the graded warning decision is finally generated based on the risk assessment results.

8.1.4.1   *Eigenvalue calculation of drilling anomaly warning indicators*

(1) Drilling process parameter data preprocessing

    In this example, 600 short-term and 3500 long-term drilling process parameters data (taking relative export flow rate as an example) are given. Through the experimental study on the preprocessing of parameter data, when the sliding window length is taken as 80–200,

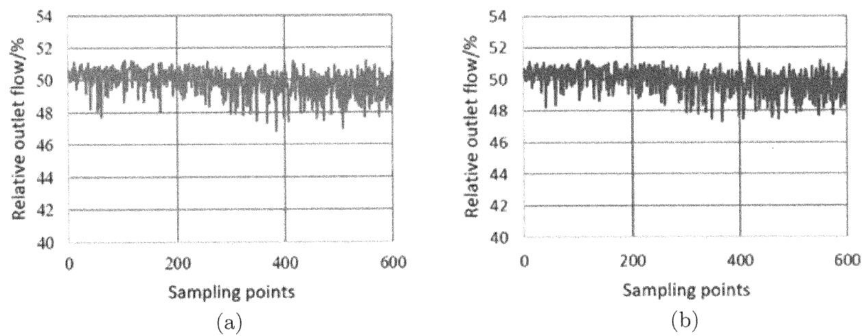

Fig. 8.2. Comparison of short-term data preprocessing (a) before and (b) after processing.

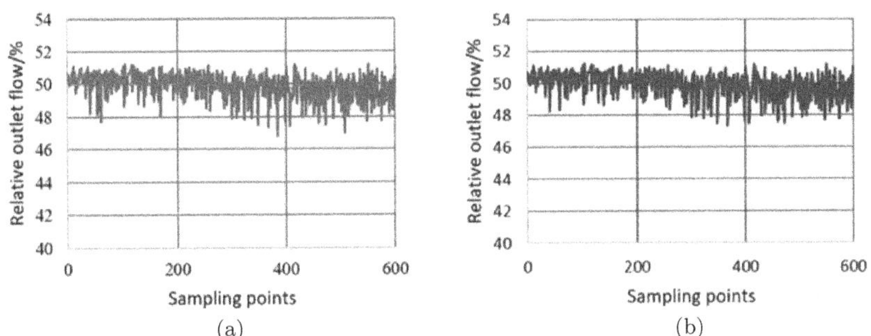

Fig. 8.3. Comparison of long-term data preprocessing (a) before and (b) after processing.

the effect of coarse error value rejection is more desirable. Therefore, the short-term data sliding window length is set to 80 and the long-term sliding window length is set to 200. The corresponding processing results are shown in Figs. 8.2 and 8.3.

(2) Calculation and analysis of eigenvolume

The warning indicators characterized by downhole anomalies vary according to different drilling wells. In this example, the main early-warning indicators characterizing the well leakage are selected for the well leakage event, which are relative outlet flow rate, inlet flow rate, riser pressure, and mud pool volume, and the trends of these indicators are analyzed.

Part of the data collected from 19:49 to 19:50 is selected. From Fig. 8.4, it can be seen that the outlet flow rate fluctuates up and down at 50% and decreases to 40%. The inlet flow rate decreases from 508 gpm to 433 gpm and 108 gpm. The riser pressure decreases

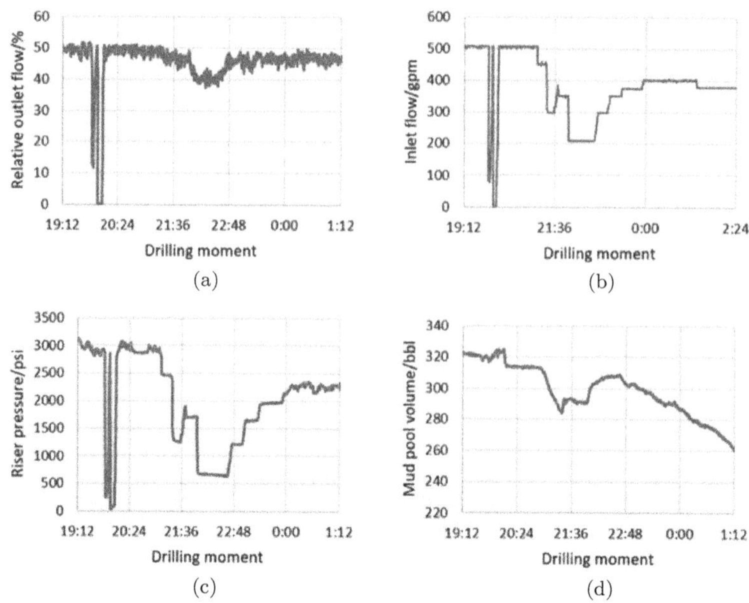

Fig. 8.4. Trends in different early-warning indicators: (a) relative outlet flow, (b) inlet flow, (c) riser pressure, and (d) mud pool volume.

from 2734.962 psi to 1671.373 psi and 623.75 psi; and the change of the mud pool volume fluctuates up and down at around 321 bbl.

The parameter data collected from the 19:12 to 01:12 time period of the next day were selected and the change trend was statistically analyzed as shown in Fig. 8.4. The analysis of the parameter changes shows that in the early stage of well leakage, the changes of each parameter are different. The outlet flow rate and inlet flow rate show an obvious downward trend, while the riser pressure changes lag behind slightly, and the mud pool volume changes are relatively smooth. It indicates that the mud pool volume change is not sensitive to the diagnosis of early well anomalies. During the plugging process, the outlet flow rate, inlet flow rate, and riser pressure show a trend of decreasing and then increasing, while the mud pool volume shows a slow decreasing trend.

During the drilling process, there are certain fluctuations in the changes of drilling engineering index parameters, but not all fluctuations are signs of abnormality. It is necessary to calculate the eigenvalue that characterizes the warning indicators. Here, the rate of change that indicates the trend of change is used as the eigenquantity,

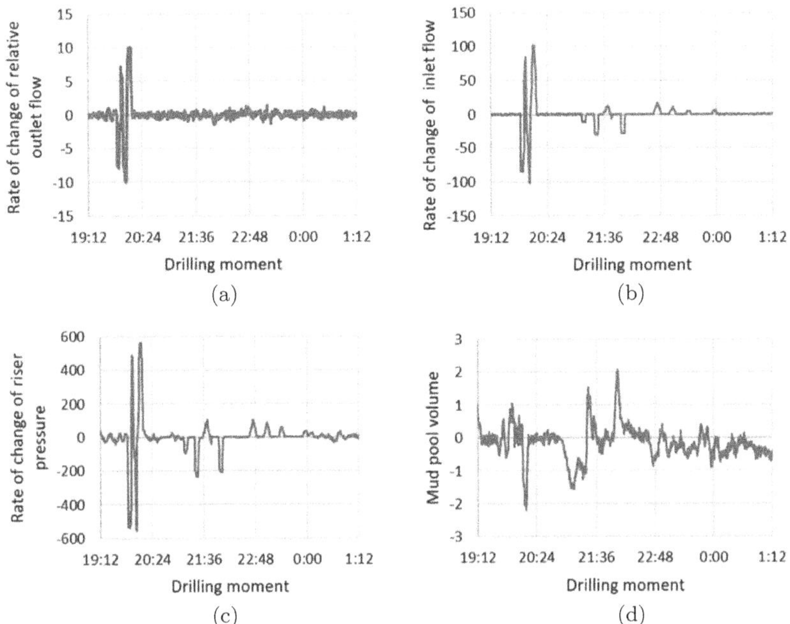

Fig. 8.5. Rates of change of different warning indicators: (a) relative outlet flow, (b) inlet flow, (c) riser pressure, and (d) mud pool volume.

and the time interval for calculating the rate of change is set as 5 minutes. In actual monitoring, if there is no historical value corresponding to the current monitoring moment; it is considered normal. Taking the dynamic leakage rate $> 10\,\text{bbl/h}$ as the leakage criterion, by comparing the rate of change of relative outlet flow rate, inlet flow rate, riser pressure, and mud pool volume as shown in Fig. 8.5, the set safety threshold can be used to obtain high-precision well leakage time period localization information in a short period of time. Different variable safety thresholds generally need to be set according to the actual formation conditions and well depth conditions, and the values are different as the conditions change.

8.1.4.2　*Analysis of real-time warning results*

(1) Judgment based on dynamic safety thresholds

Based on the offline dataset of historical well leakage drilling parameters, the dynamic moving average of each indicator is calculated by the sliding window algorithm. By statistically analyzing

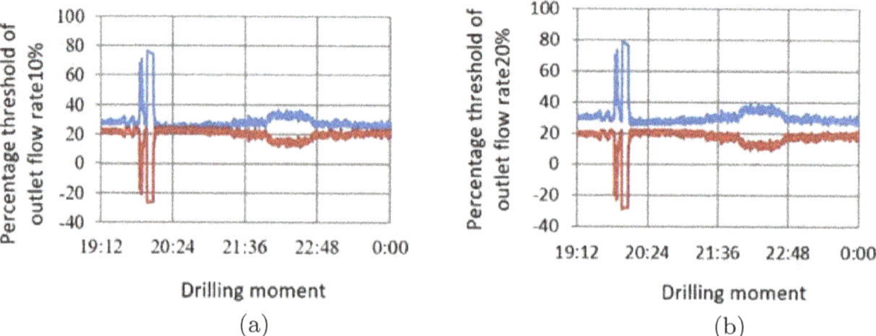

Fig. 8.6. Upper and lower dynamic safety thresholds at different percentile values: (a) 10% and (b) 20%.

the distribution characteristics conditions, the upper and lower limits of the dynamic safety thresholds of each indicator are determined by the percentile method (usually the 5th and 95th percentiles are selected), and the quantitative warning boundary is established. For example, if the average value of the relative export flow rate at a certain well depth for a certain period of time is 50.25%, the corresponding upper and lower thresholds are obtained by taking the percentile values of 10% and 20% as follows: 55.26%, 45.216%, and 60.28%, 40.192%, respectively. Based on the actual data observation and the comparison of the two cases herein, it is found that the percentile value of 10% is more suitable for the indicator of the relative export flow rate. The percentile values are different for different drilling metrics. A drop in outlet flow rate signals the possibility of a well leak, so a lower limit on the dynamic safety threshold needs to be determined. From Fig. 8.6, it can be seen that $F(R_{\mathrm{DL}}) < 0$ indicates the risk of well leakage. Relying on dynamic thresholds alone to determine well leakage can easily lead to false alarms during drilling. For example, if $F(R_{\mathrm{DL}}) > 0$ after 21:36, the well leakage is actually being plugged at that time.

(2) Comprehensive multi-indicator early warning based on Bayes parameter estimation

In order to improve the accuracy and reduce the false alarm rate of drilling anomaly early warnings, a priori distribution and likelihood function of relevant sign parameters are analyzed, then a posteriori distribution of anomalies is determined, and the early warning of multi-indicator fusion is carried out through the introduction

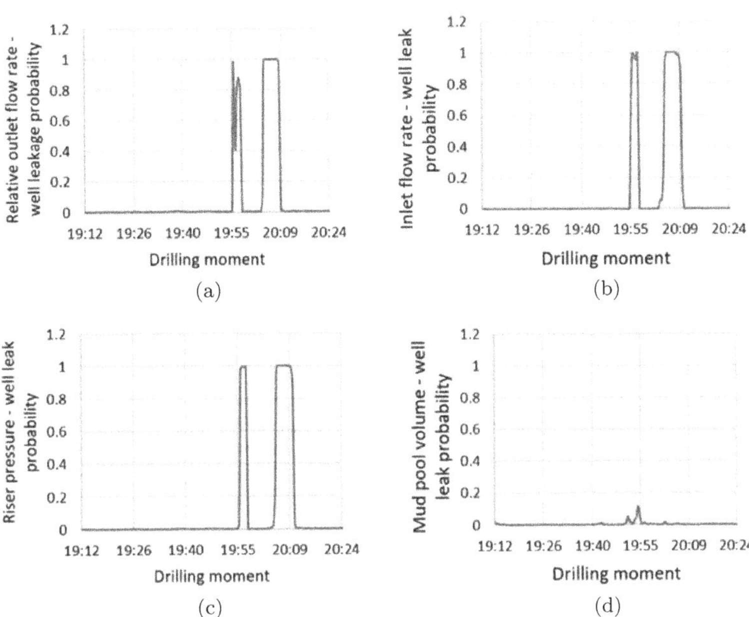

Fig. 8.7. Early well leakage probability distribution for different early-warning indicators: (a) relative outlet flow, (b) inlet flow, (c) riser pressure, and (d) mud pool volume.

of weighting coefficients and Gaussian probability density distribution. After preprocessing the data corresponding to these indicators, eigenvalue extraction is carried out to determine the corresponding short-term and long-term mean and variance by taking the rate of change as an example. The probability of well leakage and plugging corresponding to different indicators at different drilling moments is calculated as shown in Figs. 8.7 and 8.8. From Fig. 8.7, it can be seen that the probability of well leakage occurrence under relative outlet flow rate, riser pressure, and inlet flow rate rises from close to 0 to 99.9% in 3 min, close to 1, which can be used as a reference for early warning. And the change of well leakage probability corresponding to the mud pool volume variable is not obvious, which cannot be used as an early warning indicator. As can be seen from Fig. 8.8, compared with other early warning indicators, the change of well leakage probability corresponding to the mud pool volume variable is more obvious, and it can be used as an early warning for the well leakage in the plugging process.

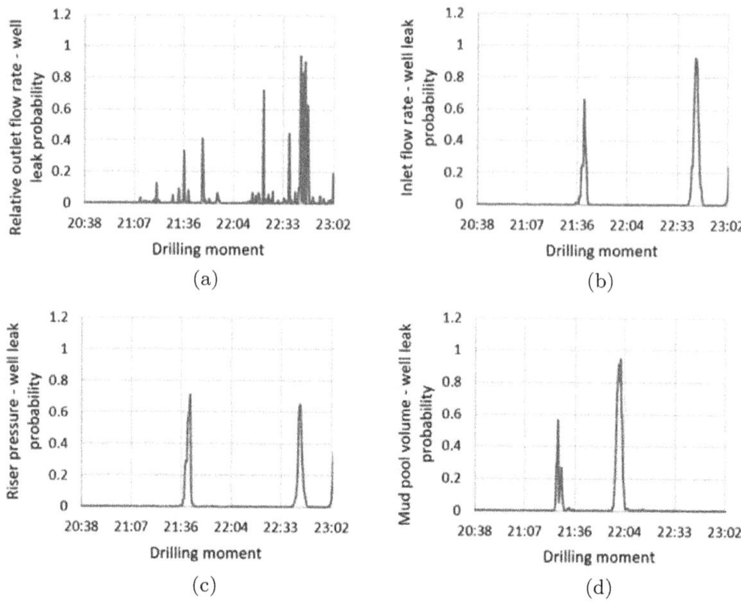

Fig. 8.8. Probability distribution of well leakage during plugging with different warning indicators: (a) relative outlet flow rate, (b) inlet flow rate, (c) riser pressure, and (d) mud pool volume.

Real-time early warning of well leakage is carried out by extracting and combining the eigenvolume of drilling metrics into the computation. Within the time period 19:43–20:24, while drilling, given the weights of $\omega_m = \{0.4, 0.2, 0.3, 0.1\}$, a multi-indicator eigenvalue fusion of well leakage probability estimation is carried out. The multi-indicator fusion warning model can improve the warning rate of process anomalies and can avoid leakage due to the delay of the probabilistic warning of the mud pool volume, i.e., to reduce the leakage rate. At the anomaly correction factor $\varphi_m = 0.05 = 0.05$, the actual probability value of a single indicator, when well leakage probability occurs, is obtained. The multi-indicator probability value before correction and the multi-indicator probability value after correction are shown in Fig. 8.9. After correction, compared with before correction, its average relative error is reduced by 3.9% and the maximum relative error is reduced by 22.7%. The application of the anomaly correction coefficient reduces the maximum relative error and average relative error values, and the warning effect is significantly improved.

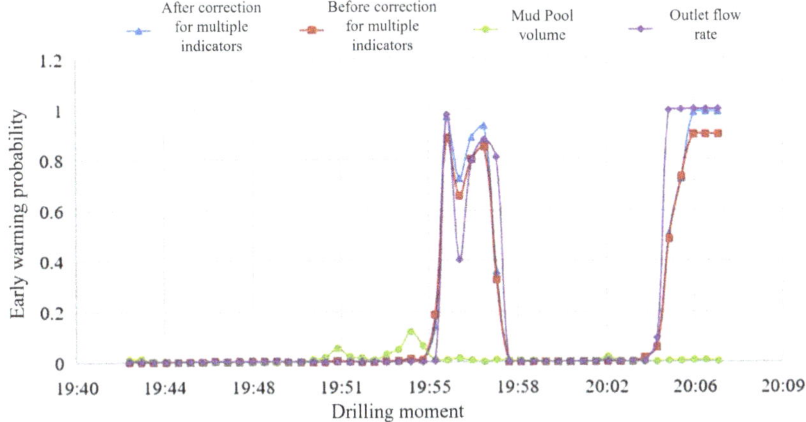

Fig. 8.9. Comparison of the probability of occurrence of well leakage before and after correction.

8.2 Early Warning Method for Offshore Drilling Overflow Based on Pattern Recognition

Drilling overflow early warning is mostly realized by the amount of drilling fluid return and the volume increase of drilling fluid in the circulation pool, and this threshold exceeding alarm method often produces a large number of false alarms and omissions due to the improper selection of the threshold limit, and it is difficult to adapt to the characteristics of unstable parameter changes caused by variable drilling conditions. For this reason, this book proposes an overflow warning method based on parameter trend pattern recognition, which recognizes drilling anomalies by parameter change trend characteristics, avoids the calculation of threshold limits, and has good dynamic adaptability.

This section firstly analyzes the causes and manifestations of gas intrusion overflow and selects warning indicators based on the characteristics of the gradual development of the accident. Secondly, the overflow early-warning method based on parameter trend pattern recognition is introduced: Based on the wellbore dynamic multiphase flow simulation theory, this study initially constructs a standard pattern library of overflow characteristics through the fusion of adjacent wellbore influx monitoring data. Subsequently, wavelet filtering denoising and manifold learning-based dimensionality reduction techniques are applied to offline logging signals to extract temporal evolution features. Finally, a sparse coding algorithm is employed to achieve pattern matching and cluster diagnosis of drilling anomaly conditions. Finally, the probability value of overflow risk is calculated based on the fuzzy logic method to realize early warning. In the case study, the method is applied to the logging data of a well in Ledong, South China Sea, and a comparison between the early-warning results and the drilling daily report proves that the method can realize accurate early warning at the early stage of overflow.

8.2.1 *Overflow early-warning model*

8.2.1.1 *Characterization of gas overflow drilling process parameters*

Gas overflow occurs for a number of reasons, the most fundamental of which is an imbalance of pressure in the well, i.e., the pressure in the well is less than the formation pressure. Natural gas in the pore space of the formation enters the wellbore through rock chip gas intrusion, replacement gas intrusion, diffusion gas intrusion, and gas overflow. The reasons for the pressure in the well being less than the formation pressure are inaccurate prediction of formation pressure, low density of drilling fluid, and the well not being filled with drilling fluid when starting the drilling, which leads to a reduction of hydrostatic pressure. Excessive suction pressure will make the hydrostatic pressure lower than the formation pressure, resulting in overflow. When the drilling fluid leaks, the drilling fluid enters the formation, causing the hydrostatic pressure of the fluid column in the well to drop, and overflow occurs once it is less than the formation pressure; in addition, drilling encounters an abnormally high pressure formation, which results in overflow because the design density of the drilling fluid is unable to satisfy the pressure demand in the well.

After the natural gas intrusion into the borehole, it is in the state of gas–liquid two-phase flow, forming bubble flow, segment plug flow, and other flows. The gas slips off and rises with the circulating drilling fluid in the process of upward return in the annulus. The gradual decrease of the liquid column pressure in the well from the bottom to the top leads to an increase in the expansion of the gas volume in the process of upward transportation. For the wells with shallow

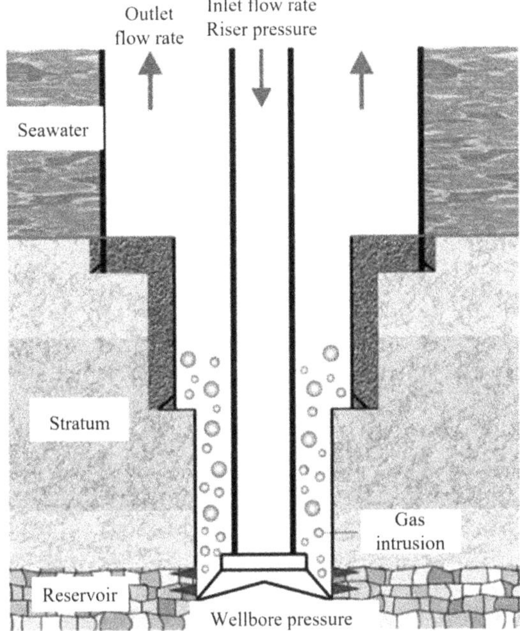

Fig. 8.10. Schematic diagram of gas intrusion overflow during deepwater drilling.

depths, the expansion of the gas will cause a significant reduction in the static liquid column pressure in the well, which will also cause the density value of the returned drilling fluid to decrease, as shown in Fig. 8.10.

After the gas intrusion occurs, the recorded well data monitored in real time will change significantly, and the drilling anomalies of the spill incident will be as follows:

(1) If the inlet flow rate of drilling fluid is unchanged, the outlet flow rate increases gradually, the difference between the inlet and outlet flow rates increases gradually, and the drilling fluid spills out of the wellhead autonomously after the pump is stopped.
(2) There is increased volume of drilling fluid in the mud circulation basin.
(3) The standpipe pressure slowly decreases and the pump speed (number of pump strokes) increases.
(4) The density of the exported drilling fluid is reduced.

Therefore, the recording parameters of drilling fluid flow rate, mud pool volume, and riser pressure, are selected as the early-warning indicators of overflow accidents. The overflow information of multiple early-warning indicators is integrated to make a comprehensive judgment.

According to the Specification for Interpretation of Recorded Data of Comprehensive Recorder (SY/T 5977-94), the criteria for abnormal parameters related to overflow under specific conditions and specified requirements are as follows:

(1) The pressure of the standpipe gradually decreases by 0.5 MPa, or suddenly increases or decreases by more than 2 MPa.
(2) Relative change in total pool volume of drilling fluid exceeds 2m^3.
(3) The density of the drilling fluid outlet suddenly decreases by more than 0.04 g/cm^3, or decreases or increases in trend.
(4) The drilling fluid outlet discharge is significantly larger or smaller than the inlet discharge.

8.2.1.2 *Pattern recognition of drilling anomalies*

Pattern recognition of drilling anomalies involves finding an automatic recognition algorithm that can be applied to computers. It can transform the process of drilling anomaly discrimination based on experts' experience into an automatic discrimination process through computer processing of logging signals, so as to improve the timeliness and accuracy of the monitoring of drilling anomalies, avoid the lagging and miscalculation effects brought about by the manual interpretation of drilling anomalies, increase the efficiency of the production monitoring, and realize the accurate warning of drilling anomalies.

In this section, a new method for drilling anomaly identification, a pattern recognition algorithm based on parameter trend features, is proposed. The algorithm gives the standard pattern of drilling anomalies in advance by fusing the simulation results of the dynamic wellbore multinomial flow model and the training results of neighboring well data, which reduces the workload of sample training. In addition, the algorithm simplifies the data pattern through the segmented approximation method of dimensionality reduction processing, so that its pattern features become easy to extract. By extracting the shape features, trend features and similarity calculation results

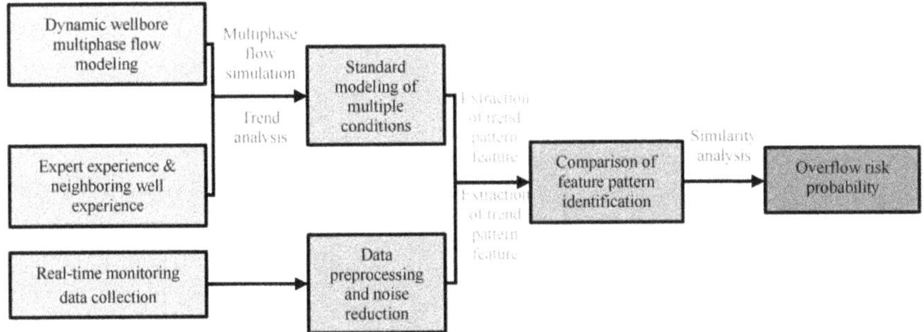

Fig. 8.11. Overflow pattern recognition model structure.

of the signal sequence as the auxiliary feature set for pattern classification, the method has the advantages of feature intuition and small data volume, which not only reduces the computational complexity, but also avoids the need for the design of classifiers, thus simplifying the implementation of pattern clustering. The model structure of this pattern recognition algorithm is given in Fig. 8.11.

Based on the dynamic multiphase flow model of a gas intrusion wellbore, the standard pattern of overflow parameters is extracted by combining the surge data of neighboring wells. The trend features of the signal sequence are extracted by filtering and preprocessing the recorded well data collected offline. The clustering identification of drilling anomalies is realized by comparing the trend eigenvalues of the standard pattern of overflow parameters and the recorded well parameters. Finally, the trend eigenvalues are analyzed for similarity, and the overflow risk probability is calculated based on the fuzzy logic method.

8.2.1.3 *Warning parameter overflow standard model*

The amount of mud pool volume change during gas intrusion calculated by the dynamic wellbore flow model simulation is shown in Fig. 8.12, which shows that the mud pool level in the overflow mode exhibits a quadratic upward trend.

The relationship between inlet flow and outlet flow, liquid level, and standpipe pressure during gas intrusion overflow is as follows:

$$Q_{\text{out}} = [A(\nu_l + \nu_g)]|_{s=0}, \tag{8.19}$$

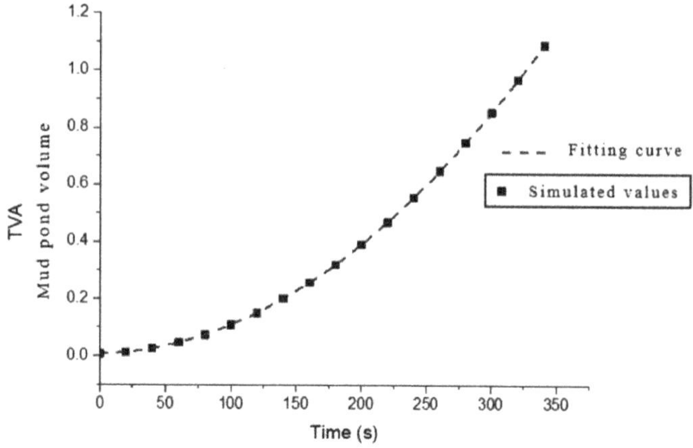

Fig. 8.12. Mud pool level change during a typical gas intrusion.

$$V_{\text{pit}}(t) = \int_0^t Q_{\text{out}}(\tau)d\tau - Q_{\text{in}}t, \tag{8.20}$$

$$p'_{\text{tn}}(t) = p'_{\text{wf}}(t) \approx -\{[(Q_{\text{out}}(t) - Q_{\text{in}}] + AR_p\}\frac{\rho_w - \rho_g}{A}$$

$$+ \rho_w R_p\left(g + \frac{2f\nu^2}{D}\right), \tag{8.21}$$

where Q_{out} denotes the outlet flow rate, m^3/s; Q_{in} denotes the inlet flow rate, m^3/s; A denotes the annulus cross-sectional area, m^2; and p_{wf} denotes the bottomhole pressure, p_a.

According to Eqs. (8.19)–(8.21), the flow differential ($Q_{\text{out}} - Q_{\text{in}}$) and riser pressure will tend to increase linearly and decrease quadratically, respectively. The qualitative characteristics of the trend changes of these parameters remain consistent with the neighboring well overflow data.

Simplifying the overflow model yields a linear trend for the three early-warning indicators (Fig. 8.13), using the angle α to characterize the degree of data trending.

Well leaks, drilling anomalies and spills usually show opposite patterns in the response characteristics of drilling fluid parameters, while riser pressures also reduced slowly due to leakage of drilling fluid into the formation, resulting in a reduction in the pressure of the drilling fluid column wells. Therefore, the standard patterns of

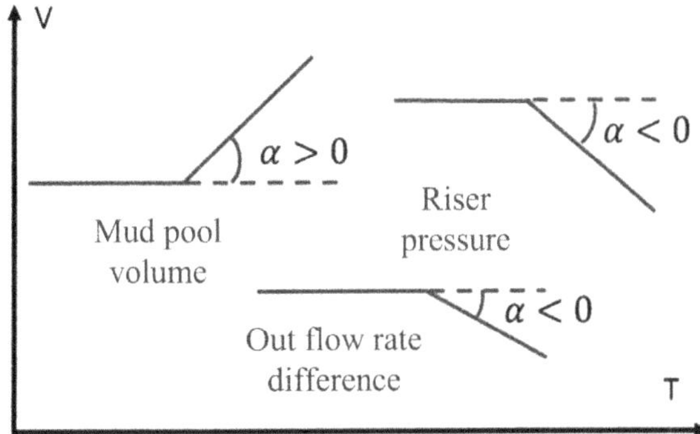

Fig. 8.13. Trends in warning indicators for overflow events.

drilling abnormality (overflow, well leakage) and normality are summarized as follows:

$$\text{overflow}: \quad \alpha_{\text{mfop}} < 0; \quad \alpha_{\text{tva}} > 0; \quad \alpha_{\text{sppa}} < 0.$$

$$\text{well leakage}: \quad \alpha_{\text{mfop}} > 0; \quad \alpha_{\text{tva}} < 0; \quad \alpha_{\text{sppa}} < 0.$$

$$\text{normal}: \quad \alpha_{\text{mfop}} \approx 0; \quad \alpha_{\text{tva}} \approx 0; \quad \alpha_{\text{sppa}} \approx 0.$$

8.2.2 *Logging signal pattern feature extraction methods*

The pattern feature extraction method of logging signal has 3 main steps: data preprocessing, dimension reduction, and pattern feature extraction. Data preprocessing means to eliminate the noise of the original signal and standardize the data for subsequent multi-indicator fusion warning. Data dimensionality reduction means to simplify the expression of the shape change of the signal sequence for pattern feature extraction, i.e., the shape feature and the trend change feature of the signal sequence. Finally, the pattern feature of the signal sequence is compared with the standard pattern feature of the overflow to realize the pattern recognition of the drilling anomaly.

8.2.2.1 *Data preprocessing*

The logging data collected by the logging instrument is a continuous signal time series, and a set of effective data series (including drill bit sounding, borehole depth, mechanical drilling speed, large hook load, rotary table torque, drilling pressure, riser pressure, pump speed, mud pool volume, mud flow rate, inlet and outlet mud temperatures, and inlet and outlet mud resistivity) is collected every 5 s. The overflow accident is an accident of gradual development, and it often needs the early-warning indicator to show an abnormal trend of change in a period of time before it can be judged, so it is necessary to analyze the signal sequence of the early-warning indicator for a length of time.

The signal sequence of drilling parameters with m data sampling points is denoted as S:

$$S = \{(s_1, t_1), (s_2, t_2), \ldots, (s_m, t_m)\}. \tag{8.22}$$

Flow data obtained from real-time measurements in deepwater drilling usually contain high noise due to the high influence of sensors by wind, ocean currents, and electric waves. The influence of noise on overflow diagnosis can be significantly reduced by signal filtering. The signal sequence S is first filtered by wavelet noise reduction to obtain the signal sequence Q^*:

$$Q^* = \{(q_1, t_1), (q_2, t_2), \ldots, (q_m, t_m)\}. \tag{8.23}$$

In this book, with the help of MATLAB wavelet toolbox, 6-layer decomposition of the signal is performed using the db6 wavelet basis, and the 6th layer approximation component is reconstructed as the filtered signal sequence after noise reduction. An example of noise reduction is shown in Fig. 8.14.

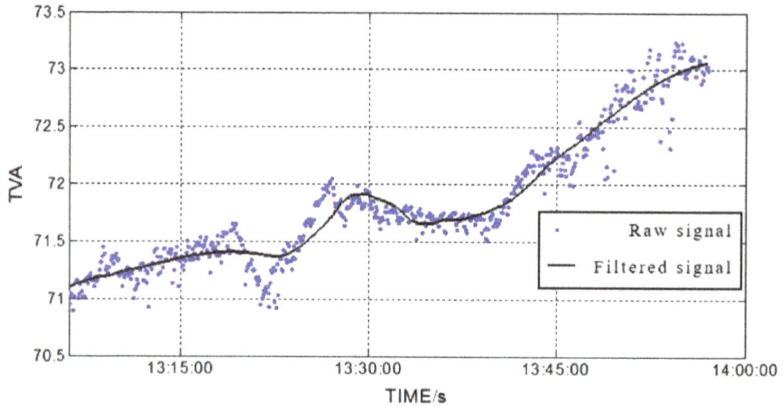

Fig. 8.14. Comparison of filtered signal and original signal.

Early-warning indicators have different units and different sets of quantities. In order to improve the accuracy of overflow identification, it is necessary to fuse multiple indicators for identification. So, it is necessary to standardize multiple early-warning indicators first and transform the signal sequence within $[0, 1]$ in order to standardize different early-warning indicators to have a uniform range. The standardized signal sequence is denoted as Q.

8.2.2.2 *Downscaling*

The signal sequence after smoothing has obvious trend change characteristics. The analysis of the manifestations of drilling anomaly recording parameters that characterize trend changes useful for early warning is only increasing or decreasing. In order to simplify the difficulty of identification, the curve change of the high power of the signal sequence reduces the dimensionality to one linear change. A segmented approximation algorithm is used for dimension reduction.

The key to the segmentation approximation algorithm is to find the segmentation interruption point of the signal sequence. The steps of the interruption point search method established in this paper are as follows (Fig. 8.15):

Step 1: Input a fixed-length time series $Q(i : j)$ of size m, and initialize the following variables: Minimum fitting error $\sigma_{\min} = +\infty$ (used to store the optimal fitting error); Segment boundary vector $B_D = [Q(i), Q(j)]$ (used to record breakpoints between segments); Segment count $N = 1$ (indicating the current number of subsequences, intially 1).

Step 2: Find the discontinuity point by calculating the distance from the sequence point to the fitted line:

(2-1) Establish the fitting equation for the linear fitting line L_{ij} with $Q(i), Q(j)$ as endpoints.
(2-2) Calculate the distance d_k from each point of the time series $Q(i : j)$ to the fitting line.
(2-3) Find the point of maximum value of d_k as the newly found discontinuity point $Q(k)$, $\boldsymbol{B}p = (N + 1) = Q(k)$, and $N = N + 1$.
(2-4) Judge $k < j$; if it is valid, return to (2-2) and repeat the execution; if not, execute step 3.

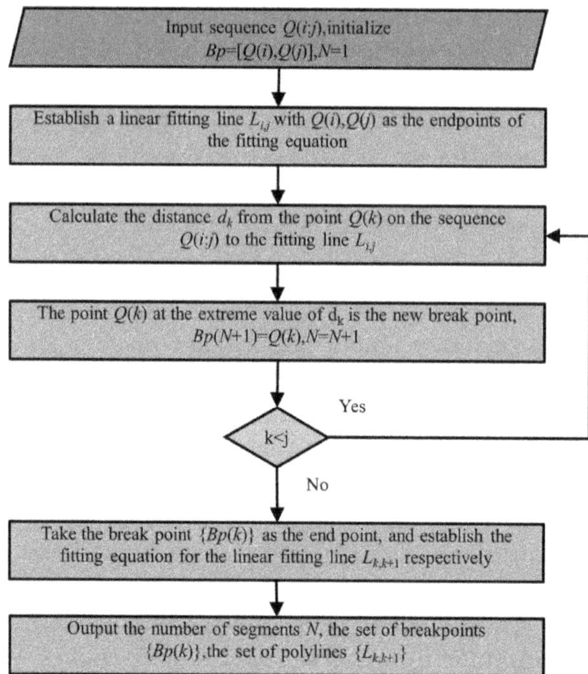

Fig. 8.15. Steps of segmented approximation breakpoint searching.

Step 3: The time series $Q(i:j)$ is divided into N segments, and the equation for fitting the linear fit line $\{L_{k,k+1}\}$ is established in segments with the breakpoint $\{\boldsymbol{Bp}(k)\}$ as the endpoint, and the segments are recalculated $\sigma_{\min}$.

Step 4: Output the number of segments N, the set of discontinuities $\{\boldsymbol{Bp}(k)\}$, the polyline $\{L_{k,k+1}\}$, and the fitting error $\sigma_{\min}$.

Segmental approximation breaks for a given sequence can be searched automatically by MATLAB programming.

8.2.2.3 *Pattern feature extraction*

The segmentation approximation divides the original signal sequence Q into polyline signal sequences of varying sequence lengths, denoted as P:

$$P = (p_1, p_2, \ldots p_N). \tag{8.24}$$

Each p_i is a line segment represented by a primary function, and k_i is the slope of p_i, then Q has the shape feature set K:

$$K = \{k_1, k_2, \ldots k_N\}. \tag{8.25}$$

The angle α is introduced to represent the trend characteristics of the shape change of the signal sequence after dimensionality reduction, and α is the pattern characteristics of the signal sequence, which characterizes the changing trend of the parameters of the warning indicator. Q has the trend feature set $A = \{\alpha_1, \alpha_2, \ldots, \alpha^N\}$. α_i is calculated by Eq. (8.26), where $k_0 = 0$:

$$\alpha_i = \arctan(k_i) - \arctan(k_{i-1}). \tag{8.26}$$

8.2.3 *Clustering methods based on trend characteristics and similarity measures*

Pattern recognition realizes fault recognition and diagnosis through clustering of pattern features. After extracting the shape features and trend features of the warning parameter signal sequence, the pattern matching algorithm is used to realize signal classification by calculating the feature similarity between the signal to be recognized and the standard overflow pattern, and finally complete the intelligent recognition of the overflow condition. In this chapter, two methods, trend eigenvalue classification and shape similarity measurement, are used for the identification of overflow. Overflow is considered to occur when both methods are judged as overflow at the same time. Among them, trend eigenvalue has good sensitivity and gives a quick response when the parameters deviate from the normal value or deviate from the original state with abnormal changes. Shape similarity measurement has good error tolerance and gives a robust response based on the trend of the changes in a short period of time, taking into account the fluctuation amplitude of the signal sequence and time span, which compensates for false alarms caused by the high sensitivity of the trend eigenvalue classification.

Since anomalous changes in a single-warning indicator often indicate multiple complex situations, both methods need to fuse multiple indicators to identify drilling anomalies. The method outputs a spill warning when multiple warning indicators point to spill anomalies at the same time.

8.2.3.1   *Classification of trend eigenvalue patterns*

First, the number of sampling points m of the signal sequence is determined, and the wellbore dynamic multiphase flow model is applied to simulate and calculate the standard pattern of the amount of change of the warning parameter over time for a 5-min time period. Second, the same normalization rule is applied to each warning indicator. The acquired signal sequences and wellbore dynamics after wavelet noise reduction, as well as the standard pattern of drilling anomalies obtained from multiphase flow simulations, are normalized so that the resulting time series have the same range. Herein, the standard pattern of α for negative values is normalized to $[-1, 0]$, that of α for positive values is normalized to $[0, 1]$, and that of the collected signal sequence is normalized to $[0, 1]$. Then, the standard pattern and the acquired signal sequence are segmented and approximated for dimensionality reduction, respectively, to obtain the polyline sequences P^{mfop}, P^{tva}, P^{sppa} and P_Q^{mfop}, P_Q^{tva}, P_Q^{sppa} that can be used for pattern feature extraction. Finally, the shape features and trend features of the standard mode polyline sequence and the acquired signal polyline sequence are extracted, respectively, and the feature values of the standard mode are listed in Table 8.3.

Among them, indicators such as $a_{\mathrm{kick}}^{\mathrm{sppa}}$ are related to the well location and drilling stratum, which are obtained through the drilling anomaly data statistics of neighboring wells. Firstly, combined with the drilling daily report, the drilling anomaly occurrence moment is judged by the expert's experience based on the logging data. Secondly, m signals of the same length are collected from the anomaly

Table 8.3.   Drilling anomaly standard model eigenvalues.

Early warning parameter	Drilling anomaly	N	Shape characteristics k	Trend characteristics α
Riser pressure	Normal 0	1	$k_1 \approx 0$	$\alpha \approx 0$
	Well surge/Overflow 1	2	$k_1 \approx 0, k_2 < 0, k_1 > k_2$	$a_{\mathrm{kick}}^{\mathrm{sppa}} < \alpha < 0$
	Well leakage 2	2	$k_1 \approx 0, k_2 < 0, k_1 > k_2$	$a_{\mathrm{kick}}^{\mathrm{sppa}} < \alpha < 0$
Mud pool volume	Normal 0	1	$k_1 \approx 0$	$\alpha \approx 0$
	Well surge/Overflow 1	2	$k_1 \approx 0, k_2 > 0, k_1 < k_2$	$0 < \alpha < \alpha_{\mathrm{kick}}^{\mathrm{tva}}$
	Well leakage 2	2	$k_1 \approx 0, k_2 < 0, k_1 > k_2$	$\alpha_{\mathrm{kick}}^{\mathrm{tva}} < \alpha < 0$
Mud flow rate	Normal 0	1	$k_1 \approx 0$	$\alpha \approx 0$
	Well surge/Overflow 1	2	$k_1 \approx 0, k_2 < 0, k_1 > k_2$	$\alpha_{\mathrm{kick}}^{\mathrm{mfop}} < \alpha < 0$
	Well leakage 2	2	$k_1 \approx 0, k_2 > 0, k_1 < k_2$	$0 < \alpha < \alpha_{\mathrm{kick}}^{\mathrm{mfop}}$

occurrence moment, and the trend eigenvalues of the empirical data are extracted using the same method through the data filtering, dimensionality reduction, and feature extraction process. Lastly, the mean μ and the standard deviation σ of the trend eigenvalues of all the collected neighboring wells' data are calculated. Based on the 3σ rule, in general, μ is chosen as the value of indicators such as $a_{\text{kick}}^{\text{sppa}}$, $\mu - 3\sigma$ is chosen if the recognition sensitivity is to be improved, and $\mu + 3\sigma$ is chosen if better fault tolerance is required.

When the shape features and trend features of the three warning parameter signal sequences satisfy the standard pattern, the momentary signals are classified into the corresponding drilling anomalies, thus completing the pattern recognition of the trend eigenvalues.

8.2.3.2 *Shape similarity measurement*

Similarity calculations are utilized to evaluate the similarity between recorded signal sequences and drilling anomaly standard sequences. For two sequences from the same time domain, the Euclidean distance is an effective and simple similarity calculation method. However, the calculation results are easily disturbed by sequence anomaly data. In addition, the traditional Euclidean distance calculation method recognizes unfavorable shapes due to the lack of trend information.

Considering the fluctuation amplitude, trend, and time span of the time series, the similarity between the filtered signal sequence and the standard model is measured using the "morphological distance" combining the Euclidean distance and the improved slope distance Cr see Eq. (8.27):

$$Cr = (D_0 \times D_{\text{KM}})^{1/2}, \tag{8.27}$$

where D_0 is the Euclidean distance, m; and D_{KM} is the improved slope distance, m:

$$D_{\text{KM}}(S, S_b) = \left| \sum_{i=1}^{n} \Delta t_i W_i (k_i - k_{bi})/t_n \right|, \tag{8.28}$$

$$W_i = \frac{(S_i - S_{\text{min}}) \times a}{S_{\text{max}} - S_{\text{min}}} + (1 - a), \tag{8.29}$$

$$S_i = \max(|q_1 - q_{b1}|, \ldots, |q_i - q_{bi}|, \ldots, |q_m - q_{bm}|),$$

$$(8.30)$$

where S_b denotes the standard model; k_i denotes the slope of segment i; W_i denotes the weighting factor; and $\alpha \in [0,1]$. The improved slope distance takes into account the effect of the trend change and time span at the same time. If $\alpha = 0$, Equation 8.30 becomes the traditional slope distance, which can reflect the trend change well and has a good noise tolerance.

The similarity Cr of different drilling anomaly standard patterns is obtained from calculating the signal sequences of the three warning parameters separately. The drilling pattern with the smallest value is the anomaly pattern of the signal at that moment.

8.2.4   *Early warning of spill risk based on trend characteristics*

When the parameter X is smaller than v_0, it does not deviate from the normal range, and 0 is used to indicate normal; when X is larger than v_1, it is abnormal, and 1 is used to indicate it; when the parameter X is between normal v_0 and abnormal v_1, it indicates that there is a certain probability of abnormality, and it is easy to think of mapping the interval (v_0, v_1) to $(0, 1)$ to indicate the level of parameter deviation from normal. The closer the value of the map is to 1, the further it is from the normal value, and the higher the probability of abnormality that it is. The closer the mapping value is to 0, the more likely it is that the parameter deviates slightly from the normal value and an abnormality is less likely to occur. The fuzzy logic method is based on this idea and calculates the probability of abnormality by judging the degree of deviation of the parameter from the normal level. The mathematical expression formula for calculating the abnormality of parameter X by the fuzzy logic method is

$$P(X) = \begin{cases} 0 & x < v_0 \\ (x - v_0)/(v_1 - v_0)^* & v_0 < x < v_1 , \\ 1 & x > v_1 \end{cases} \quad (8.31)$$

where v_0 is called the lower threshold limit of the parameter and v_1 is called the upper threshold limit of the parameter.

The occurrence of an anomalous event often requires the joint decision of multiple parameters. The contribution of each parameter to the anomalous event is expressed by the weight w_i. Assuming that the event A anomaly has a total of n identifying parameters, denoted as $\{x_1, x_2, \ldots x_n\}$, and the resulting probability of each parameter anomaly is p_i, then the probability of occurrence of the event A anomaly is

$$P(A) = (p_1, p_2, \ldots p_n) \cdot \begin{pmatrix} w_1 \\ w_2 \\ \vdots \\ w_n \end{pmatrix} = p_1 \cdot w_1 + p_2 \cdot w_2 + \cdots + p_n \cdot w_n.$$

$$(8.32)$$

Among them, the matrix composed of parameter weight values $(w_1, w_2, \ldots w_n)^T$ is called the fuzzy relationship matrix, which is usually denoted by $\boldsymbol{R}$. The vector $(p_1, p_2, \ldots p_n)$ composed of parameter abnormal probability values is called the abnormal sign domain $\boldsymbol{P}$.

In short, the fuzzy logic method calculates the probability of an event being abnormal based on the fuzzy vector and the fuzzy relationship matrix $\boldsymbol{R}$ after obtaining the domain of abnormal signs $\boldsymbol{P}$ (fuzzy vector) based on the parameter monitoring data and the threshold limit calculation.

In the overflow anomaly events occurring during drilling operations, warning indicators (mud pool volume, drilling fluid flow rate, riser pressure, drilling fluid outlet density, etc.) are parameters characterizing overflow anomalies. The trend eigenvalue α of the signal sequence of the warning parameter is the threshold for the probability of parameter anomalies. Since overflow accidents are gradual accidents, the diagnosis of the overflow is performed using the trend pattern recognition method. So, the threshold lower limit of all the warning indicators is 0. The threshold upper limit of different warning indicators is not only the same but the upper thresholds are related to the well location and drilling stratum, which are obtained through the statistics of drilling anomalies of neighboring wells. The fuzzy relationship matrix of the warning indicators is calculated by expert experience or the AHP hierarchical analysis method.

8.2.5 *Applications*

In this section, the proposed overflow early-warning method based on parameter trend pattern recognition is applied to the four-opening drilling process of well LD10 in Ledong, South China Sea. According to the drilling journal, when the LD10 well was drilled to 4023 m on March 21, the mechanical drilling speed was accelerated. The drilling cycle was stopped and observed, and the circulating gas measurement value of 13%–19% was found to be abnormal. Analysis shows that gas intrusion overflow occurred at this time, but the drilling daily report only notes that the abnormality occurred in the afternoon of that day and does not indicate the specific time of occurrence. Therefore, the three recording signal sequences of mud pool volume, riser pressure (pump pressure), and mud flow rate before and after about 20 minutes of drilling to 4023 m are selected as the drilling abnormality early-warning parameters. The early-warning method based on parameter trend is applied to determine the specific time of overflow occurrence.

8.2.5.1 *Trend pattern recognition of well recording parameters*

The integrated logger collected a set of data every 5 s, with a total of 250 sampling points for the selected time period. Some of the raw parameter data are shown in Fig. 8.16.

According to the trend of parameter curves, it can be seen that during the selected 20-minute drilling process, with the passage of time, the riser pressure first showed a smooth trend and then a stepwise decrease. What's more, the mud pool volume gradually increased, and the fluctuation of mud flow rate was large. And the overall trend of growth was small. From the a priori information of the expert library, it can be seen that gas intrusion overflow is suspected to have occurred during this period of time.

Based on the pattern recognition algorithm, the MATLAB programming language is used to perform wavelet noise reduction, normalization, and segmental approximation dimensionality reduction for the three parameters. Aiming at the noise characteristics of different parameters, differentiated wavelet noise reduction strategy is adopted: for the riser pressure signal with high signal-to-noise ratio, the db4 wavelet base is selected for 4-layer decomposition and its 4-layer approximation components are reconstructed; while for the mud pool volume and mud flow rate signals with large noise interference,

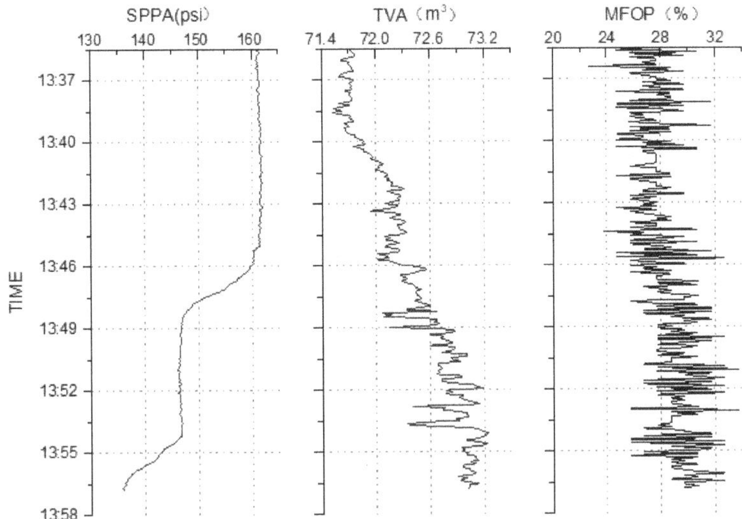

Fig. 8.16. Monitoring parameters at the onset of gas intrusion (SPPA pump pressure, TVA slurry pool volume, MFOP slurry flow).

the same wavelet base is adopted but the number of decomposition is increased to 5-layer decomposition, and the corresponding reconstruction of the 5-layer approximation components are carried out to achieve a more thorough noise suppression.

Figure 8.17 demonstrates the pattern recognition results after segmented approximate dimensionality reduction of the signal sequence for three groups of sampling points (with different time lengths) of the mud pool volume parameter. The signal sequences of 6.5 min, 14 min, and 20.5 min were selected for pattern recognition. Comparing the recognition results with the filtered signal sequences, it can be seen that the multi-segmented line, after dimensionality reduction, fits the curve change trend of the signal sequences very well. Comparing the overlapping sequence segments of the three groups of signals, the trend identified by the pattern has high consistency, which indicates that the length of the signal sequence has little influence on the similarity of the recognition results. So, the proposed pattern recognition method can extract the trend change characteristics of the signal sequence with good time continuity and time adaptability.

Based on the identification of a single early-warning indicator for the slurry pool volume, the parameter shows an increasing trend at 13:39, indicating that an overflow may have occurred. However,

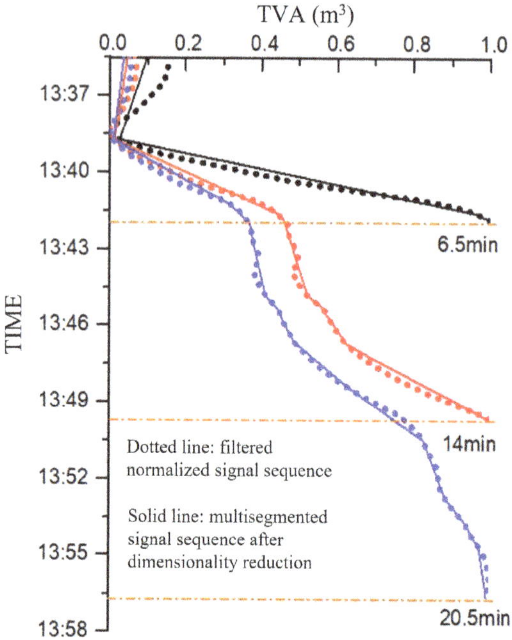

Fig. 8.17. Pattern recognition results for TVA signals of different time lengths.

affected by different flow data quality, the time point at which overflow is monitored may be very different. For example, the riser pressure parameter is more sensitive, which can be reflected by the decreasing parameter trend shortly after the gas intrusion occurs. But there are various reasons for riser pressure abnormality, and a single indicator is not sufficient to indicate that the abnormality is due to the overflow. However, the increase in the volume of the mud pool is the most accurate parameter for characterizing the overflow. The mud pool volume is often characterized by a hysteresis effect. Therefore, it is necessary to integrate the trend change patterns of multiple warning parameters to improve the reliability of the warning results.

Figure 8.17 identifies the pattern characteristics of the three parameters of riser pressure, mud pool volume, and mud flow rate under the 20-min (252 sampling points) logging data. According to the segmented approximate dimension reduction results, there are 7, 17, and 11 intermittent points searched for the three parameters (as shown in Fig. 8.18), and the filtered signal sequence

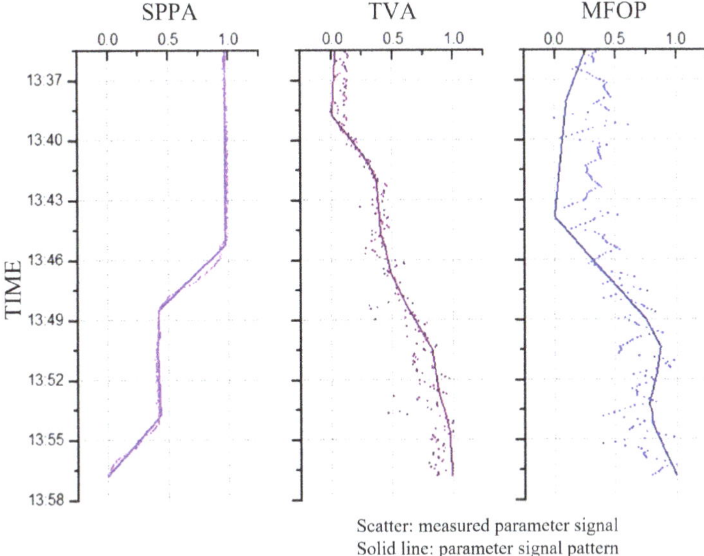

Fig. 8.18. Pattern recognition results for pump pressure, mud pond volume, and mud flow rate.

$Q\{q_{\mathrm{SPPA}}, q_{\mathrm{TVA}}, q_{\mathrm{MFOP}}\}$ is divided into 6, 16, and 10 segments of the primary linear signal sequence, respectively:

$$
P_{\mathrm{SPPA}} \left\{
\begin{array}{c}
p_1(13:36, 13:45), p_2(13:45, 13:48), \\
p_3(13:49, 13:50), \\
p_4(13:50, 13:50), p_5(13:50, 13:53), \\
p_6(13:53, 13:56)
\end{array}
\right\}, \quad (8.33)
$$

$$
P_{\mathrm{TVA}} \left\{
\begin{array}{c}
p_1(13:36, 13:39), p_2(13:39, 13:41), \\
p_3(13:42, 13:42), p_4(13:42, 13:43), \\
p_5(13:43, 13:45), p_6(13:45, 13:45), \\
p_7(13:46, 13:46), p_3(13:46, 13:47), \\
p_9(13:47, 13:48), p_{10}(13:48, 13:50), \\
p_{11}(13:50, 13:53), p_{12}(13:53, 13:54), \\
p_{13}(13:54, 13:54), p_{14}(13:54, 13:54), \\
p_{15}(13:54, 13:54), p_{16}(13:54, 13:56)
\end{array}
\right\}, \quad (8.34)
$$

$$P_{\text{MFOP}} \begin{cases} p_1(13:36,13:39), p_2(13:39,13:39), \\ p_3(13:39,13:44), p_4(13:44,13:49), \\ p_5(13:49,13:50), p_6(13:50,13:52), \\ p_7(13:52,13:53), p_3(13:53,13:53), \\ p_9(13:53,13:54), p_{10}(13:54,13:56) \end{cases}. \quad (8.35)$$

The shape eigenvalues and trend eigenvalues are extracted for the polyline sequence to obtain the feature sets K and A for the three parameters:

$$K_{\text{SPPA}} = \{1.1744e-05, -0.0146, 6.6473e-04, -0.0055,$$
$$6.1266e-04, -0.0125\}, \quad (8.36)$$

$$K_{\text{TVA}} = \begin{cases} -7.12e-04, 0.0106, 6.25e-03, 1.33e-03, \\ 1.19e-03, 5.13e-03, \\ 2.77e-03, 3.92e-03, 3.27e-03, 6.85e-03, \\ 8.39e-03, 2.20e-03 \\ 4.50e-03, 3.29e-03, 4.00e-03, 7.90e-04, 0.0106 \end{cases},$$
$$(8.37)$$

$$K_{\text{MFOP}} = \begin{cases} -5.51e-03, 3.07e-03, -1.46e-03, \\ 1.27e-02, 7.55e-03, \\ -2.05e-03, -3.47e-03, 4.89e \\ -031.43e-036.65e-03 \end{cases}, \quad (8.38)$$

$$A_{\text{SPPA}} = \{6.7291e-04, -0.8389, 0.8772, -0.3544, 0.3514, -0.7487\},$$
$$(8.39)$$

$$A_{\text{TVA}} = \begin{cases} -0.0408, 0.6493, -0.2507, -0.2818, -0.0080, 0.2260, \\ -0.1355, 0.0660, -0.0374, 0.2056, 0.0879, \\ -0.3547, 0.1319, -0.0691, 0.0409, -0.1842 \end{cases},$$
$$(8.40)$$

$$A_{\text{MFOP}} = \begin{cases} -0.3159, 0.4918, -0.2595, 0.8091, -0.2931, \\ -0.5495, -0.0818, 0.4791, -0.1980, 0.2991 \end{cases}. \quad (8.41)$$

8.2.5.2 *Early warning of overflow risks*

After obtaining the feature set of the parameter trend pattern $C = \{K, A\}$, the probability value of the overflow risk of this segment of the signal sequence is calculated using the fuzzy logic method, and when the probability exceeds 0.5, an alarm warning is given.

First, the same length of signal sequence is collected from the overflow recording data of the same layer in the neighboring wells for pattern recognition, the pattern eigenvalues are extracted, and the mean value of the statistically obtained eigenvalues is taken as the critical value of the anomaly, i.e., $K = [-0.015, 0.0108, 0.013]^T$, $A = [-0.84, 0.66, 0.82]^T$. The average value of the statistically obtained eigenvalues is taken as the critical value of the anomaly. The probability of abnormality of a single indicator at a given moment is calculated $p_i = \{k_i, a_i\}$.

Then, the weights of the warning indicators $w_{id} = [0.1738, 0.2686, 0.5576]$ and the weights of the eigenvalues $w_c = [0.4, 0.6]^T$ are given as a fuzzy relationship matrix. Calculate the probability value of overflow abnormality P_i at a certain moment of time with the fusion of multiple indicators. The result is shown in Fig. 8.19.

$$
P_i = [w_{\text{SPPA}}, w_{\text{TVA}}, w_{\text{MFOP}}] \cdot
\begin{bmatrix}
k_i^{\text{SPPA}} & a_i^{\text{SPPA}} \\
k_i^{\text{TVA}} & a_i^{\text{TVA}} \\
k_i^{\text{MFOP}} & a_i^{\text{MFOP}}
\end{bmatrix}
\cdot [w_k, w_a]
$$

$$
= w_{id} \times
\begin{bmatrix}
p_i^{\text{SPPA}} \\
p_i^{\text{TVA}} \\
p_i^{\text{MFOP}}
\end{bmatrix}. \tag{8.42}
$$

It can be seen that the calculated probability value is a step change. To 14:45, the integrated probability is more than 0.5. The length of probability value over the alarm limit is about 15 min, to 14:50, when the probability value reduces to less than 0.3. This long period of time exceeding the limit of the anomaly is judged to be the occurrence of overflow.

Analyzing the warning method, it can be seen that the process of anomaly identification and risk probability calculation is realized on the basis of the segmentation of the signal sequence. Therefore, the sensitivity of the warning method has a positive correlation with the degree of refinement of the segmentation of the signal sequence.

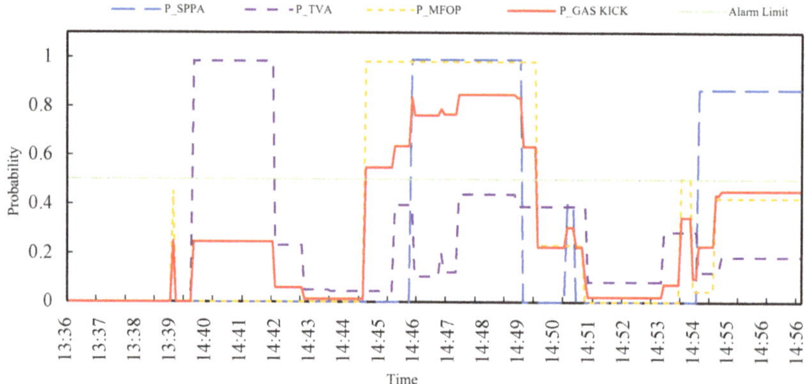

Fig. 8.19. Overflow anomaly probability values for multiple metrics fusion.

The finer the segmentation, the more sensitive the warning method, and at the same time, the higher the requirements for the data processing operation. The focus of the segmentation is to find the interruption point, which is related to the trend of the signal itself. Under the same filtering method, the larger the signal fluctuation, the more the interruption points, and the more the segments; on the contrary, the relatively smooth signal segments are fewer. Therefore, when applying this method to realize overflow warnings, the degree of filtering for each warning indicator can be set according to the demand of alarm sensitivity and the fluctuation characteristics of the parameter itself. For the parameter with a large fluctuation amplitude, the filtering is appropriately enhanced to make the signal sequence smoother. For the smoothly changing parameter, the filtering is weakened to retain the characteristics of the signal change.

8.3 Classification and Prediction of Stuck Drill Accidents Based on Generalized Fuzzy Neural Network Methods

8.3.1 *Generalized neural network prediction models*

GRNN prediction was proposed by American scholar Donald F. Specht in 1991; it is a variant form of a radial neural network and is an effective new feed-forward neural network. GRNN has strong non-linear mapping ability and a flexible grid structure, which can realize approximation of any continuous function with arbitrary accuracy. At the same time, it has high fault tolerance, robustness, and has good applicability in the solutions of nonlinear problems. Compared with the traditional radial-basis function neural network, GRNN has a stronger approximation ability and a faster learning rate. The lattice finally converges to the optimized regression surface with more aggregated data volume. In addition, when the amount of sample data is small and the data are unstable, better prediction results can also be obtained, so it has been widely used in various fields.

Nonlinear regression analysis is the mathematical and theoretical basis of GRNN implementation. With random variables x and y, their observations are X and Y, respectively. Then, the regression analysis of the non-independent variable Y with respect to the independent variable x is implemented by solving for y with the largest probability value. Let the joint probability density function of x and y be $f(x, y)$; then, the regression of y relative to X (also known as the conditional mean) is shown in Eq. (8.43):

$$\hat{Y} = E(y|X) = \frac{\int_{-\infty}^{\infty} y f(X, y) dy}{\int_{-\infty}^{\infty} f(X, y) dy}, \tag{8.43}$$

where $\hat{Y}$ is the predicted output of Y given the input as X.

The estimate $\{x_i, y_i\}_{i=1}^n$ of the density function $f(X, y)$ in Eq. (8.44) can be found by computing the non-parametric estimate of the sample dataset $\hat{f}(X, y)$:

$$\hat{f}(X, y) = \frac{1}{n(2\pi)^{\frac{s+1}{2}} \sigma^{s+1}} \sum_{i=1}^n \exp\left[-\frac{D_i^2}{2\sigma^2}\right] \exp\left[\frac{(X - Y_i)^2}{2\sigma^2}\right], \tag{8.44}$$

$$D_i^2 = (X - X_i)^T (X - X_i) \tag{8.45}$$

where X_i and Y_i are the observations of x and y, respectively. n is the number of samples. s is the dimension of x. σ is the smoothing factor. D_i^2 is the square of the Euclidean distance between the sample X and its corresponding variable input value X_i.

Replacing $f(X, y)$ with $\hat{f}(X, y)$ and substituting into Eq. (8.43) yields

$$\hat{Y}(X) = \frac{\sum_{i=1}^n \exp\left(\frac{-D_i^2}{2\sigma^2}\right) \int_{-\infty}^{\infty} y \exp\left[-\frac{(Y-Y_i)^2}{2\sigma^2}\right] dy}{\sum_{i=1}^n \exp\left(\frac{-D_i^2}{2\sigma^2}\right) \int_{-\infty}^{\infty} \exp\left[-\frac{(Y-Y_i)^2}{2\sigma^2}\right] dy}. \tag{8.46}$$

The grid output $\hat{Y}(X)$ is computed for two integrals:

$$\hat{Y}(X) = \frac{\sum_{i=1}^{n} Y_i \exp\left(\frac{-D_i^2}{2\sigma^2}\right)}{\sum_{i=1}^{n} \exp\left(\frac{-D_i^2}{2\sigma^2}\right)}. \tag{8.47}$$

From the calculations, it can be seen that the output of the network is a weighted average of all the sample observations Y_i, and the weight factor of each Y_i is the exponential form of the square of the Euclidean distance between the corresponding X_i and X. If σ is large, the output of the network is very close to the mean of the sample-dependent variable; if σ is very small and close to zero, the output of the network is very close to the training sample. Therefore, the value of σ should be reasonable in the calculation process, so that the training results can take into account the effects on the neighboring samples and all samples.

8.3.1.1 *Structure of generalized neural networks*

GRNN is structurally more similar to a radial-basis function neural network with four layers, which are the input layer, pattern layer, summation layer, and output layer:

(1) Given an input X, the number of neurons in the input layer is equal to the dimension of X. Neurons of simple distribution units are passed to the pattern layer via the input layer.
(2) The pattern layer takes the number of samples n in the input X as the number of neurons in the layer, each corresponding to a different sample, and its delivery results in

$$p_i = \exp\left(\frac{-D_i^2}{2\sigma^2}\right), \quad i = 1, 2, \ldots, n. \tag{8.48}$$

(3) The summation layer is for the summing calculation of the two neurons it contains.

The first category is to sum the outputs of the neurons of all the pattern layers, i.e., the denominator of Eq. (8.47), with a weight of 1 between the pattern layer and each neuron, and a transfer function of

$$S_D = \sum_{i=1}^{n} P_i. \tag{8.49}$$

The second category is a weighted sum of the outputs of all neurons in the pattern layer, i.e., the numerator of Eq. (8.47), where the weights between the ith neuron in the second layer and the jth neuron in the third layer are the jth element in the output sample Y_i, and the transfer function is

$$S_{Nj} = \sum_{i=1}^{n} y_{ij} P_i, \quad j = 1, 2, \ldots, k. \tag{8.50}$$

(4) In the output layer, the number of neurons is the dimension k of the output data vector in the learning sample, and each neuron divides the two classes of outputs from the summation layer to take the jth element of $\hat{Y}(X)$ as the output of neuron j:

$$y_j = \frac{S_{Nj}}{S_D}, \quad j = 1, 2, \ldots, k. \tag{8.51}$$

8.3.1.2 *MATLAB implementation of generalized neural networks*

The GRNN implementation in MATLAB consists of a three-layer grid structure with an input layer, a radial base layer, and a linear output layer:

(1) *Input layer:* The grid input is X (s-dimensional) and the number of training samples is q.
(2) *Radial base implicit layer:* The number of training samples q is the number of units in this layer, and the Euclidean distance metric function *dist* is used as the weight, as shown in Eq. (8.52):

$$\|dist\|_j = \sqrt{\sum_{i=1}^{n} (x_i - LW_{j,i})^2} \quad (j = 1, 2, \ldots, q), \tag{8.52}$$

The *dist* function is used to solve for the distance between the mesh input X and the weights $LW_{1,1}$ of the radial basis implicit layer. $LW_{1,1}$ is a $q \times s$ order matrix. b_1 is the implicit layer threshold, set to $0.8326/A$. The size of the threshold is usually changed by varying the value of A. The result n_1 of multiplying the threshold with the weight input is input to the radial basis function (radbas) of the implied layer through the network basis function neprod. Usually, a Gaussian function is used as the transfer function of the implied layer, then the output of the implied layer is as shown in Eq. (8.53):

$$a'_j = radbas[netprod(\|dist\|_j \cdot b_{1j})] = \exp\left[-\frac{(n'_j)^2}{2\sigma^2}\right]$$

$$= \exp\left[-\frac{(\|dist\|_j \cdot b_{1j})^2}{2\sigma^2}\right], \tag{8.53}$$

where σ is the smoothing factor and the basis functions flatten out as σ increases.

Since the Gaussian function is a nonlinear local distribution function with a value domain greater than or equal to 0, it tends to decay symmetrically about the center, and the hidden layer output increases as the distance of the input signal from the center of the basis function decreases. This local approximation capability makes this network learn at a fast rate.

(3) *Linear output layer*: The network layer uses the nprod weight function, which computes weights as follows: first, it performs element wise dot product between the hidden layer output and weight matrix $LW_{2,1}$, then divides the result by the sum of vector a1 elements. Finally, the ratios are processed by the purlin transfer function to generate network output. The transfer function's input characteristics are shown in Fig. 8.20.

8.3.2 *Card drill classification model*

8.3.2.1 *Various signs of jamming occurrence*

(1) Borehole plugging precursors from viscous adsorption

(i) The prerequisite for the occurrence of viscous suction jamming is that the drilling tools are stationary for a period of time ranging from several minutes to several tens of minutes. The specific duration is related to the drilling fluid system, the structure of

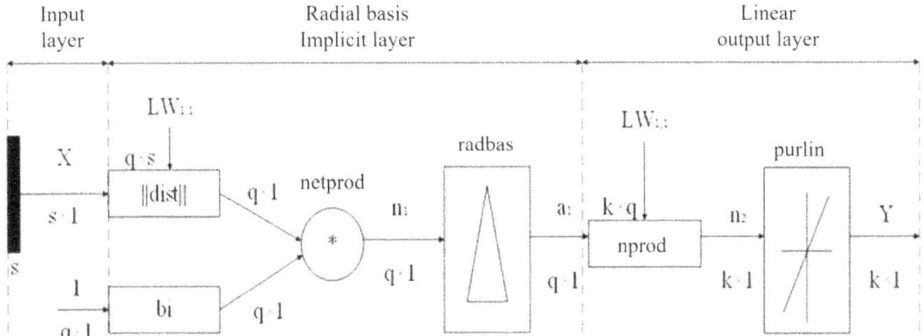

Fig. 8.20. Schematic of GRNN MATLAB implementation.

the lower drilling tools, and the borehole conditions. Differential pressure jamming occurs when normal friction is removed and the tool can be lifted, lowered, and rotated without resistance.

(ii) The part of the drill where stick-suction jamming occurs will not be the drill bit, and the most likely part to encounter obstruction is the drill collar with a large contact surface with the well wall, followed by the drill pipe.

(iii) After the occurrence of viscous suction jammed drilling, the drilling fluid is circulated as usual, the amount of drilling fluid returned is stabilized, and the riser pressure basically remains unchanged.

(iv) After the occurrence of the sticky suction jammed drill, if it is not handled in time and not immediately active, then the jam point may move up until near the casing shoe.

(2) Signs of collapsed and stuck drill

(i) Slight slumping during the drilling will make the drilling fluid performance unstable. Its viscosity, density, and sand content will increase, and there will be a large amount of angular flaky rock chips in the return volume. The location of the collapse is divided into the positive drilling formation and the upper part of the positive drilling formation. In the former case, it will be difficult to drill. The riser pressure and rotary torque will increase during the drilling process. The pressure of the riser of the lifting tool will be reduced to normal, but the drill bit will not be able to be lowered to the bottom of the well. In the latter case, the pressure in the riser increases. The pressure in the riser does not decrease when lifting the drilling tools. The drilling tools are blocked, and the amount of drilling fluid returned is reduced or even not returned.

(ii) If the well collapses during drilling, the amount of wellhead return decreases or does not return after entering the collapsed layer, and the drilling fluid may be back-sprayed through the drill pipe. In the drilling process, the dispersed wall slump rock chips can be kept in the transportation state under the buoyancy of the drilling fluid to avoid plugging, while the concentrated slump produces a high on centration of rock chip clusters that are prone to form flow obstacles, leading to the occurrence of wellbore plugging. The riser pressure is not abnormal when the drill bit is down in the collapsed layer before the pump is open. The riser pressure rises and the suspended weight decreases when it bites down into the collapsed layer after the pump is opened. The drill bit is lifted away from the collapsed layer and then is back to normal. The resistance and rotary torque do not increase significantly during the process of drilling down the collapsed hole, but the riser pressure and the amount of return from the wellhead fluctuate greatly. Sometimes, the riser pressure suddenly rises and the weight of the suspension decreases, and sometimes the amount of return decreases to the point of cutting off the flow. The return of the rock debris is either due to the newly collapsed prismatic rock or the non-prismatic rock that had been milled for a long time.

(iii) Failure to fill drilling fluid in time or an insufficient amount of drilling fluid during drilling and the occurrence of well leakage may lead to well collapse during drilling. When the wall collapses during drilling, the up-and-down drilling operation will face a significant jamming problem: the resistance to the drilling tools shows a large fluctuation and is difficult to be stabilized, but the trend of change is generally monotonically increasing; at the same time, the rotating torque of the tools is also continuously increasing. The pressure of the open pump riser increases and the suspended weight falls. There is back pressure when the circulation stops, and the starting drilling fluid is back-blasted through the drill pipe. The amount of wellhead return decreases or even does not return.

(iv) In addition to shrinkage-type stuck drilling caused by rock creep (this type of stuck drilling can usually be lifted by itself after one complete drilling cycle), the stuck drilling accidents induced by well wall collapse are often accompanied by drilling interruptions and abnormally high pump pressures, which are significantly more difficult to deal with.

(3) Signs of sand bridge stuck drill

(i) Sand bridge jamming occurs at the beginning of the soft encounter resistance without a specific jamming point. The size of the resistance is directly proportional to the depth of the drilling tool. When the weight of the drilling tool increases to the point that the resistance of the sand bridge is canceled out, the depth of the drilling tool will not have an effect on the suspended weight.

(ii) After the drill bit is lowered into the sand bridge, the drilling fluid cannot be returned due to the sand bridge, resulting in reduced or even no return at the wellhead. The drilling fluid is either squeezed into the soft formation or can only be spewed back out through the drill pipe.

(iii) If sand bridge blockage occurs during drilling, the liquid level in the wellbore annulus does not decrease, but the liquid level in the water eye decreases rapidly.

(iv) After the drilling tool enters the sand bridge, the circulation stops and the drilling tool can be freely lifted up, lowered, and rotated. When turning on the pump, the riser pressure increases. The suspended weight decreases, and the wellhead does not return.

(v) During drilling, if the sand-carrying capacity of the drilling fluid is poor and the pump displacement is small, the drilling tools can be lifted up, lowered, and rotated freely with the pump on. But the drilling tools cannot be lifted up when the pump is stopped. This phenomenon is especially common in non-solid-phase mud.

(vi) When gas drilling, the amount of gas returned is reduced or not returned and accompanied by the return of wet mud mass. The pressure of the riser rises, and there is resistance to starting and stopping the drilling.

(4) Signs of shrinkage and drill jamming

(i) Due to lithology and other reasons, the well section with small well diameter is relatively fixed, so the depth of the well at the point of obstruction is basically fixed at one point or several points. Only one of the uplift and downlift processes encounters obstruction.

(ii) Most of the jamming occurs when the drilling tools are lifted up and lowered down, not when they are stationary. The obstruction of drilling in the creeping rock layer and water-bearing soft mud layer is likely to occur in the drilling process.

(iii) The pressure of the circulating riser is normal. The inlet and outlet flow rates are stable, and the performance of the drilling fluid is basically stable. However, the pressure of the riser gradually increases when drilling the rock formation with strong fluidity, blocks the annulus, and cannot circulate.

(iv) Leaving the jammed position, the drill bit can be lifted up, lowered, and rotated normally, but it is difficult to rotate with a slight increase in resistance.

(v) Jamming near the bottom of the well when drilling down is either caused by sinking sand at the bottom of the well or drilling a small borehole when the diameter of the drill bit becomes smaller due to wear and tear from prolonged use.

(vi) Drilling in creeping formations usually results in faster drilling speeds, higher torque, and drilling hold-up, making it difficult to lower the drill bit and cut a hole. And the pump hold-up gets worse as the creep rate increases.

(vii) The point of obstruction is a large-diameter tool, probably a drill bit but not a drill pipe or collar.

(5) Keyway jam drilling signs

(i) The large outer-diameter part cannot enter the keyway when drilling down. Therefore, the keyway will only be jammed during drilling. The weight of the drill column itself will cause it to tilt to the side of the keyway, and the larger-diameter part will be jammed if it enters the keyway.

(ii) The jammed part is a tool with a large outer diameter, such as a drill bit and a drill collar with a larger diameter than the joint of the drill pipe. When it is lifted up and touches the keyway, it will be blocked.

(iii) The jamming point of keyway is related to the lithology of the formation, well diameter, and borehole trajectory. In the case of uniform lithology and good borehole quality, the position of the starting jamming point shifts downward with the number of times of obstruction. On the contrary, in the case of mud shale sandstone interactive stratum and large change of well inclination, only the sandstone layer is scraped and pulled by the joint of the drill pipe. The jamming point is basically the same every time the drill is started.

(iv) It is difficult to rotate the turntable if the drilling tool is lifted up with extra force, but it can rotate freely if it is lowered and left. The pressure of the standpipe is unchanged when the pump is turned on. The performance of drilling fluid is stable, and the inlet and outlet flow rates are stabilized.

(6) Signs of mud pack jamming

(i) During the drilling process, the drilling speed gradually decreases and the torque gradually increases, and there may be a hold-up situation. If the drill bit or the corrector mud pack is jammed, the drilling fluid circulation path is reduced and the riser pressure increases.

(ii) There is resistance to uplift the drill bit, and the amount of resistance is determined by the severity of the mud pack.

(iii) The liquid level in the annulus of the starting wellhead drops slowly, or even does not drop, or overflows while starting drilling. The liquid level cannot be observed in the drill pipe. When mud pack jamming occurs, the drilling tools can still be raised and lowered within a certain amount of resistance and well diameter, but the resistance gradually becomes larger with the uplift process. Jamming will be encountered in the section of the well with a smaller diameter.

(7) Signs of a falling object jamming the drill

(i) When drilling, the drill will be held up when there is a falling object in the ring and the uplift will be blocked. A small falling object may fall by lifting, but a large falling object may be stuck by strong lifting.

(ii) The drill will be suddenly blocked when it meets the falling object. If the position of the falling object remains unchanged, the blocked position will remain unchanged. It is easy to lower the drill as long as it is not forced. If the position of the falling object changes when the drill is started, the drill can only be lowered, and the point of obstruction moves downward during the process. It is easy to rotate the drill without obstruction, but difficult to rotate the drill with obstruction.

(iii) If the drop is small, the blocked position is usually the large OD tool. If the drop is large, the blocked position may also be the joint of the drill pipe.

(iv) Opening the pump can make the fluid circulate normally. The riser pressure remains unchanged. The inlet and outlet flow rates are stabilized, and the performance of the drilling fluid is stable.

8.3.2.2 *Classification of jammed drills*

The results of FCM clustering depend largely on the selection of random initial clustering centers. Since the initial clustering centers will change each time the detection is run, the accuracy of the clustering results will change and the results of classification analysis using clustering are unstable. In addition, when the number of data dimensions is too large, the amount of data is too large, and the difference between the data is not obvious, the FCM clustering can easily fall into the local extremes, part of the card drilling data cannot be correctly classified, and the accuracy of the processing of the data decreases, increasing the processing time. Therefore, this card drill classification uses FCM clustering combined with generalized neural network regression to propose a fuzzy generalized neural network classification algorithm; the implementation steps of this model are as follows:

(1) The preprocessed multi-parameter card drill data were subjected to FCM clustering, where the input data were classified into m classes and the cluster centers and individual fuzzy affiliation matrices $\boldsymbol{u}$ were obtained for each class of card drill data.

(2) Optimization of training data for the GRNN network, i.e., for the results of FCM clustering is carried out. The 30 sets of samples closest to the center of each class are selected as the sample data

for the GRNN network. The selection process is to first find the intra-class mean for each class of data in the FCM clustering results $mean_i$ ($i = 1, 2, \ldots, m$). Then, find the distance from the intra-class mean for all sample data in each class $ecent_i$ ($i = 1, 2, \ldots, m$), select the n samples with the smallest distance in the distance matrix as the GRNN training data for that class, and set their corresponding outputs as i. Finally, we get the $m \times n$ groups of training data, where the input data are the parameter data of different card drills types, and the output data are the classes of the card drills.

(3) GRNN network training is conducted using $m \times n$ sets of training data.

(4) Prediction of output Y using GRNN network training results is done for all input sample data X.

(5) GRNN network prediction will reclassify the card drill data into m classes, and then use the training sample reselection module to find out the closest samples to the center value of each class as the training samples. The selection process is the same as in step 4, but the sample reselection is performed on the GRNN network training results.

(6) Return the reoptimized training data to the GRNN network for training and classify all samples. Iterate the operation repeatedly until the classification results output before and after are consistent to stop the iteration. Output the classification results. The specific implementation process of the card drill accident classification model based on a fuzzy generalized neural network is shown in Fig. 8.21.

8.3.3 *Applications*

Based on the analysis of drilling dailies and logging data, the parameters selected to characterize the stuck drill were large hook load, rotary torque, rotary speed, relative outlet flow rate, and riser pressure.

Firstly, the raw data of three kinds of jam drills were extracted, including 30 sets of mechanical jam drill data, 30 sets of viscous suction jam drill data, and 30 sets of circulation jam drill data.

Next, normal data were selected for the depths of wells adjacent to the location of the corresponding well where the stuck drill occurred.

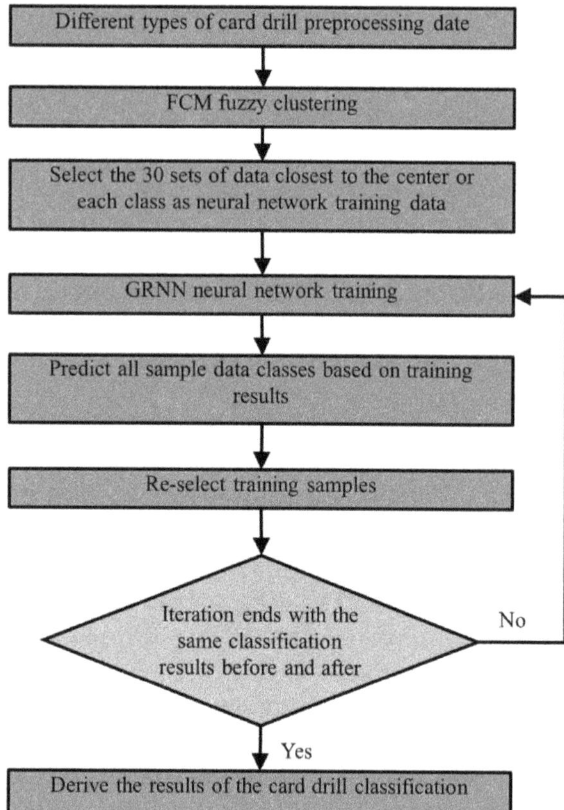

Fig. 8.21. Generalized fuzzy neural network card drill accident classification model flow.

Parameter data were averaged for the normal start-and-stop drilling or cycling period.

Finally, in order to avoid the influence of drilling depth and working conditions on the data, the difference between the raw data during the stuck drilling period and the average of the normal data of the corresponding wells with neighboring depths was taken, and these data were used for clustering.

This can avoid the influence of different well conditions on the data and also can reflect the trend of the data of different stuck drilling conditions relative to the normal data through the difference. The average value and the data after the difference are shown in Tables 8.4–8.5.

Table 8.4. Mean values of normal parameter data of proximity depth for each type of stuck drilling occurrence.

Well number	Well depth avg (m)	Drill depth avg (m)	Hook load avg (t)	Turntable torque avg (kN · m)	Turntable speed avg (r/min)	Relative outlet flow rate avg (%)	Riser pressure avg (MPa)
YC2641	3216	Circa 2684	50.73	−0.033	0	17.5	0.25
YC2721	3468	Circa 3210	103.59	0	0	0	0.42
YC2641	3216	Circa 2825	49.92	0.262	13.3	9.8	0.024

Table 8.5. Selected data on the difference between the parameter data and the average value for each card drill type.

Well number	Well depth ds (m)	Drill depth ds (m)	Hook load ds (t)	Turntable torque ds (kN · m)	Turntable speed ds (r/min)	Relative outlet flow rate ds (%)	Riser pressure ds (MPa)	Type of jammed drill
YC2641	3216	2677.03	101.46	0.0025	0	−4.5	−0.00134	1
YC2641	3216	2677.51	103.68	0.0025	0	−4.5	−0.03134	1
YC2641	3216	2674.05	106.66	0.0025	0	−7.5	−0.00134	1
YC2641	3216	2673.78	111.40	0.0025	0	−3.5	0.22866	1
YC2641	3216	2673.60	106.81	−0.001	0	0.5	0.20866	1
YC2641	3216	2676.69	98.78	−0.0075	0	−1.5	−0.004	1
YC2721	3468	3210.62	62.85	0	0	0	−0.114	2
YC2721	3468	3209.88	62.71	0	0	0	−0.114	2
YC2721	3468	3208.28	62.18	0	0	0	−0.114	2
YC2721	3468	3207.24	63.98	0	0	0	−0.114	2
YC2721	3468	3205.69	71.61	0	0	0	−0.114	2
YC2721	3468	3204.24	75.93	0	0	0	−0.114	2
YC2641	3216	2824.97	109.59	1.218	−11.3	−9.8	0.386	3
YC2641	3216	2824.97	108.83	0.988	−11.3	−9.8	0.376	3
YC2641	3216	2824.97	111.31	0.988	−11.3	−9.8	0.386	3
YC2641	3216	2825.03	111.91	1.258	−11.3	−9.8	0.386	3
YC2641	3216	2825.06	110.35	0.828	−11.3	−9.8	0.396	3
YC2641	3216	2825.08	112.62	1.088	−11.3	−9.8	0.376	3

In Table 8.4, avg stands for average. Stuck drill types 1, 2, and 3 in Table 8.5 represent viscous suction stuck drills, mechanical stuck drills, and circulation stuck drills, respectively; ds represents the difference.

8.3.3.1 *FCM clustering results*

By calculating the fuzzy affiliation degree, the center value of each category, and the objective function value for the three major

Table 8.6. FCM clustering results for each card drill category.

Type of jammed drill	Clustering for viscous suction jamming	Cluster for mechanical jamming	Clustering for cyclic jamming
Actual viscous suction jamming	26	0	4
Actual mechanical jamming	5	25	0
Actual cyclic jamming	0	0	30

Table 8.7. FCM clustering accuracy by card drill category.

Type of jammed drill	Correct rate (%)
Viscous suction jamming	86.7
Mechanical jamming	83.3
Cyclic jamming	100

categories of card drilling data, after the end of the iteration, the position of the largest element in each column of the fuzzy affiliation matrix is used to determine the category to which the element belongs, and the results of the calculation are shown in Table 8.6.

From the clustering results in Table 8.7, it can be seen that FCM clustering is able to categorize the sticky suction card drills and mechanical card drills with circulation card drills. However, the classification between mechanical card drills and sticky suction card drills is not good, and the clustering process will judge more mechanical card drill data as sticky suction card drills.

8.3.3.2 *Classification results of GRNN network prediction combined with FCM clustering*

According to the results of FCM clustering for sample optimization, training, and prediction, the predicted values are obtained as shown in Fig. 8.22, from which it can be seen that the predicted result values are very close to the actual values.

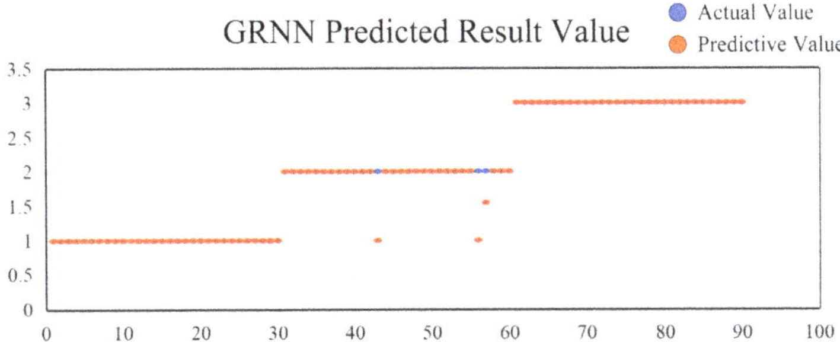

Fig. 8.22. GRNN predicted outcome values.

Table 8.8. GRNN prediction errors.

Predicted data category	Average relative error	Maximum relative error
Viscous suction jamming	0	0.001
Mechanical jamming	0.06	0.5
Cyclic jamming	0	0

Table 8.9. Generalized fuzzy neural network card drill classification results.

Type of jammed drill	Clustering for viscous suction jamming	Cluster for mechanical jamming	Clustering for cyclic jamming
Actual viscous suction jamming	30	0	0
Actual mechanical jamming	2	28	0
Actual cyclic jamming	0	0	30

The FCM clustering results were subjected to sample optimization and the GRNN training test was able to predict the card drill types well, as shown in Table 8.8. The three card drill types were classified based on the predicted values and the classification results are shown in Table 8.9.

As can be seen from the classification results in Table 8.10, the results of card drills classification based on the generalized fuzzy

Table 8.10. Generalized fuzzy neural network card drill classification accuracy.

Type of jammed drill	Correct rate (%)
Viscous suction jamming	100
Mechanical jamming	93
Cyclic jamming	100

neural network effectively improve the resolution of viscous suction card drills and mechanical card drills compared to the use of FCM clustering alone for the discrimination of card drill categories. And they increase the original correct rate of classification of viscous suction card drills from 86.7% to 100%, and increase the correct rate of classification of mechanical card drills from 83.3% to 93%. It shows that the process of optimizing the sample data by using the FCM clustering results is good at singling out the data samples with more characteristics of various types of card drills and separating the three types of card drill data with high speed and accuracy by the high efficiency and stability of GRNN.

8.4 Prediction and Early-Warning Methods of Gas Hydrate Generation in Drilling Encounters

The temperature and pressure field in the annulus of the water separator near the seafloor of a deepwater well is affected by the heat transfer from the seafloor at a low temperature and is characterized by low temperature and high pressure. The seafloor temperature at a depth of 1,500 m in the South China Sea is 3–4°C. When hydrocarbon gas overflow occurs in the annulus of the wellbore using water-based drilling fluid, the intruding hydrocarbon gases, especially methane gas, are highly susceptible to the regeneration of natural gas hydrate crystals in the annulus of the low-temperature and high-pressure section of the well and the solid hydrate particles formed under the condition of pressure fluctuation at the bottom of the well. After a period of aggregation and volume increase, the hydrate will hinder the return of drilling fluid or even block the annulus of the wellbore, seriously threatening the safety of drilling operations. In order to effectively prevent and control the hazards of gas hydrate during drilling, it is necessary to predict and analyze the scope and extent of hydrate formation in the wellbore annulus.

In this section, a numerical simulation method for gas hydrate prediction in the annulus of the deepwater wellbore is established based on the CSMHyK hydrate dynamics model using a development well in a gas field in Lingshui, South China Sea, as an example; the range of the well section with hydrate generation, the rate of hydrate generation, and the amount of hydrate generation are predicted for a number of working conditions. The sensitivity of four drilling parameters, namely, gas intrusion pressure, drilling fluid temperature, density, and displacement, was analyzed, and the role of different drilling parameters on hydrate formation was obtained. Based on the results of model calculation, several characteristic values characterizing hydrate formation were extracted, and supercooling density, hydrate generation section length, total hydrate generation, and peak generation rate were selected as risk evaluation indicators; the weights of the indicators were assigned based on the idea of fuzzy comprehensive evaluation to realize the quantitative comparison of the risk level of hydrate formation in wellbores under different working conditions.

8.4.1 *Natural gas hydrate formation mechanism and phase equilibrium conditions*

When gas overflow occurs in the deepwater drilling process, shallow gas, primary hydrate decomposition gas, and reservoir gas within the formation enter the wellbore annulus by means of pressure gas intrusion, replacement gas intrusion, etc., and then there is an adequate hydrocarbon gas source within the annulus; the water-based drilling fluid provides a large amount of free water, and the flow of the drilling fluid and the disturbance of the gas flow under the pressure of gas intrusion during the circulation period provide the auxiliary dynamic conditions for pressure fluctuations. In the high-pressure, low-temperature well section near the seafloor, free gas and water molecules combine to form hydrate nuclei at the gas–liquid interface and on the surface of gas bubbles in the circulation channel, and the nuclei grow to become hydrate particles that build up in the roughness of the tubing wall and at locations such as elbows and congregate to form plugs.

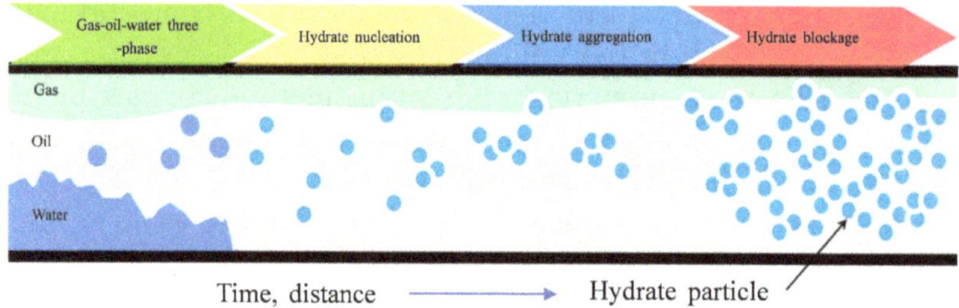

Fig. 8.23. Schematic diagram of natural gas hydrate formation blockage.

Experiments found that the degree of hydrate blockage circulating channel and hydrocarbon gas transport in the channel by the obstruction of the strength of the gas transport process is very weak, it is difficult to have hydrate aggregation to form a blockage; on the contrary, when the flow resistance increases, hydrate nuclei are easily formed locally and continue to grow, which ultimately leads to plugging. It is worth noting that the hydrate deposits formed by free gas in the wellbore annulus are usually characterized by inhomogeneous porous media, and their spatial distribution has obvious anisotropy. Figure 8.23 is a schematic diagram of natural gas hydrate formation blockage.

The hydrate phase curve describes the hydrate phase equilibrium condition, i.e., the critical pressure-temperature values for hydrate generation from specific components of hydrocarbon gases, which is its inherent property and is not affected by changes in environmental conditions. The hydrate phase equilibrium condition is the main basis for predicting hydrate generation in the wellbore annulus. According to the hydrate formation mechanism, it is known that the phase equilibrium conditions of natural gas hydrates are mainly related to the components of natural gas, and a gas field in Lingshui is located in the central canyon of the Lingshui depression in the Qiongdongnan basin. The reservoir fluid components of the gas field obtained by sampling are shown in Table 8.11.

Based on the natural gas component data, combined with the formation water mineralization and water–gas ratio data, PVTsim software was applied to analyze the natural gas component, the water component was processed by using the EOS equation of state, and the natural gas hydrate phase equilibrium conditions of this component were numerically calculated to obtain the hydrate phase curve and

Table 8.11. Temperature and pressure of gas hydrate formation in a gas field at Lingshui gas field.

Temperature/°C	Pressure/bar	Temperature/°C	Pressure/bar
2.00	12.10	18.28	111.54
4.00	15.39	19.45	141.54
6.00	19.58	20.39	171.54
8.00	24.98	21.47	211.54
10.00	32.06	22.20	241.54
12.00	41.54	23.09	281.54
14.81	61.54	23.92	321.54
16.58	81.54	26.94	491.51

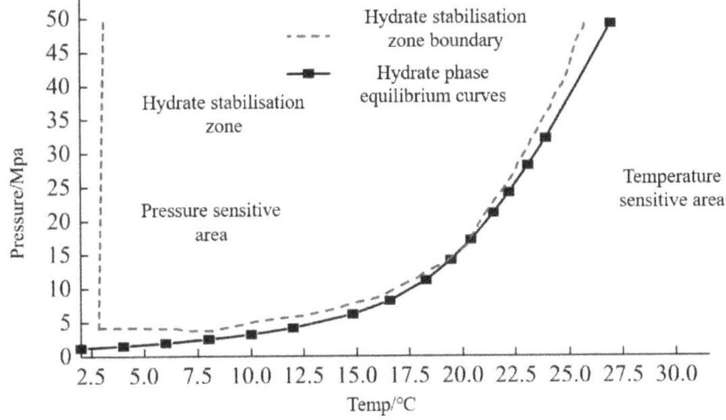

Fig. 8.24. Phase equilibrium curve of natural gas hydrate in a gas field in Lingshui.

the prediction data of hydrate phase equilibrium temperature and pressure, as shown in Fig. 8.24.

The region to the left of the hydrate phase equilibrium curve is the stable region for hydrate formation, and if the temperature and pressure are in this region, there is a risk of solid hydrate formation. Analyzing the hydrate phase equilibrium curves, it can be seen that the correlation between temperature and pressure of hydrate formation grows with an approximate exponential function, and the low temperature is more favorable for the stable existence of hydrate compared with the high-pressure condition.

Under high-pressure conditions, the hydrate phase equilibrium is more sensitive to temperature, and a slight increase in temperature can cause the decomposition of solid hydrates; under low-pressure conditions, the hydrate phase equilibrium is more sensitive to pressure, and a slight increase in pressure can cause the combination of natural gas and free water to form solid hydrates. Accordingly, the temperature-sensitive and pressure-sensitive zones of hydrate phase equilibrium can be roughly classified. In the temperature-sensitive zone, increasing the temperature field in the wellbore annulus is more effective in preventing the regeneration of hydrate in the wellbore, whereas in the pressure-sensitive zone, lowering the pressure field in the wellbore annulus can help to reduce the risk of regeneration of hydrate.

According to the low-temperature and high-pressure characteristics of deepwater operations, the water separation pipe section is affected by the low temperature of deepwater, has a low temperature in the annulus, and is in the pressure-sensitive zone of hydrate phase equilibrium, so lowering the density of drilling fluids can reduce the hydrostatic pressure, thus reducing the risk of hydrate regeneration in the annulus.

8.4.2 *Deepwater wellbore gas hydrate prediction modelling*

OLGA is the world's leading full-dynamic multiphase flow simulation and computation program for modelling the movement of oil, gas, and water in oil wells and pipelines. The hydrate dynamics module in it uses the CSMHyK kinetic model and multiphase flow equations to solve for the location of the pipeline section that satisfies the hydrate phase equilibrium condition and the hydrate generation. In this section, based on the kinetic theory of hydrate formation, a hydrate prediction model for the deepwater wellbore annulus is established with the help of OLGA software to numerically simulate the scope and extent of hydrate generation and predict the risk of hydrate formation in the annulus.

The integration of the CSMHyK computational model in the OLGA software is shown in Fig. 8.25.

8.4.2.1 *Theoretical model for predicting gas hydrate formation*

The CSMHyK kinetic model uses the first-order kinetic equation with adjustable rate constants Eq. (8.54) proposed by Boxall to calculate

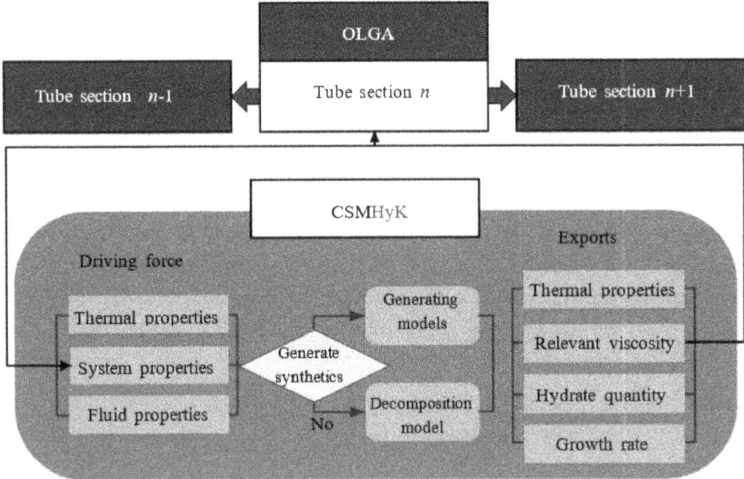

Fig. 8.25. CSMHyK computational model integration in OLGA.

hydrate growth rates:

$$\frac{dm_{\text{gas}}}{dt} = -uk_1 \exp\left(\frac{k_2}{T_{\text{sys}}}\right) A_s(T_{\text{hyd\_eq}} - T_{\text{system}}). \tag{8.54}$$

The kinetic equation gives the calculation of the gas consumption rate dm_{gas}/dt during hydrate formation, where k_1 and k_2 denote the intrinsic rate constants, taking values based on the experimental data obtained by Englezos *et al.* for measuring the gas rates consumed by methane and ethane hydrates at different temperatures, which do not take into account the resistance to mass and heat transfer, $k_1 = 7.3548 \times 1017\,\text{kg}/(\text{m}^2 \cdot \text{s} \cdot \text{K})$, $k_2 = -13600\,\text{K}$; A_s denotes the surface area between the water molecule and the hydrocarbon phase, m^2/m; $T_{\text{hyd\_eq}} - T_{\text{system}}$ denotes the degree of supercooling as a driving force for the formation of the hydrates, °C; $T_{\text{hyd\_eq}}$ denotes the hydrate phase equilibrium temperature; T_{system} denotes the system temperature; and u denotes the scaling factor for the transport resistance, taking the value $1/500$.

After the formation of hydrate particles, the size of the hydrate particle clusters was calculated using a relational model based on the steady-state equilibrium of inter-particle cohesion and shear proposed by Camargo and Parmo, assuming that the inter-particle cohesion is a constant, and the diameter of the hydrate particle clusters was calculated by employing the nonlinear equation Eq. (8.55):

$$\left(\frac{d_A}{d_P}\right)^{4-f} = \frac{F_a[1 - (\phi/\phi_{\max})(d_A/d_P)^{3-f}]^2}{d_P^2 \mu_0 \gamma [1 - \phi(d_A/d_P)^{3-f}]}, \tag{8.55}$$

where d_A is the hydrate particle cluster diameter, m; d_P is the hydrate particle diameter, m; ϕ is the hydrate particle volume fraction; $\phi_{\max}$ represents the maximum volume fraction in which particles can aggregate, which is assumed to be 4/7; f is the fractal dimension, which is assumed to be 2.5; F_a is the inter-particle cohesion, which is assumed to be 50 Mn/m; μ_0 is the oil phase viscosity, cp; and γ is the shear rate, 1/s.

Based on the fractal structure of the polymer, the effective volume fraction of hydrate particle aggregates, ϕ_{eff}, is as shown in Eq. (8.56):

$$\phi_{\mathrm{eff}} = \phi \left(\frac{d_A}{d_P}\right)^{3-f}. \tag{8.56}$$

The relative viscosity μ_r of the hydrate slurry was calculated using the modified Mills model Eq. (8.57), which takes into account the effective volume fraction of hydrate particle aggregates:

$$\mu_r = \frac{1 - \phi_{\mathrm{eff}}}{[1 - (\phi_{\mathrm{eff}}/\phi_{\max})]^2}. \tag{8.57}$$

The Van der Waals–Platteeuw molecular thermodynamic model was used to calculate the hydrate phase equilibrium conditions Eq. (8.58):

$$\frac{\Delta\mu_0}{RT_0} - \int_{T_0}^{T} \frac{\Delta h_0 + \Delta c_p(T - T_0)}{RT^2} dT + \int_{p_0}^{p} \frac{\Delta V}{RT} dp$$

$$= \ln\left(\frac{f_w}{f_w^0}\right) - \sum_{i=1}^{2} v_i \ln\left(1 - \sum_{j=1}^{Nc} \theta_{ij}\right). \tag{8.58}$$

Among them,

$$\ln\left(\frac{f_w}{f_w^0}\right) = \begin{cases} \ln x_w \\ \ln(y_w x_w)(\text{addition of inhibitors}) \end{cases},$$

where $\Delta\mu_0$ is the chemical potential difference between water in an empty hydrate lattice and pure water in a standard state, J/mol; R is the universal gas constant, J/(mol · K); T_0 and p_0 are the temperature and pressure in standard state, $T_0 = 273.1\,\text{K}$, $p_0 = 0$, respectively; T and p are the temperature and pressure at the point of hydrate-generating phase, K, Pa; Δh_0, ΔV, and ΔC_p are the specific enthalpy difference, specific volume difference, and specific heat capacity difference between the empty hydrate lattice and pure water, J/mol, m^3/mol, and J/(mol · K), respectively; f_w and f_w^0 are the water fugacity in the water-rich phase and at the hydrate-generating phase point, Pa, respectively; v_i is the ratio of the number of i-type cavities to the number of water molecules in the hydrate phase; θ_{ij} is the probability of the number of i-type cavities to be occupied by the molecules of the j-type gas; and x_w and y_w are the water the substance fraction and activity coefficient of the water-rich phase, respectively, with no causality.

8.4.2.2 Construction of physical model of deepwater drilling wellbore annulus

The model assumes that gas overflow occurs when the drilling encounters shallow gas at 670 m below the mudline, the shallow gas with pressure is regarded as a small gas reservoir, and the hydrocarbon gas enters the annulus of the wellbore and then passes through the annulus of the casing, the blowout preventer, and the watertight pipe with the upward returning drilling fluids in turn. The modelling process and model sketch are shown in Figs. 8.26 and 8.27, respectively.

Considering the drill pipe and the wellbore annulus as a connected "U" pipe structure, the drilling fluid enters the pipe from the inlet of the drill pipe, and the shallow gas enters the wellbore annulus from the bottom of the well. According to the well structure, the wellbore routing is set up and the boundary conditions of the inlet and outlet nodes are defined. According to the literature research and field research, the drilling fluid outlet temperature is taken as 17°C, and the outlet pressure is 40 bar; the drilling fluid inlet temperature is taken as 25°C, and the displacement is 45 L/s; the gas intrusion pressure is taken as 200 bar, and the reservoir temperature is taken as 22.5°C. The natural gas fluid file in the pipeline applies the fluid component file generated by PVTsim software.

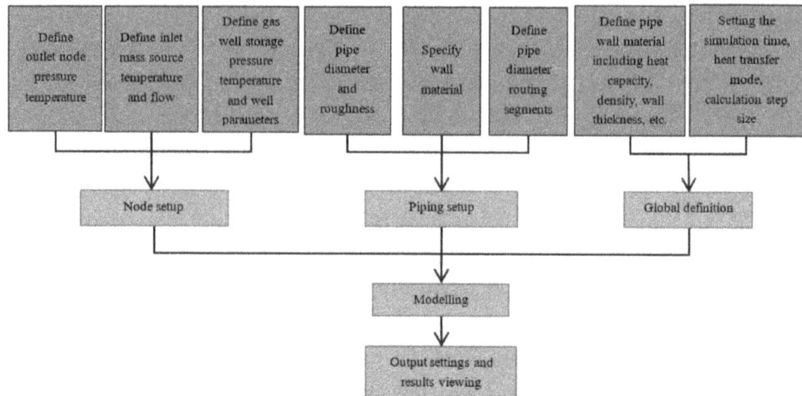

Fig. 8.26. OLGA deepwater gas intrusion wellbore annulus modelling flowchart.

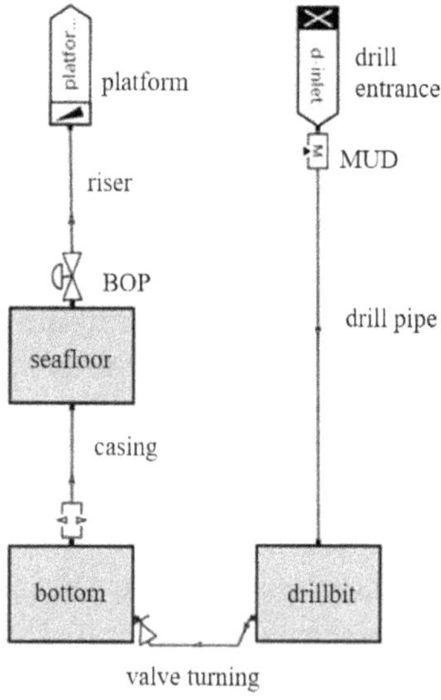

Fig. 8.27. Sketch of deepwater gas intrusion wellbore annulus modelling.

A gas field in Lingshui is located in the central canyon of the Lingshui Depression in the Qiongdongnan Basin, with a geothermal gradient of $3.87°C/100\,m$. The water depth in the area is 1,336–1,531 m. According to the follow-up records of the evolution of the water temperature in the South China Sea waters, the surface temperature of seawater in the summer is about 24–26°C, the surface temperature of seawater in the winter is about 18–22°C, and the seafloor temperature is about 3°C. The temperature of the sea bottom is about 3°C.

In this chapter, a well with a water depth of 1500 m is selected as an example for the simulation study. The total length of the water separator is 1500 m, the outer diameter is 533.4 mm, the wall thickness is 16 mm, the total length of the surface casing is 650 m, the wall thickness is 20 mm, the outer diameter of the drilling pipe is 168.3 mm, the inner diameter is 148.3 mm, and the equivalent diameter of the water hole of the drilling bit is taken as 40 mm; the thermal conductivity of the tubing materials is all 45 W/m · K, the roughness of pipe wall is 4.5×10^{-5} m; the thermal conductivity of seawater is 0.6 W/(m · °C), the thermal conductivity of drilling fluid is taken as 1.73 W/(m · °C), and the thermal conductivity of steel is 46 W/(m · °C).

The formation of natural gas hydrate requires four conditions: (a) temperature lower than the hydrate formation temperature, (b) pressure higher than the hydrate formation pressure, (c) sufficient hydrocarbon gas source and liquid water, and (d) high gas flow rate and the existence of gas flow disturbance or pressure fluctuation. The wellbore annulus drilled with water-based drilling fluid can ensure the existence of a large amount of free liquid water, and after the occurrence of gas intrusion, the annulus has a sufficient hydrocarbon gas source. Therefore, the key factors affecting hydrate formation in the wellbore annulus are the annulus temperature, pressure field, and gas flow rate.

In order to analyze the sensitivity of drilling parameters for gas hydrate formation, this section simulates the drilling parameters affecting the temperature field, pressure field, and gas flow rate in the annulus by setting different model boundary conditions.

According to the drilling operation process, it is known that the distribution of the temperature field in the annulus of the wellbore is related to the inlet temperature of the drilling fluid and the ambient temperature; the distribution of the pressure field in the annulus of the wellbore is related to the density of the drilling fluid, the discharge volume, and the depth of the well; the gas flow rate in the annulus is related to the discharge volume of the drilling fluid and the gas intrusion pressure. The ambient temperature and well depth of a fixed well location are fixed values and are not taken into consideration. In this study, only the formation of hydrates in the wellbore annulus under different drilling fluid inlet temperatures, discharge rates, densities, and gas intrusion pressures is considered.

According to the results of deepwater drilling site research, the drilling fluid inlet temperature is affected by the sea surface temperature, which is usually 24–28°C in summer and 18–22°C in winter; the commonly used drilling fluid displacements are 46 L/s, 37 L/s, and 28 L/s. The drilling fluid densities used for shallow drilling are selected as 1080 g/cm^3 and 1030 kg/s. The shallow gas pressures are selected as 180 bar, 200 bar, and 220 bar, based on the above; the following seven groups of working conditions are set as model boundary conditions (Table 8.12), and the prediction model of hydrate in the annulus of deepwater drilling wellbore is established to numerically simulate the process of hydrate formation by gas intrusion into the wellbore during deepwater drilling.

By means of numerical simulation methods, this section firstly establishes a distribution model of the coupled temperature-pressure field in the wellbore annulus. Based on the depth-pressure relationship curve, the hydrate phase equilibrium conditions are transformed into depth-temperature function relationship, and then combined with the measured circumferential temperature profile, a template

Table 8.12. Simulated boundary conditions.

Group	Gas intrusion pressure (bar)	Drilling fluid temperature (°C)	Drilling fluid displacement (L/s)	Density of drilling fluid (g/cm^3)
1#	200	25	46	1080
2#	200	20	46	1080
3#	200	25	46	1030
4#	200	25	37	1080
5#	220	25	46	1080
6#	200	25	28	1080
7#	180	25	46	1080

for predicting hydrate-generating regions is constructed. Secondly, the hydrate generation rate and hydrate generation per unit time in the corresponding well section range were simulated, and the hydrate generation under different working conditions was simulated by setting multiple sets of model boundary conditions to analyze the sensitivity of drilling parameters. Finally, based on the results of the hydrate kinetic model, four eigenvalues were extracted, namely, supercooling density, hydrate generation range, hydrate generation peak rate, and hydrate generation total, which characterized the hydrate generation range and degree under a certain condition to establish a set of eigenvalues for evaluating the hydrate formation situation and to realize the quantitative analysis of the hydrate risk level in the annulus of the wellbore under different conditions.

8.4.3 *Applications*

8.4.3.1 *Pressure-temperature analysis of hydrate formation in the wellbore annulus*

The temperature-pressure field distribution of the wellbore annulus under 7 groups of working conditions was simulated and calculated with the help of OLGA software; the software used the calculation model based on the equation of state and the coupling effect of temperature and pressure to achieve the temperature-pressure value of the well section. The calculation formula is shown as follows, the results of the pressure field calculations are shown in Fig. 8.28, and the results of the temperature field calculations are shown in Fig. 8.29.

Wellbore pressure field equation Eq. (8.59):

$$\frac{\partial}{\partial t}\left(\sum_{i=1}^{2} A\rho_i E_i \nu_i\right) + \frac{\partial}{\partial s}\left(\sum_{i=1}^{2} A\rho_i E_i \nu_i^2\right) + Ag\cos\alpha\left(\sum_{i=1}^{2} \rho_i E_i\right)$$

$$+\frac{\mathrm{d}(Ap)}{\mathrm{d}s} + \frac{\mathrm{d}(AF_r)}{\mathrm{d}s} = 0, \tag{8.59}$$

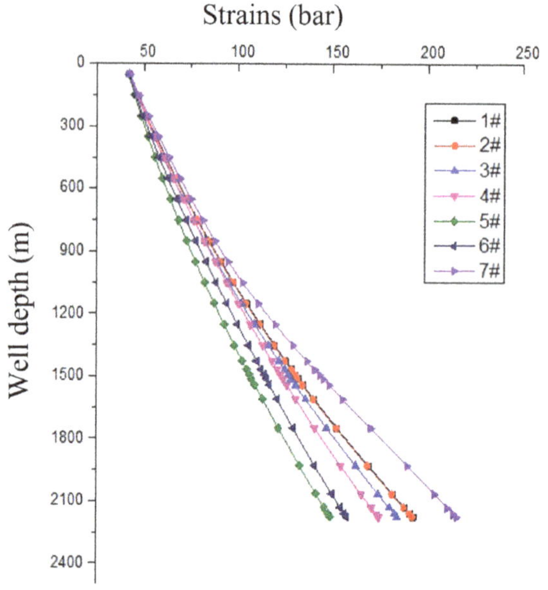

Fig. 8.28. Well depth–circumferential pressure curve.

Wellbore temperature field equations Eq. (8.60):

$$\frac{2}{vr_n^2}\left(\frac{r_0 U_0 k_e}{k_e + T_D r_0 U_0}\right)(T_e - T_f)$$

$$+ \frac{\partial}{\partial z}\left[\rho(H + gz\cos\alpha) + \frac{1}{2}v^2 + \frac{fv^2}{2d}\right]$$

$$= \frac{\partial}{\partial t}\left[\rho\left(C_f T_f + gz\cos\alpha + \frac{1}{2}v^2\right)\right], \tag{8.60}$$

where ρ_i is the density of gas and liquid phases, kg/m^3; ν_i is the velocity of each phase, m/s; E_i is the volume fraction of each phase; α is the well inclination angle, °; F_r is the friction pressure drop, Pa; p is the annulus pressure, Pa; r_n is the inner radius of the test concern, m; r_0 is the outer radius of the test tubing column, m; U_0 is the total heat transfer coefficient with the outer surface of the tubing as the datum surface, W/(m$^2 \cdot$K); k_e is the thermal conductivity of the formation, W/(m$\cdot$K); H is the enthalpy of the gas, J; T_e is the temperature of the formation, K; and T_f is the temperature of the fluid, K.

According to the model simulation results, comparing the wellbore annulus depth–pressure curves for the seven groups of conditions (Fig. 8.28), it can be seen that the pressure field inside the wellbore annulus is different for different drilling fluid discharges, densities, and gas intrusion pressures. The smaller the gas intrusion pressure (1#, 7#, and 5#), the larger the drilling fluid displacement (1#, 4#, and 6#), and the larger the drilling fluid density (1# and 3#), the larger the pressure field in the wellbore annulus, whereas the drilling fluid temperature has a negligible effect on the pressure field in the wellbore annulus (1# and 2#). In addition, the pressure field in the wellbore annulus was linearly distributed along the well depth for all 7 groups of conditions, and the annulus pressure gradually decreased from the bottom of the well to the wellhead.

Based on the well depth–annular pressure curve, the hydrate phase equilibrium pressure–temperature curve is transformed into the well depth–hydrate generation temperature curve (Fig. 8.29), which is plotted in the same coordinate system with the well depth–temperature curve of the wellbore annulus obtained by simulation, and the hydrate generation range prediction graphic plate is obtained (Fig. 8.30); the area surrounded by the two curves is the region that

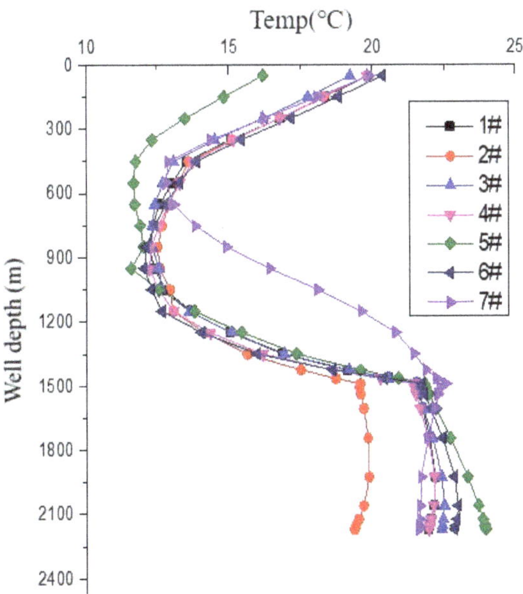

Fig. 8.29. Well depth–circumferential temperature curve.

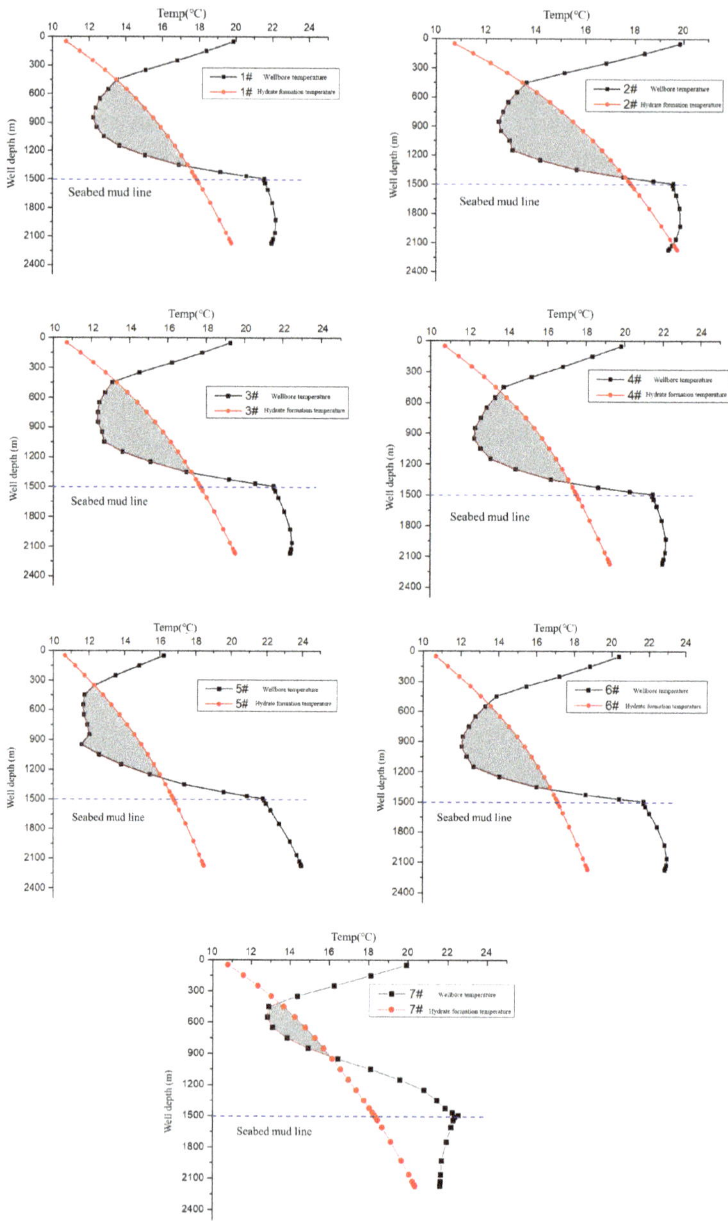

Fig. 8.30. Well depth–wellbore annulus temperature–hydrate phase equilibrium temperature curve.

meets the temperature-pressure conditions of gas hydrate generation. When the pressure fluctuation in the annulus also meets the hydrate formation conditions, it is very likely that gas hydrate will be generated in this well section.

Introducing subcooling to represent the difference between the hydrate phase equilibrium temperature at a given well depth and the annulus temperature of the wellbore at that depth, i.e., the difference between the region of overlap of the two curves in (g) in Fig. 8.30, and supercooling degree > 0 indicates that the temperature of the tubing section meets the conditions of hydrate generation.

There are two overlapping curves in all seven simulated conditions, i.e., the well section with supercooling greater than 0. The overlapping area is located in the watertight tubing annulus with a depth of less than 1500 m, while there is no overlapping area in the surface casing annulus with a depth of less than 1500 m. This indicates that the temperature and pressure conditions for hydrate formation do not exist in the surface casing annulus under the seven simulated conditions. This indicates that under the simulated seven working conditions, the temperature and pressure conditions for hydrate formation are not available in the surface casing annulus, while some sections of the water separator annulus meet the temperature and pressure conditions for hydrate formation, and there is a possibility of hydrate generation.

8.4.3.2 *Prediction of hydrate formation in wellbore annulus and parameter sensitivity analysis*

The OLGA hydrate module CSMHyK kinetic model was used to simulate and calculate the hydrate generation in the watertight pipe annulus with supercooling greater than 0. The trends of the two characteristic quantities, PSIHYD (hydrate generation rate per unit volume) and HYDMASS (hydrate generation per unit time), were obtained for both the temporal (simulation duration) and spatial (pipe length) dimensions (Fig. 8.31).

In terms of the hydrate generation rate, the common pattern presented by multiple groups of working conditions is as follows: Along the pipe length: The hydrate generation rate shows a gradual decrease along the pipe length, i.e., the generation rate is large in the well section near the seafloor, and the closer it is to the sea

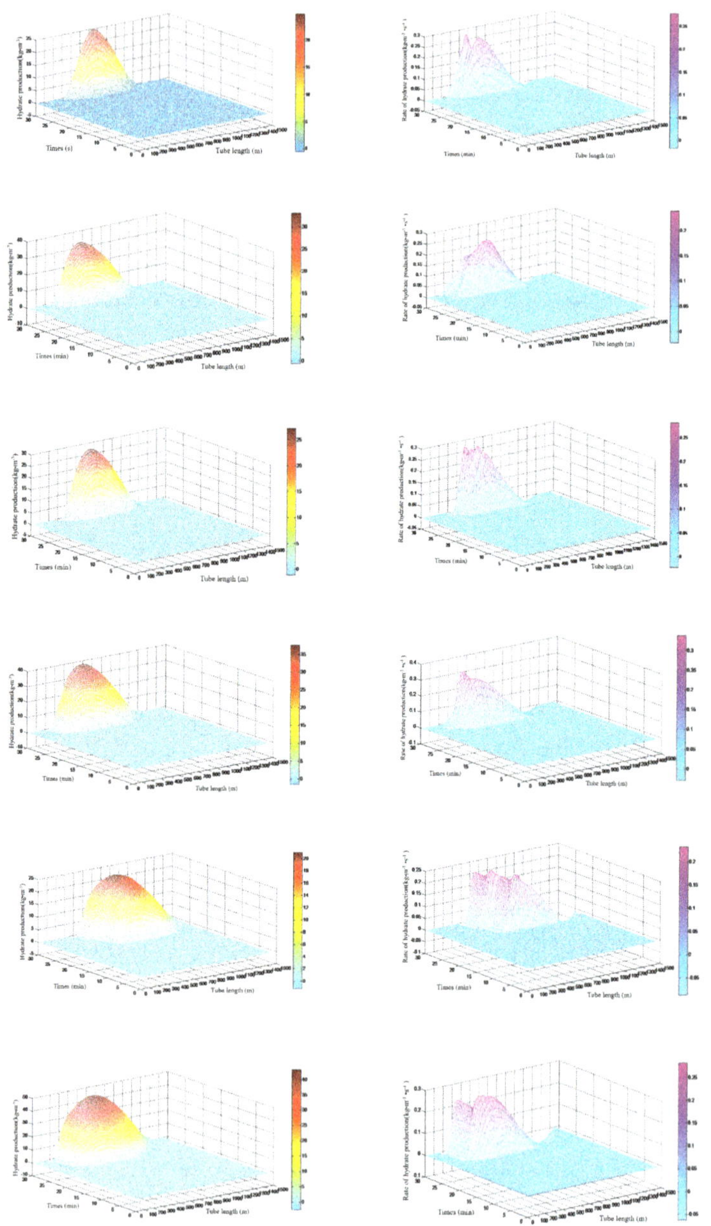

Fig. 8.31. Trends in hydrate production and production rates over time-space.

level, the smaller the generation rate is; the generation rate is negative in the well section close to the sea level, which indicates that part of hydrate decomposition has taken place. Along the time axis: From the moment of hydrate generation, the peak rate of hydrate generation gradually increases, and after the peak value increases to a certain value, it tends to stabilize and does not continue to increase. In addition, it can be observed in the combined pipe and time dimensions that the location of the pipe section where the peak hydrate generation rate occurs moves closer to the seafloor as the simulation time progresses. This suggests that, under the condition of constant gas intrusion pressure, the location of rapid and large amounts of hydrate generation may eventually appear in the blowout preventer near the seafloor, and the structure of the elbow, orifice plate, and valve of the blowout preventer's gate cavity is very favorable for hydrate aggregation; so, it can be deduced that, if the downhole gas overflow is not suppressed in a timely manner, the blowout preventer will be ice-blocked and lose the ability to control the well.

In terms of the hydrate generation amount, the common pattern presented by multiple groups of working conditions is that the hydrate generation amount increases gradually and cumulatively with the increase of time. The differences are the location of the pipe section with hydrate generation, the range of the pipe section, the amount of hydrate generation, and the moment of hydrate generation. 1#, 2#, 3#, and 4# started to generate hydrate after 25 min of gas intrusion, whereas the time of hydrate generation of 5# and 6# was earlier than the time of 20 min of gas intrusion.

Among them, 1#, 2#, 3#, 4#, and 6# are all located in the annular section of the bulkhead near the seafloor, while 5# is near the sea level. Analyzing the hydrate generation mechanism, we can see that the gas intrusion pressure of 5# is large, the gas flow rate is high in the lower pipe section, and the gas transport is weakly obstructed under the dynamic water–gas system, which is not conducive to the generation and aggregation of hydrate, whereas the gas flow rate is slowed down and the pressure inside the wells decreases in the upper pipe section close to sea level, so that the gas slides out of the liquid phase and becomes bubbles with gradually increasing volume. According to the hydrate nucleation interface hypothesis, the bubble surface provides a good interface for hydrate formation, thus providing conditions for rapid hydrate generation.

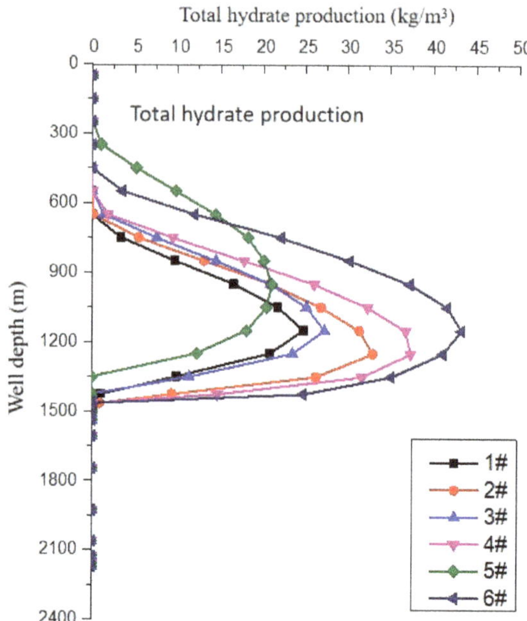

Fig. 8.32. Distribution of simulated 30-min hydrate generation along the pipe length.

In order to better compare the differences in hydrate generation under multiple sets of conditions, the distribution of total hydrate generation along the pipeline at the end of the simulation was obtained by integrating over the time axis based on the hydrate generation rate values of the well sections (Fig. 8.32). The total hydrate production in the whole pipeline was obtained by quadratic integration over the pipe axis. In addition, the peak hydrate production rate and the location of the pipe section with the peak rate were recorded for the whole simulation length (Table 8.13). The parameter sensitivity of drilling parameters to hydrate generation can be derived by comparing the variability of hydrate formation characteristics under different conditions.

Based on the calculation results of PSIHYD and HYDMASS, a total of five characteristic values were obtained, namely, the range of the hydrate-generating well section, the length of the well section, the peak rate of hydrate formation, the location of the peak rate, and the total amount of hydrate generation in the whole well

Table 8.13. Hydrate formation characteristic quantities.

Group	Range of well section (m)	Length of well section (m)	Peak rate position (m)	Peak rate ($kg \cdot m^{-3} \cdot s^{-1}$)	Total for all well sections (kg/s)	Subcooling density ($^{\circ}C \cdot m$)
1#	850–1350	500	1050	0.2749	107.82	2138.30
2#	750–1450	700	950	0.2367	167.10	2241.22
3#	750–1350	600	1050	0.2804	131.31	2050.92
4#	750–1450	700	1150	0.3737	207.81	2084.37
5#	450–1250	800	950	0.2199	140.43	1792.11
6#	650–1450	800	1050	0.2715	291.08	1956.66
7#	—	0	—	0	0	576.93

section. The longer the hydrate-generating well section, the larger the annulus affected by hydrate; the larger the peak rate of hydrate generation, the greater the probability of generating a large amount of hydrate in a short period of time in the corresponding well section, and the higher the risk of blockage of the circulation channel due to the gathering of a large amount of hydrate. The larger the total amount of hydrate generation, the more serious the loss of water in the annulus of drilling fluid, and the larger the alteration of the drilling fluid rheological properties; the more hazardous the normal return of drilling fluid to the annulus of the annulus, the higher the risk of clogging.

From the simulation results of several groups of working conditions (Table 8.13), it can be seen that hydrate generation occurs in some sections of 1#–6# bulkhead annulus, and the peak generation rate occurs in the middle and lower sections of the tubing, while no hydrate generation occurs in 7# annulus, which is characterized by low density of subcooling and low disturbance of the gas flow. 6#, which has the smallest drilling fluid discharge, has the longest section with hydrate generation, at a depth of about 350–1200 m, and generates the largest amount of hydrate, which is close to 300 kg per unit time. 4#, with the second smallest drilling fluid discharge, has the largest peak hydrate rate of $0.3737\,kg \cdot m^{-3} \cdot s^{-1}$, which occurs at $kg \cdot m^{-3} \cdot s^{-1}$ at a depth of 1150m, and the rest of the conditions have the peak hydrate generation rate at the section near the seafloor in the annulus of the water separator. Comparing 1# and 2# with different inlet temperatures of drilling fluid, it can be seen that 2# with low inlet temperature has a large amount of hydrate generation and a wide range of affected well sections, and although the peak rate is smaller than that of 1#, the difference is not great, and in general, the risk of hydrate generation in 2# is higher than that of 1#.

The simulation results show that the drilling fluid displacement and drilling fluid inlet temperature are the main factors affecting the hydrate generation, and the smaller the drilling fluid displacement and the lower the drilling fluid inlet temperature, the more favorable the wellbore annulus is to generate a large amount of hydrates. The drilling fluid displacement is the most sensitive drilling parameter. The drilling fluid displacement makes the gas–water transport in the wellbore annulus closer to the static system, and the overflow gas encounters stronger obstruction during transportation, which is more favorable for hydrate formation and aggregation. In addition, the gas intrusion pressure is the main factor affecting the location of the hydrate generation section. The larger the gas intrusion pressure, the easier it is to generate hydrate in the downstream tubular section of the drilling fluid circulation, but when the gas intrusion pressure is not small enough to generate gas flow disturbance in the circulation channel, there is no more hydrate generation. The simulation results also further verify that hydrate formation requires not only low-temperature and high-pressure environmental conditions but also appropriate pressure fluctuations.

8.4.3.3 *Early warning of hydrate formation risk in wellbore annulus*

Analyzing the microscopic process of hydrate formation, it can be seen that the nucleation of natural gas hydrate crystals needs to be formed after a period of induction time on the basis of subcooling to meet the conditions. After the formation of crystal nuclei, under suitable conditions of temperature and pressure, the crystal nuclei undergo a rapid growth period, the size of the crystal nuclei gradually increases and tends to be more or less stable, and they finally form solid gas hydrate particles. Therefore, the thermodynamic conditions for hydrate formation not only require the presence of subcooling ($\Delta T > 0$) in the wellbore annulus, but also the kinetic characteristics of hydrate generation are significantly controlled by the magnitude and duration of subcooling. Specifically, the larger the subcooling, is maintained, the more sufficient time is available for the growth degree > 0, the longer the time available for hydrate nuclei to grow and stabilization of hydrate nuclei, which ultimately leads to a significant increase in hydrate production.

Since the drilling fluid is in real-time flow state during drilling, when the natural gas molecules and water molecules flow with the drilling fluid into the tubing section with supercooling > 0 to form hydrate nuclei, they will continue to flow upward with the drilling fluid, and the nuclei will only have enough induced time to grow and stabilize if they are always in the tubing section with super-cooling > 0. Therefore, the time that the hydrate nucleus is in the subcooling degree > 0 pipe section can be translated into the range of subcooling degree > 0 within the pipe section. For this reason, the subcooling density eigenvalue is reintroduced, and the subcooling density is defined as the area enclosed by the hydrate formation phase equilibrium curve and the wellbore annulus temperature curve, i.e., the area of the shaded portion in Fig. 8.30, in the range of tubing segments with subcooling > 0. The subcooling density examines the hydrate formation potential in terms of both subcooling values and the length of tubing segments with subcooling > 0 as shown in Eq. (8.61):

$$\rho_{\text{scd}} = \int_{h}^{h+\Delta h} (T_{\text{ba}} - T_{\text{te}})\mathrm{d}h, \tag{8.61}$$

where ρ_{scd} is the density of subcooling degree, $°\text{C} \cdot \text{m}$; T_{ba} is the phase equilibrium temperature of natural gas hydrate at a certain depth, $°\text{C}$; T_{te} is the annulus temperature of the wellbore at a certain depth, $°\text{C}$; h is the starting point of tubing section with subcooling degree > 0; and Δh is the length of tubing section with subcooling degree > 0, m.

From the supercooling density values in the annulus of the well-bore under 7 groups of working conditions (Table 8.13), it can be seen that the supercooling density values of 1#–6# are all in 2000°C-m, among which the supercooling density of 2# is the largest, and the supercooling density of 7# is only 576.93°C $\cdot$ m, which indicates that the possibility of hydrate generation in 2# is the largest and the possibility of hydrate generation in 7# is the smallest. This is related to the mechanism of hydrate generation, which requires gas flow disturbance, and the gas intrusion pressure of 7# is too small to generate enough pressure fluctuation, so although the supercooling degree meets the low-temperature and high-pressure environmental conditions, there is still no hydrate generation.

In order to further compare and analyze the risk of hydrate formation in the wellbore annulus under different working conditions, four eigenquantities, namely, supercooling density, hydrate-generating well section length, total hydrate generation in the whole pipeline, and peak rate of hydrate generation, are selected as the evaluation indexes of the risk of hydrate generation, and based on the idea of fuzzy comprehensive evaluation, weights are assigned to the indexes by applying the Delphi method, which are 0.3, 0.3, 0.2, and 0.2, respectively. It indicates the contribution rate of the four eigenvalues to the total risk of hydrate formation, and the total risk is calculated according to Eq. (8.62). The risk value combines the dangers of the four eigenvalues, which can visually and clearly compare the severity of hydrate formation in the wellbore annulus under different working conditions. It should be noted that this risk value is only used to compare the risk level of hydrate formation in the annulus of the wellbore under different working conditions. Since the four eigenvalues have different units and the range of values is not uniform, it is necessary to normalize the data first (Table 8.14), and the calculation of risk values is shown in Eq. (8.6):

$$R = \sum_{i=1}^{4} w_i \times E_i, \tag{8.62}$$

where R denotes the value at risk; w_i denotes the weight of the ith eigenvalue; and E_i denotes the eigenvalue.

As shown in Fig. 8.33, comparing multiple groups of working conditions, 6# has the highest risk, followed by 2# and 4#, and the risk levels of 1#, 3#, and 5# are comparable. Among the drilling parameters affecting the hydrate formation, the drilling fluid discharge plays the most prominent role, followed by the drilling fluid inlet temperature, while the drilling fluid density and the gas intrusion pressure do not have a significant effect on the risk of hydrate generation. The total risk values calculated based on the simulation results can be used to compare the hydrate formation risk under different drilling

Table 8.14. Normalized four eigenvalues and risk values for hydrate formation.

Group	Length of well section/m	Peak rate/ $(kg \cdot m^{-3} \cdot s^{-1})$	Total amount generated/ $(kg \cdot s^{-1})$	Subcooling density/ $(°C \cdot m)$	Exposures
1#	0.625	0.7356	0.3704	0.9541	0.6799
2#	0.875	0.6334	0.5741	1	0.8557
3#	0.75	0.7503	0.4511	0.9151	0.7062
4#	0.875	1	0.7139	0.9300	0.8843
5#	1	0.5884	0.4824	0.7996	0.6784
6#	1	0.7265	1	0.8730	0.9146
Weights	0.3	0.3	0.2	0.2	

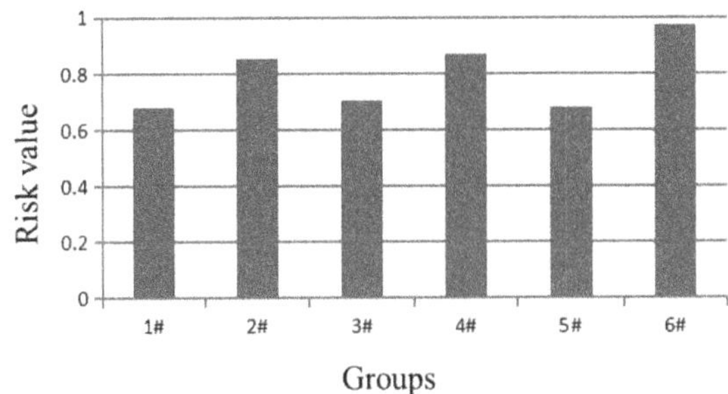

Fig. 8.33. Comparison of hydrate formation risk for multiple conditions.

scenarios and provide a theoretical basis for the optimal design of drilling parameters.

8.5 MMPR-based Availability Prediction Method for Deepwater Blowout Preventer System

A blind shear ram preventer (BSRP) is the last important safety barrier in a subsea blowout preventer system. In case of emergency, the BSRP prevents serious accidents such as a blowout or uncontrolled blowout by cutting off the drill pipe and sealing the wellhead. Therefore, it is important to analyze the availability of the BSRP and perform functional testing and maintenance to ensure that it is in a usable condition during its service life. Conventional methods for analyzing the availability of BSRPs ignore the dynamic characteristics of the testing process, such as the non-periodicity of testing and maintenance due to incidents such as well surges and blowouts, and the false alarms or errors that may occur during the testing process. In addition, traditional blowout preventer availability analysis methods do not consider the impact of maintenance status on unavailability. Unscheduled lifting of blowout preventers for maintenance increases the risk of blowouts due to the subsea environment, such as the difficulty of accessing subsea equipment and the delay in subsea maintenance.

The Multiphase-Markov process (MMP) is a stochastic process in which a Markov chain is transferred from one state to another at different times in continuous time. Each test maintenance cycle can be regarded as a phase (stage) of MMP. Therefore, in this section, the MMP method is chosen to study the change rule of multiple states in different test and maintenance cycles. Using MMP to analyze the unavailability of underwater BSRP, a preventive testing and maintenance method based on the Multiphase-Markov process with repair (MMPR) is proposed. The time-dependent unavailability and average unavailability of underwater BSRP are analyzed and modeled, and the effects of non-periodicity of preventive testing and maintenance, testing errors, and repair delays on the unavailability of the system are predicted and evaluated.

8.5.1 *Unavailability prediction method based on MMPR*

Based on the relevant theories, definitions, and assumptions of MMP process modelling, the mathematical formulas of unavailability and average unavailability for different test and maintenance cycles based on MMP are derived, and the multiphase Markov unavailability analysis model is constructed to consider different scenarios (off-cycle test and maintenance and test failures) of the state to be repaired.

8.5.1.1 *MMP unavailability derivation process*

Modelling based on MMP allows the analysis of the unavailability of a real system during the test and maintenance cycle. The usability state of a system is closely related to the realization of its basic functions, and a system in a usable state means that the system is able to realize its basic functions, while on the contrary, the system is in an unavailable state. We can evaluate and analyze the system unavailability state (faulty state and pending maintenance state) in different test cycles, and the analysis indexes include instantaneous unavailability and average unavailability, and the derivation process of instantaneous unavailability and average unavailability is described in detail in the following.

(1) Derivation of the instantaneous unavailability analytic formula

In MMP, $C(i, j)$ is defined as the transfer rate matrix from one state i to another state j during the test cycle. $P_t(i)$ denotes the probability of state i at moment t, $P_t = [P_t(1), P_t(2), \ldots, P_t(i)]$. $C(i, j)$ determines the state transfer probability P_t, if $C(i, j)$ is constant during the test cycle, the system state can be expressed by the Chapman–Kolmogorov equation, see Eqs. (8.63) and (8.64):

$$\frac{\mathrm{d}P_t}{\mathrm{d}t} = P_t \cdot C, \tag{8.63}$$

$$P_t = \exp(C \cdot t). \tag{8.64}$$

In an MMP-based model, the k test cycles are denoted as $[T_0 = 0, T_1]$, $[T_1, T_2], \ldots, [T_{k-1}, T_k]$. If the system is assumed to be within the first test cycle $t \in [T_0 = 0, T_1]$, the system state probabilities are as in Eqs. (8.65) and (8.66).

In an MMP-based model, the k test cycles are denoted as $[T_0 = 0, T_1]$, $[T_1, T_2], \ldots, [T_{k-1}, T_k]$. If it is assumed that the system at time t is in the first test cycle $t \in [T_0 = 0, T_1]$, then the system state probabilities are shown in Eqs. (8.65) and (8.66):

$$P_t = P_0 \cdot \exp(C_1 \cdot t), \tag{8.65}$$

$$P_{T_1} = P_0 \cdot \exp(C_1 \cdot T_1), \tag{8.66}$$

where $C_1, C_2, \ldots, C_k$ denote the transfer rate matrices for different test cycles.

The state probabilities are calculated in Eq. (8.67) assuming $t \in [T_1, T_2]$:

$$P_t = P_{T_1} \cdot M_1 \cdot \exp(C_2 \cdot (t - T_1)), \tag{8.67}$$

where M_1 is the probability transfer matrix for the different states in a new test cycle following the last test repair action, and M_1 may be affected by the test strategy and the repair action. The repair action at the moment of T_1 can be obtained by a linear transformation of the probability P_{T1}. $P_{T1} \cdot M_1$ denotes the probability of all the states after the completion of the repair action at the moment of T_1. The method allows the state probabilities to be linearly redistributed at the beginning of each test cycle by multiplying the probability transfer matrix. Thus, Eqs. (8.68) and (8.69) are

$$P_{T_2} = P_{T_1} \cdot M_1 \cdot \exp[C_2 \cdot (T_2 - T_1)]$$

$$= P_0 \cdot \exp(C_1 \cdot T_1) \cdot M_1 \cdot \exp[C_2 \cdot (T_2 - T_1)], \quad (8.68)$$

$$P_{T_{(k-1)}} = P_{T_{(k-2)}} \cdot M_{k-2} \cdot \exp[C_{k-1} \cdot (T_{k-1} - T_{k-2})]$$

$$= P_0 \cdot \prod_{n=1}^{n=k-2} \exp\{[C_n \cdot (T_n - T_{n-1})] \cdot M_n\}$$

$$\cdot \exp[C_{k-1} \cdot (T_{k-1} - T_{k-2})], \quad (8.69)$$

where M_{k-2} is the probability transfer matrix for the test cycle $[T_{k-1} - T_{k-2}]$.

If $t \in [T_{k-1} - T_{k-2}]$ within, then Eq. (8.70)

$$P_t = P_{T(k-1)} \cdot M_{k-1} \cdot \exp[C_k \cdot (t - T_{k-1})]$$

$$= P_0 \cdot \prod_{n=1}^{n=k-1} \exp\{[C_n \cdot (T_n - T_{n-1})] \cdot M_n\}$$

$$\cdot \exp(C_k \cdot (t - T_{k-1})). \quad (8.70)$$

According to the definition of instantaneous unavailability $UA(t)$, the system will no longer fulfill its basic functions as long as it is in one of the unavailable states. Therefore, the system's instantaneous unavailability is calculated using the system's unavailable states during the test cycle with the following expression Eq. (8.71):

$$UA(t) = P_t \cdot B, \quad (8.71)$$

where B is a vector containing 1 s and 0 s, with 1 s indicating that the system is in a functionally available state and 0 s indicating that the system is in an unavailable state.

The same methodology is applied to evaluate the system $UA_k(t)$ in a test cycle $t \in [T_{k-1} - T_{k-2}]$, as in Eq. (8.72):

$$UA_k(t) = P_0(i) \cdot \prod_{n=1}^{n=k-1} \{\exp[C_n \cdot (T_n - T_{n-1})] \cdot M_n\}$$

$$\cdot \exp(C_k \cdot (t - T_{k-1})) \cdot B. \tag{8.72}$$

In contrast to the classical Markov model, Eq. (8.72) provides a $UA_k(t)$ method capable of evaluating a system's periodicity, non-periodicity, or a particular test cycle.

(2) Derivation of analytical formula for average unavailability

The overall average unavailability can be expressed in the following equation Eq. (8.73):

$$UA_{\text{avg}} = \frac{1}{T} \int_0^T UA(t)\mathrm{d}t = \frac{1}{T} \left[\int_{T_0}^{T_1} UA_1(t)\mathrm{d}t + \int_{T_1}^{T_2} UA_2(t)\mathrm{d}t \right.$$

$$\left. + \cdots + \int_{T_{k-1}}^{T_k} UA_k(t)\mathrm{d}t \right] = \frac{1}{T} \sum_{n=1}^{n=k} \int_{T_{n-1}}^{T_n} UA_n(t)\mathrm{d}t. \tag{8.73}$$

If $t \in [T_0, T_1]$, then there is Eq. (8.74):

$$\int_{T_0}^{T_1} UA_1(t)\mathrm{d}t = \int_{T_0}^{T_1} P_0(i) \cdot \exp[C_1 \cdot (t - T_0)] \cdot B$$

$$= P_0(i) \cdot \int_{T_0}^{T_1} \exp[C_1 \cdot (t - T_0)] \, \mathrm{d}t \cdot B$$

$$= P_0(i) \cdot \sum_{l=0}^{\infty} \frac{(C_1)^l}{(l+1)!} \cdot \left(T_1^{l+1} - T_0^{l+1} \right) \cdot B \tag{8.74}$$

If $t \in [T_1, T_2]$, then there is Eq. (8.75):

$$\int_{T_1}^{T_2} UA_2(t)\mathrm{d}t = \int_{T_1}^{T_2} P_0(i) \cdot \exp[C_1 \cdot (T_1 - T_0)] \cdot M_1$$

$$\cdot \exp[C_2 \cdot (T_1 - T_0)] \cdot B \, \mathrm{d}t$$

$$= P_0(i) \cdot \exp[C_1 \cdot (T_1 - T_0)] \cdot M_1$$

$$\cdot \int_{T_1}^{T_2} \exp\left[\boldsymbol{C_2} \cdot (T_1 - T_0)\right] \mathrm{d}t \cdot \boldsymbol{B}$$

$$= P_0(i) \cdot \exp\left[\boldsymbol{C_1} \cdot (T_1 - T_0)\right] \cdot M_1 \cdot \sum_{l=0}^{\infty} \frac{(\boldsymbol{C_2})^l}{(l+1)!}$$

$$\cdot \left(T_2^{l+1} - T_1^{l+1}\right) \cdot \boldsymbol{B}. \tag{8.75}$$

Similarly, if $t \in [T_{k-1}, T_k]$, then there is

$$\int_{T_{k-1}}^{T_k} UA_k(t)\mathrm{d}t = \int_{T_{k-1}}^{T_k} P_0(i) \cdot \prod_{n=1}^{n=k-1} \{\exp[\boldsymbol{C_n} \cdot (T_n - T_{n-1})] \cdot M_n\}$$

$$\cdot \exp\left[\boldsymbol{C_k} \cdot (t - T_{k-1})\right] \cdot \boldsymbol{B}\, \mathrm{d}t$$

$$= P_0(i) \cdot \prod_{n=1}^{n=k-1} \{\exp\left[\boldsymbol{C_n} \cdot (T_n - T_{n-1})\right] \cdot M\}_n$$

$$\cdot \int_{T_{k-1}}^{T_k} \exp\left[\boldsymbol{C_k} \cdot (t - T_{k-1})\right] \cdot \boldsymbol{B}\mathrm{d}t$$

$$= P_0(i) \cdot \prod_{n=1}^{n=k-1} \exp\left[\boldsymbol{C_n} \cdot (T_n - T_{n-1})\right] \cdot M_n\}$$

$$\cdot \sum_{l=0}^{\infty} \frac{(\boldsymbol{C_k})^l}{(l+1)!} \cdot \left(T_k^{l+1} - T_{k-1}^{l+1}\right) \cdot \boldsymbol{B} \tag{8.76}$$

System failure occurs randomly during any test cycle, so evaluating UA_{avg} for a test cycle can be achieved by using the treatment of the analytical formula as shown in Eq. (8.77):

$$UA_{\mathrm{avg}} = \frac{1}{T} \int_0^T UA(t)\mathrm{d}t = \frac{1}{T} P_0(i) \cdot \left(\sum_{l=0}^{\infty} \frac{(\boldsymbol{C_2})^L}{(l+1)!} \cdot \left(T_2^{l+1} - T_1^{l+1}\right)\right.$$

$$+ \exp\left[\boldsymbol{C_1} \cdot (T_1 - T_0)\right] \cdot M_1 \cdot \sum_{l=0}^{\infty} \frac{(\boldsymbol{C_2})^l}{(l+1)!} \cdot \left(T_2^{l+1} - T_1^{l+1}\right)$$

$$+ \cdots + \prod_{n=1}^{n=k-1} \left\{ \exp\left[\boldsymbol{C}_n \cdot (T_n - T_{n-1}) \right] \cdot \boldsymbol{M}_n \right\}$$

$$\cdot \sum_{l=0}^{\infty} \frac{(\boldsymbol{C}_k)^l}{(l+1)!} \cdot \left(T_k^{l+1} - T_{k-1}^{l+1} \right) \Bigg) \cdot \boldsymbol{B}. \tag{8.77}$$

8.5.1.2 *MMPR modelling based on off-cycle functional test maintenance*

(1) Off-cycle MMPR model

During the drilling process, functional verification testing of the blowout preventer system (hereinafter collectively referred to as functional testing) is required to ensure system availability. It is assumed that the functional test will be able to detect all failure modes during the test cycle. The test cycle is dependent on the drilling conditions and drilling operations, and in accident-prone well conditions, the functional test is non-cyclical. In order to illustrate the unavailability of the system during these test cycles, an MMPR model based on off-cycle functional test maintenance (including pending repair status) is developed. The 1oo1 and 1oo2 configurations of the BSRP are selected as examples, and the modeling steps are as follows:

(i) Determine all possible states of the different configurations for building the MMPR model. Due to the particular subsea environment and operating conditions, test maintenance may often be delayed, and therefore a "pending maintenance" state is introduced to characterize such delays, taking into account the maintenance time after each test cycle.

(ii) Transition rate matrices are constructed by determining equipment failure rates and repair rates based on empirical data from accident databases, accident reports, or expert judgment. The probabilistic transfer matrix for each state is determined by the operator at the site; in the initial stage, the system is in the best state of functioning and the corresponding transfer probability is 1.

(iii) Construct the mathematical model of MMPR under different cases.

Table 8.15. Possible states for 1oo1 configuration.

Statuses	Acronyms	Descriptions	Unavailability
1	FU	Functional status	Usability
2	DU	DUF status	Unavailable
3	IR	Pending repair status	Unavailable

Table 8.16. Possible states of 1oo2 configuration.

Statuses	Acronyms	Descriptions	Unavailability
1	FU-FU	Both components are in functional state	Usability
2	DU-FU	One of the components DUF status	Usability
3	DU-DU	Both parts are in DUF state	Unavailable
4	DU-IR	One of the components is in a state of repair	Usability
5	IR-IR	Both parts are in a state of repair	Unavailable

It is known that there are m functional test cycles as $[T_0 = 0, T_1]$, $[T_1, T_2], \ldots, [T_{m-1}, T_m]$. Depending on the particular working conditions and specific operational requirements, either cyclic or non-cyclic test phases are considered. The MMPR models for the 1oo1 configuration and the 1oo2 configuration are shown in Fig. 8.34. During the functional test cycle, there are three possible states for the 1oo1 configuration (Table 8.15) and five possible states for the 1oo2 configuration (Table 8.16). The green circle represents that the component is available, while the yellow circle represents that the component is not available. As an example, in Fig. 8.34(a), if a fault is detected during a certain test cycle but the repair program is not immediately responsive, the arc from the pending repair state does not go directly back to the DU, and therefore the IR state is taken into account in the MMPR model. The dotted line in the model represents the pending repair and there is no Markov chain process transfer between the two states. It can be seen that the test and repair behavior of the system can be integrated into the MMP process, and this model can

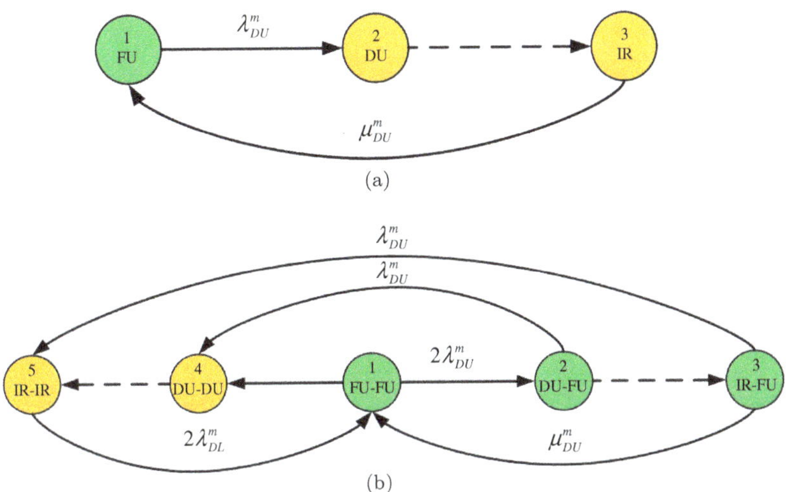

Fig. 8.34. MMPR models for different functional test phases for (a) 1oo1 configuration and (b) 1oo2 configuration.

more fully and reasonably simulate the actual state of system safety, especially for systems operating in subsea environments.

(2) Determination of parameters of the analytical formula

From Eqs. (8.71) and (8.76), the parameters to be determined for the analytical formula are the initial state probability P_0, the state transfer rate matrix $\boldsymbol{C_m}$, the probability transfer matrix $\boldsymbol{M_m}$, and the vector $\boldsymbol{B}$.

At the beginning of the first test cycle, the system is in the "FU" or "FU-FU" state. Thus, the initial state probabilities of the two types of configurations at the moment $t = 0$ are as in Eqs. (8.78) and (8.79), respectively:

$$P_{01}(i) = [1, 0, 0], \tag{8.78}$$

$$P_{02}(i) = [1, 0, 0, 0, 0], \tag{8.79}$$

where $P_{01}(i)$ represents the initial probability of the 1oo1 configuration and $P_{02}(i)$ represents the initial probability of the 1oo2 configuration.

From Fig. 8.34, it can be seen that the state transfer rate matrix can be composed of a failure rate λ_{DU}^m m and a repair rate μ_{DU}^m. The state transfer rate matrix is shown in Eqs. (8.80) and (8.81) for any test cycle $[T_{m-1}, T_m]$:

$$C_{m1} = \begin{bmatrix} -\lambda_{DU}^m & \lambda_{DU}^m & 0 \\ 0 & 0 & 0 \\ \mu_{DU}^m & 0 & -\mu_{DU}^m \end{bmatrix}, \tag{8.80}$$

$$C_{m2} = \begin{bmatrix} -2\lambda_{DU}^m & 2\lambda_{DU}^m & 0 & 0 & 0 \\ 0 & -\lambda_{DU}^m & 0 & \lambda_{DU}^m & 0 \\ \mu_{DU}^m & 0 & -(\mu_{DU}^m + \lambda_{DU}^m) & 0 & \lambda_{DU}^m \\ 0 & 0 & 0 & 0 & 0 \\ 2\mu_{DU}^m & 0 & 0 & 0 & -2\mu_{DU}^m \end{bmatrix}, \tag{8.81}$$

where C_{m1} denotes a state transfer rate for the 1oo1 configuration and C_{m2} denotes a state transfer rate for the 1oo2 configuration.

The state probabilities in the probabilistic transfer matrix M_m can be obtained by expert judgment in field operations. It is assumed that M_m is the same for different test cycles, and errors (e.g., false alarms) are not considered in each test cycle, i.e., $M_1 = M_2 = \cdots = M_m = M$. Thus, Eqs. (8.82) and (8.83) are as follows:

$$M_{m1} = \begin{bmatrix} 1 & 0 & 0 \\ 0 & 0 & 1 \\ 0 & 0 & 1 \end{bmatrix}, \tag{8.82}$$

$$M_{m2} = \begin{bmatrix} 1 & 0 & 0 & 0 & 0 \\ 0 & 0 & 1 & 0 & 0 \\ 0 & 0 & 1 & 0 & 0 \\ 0 & 0 & 0 & 0 & 1 \\ 0 & 0 & 0 & 0 & 1 \end{bmatrix}, \tag{8.83}$$

where M_{m1} denotes the probability transfer matrix for the 1oo1 configuration and M_{m2} denotes the probability transfer matrix for the 1oo2 configuration.

If the system is available in state i, then $B_j(1, i) = 0$, if the system is not available in state i, $B_j(1, i) = 1$, $j = 1, 2$ denotes the different configurations of the system, and Vector B is shown in Eqs. (8.84) and (8.85):

$$B_1(1, i) = [0, 1, 1]^T, \tag{8.84}$$

$$B_2(1, i) = [0, 0, 0, 1, 1]^T. \tag{8.85}$$

Typically, the instantaneous unavailability $UA(t)$ and average unavailability UA_{avg} for different configurations at any test cycle $t \in [T_{m-1}, T_m]$ can be obtained from Eqs. (8.71) and (8.77).

8.5.1.3 *MMPR modelling based on incomplete test maintenance*

(1) MMPR Modelling for Incomplete Test Maintenance

The actual testing and maintenance strategy for underwater operational systems is usually a combination of functional verification testing and incomplete testing. Incomplete testing can only detect partial faults and is usually used to verify whether the actual system is usable under specific conditions, so errors (e.g., false alarms) may occur during the testing process. In addition, the test cycles are shorter than functional tests and it can be assumed that the failure rate is the same for all cycles.

It is known that there are k incomplete test cycles as $[T_0 = 0, T_1]$, $[T_1, T_2], \ldots, [T_{k-1}, T_k]$. In accordance with the requirements of specific operating conditions and specific operational requirements, periodic incomplete testing is considered. The MMPR model for the incomplete testing phase for the 1oo1 configuration and the 1oo2 configuration is shown in Fig. 8.35. The MMPR model during the incomplete test cycle considers two types of failures, i.e., DUF1 and DUF2. The 1oo1 configuration has 4 possible states (Table 8.17) and 1oo2 configuration has 10 possible states (Table 8.18). Green circles represent parts available, while yellow circles represent parts unavailable. As an example, in Fig. 8.35(a), the dotted line represents waiting for repair, and there is no Markov chain process transfer between the two states, but the transfer after incomplete testing is considered valid. The test and repair behavior of the system can be integrated into the MMP process.

(2) Incomplete test analytical formula to determine the parameters

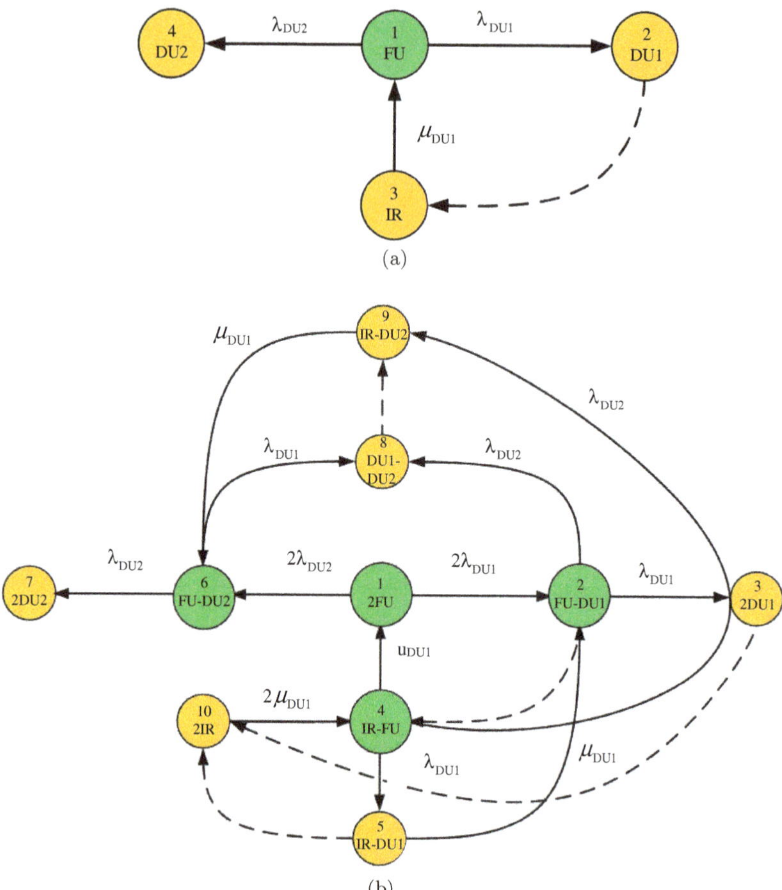

Fig. 8.35. Two configuration models: (a) 1oo1 configure and (b) Incompletely tested MMPR model for 1oo2 configurations.

At the moment $t = 0$, the system is in the "FU" or "2FU2" state, the initial state probabilities for the two types of systems are as in Eqs. (8.86) and (8.87), respectively:

$$P_{01}(i) = [1, 0, 0, 0], \tag{8.86}$$

$$P_{02}(i) = [1, 0, 0, 0, 0, 0, 0, 0, 0, 0], \tag{8.87}$$

where $P_{01}(i)$ represents the initial probability of the 1oo1 configuration and $P_{02}(i)$ represents the initial probability of the 1oo2 configuration.

Table 8.17. Possible states for incomplete testing of 1oo1 configurations.

Statuses	Acronyms	Descriptions	Unavailability
1	FU	Functional status	Usability
2	DU1	Incompletely tested DUF1 status	Unavailability
3	IR	Pending repair status	Unavailability
4	DU2	Completely tested DUF2 status	Unavailability

Table 8.18. Possible states of an incomplete test 1oo2 configuration.

Statuses	Acronyms	Descriptions	Unavailability
1	2FU	Both parts in functional state	Availability
2	FU-DU1	One part in DUF1 state, one in functional state	Availability
3	2 DU1	Both parts in DUF1 state	Unavailability
4	IR-FU	One part in maintenance state, one in functional state	Usability
5	IR-DU1	One component pending repair status, one DUF1 status	Unavailability
6	FU-DU2	One part in DUF2 state, one functional state	Availability
7	2DU2	Both parts in DUF2 state	Unavailability
8	DU1-DU2	One part in DUF2 state, one in DUF1 state	Unavailability
9	IR-DU2	One component pending repair status, one DUF2 status	Unavailability
10	2IR	Both parts are in a state of repair	Unavailability

As shown in Fig. 8.35, the state transfer rate matrix can be composed of the first type of failure rate λ_{DU1}, the second type of failure rate λ_{DU2}, and the maintenance rate μ_{DU}. In any detection cycle $[T_{k-1}, T_k]$, the detected faults cannot be repaired immediately and the state transfer rate matrix is shown in Eqs. (8.88) and (8.89):

$$C_{k1} = \begin{bmatrix} -(\lambda_{\mathrm{DU1}} + \lambda_{\mathrm{DU2}}) & \lambda_{\mathrm{DU1}} & 0 & \lambda_{\mathrm{DU2}} \\ 0 & 0 & 0 & 0 \\ \mu_{\mathrm{DU}} & 0 & -\mu_{\mathrm{DU}} & 0 \\ 0 & 0 & 0 & 0 \end{bmatrix}, \tag{8.88}$$

$$C_{k2} = \begin{bmatrix}
-2(\lambda_{DU1}+\lambda_{DU2}) & 2\lambda_{DU1} & 0 & 0 & 0 & 2\lambda_{DU2} & 0 & 0 & 0 & 0 \\
0 & -(\lambda_{DU1}+\lambda_{DU2}) & \lambda_{DU1} & 0 & 0 & 0 & 0 & 0 & \lambda_{DU2} & 0 \\
0 & 0 & 0 & 0 & 0 & 0 & 0 & 0 & 0 & 0 \\
\lambda_{DU1} & 0 & 0 & -(\mu_{DU1}+\lambda_{DU1}+\lambda_{DU2}) & \lambda_{DU1} & 0 & 0 & 0 & \lambda_{DU2} & 0 \\
0 & \mu_{DU1} & 0 & 0 & -\mu_{DU1} & 0 & 0 & 0 & 0 & 0 \\
0 & 0 & 0 & 0 & 0 & -(\lambda_{DU2}+\lambda_{DU1}) & \lambda_{DU2} & \lambda_{DU1} & 0 & 0 \\
0 & 0 & 0 & 0 & 0 & 0 & 0 & 0 & 0 & 0 \\
0 & 0 & 0 & 0 & 0 & 0 & 0 & 0 & 0 & 0 \\
0 & 0 & 0 & 0 & \mu_{DU1} & 0 & 0 & 0 & -\mu_{DU1} & 0 \\
0 & 0 & 0 & 2*\mu_{DU1} & 0 & 0 & 0 & 0 & 0 & -2*\mu_{DU1}
\end{bmatrix}, \tag{8.89}$$

where C_{k1} denotes the state transfer rate for the 1oo1 configuration and C_{k2} denotes the state transfer rate for the 1oo2 configuration.

The state probabilities in the probabilistic transfer matrix $\boldsymbol{M_k}$ can be obtained by the expert judgment of the field operation, assuming that $\boldsymbol{M_k}$ is the same for different test cycles. The characteristics of incomplete testing are modelled by introducing defect factors ω and η. The parameter ω is the false alarm probability and η is the detection failure probability. There are two scenarios:

Scenario 1: When testing and maintenance are normal, $\omega = \eta = 0$ and the state probability is 1.

Scenario 2: When testing and maintenance are imperfect, $\omega \neq 0$ or $\eta \neq 0$ and the state probability is not equal to 1.

Therefore, $\boldsymbol{M}$ can be given by Eqs. (8.90) and (8.91):

$$\boldsymbol{M_{k1}} = \begin{bmatrix}
1-\omega & 0 & \omega & 0 \\
0 & \eta & 1-\eta & 0 \\
0 & 0 & 1 & 0 \\
0 & 0 & 0 & 1
\end{bmatrix}, \tag{8.90}$$

$$
M_{k2} =
\begin{bmatrix}
1-\omega-\omega & 0 & 0 & \omega & \omega & 0 & 0 & 0 & 0 & 0 \\
0 & \eta & 0 & 1-\eta & 0 & 0 & 0 & 0 & 0 & 0 \\
0 & 0 & \eta & 0 & 0 & 0 & 0 & 0 & 0 & 1-\eta \\
0 & 0 & 0 & 1 & 0 & 0 & 0 & 0 & 0 & 0 \\
0 & 0 & 0 & 0 & 0 & 0 & 0 & 0 & 0 & 1 \\
0 & 0 & 0 & 0 & 0 & 1 & 0 & 0 & 0 & 0 \\
0 & 0 & 0 & 0 & 0 & 0 & 1 & 0 & 0 & 0 \\
0 & 0 & 0 & 0 & 0 & 0 & 0 & 0 & 1 & 0 \\
0 & 0 & 0 & 0 & 0 & 0 & 0 & 0 & 1 & 0 \\
0 & 0 & 0 & 0 & 0 & 0 & 0 & 0 & 0 & 1
\end{bmatrix},
\tag{8.91}
$$

where M_{k1} denotes the probability transfer matrix for the 1oo1 configuration and M_{k2} denotes the probability transfer matrix for the 1oo2 configuration. Taking the 1oo1 configuration as an example, the transfer probability from state i to state j is shown in the following:

$P(\mathrm{FU}_k \to \mathrm{FU}_{k+1}) = 1 - \omega = M_{k1}(1,1)$, which represents the probabilistic transfer of the system state FU from the current $\mathbf{k}$-cycle to the next $(k+1)$-cycle, as in the first row and the first column of the first value corresponding to the first row and the first column in Eq. (8.91).

$P(\mathrm{FU}_k \to \mathrm{IR}_k) = \omega = M_{k1}(1,3)$, representing the probabilistic transfer of the system from the state FU to the IR state at the current k-cycle, as the values corresponding to the first row and the third column in Eq. (8.91).

$P(\mathrm{DU1}_k \to \mathrm{DU1}_{k+1}) = \eta = M_{k_1}(2,2)$, representing the probabilistic transfer of the system's state DU1 from the current k-cycle to the next $(k+1)$-cycle, as shown by the values corresponding to the second row and the second column in Eq. (8.91).

$P(\mathrm{DU1}_k \to \mathrm{IR}_k) = 1 - \eta = M_{k1}(2,3)$, representing the probabilistic transfer of the system from the state DU1 to the IR state at the current k-cycle, as the values corresponding to the second row and the third column in Eq. (8.91).

Similarly, for incomplete test cycles, the B vector is defined as shown in Eqs. (8.92) and (8.93):

$$
B_1(1, i) = [0, 1, 1, 1]^T,
\tag{8.92}
$$

$$
B_2 = (1, i) = [0, 0, 1, 0, 1, 0, 1, 1, 1, 1]^T.
\tag{8.93}
$$

The instantaneous unavailability $UA(t)$ and the average unavailability UA_{avg} for different configurations at the incomplete test cycle $t \in [T_{k-1}, T_k]$ can also be obtained from Eqs. (8.71) and (8.77).

8.5.2 *Monte Carlo Simulation validation model construction*

Monte Carlo Simulation (MCS) is a simulation methodology for random sampling or statistical testing, which is often used to solve risky problems for complex decision-making systems with a large number of uncertainties. The MCS methodology provides a flexible description of system-related operations (equipment failures and subsequent maintenance) and simulates the entire life cycle scenario of a real system by modeling the actual processes and stochastic behavior of the system. MCS is widely used in reliability analysis and studies of real systems, and the IEC 61508 standard also recommends the application of MCS to validate the results obtained from other relevant analysis methods. Therefore, in this section, a series of statistical experimental studies are conducted on the problem to be solved using the MCS methodology to estimate the availability or unavailability of the safety barriers by counting the number of times an event occurs or the time spent in the simulation and analyzing the results in comparison with the MMPR methodology.

(1) MCS calculation flow

In this section, MATLAB software is used to write the program code for the simulation to calculate the UA_{avg} of the safety barrier, and the MCS validation process is shown in Fig. 8.36.

(i) Based on the unavailability analysis of the safety barriers, the possible states (normal, faulty, and awaiting repair) of the components during a given functional test or incomplete test phase are determined. Model parameters to be entered include mean time to failure, probability density function, failure rate, repair rate, and test or maintenance time.

(ii) Over time, the states of the system will replace each other, i.e., transfer from one state to another. The MCS can obtain information about these state changes, and from this information, one can know when the next transfer of the system state occurs

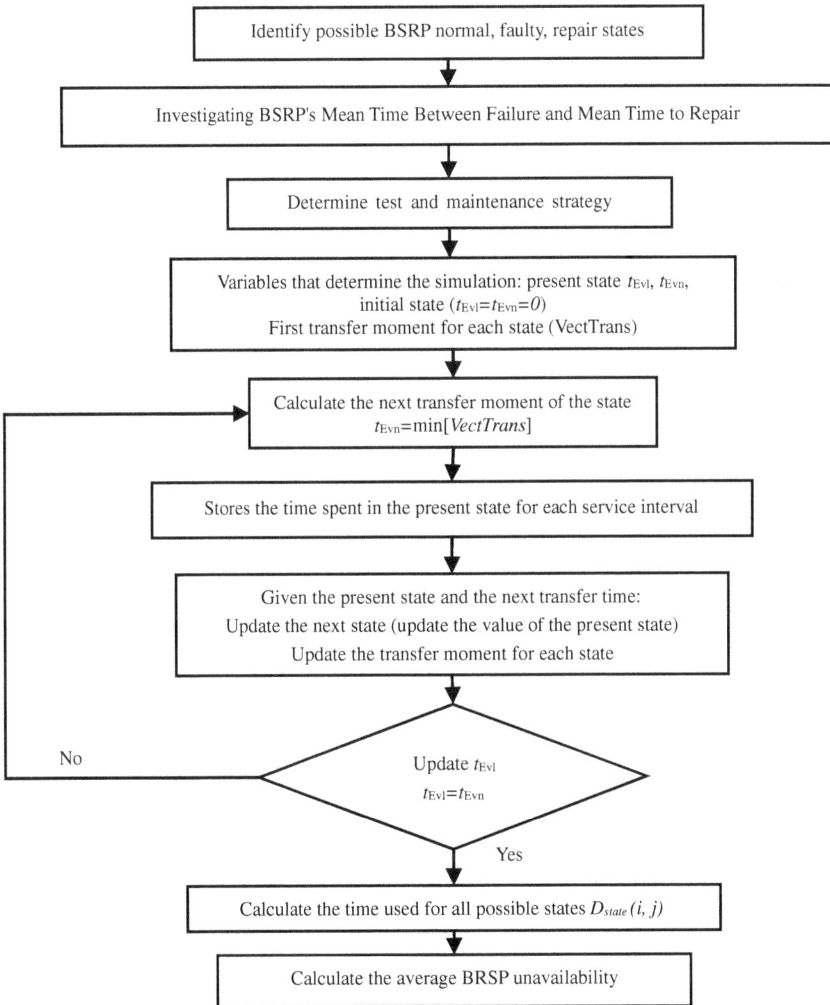

Fig. 8.36. Schematic diagram of the MCS validation process.

(fault, pending maintenance, or normal) and how the states change from one state to another at the end of the transfer.

(iii) The simulation is carried out for the transfer moment of the last failure (t_{Evl}) and the next transfer moment (t_{Evn}), depending on the test phase, the failure rate, and the repair rate. The durations of the available and unavailable states are stochastic and depend mainly on the probability density function of the failure time and the repair time. The duration $D_{\text{state}}(i,j)$ corresponding to state i in the jth test cycle can be used to analyze the unavailability of the system. Since the system is in a normal functional state in the initial phase, the initial transfer time is set to 0, i.e., $t_{\text{Evl}} = t_{\text{Evn}} = 0$, then $D_{\text{state}}(i,j)$ can be calculated by the following equation:

$$D_{\text{state}}(i,j) = D_{\text{state}}(i,j) + (t_{\text{Evn}} - t_{\text{Evl}}). \qquad (8.94)$$

(iv) When t_{Evn} is minimum, the system state is changed. $D_{\text{state}\,UA}(i,j)$ denotes the corresponding duration in the unavailable state of the system; from Eq. (8.95), this yields the average system unavailability UA_{avg}:

$$UA_{\text{avg}} = \frac{\sum_1^i \sum_1^j D_{\text{state}UA}(i,j)}{\sum_1^i \sum_1^j D_{\text{state}}(i,j)}. \qquad (8.95)$$

(v) After all MCS tests are completed, the simulation results are compared and analyzed with the MMPR resolution results.

(2) Comparative analysis of off-cycle validation

The BSRP parameters in the functional test phase are shown in Table 8.19. The 1oo1 configuration and 1oo2 configuration of the BSRP are simulated under different scenarios of constant failure rate and cyclic incremental failure rate, respectively. The MCS simulation results and MMPR analysis results are shown in Table 8.20.

As can be seen from the average unavailability calculations in Table 8.20, the MCS simulation results have the same gradient as

Table 8.19. BSRP parameters for the functional test phase.

Parameter	Symbol	Parameter value
Failure rate of the first cycle	λ_{DU}^1	1.8×10^{-6}
Failure rate of the second cycle	λ_{DU}^2	3.6×10^{-6}
Failure rate of the third cycle	λ_{DU}^3	7.2×10^{-6}
Repair rate	$\mu_{\text{DU}}^1 = \mu_{\text{DU}}^2 = \mu_{\text{DU}}^3$	0.0417
Non-cyclic functional test interval	T_1, T_2, T_3	720 h, 1440 h, 2160 h
Cyclic sexual function test intervals	$T_1 = T_2 = T_3$	1440 h
periodic number	m	3

Table 8.20. Average unavailability of off-cycle functional tests.

Cases	Test cycle/h	Constant λ (UA_avg)			Cyclic increment λ (UA_avg)		
		MMPR	MCS	relative error	MMPR	MCS	relative error
1oo1 configuration	720	1.29×10^{-3}	1.31×10^{-3}	1.55%	6.48×10^{-4}	6.70×10^{-4}	3.40%
	1440	2.63×10^{-3}	2.65×10^{-3}	0.76%	2.61×10^{-3}	2.65×10^{-3}	1.53%
	2160	3.93×10^{-3}	3.90×10^{-3}	0.76%	7.79×10^{-3}	7.88×10^{-3}	1.16%
1oo2 configuration	720	2.24×10^{-6}	2.28×10^{-6}	1.79%	5.56×10^{-7}	5.80×10^{-7}	4.32%
	1440	9.04×10^{-6}	9.22×10^{-6}	1.99%	8.95×10^{-6}	9.22×10^{-6}	3.02%
	2160	2.03×10^{-5}	2.08×10^{-5}	2.46%	8.00×10^{-5}	7.66×10^{-5}	4.25%

Table 8.21. BSRP parameters for the incomplete test phase.

Parameter	Symbol	Parameter value
Failure rate of the first type	λ_{DU1}	1.8×10^{-6}
Type II failure rate	λ_{DU2}	1.8×10^{-6}
Repair rate	μ_{DU1}	0.0417
Test cycle	Δ	168 h
Number of tests	k	4
Failure probability	$\eta = \omega$	0.01

the MMPR parsing results. The absolute values of the relative errors of the MCS simulation results of the 1oo1 configuration are smaller than those of the 1oo2 configuration, which is due to the fact that the system of the 1oo1 configuration is simple and has fewer states, which has less influence on the simulation results, and the differences between the results of the two methods are smaller. Comparing the calculation results of the two configurations obeying the cycle increment λ and the constant λ, the former calculation results are on the high side, and the relative error is on the high side. Using MCS to calibrate the results of the off-cycle function test, the relative error of the results corresponding to different cases is less than 4.32%, which tests the applicability and reasonableness of the MMPR model. Therefore, the MMPR analysis method can be used to evaluate the usability of the actual system.

(3) Incomplete test maintenance time validation comparison

The MCS method in this section is mainly used to test whether the results of different maintenance times in the incomplete test phase are reasonably feasible. Usually, it is more difficult to apply MCS to test the defective factors in incomplete test cycles, so the test failure factor ($\omega = \eta = 0$) is no longer considered in the validation simulation. The BSRP parameters for the incomplete test phase are shown in Table 8.21. Table 8.22 shows the comparative analysis of the MCS simulation results with the MMPR parsing results for different repair times.

As can be seen from the results of the mean unavailability calculations in Table 8.22, the MCS simulation results have the same gradient as the MMPR parsed results. The MCS simulation results

Table 8.22. Incomplete testing of UA_{avg} at different repair times.

Maintenance time/h	1oo1 configuration (UA_{avg})			1oo2 configuration (UA_{avg})		
	MMPR	MCS	relative error	MMPR	MCS	relative error
24	3.26×10^{-3}	3.20×10^{-3}	1.84%	1.33×10^{-5}	1.29×10^{-5}	3.10%
48	3.30×10^{-3}	3.24×10^{-3}	1.82%	1.35×10^{-5}	1.30×10^{-5}	3.85%
72	3.33×10^{-3}	3.27×10^{-3}	1.80%	1.37×10^{-5}	1.32×10^{-5}	3.79%
96	3.36×10^{-3}	3.30×10^{-3}	1.79%	1.39×10^{-5}	1.34×10^{-5}	3.73%
120	3.39×10^{-3}	3.33×10^{-3}	1.77%	1.41×10^{-5}	1.36×10^{-5}	3.68%

for both configurations are smaller than the MMPR parsed results, which may be due to the fact that the repair time is less sensitive to the test time in the MCS model. The absolute value of the relative error of the mean unavailability for both 1oo2 configurations is larger than that for the 1oo1 configuration. As the repair time increases, the MCS simulation results also increase, which is consistent with the change rule of the parsing results. The relative errors of the results corresponding to different maintenance times are less than 3.85%, which test the applicability and reasonableness of the MMPR model for the incomplete test maintenance phase, and the method can be used to evaluate the maintenance optimization characteristics of real systems.

8.5.3 *Application*

In this chapter, the BSRP subsystem of the underwater wellhead blowout preventer system is selected as the research object, including the BSRP of the 1oo1 configuration and the BSRP of the 1oo2 configuration, and the MMPR model proposed in Section 8.5.2 is applied to simulate the dynamic characteristics of the functional testing process, the non-periodicity of the test and maintenance, the impact of the test process errors, and the delayed phenomenon of the underwater maintenance to provide effective decision-making for the preventive testing and the optimization of the maintenance. The example analysis process of BSRP is shown in Fig. 8.37.

(1) BSRP failure analysis

Subsea shear gate blowout preventers remain inactive during routine downhole operations and therefore require periodic performance of functional verification test maintenance and barrier system integrity testing maintenance to identify potential Dangerous Undetected Failures (DUFs) in the barrier assembly. For shear gate blowout preventers, DUFs are the primary failure factors affecting their availability. When it comes to incomplete test maintenance, DUFs are divided into two categories: those that can be detected through incomplete testing, or DUF1, such as failure to close the shear gate blowout preventer during the bare eye phase and leakage after closure without a drill pipe, and those that can only be detected through functional verification testing, or DUF2, such as failure to

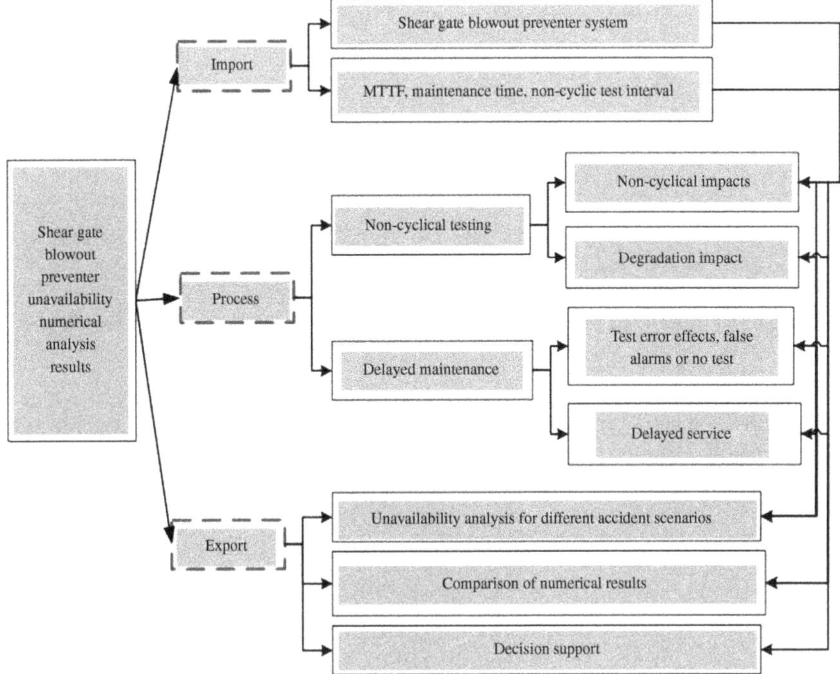

Fig. 8.37. Schematic diagram of the process of analyzing the example.

shear in the presence of a drill pipe, failure to shut down the shear, or delayed shutdown.

Based on the proposed MPPR model and unavailability calculation formula, the relevant failure parameters of the BSRP are counted. Based on the relevant literature (PSA2014), accident report (Holand report), NORSOK D-010 standard and SINTEF, and WEST E.S well blowout database, etc., the mean time to failure (MTTF) and mean downtime, MDT) statistics are given. The statistics given by different studies and databases are different: In the case of BOP, the MTTF is 5564 days and 22256 days; in the case of Item, the MTTF is 6276 days and 25104 days; in the SINTEF database, the MTTF is 1012 days; and in WEST E.S, the MTTF is 7770 days where the MDT is 189 days. Based on the researched failure data, the relevant parameters of BSRP are obtained. According to the NORSOK D-010 standard, it is known that for other blowout preventer subsystems, the functional test is 7/14 days, while for BSRP it is 30 days.

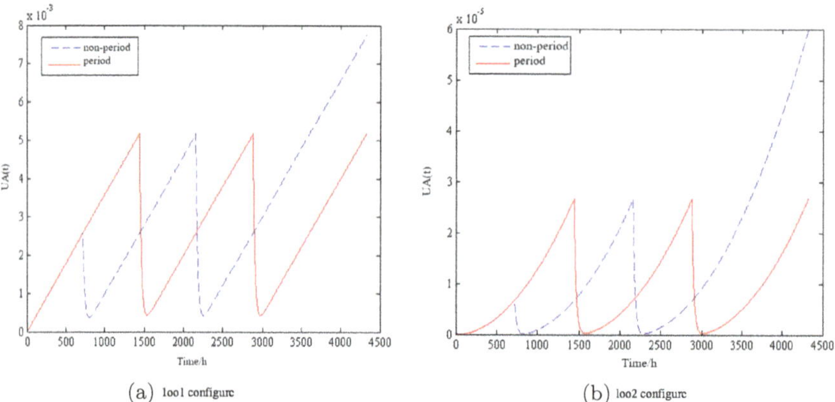

Fig. 8.38. Change in unavailability during cyclic and non-cyclic functional test phases for two configurations.

(2) Analysis of the impact of non-periodic functional testing

Due to the underwater environment and drilling events, functional tests of the BSRP are not always performed as specified, i.e., there is a non-periodicity of testing. In this section, three cyclic and three non-cyclic functional tests are used to carry out a study on the effect of cyclicity and non-cyclicity on system unavailability. Failure rates are not always constant across test cycles and may show a cyclical increasing trend. The effects of three different failure rates and three identical failure rates on off-cycle test unavailability are considered. The corresponding failure rates, repair rates, cyclic and off-cycle test intervals, and numbers are obtained from the BSRP statistics as shown in Table 8.19, which are used to construct the corresponding transfer rate matrices. The trend of unavailability for the two configurations is calculated by applying Eqs. (8.56) and (8.61)–(8.68) as shown in Fig. 8.38(a) and (b).

Obviously, for the periodic test, the trend and gradient of the two configurations are the same in each cycle of $UA(t)$. 1oo1 configuration changes in a linearly increasing and then rapidly decreasing trend, and 1oo2 configuration shows a nonlinearly increasing and then rapidly decreasing trend. The non-periodic test shows an increasing trend in each cycle, and the maximum value of $UA(t)$ also shows an increasing trend in each cycle. It can be seen that non-periodic functional testing leads to an irregular increase in $UA(t)$

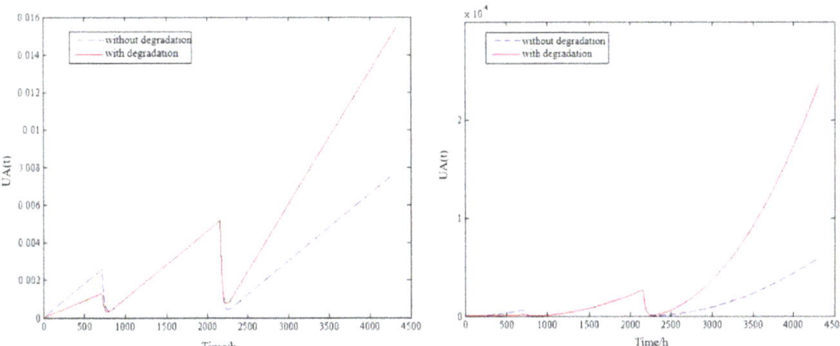

Fig. 8.39. Variation in unavailability with constant and variable failure rates.

and an increase in the unavailability of the system for the operational process.

Due to the difficulty of replacing underwater BSRP components, there is a degradation of BSRP performance over time at different test cycles. Therefore, it is assumed that the failure rate is constant for the same cycle and increases for each subsequent cycle. Based on the parameters given in Table 8.19, the effect of constant failure rate (λ_{DU}^2) and variable failure rates ($\lambda_{\mathrm{DU}}^1, \lambda_{\mathrm{DU}}^2$ and λ_{DU}^3) on system unavailability is analyzed. Given the off-cycle test times of 720 h, 1440 h, and 2160 h, the comparison between the two is shown in Fig. 8.39. The results show that the $UA(t)$ of the system with a constant failure rate increases at the same rate in each cycle, while the variable failure rate increases at an incremental rate. For example, at the end of the third test cycle ($t = 4320\,\mathrm{h}$), the unavailability value $UA(t)$ of the 1oo1 configuration rises from 8×10^{-3} to 1.5×10^{-2}, while the unavailability value $UA(t)$ of the 1oo2 configuration rises from 6×10^{-5} to 2.4×10^{-4}, indicating that degradation of the system causes its unavailability to take on a higher value and to grow at a faster rate within the off-cycle test. Combined with actual testing and maintenance, the off-cycle and degradation effects of preventive testing are important factors affecting system unavailability.

The average unavailability (UA_{avg}) of the system for the four cases is calculated by Eqs. (8.72) and (8.77)–(8.84), and the results are shown in Table 8.23:

Case 1: Off-cycle testing of the BSRP with a 1oo1 configuration with constant/variable failure rates;

Table 8.23. Average unavailability of BSRP across cases.

Case	Test cycle/h	UA_{avg} Constant failure rate	Variable failure rate	Specific value
1	720	1.29×10^{-3}	6.48×10^{-4}	0.502326
	1440	2.63×10^{-3}	2.61×10^{-3}	0.992395
	2160	3.93×10^{-3}	7.79×10^{-3}	1.982188
2	1440	2.59×10^{-3}	1.30×10^{-3}	0.501931
	1440	2.63×10^{-3}	2.61×10^{-3}	0.992395
	1440	2.67×10^{-3}	5.25×10^{-3}	1.966292
3	720	2.24×10^{-6}	5.56×10^{-7}	0.248214
	1440	9.04×10^{-6}	8.95×10^{-6}	0.990044
	2160	2.03×10^{-5}	8.0×10^{-5}	3.940887
4	1440	8.92×10^{-6}	2.24×10^{-6}	0.251121
	1440	9.04×10^{-6}	8.95×10^{-6}	0.990044
	1440	9.39×10^{-6}	3.6×10^{-5}	3.833866

Case 2: BSRP cyclic testing of 1oo1 configurations with constant/variable failure rates;

Case 3: BSRP non-cyclic testing of 1oo2 configurations with constant/variable failure rates;

Case 4: BSRP cyclic testing of 1oo2 configurations with constant/variable failure rates.

From column 3 of Table 8.23, it can be seen that the UA_{avg} of Case 1 has a substantial increase with the increase of cycles, the UA_{avg} of Case 2 has a relatively small change and can be considered to be essentially constant; the UA_{avg} of Cases 3 and 4 have the same trend compared to Cases 1 and 2, but with a decrease of the order of magnitude of 10^{-3}. From the above analysis, it can be seen that off-cycle testing also makes the BSRP's UA_{avg} values increase. By finding the ratio of the UA_{avg} of the system with constant failure rate to the UA_{avg} of the system with variable failure rate, it can be seen that the ratio keeps on increasing for both periodic and non-periodic tests, and the ratio is maximum in the third cycle, which indicates that the variable failure rate makes the UA_{avg} of the system increase in the subsequent cycles.

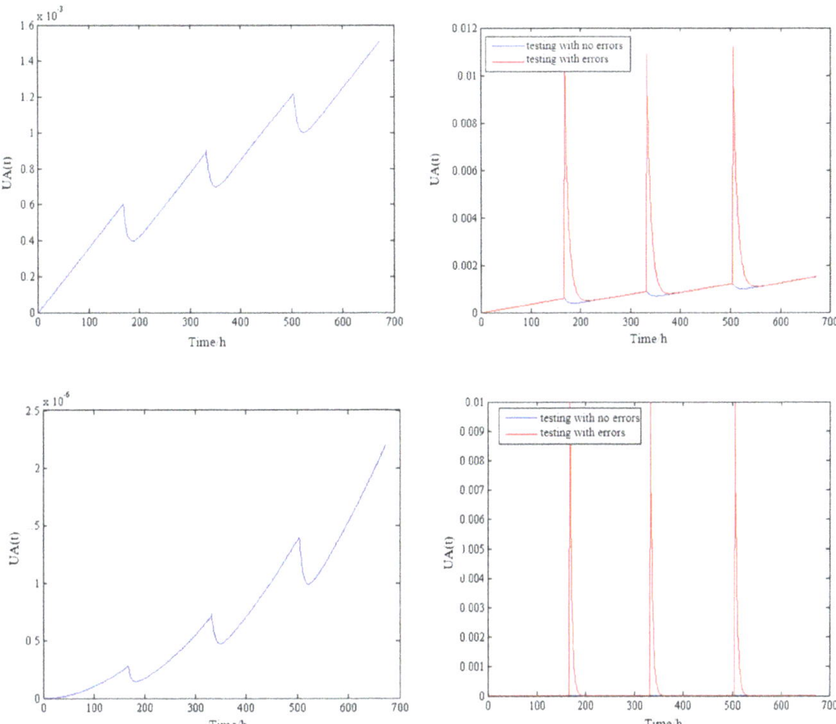

Fig. 8.40. Impact of test errors on unavailability for incomplete tests.

(3) Analysis of the impact of test errors in incomplete testing

According to the relevant standards, incomplete testing of BPSR is required to improve the system availability, but incomplete testing is susceptible to defect factors, so this section analyzes the impact of test failures on system availability while considering two types of failures DUF1 and DUF2 of the incomplete testing process. Due to the short cycle of incomplete testing, it is assumed that the two types of system failure rates are constant in each cycle, and the incomplete testing is periodic, i.e., $\Delta = 168\,$h, and the number of incomplete testing cycles $k = 4$. The defect factor has a small error, and is set to have a probability of $\omega = \eta = 0.01$. The parameters of the two types of failure rates and the repair rate are shown in Table 8.23. According to Eqs. (8.72), (8.89), and (8.90), the results of the test failures of incomplete testing on unavailability are shown in Fig. 8.40.

Figure 8.40 shows the variation of $UA(t)$ during incomplete test cycles with different configurations and with test error factors. The $UA(t)$ increases and then decreases rapidly in the first test cycle, and its value decreases from 0.6×10^{-3} to 0.4×10^{-3} instead of 0, which is due to the effect of DUF2, as shown in Fig. 8.40(a) and (c). Figure 8.40(b) and (d) compare the $UA(t)$ effect on the system with and without test errors. The results show that small errors in incomplete testing can lead to an instantaneous increase in the system's unavailability, but a rapid decrease after repair.

The average unavailability (UA_{avg}) of the system is evaluated by calculating four cases through Eqs. (8.72) and (8.77)–(8.90), and the results are shown in Table 8.24:

Case 1: 1oo1 configured BSRP incomplete test does not consider the effect of test errors;

Case 2: 1oo1 configured BSRP incomplete test considering the effect of test errors;

Case 3: 1oo2 configured BSRP incomplete test does not consider the effect of test errors;

Case 4: 1oo2 configured BSRP incomplete test considering the effect of test errors.

From Table 8.24, it can be seen that the UA_{avg} of Case 1 increases with the incomplete cycle; the UA_{avg} of Case 2 increases faster because of the effect of having test errors; and the pattern of change of Cases 3 and 4 is similar to that of Cases 1 and 2. The analyses show that the test errors increase the UA_{avg} value of the BSRP. By finding the ratio of UA_{avg} with test errors to UA_{avg} without test errors, it can be obtained that the ratio is 1 in the first cycle, which is due to the fact that it is not affected by test errors; the ratio gradually decreases from the second test cycle onward, and the effect of test errors on the UA_{avg} in the later period gradually decreases because it has been increasing. Especially for the 1oo2 configuration, the ratio increases by 596.99 times from the second cycle onward, and it is obvious that the test errors have the greatest impact on the unavailability of the 1oo2 configuration, so it is important to avoid this situation in the test.

Table 8.24. Comparison of test failure and no test failure UA_{avg}.

Test cycle/h	UA_{avg}					
	Case 1	Case 2	Specific value	Case 3	Case 4	Specific value
$[0, 168]$	3.02×10^{-4}	3.02×10^{-4}	1.00	1.01×10^{-7}	1.01×10^{-7}	1.00
$[168, 336]$	5.97×10^{-4}	1.08×10^{-3}	1.81	3.99×10^{-7}	2.382×10^{-4}	596.99
$[336, 504]$	9.50×10^{-4}	1.43×10^{-3}	1.51	8.82×10^{-7}	2.385×10^{-4}	270.41
$[504, 672]$	1.22×10^{-3}	1.70×10^{-3}	1.39	1.34×10^{-6}	2.39×10^{-4}	178.36

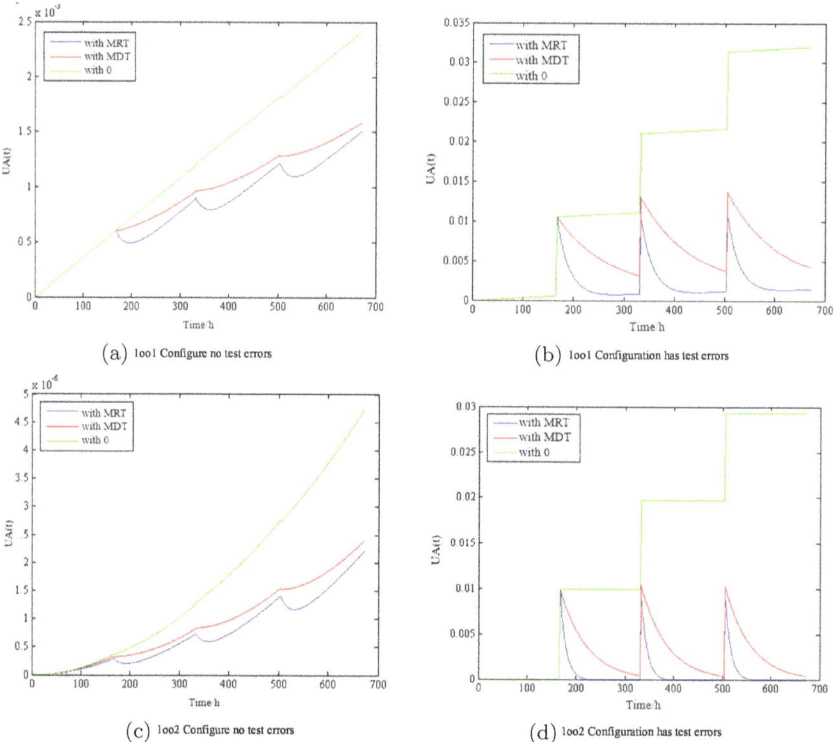

(a) 1oo1 Configure no test errors

(b) 1oo1 Configuration has test errors

(c) 1oo2 Configure no test errors

(d) 1oo2 Configuration has test errors

Fig. 8.41. Incomplete testing of maintenance delay impact on unavailability.

(4) Maintenance time optimization analysis

Due to the complex application environment of the seabed, it is difficult to activate the maintenance plan immediately even if a fault is detected. Therefore, the impact of repair delays should be considered in preventive test maintenance. Figure 8.41 shows the assessment of the $UA(t)$ impact of maintenance delays on the system during incomplete test cycles considering maintenance time and test failures included. In this section, three scenarios are considered to calculate the repair rate of the system, and the parameters for the calculation are shown in Table 8.21, where the repair rate needs to be recalculated.

Calculate the repair rate μ based on the non-negligible mean repair time (MRT);

Calculate the repair rate μ based on the non-negligible mean downtime (MDT);

Negligible repair time, i.e., no repair response, $\mu = 0$.

Figure 8.41(a) and (c) show that the 1oo1 configuration shows an approximately linear increase in $UA(t)$ when there are no test failures and no maintenance response, while the 1oo2 configuration shows an approximately exponential increase; both show a stepwise increase if test failures are considered and there is no maintenance response (Fig. 8.41(b) and (d)). This shows that not carrying out maintenance actions leads to a significant increase in the unavailability of the system. Furthermore, the instantaneous value of $UA(t)$ for considering MRT during any incomplete test cycle is lower than that of $UA(t)$ for the other two scenarios. This is due to the fact that estimating the repair rate by the MDT method assumes that the failure is found during the test cycle, i.e., MDT $= \Delta/2$. It is evident that the method of estimating the repair rate also affects the unavailability of the system, and the decision-maker needs to choose the appropriate estimation method according to the actual situation.

In this section, four cases are given to evaluate the average unavailability (UA_{avg}) of the system, as shown in Table 8.25:

Case 1: calculation of maintenance rate for 1oo1 configuration based on MRT;

Case 2: calculation of maintenance rate for 1oo1 configuration based on MDT;

Case 3: calculation of maintenance rate for 1oo2 configuration based on MRT;

Case 4: calculation of maintenance rate for 1oo2 configuration based on MDT.

From the results of the mean unavailability calculations in Table 8.25, it can be seen that the UA_{avg} values under different cases

Table 8.25. Comparison of UA_{avg} in MRT and MDT mode.

Test cycle/h	Case 1	Case 2	Specific value	Case 3	Case 4	Specific value
$[0, 168]$	3.02×10^{-4}	3.02×10^{-4}	1.00	1.01×10^{-7}	1.13×10^{-7}	1.12
$[168, 336]$	2.05×10^{-3}	5.75×10^{-3}	2.80	7.133×10^{-4}	3.06×10^{-3}	4.29
$[336, 504]$	2.41×10^{-3}	7.31×10^{-3}	3.03	7.134×10^{-4}	3.21×10^{-3}	4.50
$[504, 672]$	2.68×10^{-3}	7.74×10^{-3}	2.89	7.136×10^{-4}	3.20×10^{-3}	4.48

tend to increase in different test cycles due to the influence of the second type of hidden faults. The UA_{avg} value is the same in the first test cycle due to the effect of no maintenance delay. By comparing the UA_{avg} estimated by the MDT method with the UA_{avg} estimated by the MRT method, it can be seen that from the second test cycle onward, the ratio shows an increasing trend, and the trend gradually becomes slower. That is, the UA_{avg} of the two methods decreases with the increase of the repair time. The ratio of the 1oo1 configuration is smaller than the ratio of the 1oo2.

In order to examine the effect of different repair times on the UA_{avg} of the BSRP, the MRT and MDT methods were applied to estimate the repair rate and analyzed comparatively, as shown in Fig. 8.42. Figure 8.42(a) and (b) show the trend of UA_{avg} for different repair times for the two configurations without considering test errors. It is clear that the UA_{avg} decreases rapidly and then grows slowly as the repair time increases, and the UA_{avg} value estimated based on MDT is larger than the UA_{avg} value estimated by MRT, and the difference is decreasing. Figure 8.42(c) and (d) show the trends of UA_{avg} for different repair times for 1oo1 and 1oo2 configurations, respectively, without considering test errors, and the difference between the UA_{avg} value estimated based on MDT and the UA_{avg} estimated based on MRT shows a trend of increasing and then decreasing. In practical applications, both test errors and maintenance delays increase the unavailability of the system, and the appropriate maintenance time can be selected based on different estimation methods and requirements for unavailability.

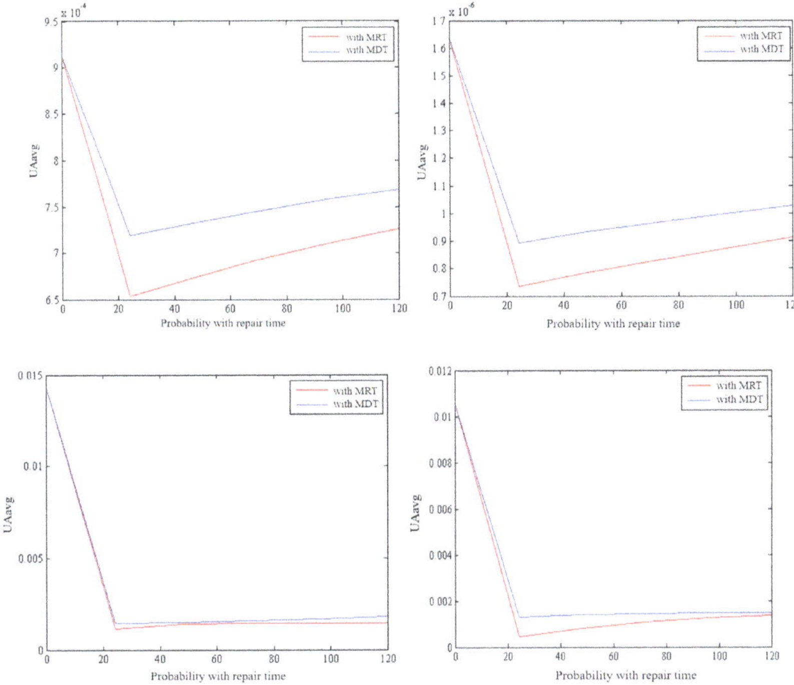

Fig. 8.42. Effect of MRT and MDT based repair time on UA_{avg}.

References

[1] Wu S, Zhang L, Barros A, *et al.* Performance analysis for subsea blind shear ram preventers subject to testing strategies. *Reliability Engineering and System Safety*, 2018, 169: 281–298.

[2] Wu S, Zhang L, Zheng W, *et al.* Reliability modeling of subsea SISs partial testing subject to delayed restoration. *Reliability Engineering and System Safety*, 2019, 191: 106546.

[3] Wu S, Zhang L, Fan J, *et al.* Dynamic risk analysis of hydrogen sulfide leakage for offshore natural gas wells in MPD phases. *Process Safety and Environmental Protection*, 2019, 122: 339–351.

[4] Zhang L, Wu S, Zheng W, *et al.* A dynamic and quantitative risk assessment method with uncertainties for offshore managed pressure drilling phases. *Safety Science*, 2018, 104: 39–54.

[5]   Wu S, Zhang L, Lundteigen MA, *et al.* Reliability assessment for final elements of SISs with time dependent failures. *Journal of Loss Prevention in the Process Industries*, 2017, 51: 186–199.

[6]   Wu S, Zhang L, Fan J, *et al.* Real-time risk analysis method for diagnosis and warning of offshore downhole drilling incident. *Journal of Loss Prevention in the Process Industries*, 2019, 62: 103933.

[7]   Wu S, Zhang L, Fan J, *et al.* A leakage diagnosis testing model for gas wells with sustained casing pressure from offshore platform. *Journal of Natural Gas Science and Engineering*, 2018, 55: 276–287.

[8]   Wu S, Zhang L, Fan J, *et al.* Prediction analysis of downhole tubing leakage location for offshore gas production wells. *Measurement*, 2018, 127: 546–553.

[9]   Wu S, Zhang L, Zheng W, *et al.* A DBN-based risk assessment model for prediction and diagnosis of offshore drilling incidents. *Journal of Natural Gas Science and Engineering*, 2016, 34: 139–158.

[10] Zhou Y, Wu S, Fan J, *et al.* A safety-barrier-based risk analysis model for offshore oil and gas leakage incidents. In: *2020 European Safety and Reliability Conference*, November 2–5, 2020, Venice, Italy.

[11] Wu S, Zhang L, Zheng W, *et al.* Reliability assessment for subsea HIPPS valves with partial stroke testing. In: *2016 European Safety and Reliability Conference*, September 24–29, 2016, Glasgow, England.

[12] Wu S, Zhang L, Zheng W, *et al.* Risk prediction for o&g fire accident of offshore platform based on numerical simulation. In: *The 6th World Conference on Safety of Oil and Gas Industry*, October 26–29, 2016, Beijing, China.

[13] Wu S, Fan J, Zhang L, *et al.* Characterization of completion operational safety for deepwater wells. In: *2013 2nd International Conference on Environment, Energy and Biotechnology*, 2013, Singapore.

Index

L

localized abnormal perturbation, 20

M

Markov process, 98
multi-class support vector machine (MCSVM), 463
multi-scale, 22

N

natural language processing (NLP), 278

P

pipe spikes, 61
pipeline integrity management, 5
PLC, 93

R

RGB, 148

S

safety component functionality, 16
sand plugging, 61
sand sinking, 61
SCADA, 93
shale gas, 3
support vector machine (SVM), 461

T

Tensorboard tool, 273
TensorFlow deep learning framework, 317
Textrank algorithm, 291
TF-IDF algorithm, 289
tower rushing accidents, 60

W

Weibo, 280
well leakage, 4
well surge, 617
word embedding, 513